BIOETHANOL

Biochemistry and Biotechnological Advances

BIOETHANOL

Biochemistry and Biotechnological Advances

Edited by
Ayerim Y. Hernández Almanza, PhD
Nagamani Balagurusamy, PhD
Héctor Ruiz Leza, PhD
Cristóbal N. Aguilar, PhD

A∧P APPLE
ACADEMIC
PRESS

First edition published 2023

Apple Academic Press Inc.
1265 Goldenrod Circle, NE,
Palm Bay, FL 32905 USA
4164 Lakeshore Road, Burlington,
ON, L7L 1A4 Canada

CRC Press
6000 Broken Sound Parkway NW,
Suite 300, Boca Raton, FL 33487-2742 USA
4 Park Square, Milton Park,
Abingdon, Oxon, OX14 4RN UK

© 2023 by Apple Academic Press, Inc.

Apple Academic Press exclusively co-publishes with CRC Press, an imprint of Taylor & Francis Group, LLC

Library and Archives Canada Cataloguing in Publication

Title: Bioethanol : biochemistry and biotechnological advances / edited by Ayerim Y. Hernández Almanza, PhD, Nagamani Balagurusamy, PhD, Héctor Ruiz Leza, PhD, Cristóbal N. Aguilar, PhD.
Other titles: Bioethanol (Palm Beach, Fla.)
Names: Hernández Almanza, Ayerim Y., editor. | Balagurusamy, Nagamani, editor. | Leza, Héctor Ruiz, editor. | Aguilar, Cristóbal N., editor.
Description: First edition. | Includes bibliographical references and index.
Identifiers: Canadiana (print) 20220137609 | Canadiana (ebook) 20220137617 | ISBN 9781774638491 (hardcover) | ISBN 9781774638507 (softcover) | ISBN 9781003277132 (ebook)
Subjects: LCSH: Biomass conversion. | LCSH: Ethanol as fuel. | LCSH: Ethanol.
Classification: LCC TP248.B55 B56 2022 | DDC 662/.88—dc23

Library of Congress Cataloging-in-Publication Data

CIP data on file with US Library of Congress

ISBN: 978-1-77463-849-1 (hbk)
ISBN: 978-1-77463-850-7 (pbk)
ISBN: 978-1-00327-713-2 (ebk)

About the Editors

Ayerim Y. Hernández Almanza, PhD

Full Professor, School of Biological Sciences,
Autonomous University of Coahuila, Torreón, Coahuila, México

Ayerim Y. Hernández Almanza, PhD, is Professor (full-time) at the School of Biological Science, the Autonomous University of Coahuila, since 2019. Prof. Hernández Almanza's experience is in extraction and purification of bioactive compounds, fermentation process, microbial pigments production, and biological characterization. She is recognized as a member (candidate) of the National System of Research (SNI) by the National Council of Science and Technology of the Government of México. In 2019 she obtained the "Mujer Universitaria" award in Scientific Ambit given by the Autonomous University of Coahuila. Prof. Hernández-Almanza has published over 10 original research papers in indexed journals and 12 book chapters and has participated and contributed in more than five scientific meetings. She has realized two research stays: at Università degli Studi di Perugia, Perugia, Italia (2013), and at Gachon University, Seongnam, Gyeonggi-do, South Korea (2016).

Nagamani Balagurusamy, PhD

Full Professor, Bioremediation Lab, Faculty of Biological Sciences,
Autonomous University of Coahuila, Torreón, Coahuila, México

Nagamani Balagurusamy, PhD, is Professor (full-time) at the Autonomous University of Coahuila, Maxico, since September 2001. He is head of the bioremediation lab of the Faculty of Biological Sciences, Torreon campus, and coordinator of the graduate program on biochemical engineering, which is recognized as of the National Post-Graduate Programs of Quality (PNPC) of the Council of Science and Technology of the Federal Government of Mexico. He is also a visiting Professor of the Graduate Program on Biotechnology of the Department of Biology, West Virginia State University, WV, USA. He has published more than 70 scientific articles in international and national indexed journals, 6 books, and more than 40 chapters in books published by Springer, Elsevier, CRC, and others. He has participated in more than 75 international and national conferences. He is a member of the National Researchers Council of Mexico

(SNI-CONACYT, Level I distinction) and also a regular member of Mexican Academy of Sciences

Héctor Ruiz Leza, PhD
Full Professor, Biorefinery Group, School of Chemistry,
Autonomous University of Coahuila, Saltillo, Coahuila, México

Héctor Ruiz Leza, PhD, is currently a full Professor in the Biorefinery Group of the Faculty of Chemistry Sciences at the Autonomous University of Coahuila. He is the founder of Biorefinery Group, principal coordinator of the biorefinery pilot plant in the Food Research Department, and leader of the biomass pretreatment stage in the Cluster of Bioalcoholes in the Mexican Centre for Innovation in Bioenergy (Cemie-Bio) from the Secretary of Energy in Mexico. Dr. Ruiz obtained his PhD in Chemical and Biological Engineering from the Center of Biological Engineering at the University of Minho, Portugal, in 2011 and was a postdoctoral researcher at the University of Minho (Portugal) and University of Vigo (Spain).

Dr. Ruiz is currently working on the development of biorefinery strategies for the production of high added-value compounds and biofuels from ligno-cellulosic, micro, and macroalgal biomass in the context of the bioeconomy. He is technically responsible for several research projects (CONACYT) and a technical consultant for biomass conversion companies in Mexico and Europe. Dr. Ruiz has conducted several research stays and technical visits at the University of North Texas (USA), Federal University of Sergipe (Brazil), Brazilian Bioethanol Science and Technology Laboratory-CTBE (Brazil), Chemical and Biological Engineering Department at the University of British Columbia (Canada), CIEMAT-Renewable Energy Division, Biofuels Unit (Spain), University of Jaén (Spain), Sadar Swaran Singh National Institute of Bio-Energy (India), Tokyo Institute of Technology (Japan), University of Concepción (Chile), Umeå University (Sweden), National Laboratory of Energy and Geology-LNEG (Portugal), CSIR-National Institute for Interdisciplinary Science and Technology (India), University of Florida-Stan Mayfield Biorefinery Plant (USA), University of Kannur (India), Federal University of Rio Grande do Norte (Brazil).

He has authored or co-authored 75 publications, including papers in indexed journals and chapters, with over 2300 citations and an h-index of 28 (Google Scholar Citations). Currently, Dr. Ruiz is Editor-in-Chief of *BioEnergy Research Journal* (Springer Publishing) and Associate Editor of *Biotechnology for Biofuels* (BioMed Central-Part of Springer Nature), and he participates in the Editorial Advisory Board of *Industrial Crops and Products* (Elsevier Editorial) and *Biofuel Research Journal*. He was the editor of the book *Hydrothermal Processing in Biorefineries* published by Springer Publishing in 2017. Dr. Ruiz was awarded with the prize "Dr. Carlos Casas Campillo" of the Mexican Society of Biotechnology and Bioengineering in 2016. This award aims to give recognition and encourage young researchers for their contribution to the development of biotechnology and bioengineering in Mexico. Dr. Ruiz has been a member of the Mexican Academy of Science in the engineering section since 2020. He received a Young Scientist Award at the Autonomous University of Coahuila for the year 2019, and he is a member of the National Researchers Council in Mexico (SNI-CONACYT, Level II distinction).

Cristóbal N. Aguilar, PhD
Full Professor, Associate Editor of Heliyon (Microbiology) and
Frontiers in Sustainable Food Systems (Food Processing), Bioprocesses,
and Bioproducts Research Group, Food Research Department,
School of Chemistry, Autonomous University of Coahuila, Saltillo, Mexico

Cristóbal N. Aguilar, PhD, is a Director of Research and Postgraduate Programs at the Autonomous University of Coahuila, Mexico. His scientific impact has an index of more than 41 h. He has been awarded several awards, the most important of which are the Outstanding Scientific Award by the International Bioprocessing Association 2018; the State Prize for Science, Technology, and Innovation Coahuila 2019; the National Research Award 2010 of the Mexican Academy of Sciences; the Carlos Casas Campillo 2008 Award from the Mexican Society of Biotechnology and Bioengineering; AgroBio National Prize-2005; and the Mexican Food Science and Technology Award.

Dr. Aguilar is a member of the Mexican Academy of Sciences, the International Bioprocessing Association, the Mexican Academy of Sciences, the Mexican Society of Biotechnology and Bioengineering, and the Mexican Association of Food Science and Biotechnology. He has developed more than 50 research projects, including 20 international exchange projects.

Contents

Contributors

P. Abdeshahian
Biopolymers, Bioprocesses, Process Simulation Laboratory, Department of Biotechnology,
Engineering School of Lorena, University of São Paulo (EEL-USP), Lorena–12.602.810, SP, Brazil

Cristóbal N. Aguilar
Bioprocesses and Bioproducts Research Group, BBG-DIA, Food Research Department,
School of Chemistry, Autonomous University of Coahuila (UAdeC), Saltillo–25280, Coahuila, Mexico

Diana Laura Alva-Sánchez
School of Biological Science, Autonomous University of Coahuila, Torreón–27000, Coahuila, Mexico

Silvia Yudith Martínez Amador
Botany Department, Agronomic Division, Antonio Narro Autonomous Agrarian University, Mexico

Lorena Amaya-Delgado
Industrial Biotechnology Unit, Center for Research and Assistance in Technology and Design of the State
of Jalisco A.C., Camino Arenero–1227, El Bajio del Arenal, C.P.–45019, Zapopan, Jalisco, Mexico

Melchor Arellano-Plaza
Industrial Biotechnology Unit, Center for Research and Assistance in Technology and Design of the State
of Jalisco A.C., Camino Arenero–1227, El Bajio del Arenal, C.P.–45019, Zapopan, Jalisco, Mexico

L. G. De Arruda
Biopolymers, Bioprocesses, Process Simulation Laboratory, Department of Biotechnology,
Engineering School of Lorena, University of São Paulo (EEL-USP), Lorena–12.602.810, SP, Brazil

D. B. Arya
Department of Environmental Sciences, University of Kerala, Thiruvananthapuram, Kerala, India

Nagamani Balagurusamy
Bioremediation Laboratory, Biological Science Faculty, School of Biological Science,
Autonomous University of Coahuila (UAdeC), Torreón–27000, Coahuila, Mexico

T. R. Balbino
Bioprocesses and Sustainable Products Laboratory, Department of Biotechnology,
Engineering School of Lorena, University of São Paulo (EEL-USP), Lorena–12.602.810, SP, Brazil

F. G. Barbosa
Bioprocesses and Sustainable Products Laboratory, Department of Biotechnology,
Engineering School of Lorena, University of São Paulo (EEL-USP), Lorena–12.602.810, SP, Brazil

M. M. Campos
Bioprocesses and Sustainable Products Laboratory, Department of Biotechnology,
Engineering School of Lorena, University of São Paulo (EEL-USP), Lorena–12.602.810, SP, Brazil

A. S. Cardoso
Bioprocesses and Sustainable Products Laboratory, Department of Biotechnology,
Engineering School of Lorena, University of São Paulo (EEL-USP), Lorena–12.602.810, SP, Brazil

L. T. Carvalho
Bioprocesses and Sustainable Products Laboratory, Department of Biotechnology, Engineering School of Lorena, University of São Paulo (EEL-USP), Lorena–12.602.810, SP, Brazil

Luis Fernando Amador Castro
Tecnologico de Monterrey, Escuela de Ingenieria y Ciencias, Av. General Ramon Corona No. 2514, Zapopan–45201, Jal., Mexico

M. J. Castro-Alonso
Bioprocesses and Sustainable Products Laboratory, Department of Biotechnology, Engineering School of Lorena, University of São Paulo (EEL-USP), Lorena–12.602.810, SP, Brazil

Héctor Hugo Molina Correa
Bioremediation Laboratory, Biological Sciences Faculty, Autonomous University of Coahuila, Torreón–27000, Coahuila, México

M. L. Silva da Cunha
Bioprocesses and Sustainable Products Laboratory, Department of Biotechnology, Engineering School of Lorena, University of São Paulo (EEL-USP), 12.602.810, Lorena, SP, Brazil

Ileana Mayela María Moreno Dávila
Department of Biotechnology, School of Chemistry, Autonomous University of Coahuila, Mexico

Ahmet Demir
Department of Environmental Engineering, Yildiz Technical University, Turkey, Davutpasa Campus, Istanbul–34220, Turkey

Kakoli Dutt
Department of Bioscience and Biotechnology, Banasthali Vidyapith, Rajasthan–304022, India, Tel.: 9887398384, E-mail: kakoli_dutt@rediffmail.com

M. P. Luévanos Escareño
School of Biological Science, Autonomous University of Coahuila, Torreón–27000, Coahuila, Mexico

Norma M. De La Fuente-Salcido
Faculty of Biological Sciences, Autonomous University of Coahuila, Torreón, Coahuila, México, E-mail: normapbr322@gmail.com

Cristian Emanuel Gámez-Alvarado
Bioremediation Laboratory, Biological Sciences Faculty, Autonomous University of Coahuila, Torreón–27000, Coahuila, México

José Antonio Rodríguez de la Garza
Department of Biotechnology, School of Chemistry, Autonomous University of Coahuila, Mexico, Phone: +52 (844) 415-5752–ext 120, E-mail: antonio.rodriguez@uadec.edu.mx

Mónica Margarita Rodríguez Garza
Department of Biotechnology, School of Chemistry, Autonomous University of Coahuila, Mexico

Miriam Soledad Valenzuela Gloria
School of Biological Science, Autonomous University of Coahuila, Torreón–27000, Coahuila, Mexico

Leopoldo Javier Ríos González
Department of Biotechnology, School of Chemistry, Autonomous University of Coahuila, Mexico

Anne Gschaedler
Industrial Biotechnology Unit, Center for Research and Assistance in Technology and Design of the State of Jalisco A.C., Camino Arenero–1227, El Bajio del Arenal, C.P.–45019, Zapopan, Jalisco, Mexico

Guadalupe Gutiérrez-Soto
University of Nuevo León, Faculty of Agronomy, Francisco Villa S/N Col. Ex Hacienda
El Canadá–66415, General Escobedo, N. L., México

Ayerim Hernández-Almanza
School of Biological Science, Autonomous University of Coahuila, Torreón–27000, Coahuila, Mexico,
E-mail: ayerim_hernandez@uadec.edu.mx

Javier Ulises Hernández-Beltrán
Bioremediation Laboratory, Biological Sciences Faculty, Autonomous University of Coahuila,
Torreón–27000, Coahuila, México, Tel.: +52-871-7571-785, E-mail: ulises.hernandez@uadec.edu.mx

Héctor Hernández-Escoto
Chemical Engineering Department, University of Guanajuato, Noria Alta s/n, Guanajuato, 36050, Mexico

R. T. Hilares
Material Laboratory, Catholic University of Santa Maria (UCSM), Yanahuara, AR–04013, Peru

Subburamu Karthikeyan
Department of Renewable Energy Engineering, Tamil Nadu Agricultural University, Coimbatore,
Tamil Nadu, India, E-mail: skarthy@tnau.ac.in

David Francisco Lafuente-Rincón
Bioprocesses and Bioprospecting Laboratory, Biological Sciences Faculty,
Autonomous University of Coahuila, Torreón–27000, Coahuila, México

Juan A. León
Laboratory of Process Intensification and Hybrid System (LIPSH), Department of Chemical Engineering,
National University of Colombia, Manizales, Caldas, Colombia, E-mail: jaleonma@unal.edu.co

Inty Omar Hernández-De Lira
Bioremediation Laboratory, Faculty of Mechanical and Electrical Engineering, Biological Science
Faculty, Autonomous University of Coahuila (UAdeC), Torreón–27000, Coahuila, México

Iosvany López-Sandin
University of Nuevo León, Faculty of Agronomy, Francisco Villa S/N Col. Ex Hacienda
El Canadá–66415, General Escobedo, N. L., México

P. R. F. Marcelino
Bioprocesses and Sustainable Products Laboratory, Department of Biotechnology,
Engineering School of Lorena, University of São Paulo (EEL-USP), Lorena–12.602.810, SP, Brazil

Thelma Karina Morales Martínez
Bioprocess and Microbial Biochemistry Group, School of Chemistry,
Autonomous University of Coahuila, Mexico

Perla Araceli Meléndez-Hernández
Chemical Engineering Department, University of Guanajuato, Noria Alta s/n, Guanajuato, 36050, Mexico

E. Mier-Alba
Bioprocesses and Sustainable Products Laboratory, Department of Biotechnology,
Engineering School of Lorena, University of São Paulo (EEL-USP), Lorena–12.602.810, SP, Brazil

Miguel Angel Medina Morales
Department of Biotechnology, School of Chemistry, Autonomous University of Coahuila, Mexico

Danay Carrillo Nieves
Tecnologico de Monterrey, Escuela de Ingenieria y Ciencias, Av. General Ramon Corona No. 2514,
Zapopan–45201, Jal., Mexico

Bestami Ozkaya
Department of Environmental Engineering, Yildiz Technical University, Turkey, Davutpasa Campus, Istanbul–34220, Turkey

Alan D. Perez
Sustainable Process Technology (SPT), University of Twente, Enschede, Overijssel, The Netherland, E-mail: a.d.perezavila@utwente.nl

Laura Andrea Pérez-García
Faculty of Biological Sciences, Autonomous University of Coahuila, Torreón, Coahuila, México

César D. Pinales-Márquez
Biorefinery Group, Food Research Department, Faculty of Chemistry Sciences, Autonomous University of Coahuila, Saltillo–25280, Coahuila, Mexico

C. A. Prado
Biopolymers, Bioprocesses, Process Simulation Laboratory, Department of Biotechnology, Engineering School of Lorena, University of São Paulo (EEL-USP), Lorena–12.602.810, SP, Brazil

Javier Quintero
Department of Chemical Engineering, National University of Colombia, Manizales, Caldas, Colombia, E-mail: jaquinteroj@unal.edu.co

David Francisco Lafuente Rincón
Faculty of Biological Sciences, Autonomous University of Coahuila, Torreón, Coahuila, México

Cindy Nataly Del Rio-Arellano
Faculty of Biological Sciences, Autonomous University of Coahuila, Torreón, Coahuila, México

Rosa M. Rodríguez-Jasso
Biorefinery Group, Food Research Department, Faculty of Chemistry Sciences, Autonomous University of Coahuila, Saltillo–25280, Coahuila, Mexico, Phone: (+52)-1-844-416-12-38, E-mail: rrodriguezjasso@uadec.edu.mx

Teresa Romero-Gutiérrez
Computer Sciences Department, Exact Sciences and Engineering University Centre, Universidad de Guadalajara, Blvd. Gral. Marcelino Garcia Barragan–1421, Olimpica. Guadalajara, Mexico

D. Rubio-Ribeaux
Bioprocesses and Sustainable Products Laboratory, Department of Biotechnology, Engineering School of Lorena, University of São Paulo (EEL-USP), Lorena–12.602.810, SP, Brazil

Héctor Ruiz Leza
Biorefinery Group, Food Research Department, Faculty of Chemistry Sciences, Autonomous University of Coahuila, Saltillo–25280, Coahuila, Mexico, E-mail: hector_ruiz_leza@uadec.edu.mx

María Alejandra Sánchez-Muñoz
Bioremediation Laboratory, Biological Sciences Faculty, Autonomous University of Coahuila, Torreón–27000, Coahuila, México

S. Sánchez-Muñoz
Bioprocesses and Sustainable Products Laboratory, Department of Biotechnology, Engineering School of Lorena, University of São Paulo (EEL-USP), Lorena–12.602.810, SP, Brazil, E-mails: salvador.sanchez@usp.br; sanchezmunoz.ssm@gmail.com

Dania Sandoval-Nuñez
Industrial Biotechnology Unit, Center for Research and Assistance in Technology and Design of the State of Jalisco A.C., Camino Arenero–1227, El Bajio del Arenal, C.P.–45019, Zapopan, Jalisco, Mexico

J. C. Santos
Biopolymers, Bioreactors, and Process Simulation Laboratory, Department of Biotechnology,
Engineering School of Lorena, University of São Paulo (EEL-USP), 12.602.810. Lorena, SP, Brazil,
E-mail: jsant200@usp.br

Charu Saraf
Department of Bioscience and Biotechnology, Banasthali Vidyapith, Rajasthan–304022, India,
Tel.: 9910331924, E-mail: charusaraf17@gmail.com

Rohit Saxena
Biorefinery Group, Food Research Department, Faculty of Chemistry Sciences,
Autonomous University of Coahuila, Saltillo–25280, Coahuila, Mexico

Shiva
Biorefinery Group, Food Research Department, Faculty of Chemistry Sciences,
Autonomous University of Coahuila, Saltillo–25280, Coahuila, Mexico

S. S. Da Silva
Bioprocesses and Sustainable Products Laboratory, Department of Biotechnology,
Engineering School of Lorena, University of São Paulo (EEL-USP), 12.602.810. Lorena, SP, Brazil,
E-mail: silviosilverio@gmail.com

R. T. Terán-Hilares
Material Laboratory, Catolic University of Santa Maria (UCSM), Yanahuara–04013, AR, Perú

Kalyanasundaram Geetha Thanuja
Department of Agricultural Microbiology, Tamil Nadu Agricultural University, Coimbatore,
Tamil Nadu, India

Daniel Tinôco
Biochemical Engineering Department, School of Chemistry, Federal University of Rio de Janeiro,
Rio de Janeiro–21941909, RJ, Brazil, E-mail: dneto@peq.coppe.ufrj.br

Dogukan Tunay
Department of Environmental Engineering, Yildiz Technical University, Turkey, Davutpasa Campus,
Istanbul–34220, Turkey

Salom Gnana Thanga Vincent
Department of Environmental Sciences, University of Kerala, Thiruvananthapuram, Kerala, India

Oznur Yildirim
Department of Environmental Engineering, Yildiz Technical University, Turkey, Davutpasa Campus,
Istanbul–34220, Turkey

Francisco Zavala-García
University of Nuevo León, Faculty of Agronomy, Francisco Villa S/N Col. Ex Hacienda
El Canadá–66415, General Escobedo, N. L., México, E-mail: francisco.zavala.garcia@gmail.com

Abbreviations

AAA	aromatic amino acids
ABTS	2,20-azino-bis-(3-ethylbenzothiazoline-6-sulfonic acid)
AD	anaerobic digestion
ADH	alcohol dehydrogenase
ADH2	alcohol dehydrogenase II
AFEX	ammonia fiber expansion
ALDH	acetaldehyde by aldehyde dehydrogenase
ARP	ammonia recycling percolation
ATPS	aqueous two-phase systems
BFP	biofuture platform
C4H	cinnamate 4-hydroxylase
CBP	consolidated bioprocessing
CCCF	closed-circulating fermentation system
CEMIE-Bio	Mexican Center for Innovation in Bioenergy
CEs	carbohydrate esterases
CMFS	continuous membrane fermenter separator
CNOOC	China National Offshore Oil Corporation
CO_2	carbon dioxide
COMT	caffeic acid 3-O-methyltransferase
CRISPR	clustered regularly interspaced short palindromic repeats
CS	chitosan
CSIR	Industrial Scientific Research Council
CSTR	continuous stirred tank reactors
CWI	cell wall integrity
DDGs	distillers dried grains
DDG	dry distillation grain
DESs	deep eutectic solvents
DHAP	dihydroxyacetone phosphate
DMC	direct microbial conversion
DP	degree of polymerization
DSSF	delayed simultaneous saccharification and fermentation
DyP	dye decolorizing peroxidase
EA	ethanol adaption
ED	Entner-Doudoroff

ED	extractive distillation
EISA	energy independence and security act
EMP	Embden-Meyerhof-parnas
EPA	energy policy act
EPS	exopolysaccharides
ESR	environmental stress response
EtOH	alcohol
EU	European Union
FISH	fluorescent *in situ* hybridization
FST	fiber separation technology
G3P	glycerol-3-phosphate
GAP	glyceraldehyde-3-phosphate
GBEP	global bioenergy partnership
GHG	greenhouse gas
GRAS	generally recognized as safe
gTME	global transcription machinery engineering
GW	global warming
H_2O_2	hydrogen peroxide
HAA	3-hydroxyanthranilic acid
HAD	homogenous azeotropic distillation
HC	hydrodynamic cavitation
HCl	hydrochloric acid
HCR	hydrodynamic cavitation reactors
HCT	hydroxycinnamoyl transferase
HER	hydrogen evolution reaction
HMF	5-hydroxymethylfurfural
HOG	high osmolarity glycerol
HPLC	high-performance liquid chromatography
HR	homologous recombination
HSP	heat shock proteins
HTAD	heterogeneous azeotropic distillation
HTLR	hydrothermal liquefaction reactors
HyPol	hyperbranched polymer
iHG	integrated high gravity
IL	ionic liquids
IMHeRE	intelligent microbial heat-regulating engine
ISC	iron-sulfur clusters
ISPR	*in situ* product recovery
KDPG	2-keto-3-deoxy-6-phosphogluconic acid

LBG	liquefied biogas
LCA	life cycle assessment
LCB	lignocellulosic biomass
LDH	lactate dehydrogenase
LHW	liquid hot water
LiP	lignin peroxidase
LLX	liquid-liquid extraction
MAVS	membrane assisted vapor stripping
MDC	microbial desalination cell
Mdh	malate dehydrogenase
MEC	microbial electrolysis cell
MEDC	microbial electrodialysis cell
MES	microbial bioeletroshyntesis
MFC	microbial fuel cell
MnP	manganese peroxidase
MREC	microbial reverse electrodialysis electrolysis cell
MWIR	microwave irradiation reactor
MWR	microwave reactors
NGS	next-generation sequencing
NRC	national research council
OD	oven-dried
OPEC	Organization of Arab Petroleum Exporting Countries
PA	phosphatidic acid
PDC	pyruvate decarboxylase
PDMS	polydimethylsiloxane
PFD	prefoldin
Pgd	phosphogluconate dehydrogenase
PKA	protein kinase A
PM	particle matters
PPP	pentose phosphate pathway
PSA	pressure swing adsorption
PSSF	pre-saccharification and fermentation strategy
PTA	phosphotransacetylase
PTMSP	poly(1-trimethylsily-1-propyne)
PV	pervaporation
RDB	renewable diesel blendstock
REFINE	RNA seq examiner for phenotype-informed network engineering
RFS	renewable fuel program
RGI	rhamnogalacturonan I

RKI1	ribose-5-phosphate isomerase
RNAi	RNA-interference
ROS	reactive oxygen species
SAA	soaking aqueous ammonia
SAGA	Spt-Ada-Gcn-5 acetyltransferase
SARSH	sulfuric acid-treated rice straw hydrolysate
SBP	sugar beet pulp
SBR	sequencing batch reactor
SE	steam explosion
SHCF	separate hydrolysis and co-fermentation
SHF	separate hydrolysis and fermentation
SMT	selective milling technology
sRNA	small RNA
SSCF	simultaneous saccharification and co-fermentation
SSF	simultaneous saccharification and fermentation
STBR	stirred tank bioreactors
STR	stirred tank reactor
Tal	transaldolase
TALEN's	transcription activator-like effector nuclease
TAME	TALEN assisted multiplex editing
TEM	transmission electron microscopy
TEMPOL	4-Hydroxy-2,2,6,6-tetramethylpiperidin-1-oxyl
TF	transcription factors
TiO_2	titanium dioxide
TKL1	transketolase
TPP	trehalose-6-phosphate phosphatase
TPS	trehalose-6-phosphate synthase
VP	versatile peroxidase
WW	water washing
XDH	xylitol dehydrogenase
XGA	xylogalacturonan
XI	xylose isomerase
XOs	xylooligosaccharides
XR	xylose reductase
XYL1	xylose reductase heterologous
ZFN	zinc finger nuclease

Acknowledgment

The editors thank the researchers who participated as reviewers of the submitted chapters. Their dedicated time in the revision and the experience that each one has in the different topics contributed to the improvement and quality of this book.

- **Raúl Rodríguez Herrera**
  Autonomous University of Coahuila, Mexico

- **Arturo Sócrates Palacios**
  ESPOL Polytechnic University, Ecuador

- **Leopoldo Javier Ríos González**
  Autonomous University of Coahuila, Mexico

- **Alejandra Alvarado**
  University of Tübingen, Germany

- **Danay Carrillo Nieves**
  Tecnológico de Monterrey, Mexico

- **Aldo Ricardo Almeida Robles**
  University of Copenhagen, Denmark

- **Miguel Ángel Medina Morales**
  Autonomous University of Coahuila, Mexico

Preface

Worldwide, various industries generate waste during the production processes they implement. These wastes range from vegetable material (such as seeds, skin, bagasse, among others) to synthetic material (such as plastics). In most cases, it is a serious environmental problem; however, some plant residues, mainly generated by food industries, have been analyzed due to their high biological potential. A clear example is the use of corn and sugarcane bagasse, rich in sugars, to obtain bioethanol. Bioethanol is one of the most interesting biofuels since it has a positive characteristic on the environment.

During the last decades, there has been a great interest in the production and use of biodiesel or bioethanol as promising alternatives to replace fossil fuels. For this reason, the development of biorefineries and processes, as well as the design of bioreactors, have been fundamental issues to understand this area. Likewise, the study of metabolic and physiological processes that are carried out by some bioethanol-producing microorganisms is also necessary to know. Yeast, like *Zymomonas mobilis* and *Clostridium thermocellum,* have been demonstrated to have a prominent role in bioethanol production. However, it is necessary to delve into factors such as increasing the tolerance of these strains to produce bioethanol, improve genetic regulation and implement genetic engineering methods that favor production yields.

On the other hand, the development of tools and strategies that allow obtaining bioethanol of major quality, viable alternatives for purification of the product are factors that still have to be improved. Bioethanol production has proven to be a promising option to reduce the damage caused by the conventional processes used to obtain fuels; therefore, there are still some challenges to overcome.

This book includes the advances and perspectives within the bioethanol industry, also describes some biochemical and physiological parameters carried out by the main bioethanol producing microorganisms as well as the potential applications that this bioproduct can have and the advantage that it would generate.

—**Editors**

CHAPTER 1

Physiology of Ethanol Production by Yeasts

MIRIAM SOLEDAD VALENZUELA GLORIA,[1]
DIANA LAURA ALVA-SÁNCHEZ,[1] M. P. LUÉVANOS ESCAREÑO,[1]
CRISTÓBAL N. AGUILAR,[2] NAGAMANI BALAGURUSAMY,[1] and
AYERIM HERNÁNDEZ-ALMANZA[1]

*[1]School of Biological Science, Autonomous University of Coahuila,
Torreón–27000, Coahuila, Mexico,
E-mail: ayerim_hernandez@uadec.edu.mx (A. Hernández-Almanza)*

*[2]Bioprocesses and Bioproducts Research Group, BBG-DIA, Food Research
Department, School of Chemistry, Autonomous University of Coahuila,
Saltillo–25280, Coahuila, Mexico*

ABSTRACT

Yeasts have been gaining popularity due to their ability to produce a series
of compounds of interest such as pigments, phenolic compounds, fatty acids,
enzymes, and even under the right conditions, and they are ethanol producers.
The latter is characterized by producing ethanol and carbon dioxide (CO_2)
under anaerobic conditions using fermentable sugars as substrates. To
consider that yeast is suitable for the production of ethanol, it is desirable
that it meets some characteristics of adaptability to the use of different
sources of carbon and nitrogen, to acidity or low availability of glycerol, and
even tolerance to high levels of ethanol concentration. In addition to these
criteria, it is also important to consider that the fermenting strain must be
able to use hexose and pentose and tolerate the inhibitory by-products of the
pretreatment. In such a way that by manipulating both carbon and nitrogen
sources, as well as environmental factors, it would be possible to increase
or decrease ethanol production, which requires knowledge of the species to
be used. This is widely used in the industry, in which over time, the use of

ethanol has been diversifying more and more, so what began to be used in the food industry for the production of alcoholic beverages, today it is used for the production of sustainable alternative fuels, which have gained impact in recent years thanks to their profitable production thanks to the low cost of substrates and their efficient fermentation.

1.1 INTRODUCTION

Yeast can produce compounds of interest such as pigments, phenolic compounds, fatty acids, enzymes, among others. Also, yeasts have been used in various industrial fermentation processes due to their ability to convert high concentrations of sugars into ethanol and CO_2; for example, Saccharomyces cerevisiae has been exploited for centuries for the production of alcoholic beverages. Currently, ethanol obtained from this via is an alternative to be used as a substitute for gasoline [1–3].

The ethanol that is produced through fermentation represents a positive alternative as fuel to petroleum, as a source of energy for batteries through electrochemical effects, as energy in the generation of energy by thermal combustion, etc. [4]. Ethanol is significantly less toxic to humans than is gasoline in such a way that it reduces air pollution thanks to its low volatility, photochemical, and waste activity [5]. Ethanol obtained from waste materials from biomass or renewable sources is called bioethanol, and it can be used as fuel, chemical raw material, and solvent in various industries.

The production of ethanol as liquid fuel is obtained by fermentation from biomass, sugar cane, cereal grains, and sugar beets, and it is gaining a lot of popularity around the world [4, 6]. If you have greater control of the fermentation conditions, this can contribute to the reduction of stress towards yeast cells and decrease contamination by bacteria and wild yeasts. Therefore, having large information gaps around the ethanol production processes, leading to large investigations in order to achieve the generation of higher quality products and more optimized processes [7].

1.2 PRINCIPAL ETHANOL-PRODUCING YEASTS

Ethanol-producing yeasts are characterized for producing ethanol and CO_2 in anaerobic conditions using fermentable sugars as substrate [8]. The principal attributes for considering a good ethanol-producing yeast for their use in an industry are diverse as high tolerance levels of ethanol concentration, acidity, high temperature, low glycerol formation, capacity for use different sources of carbon and nitrogen. Also, it is important to observe the tolerance and inhibitors of the yeast in production about biomass hydrolyzed [9–12].

Saccharomyces cerevisiae is the more globally used yeast for industrial production of ethanol [13, 14]; Brazil, the United States, European Union (EU), and China being the main producers of ethanol (Renewable Fuels Association). Furthermore, S. cerevisiae is an ideal yeast for the industry due to easy manipulation with molecular methods such as genetic engineering since its genome has been extensively studied [15, 16]. The studies have been characterized for the improvement of the strains using by-products as substrate rich in sugars for the production of ethanol (Table 1.1). For example, yeasts such as *Schizosaccharomyces pombe*, *Candida krusei*, *Kluyveromyces marxianus*, *Dekkera bruxellensis*, *Pichia striptis*, *Pichia kudriavzevii*, *Wickerhamomyces anomalous*, among others; they have been isolated and identified as producing ethanol with good behavior and tolerant to high concentrations of alcohol (EtOH) [17–19].

The main factors that contribute to the low growth of yeasts in the process and that lead to high ethanol yields characterized in 90–92% of the theoretical conversion of sugar to ethanol are high cell densities, cell recycling, and high ethanol concentration [20–22].

1.3 BIOCHEMISTRY OF ETHANOL PRODUCTION

The generation of ethanol from lignocellulosic biomass (LCB) is possible through three stages. The first of them consists in allowing the biomass to have a better management through a treatment for the next stage, which consists of subjecting the treated biomass to an enzymatic hydrolysis process in order to improve the disposition of simple sugars, such as glucose and xylose. Finally, thanks to the availability of sugars, it is possible to carry out fermentation through the use of different microorganisms [9, 11, 24]. A generally simplified representation of the process for ethanol production from lignocellulosic materials by chemical hydrolysis is shown in Figure 1.1.

The treatment given to the biomass prior to enzymatic hydrolysis is nothing more than through different methods it is possible to modify its physicochemical properties in order to facilitate the enzymatic work. However, this can also bring consequences such as crystallization [24]. Parallel that, there is an increase in both the size of the internal surface and its pore volume; this being also an adjuvant for the enzymatic work [25]. In

TABLE 1.1 Ethanol Production by Yeast Under Different Conditions

Yeast	Substrate (%)	pH	Temperature (°C)	Ethanol (%, v/v)	Condition	Reference
Saccharomyces cerevisiae (CDBT2)	Glucose (5)	5.5	30	19.8	Electrochemical cell (4V[1])	[18]
Saccharomyces cerevisiae (UVNR56)	Molasses medium (28)	–	37	10.3	UV-C[2] radiation	[22]
Saccharomyces cerevisiae UAF-1	Molasses media (27)	Adjusted 4.0–4.5	–	12.2	VHG[3] technology	[23]
Wickerhamomyces anomalous (CDBT7)	Glucose (5)	5.5	30	23.7	Electrochemical cell (4V)	[18]
Kluyveromyces marxianus (YZB014)	Xylose (5)	–	45	5.2	Modified strain (recombinant)	[19]
Pichia stipitis (PXF58)	Xylose (11.4)	–	30	4.3	UV-mutagenesis[4]	[23]

[1] *V = VOLTS;*

[2] *VGH = Very high gravity;*

[3] *UV-C = Ultraviolet-C;*

[4] *UV-mutagenesis = Ultraviolet mutagenesis.*

response to all of the above, there is a significant improvement in the yield rate of monomeric sugars [11].

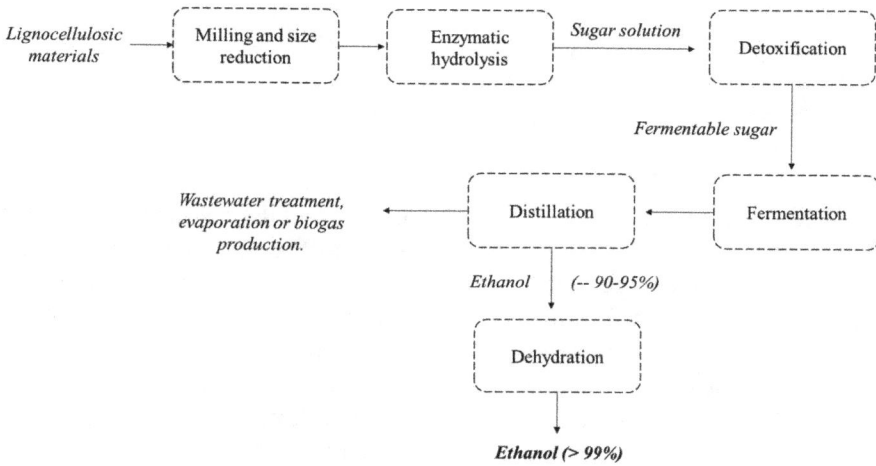

FIGURE 1.1 Rough diagram of the production of ethanol from lignocellulosic materials.

The complex and irregular reaction by which insoluble cellulose manages to defragment into solid-liquid interfaces thanks to the simultaneous action of cellobioses, exoglucanases, and endoglucanases, is better known as the enzymatic hydrolysis process. This primer reaction is accompanied by further liquid-phase hydrolysis of soluble intermediates, such as celluloligosaccharides and cellobiose, mainly, which through catalytic reactions are ungrouped in order to produce glucose through the action of β-glucosidase [27]. In the short term, enzymatic hydrolysis is in charge to convert LCB to fermentable sugars through a series of biochemical processes.

Fermentation in this context comprises the action of submitting the LCB through a catabolic process of incomplete oxidation, which does not require oxygen, and which final product is an organic compound. Across this process, the pentoses and hexoses obtained from hydrolysis get fractioned thanks to the presence of fermenting microorganisms, such as some bacteria, algae, yeasts, and even some fungi, whether natural or recombinant [28].

In order to explain this complex process in a more comprehensible way we can say that central metabolism begins with the basic conversion of sugars to pyruvate, producing energy in the form of ATP and reduced NADH cofactors, where pyruvate divergence after glycolysis acts as an essential regulatory point in metabolism [11]. As a result of this process, pyruvate manages to have the choice of either following the fermentation route or breathing. In the case of eukaryotes, this depends on the presence of oxygen. That is, under aerobic conditions, pyruvate will be converted to acetyl-CoA by the actions of a pyruvate dehydrogenase and will be directed toward the citric acid cycle. In counterpart, under anaerobic conditions, pyruvate is diverted to fermentation [13]. Where, the conversion of pyruvate to ethanol is a two-step process [1, 25, 29]. First, through the action of pyruvate decarboxylase (PDC) on pyruvate, acetaldehyde, and CO_2 are obtained, the latter as waste. In this section, three enzymes come into action, which are encoded in the yeast genome in question, which is their importance in that they act as a key point of metabolic branching between fermentation and respiration. In this process, a direct competition is generated in the action of PDC, between its use for the production of ethanol and the maintenance of the balance of pyruvate availability in the metabolic pathway. Subsequently, acetaldehyde is converted to ethanol by an alcohol dehydrogenase (ADH) [13, 25]. Due to its nature as an oxidoreductase it manages to promote the reversible interconversion of alcohols and the aldehydes/ketones themselves [25]. ADH has available a number of substrates present in the different metabolic pathways, which results in the need for a strict order in order to preserve the homeostasis of the intermediates and products [1]. Therefore, this is why eukaryotes and even humans have numerous ADH enzymes. Of which it is possible to mention the enzymes Adh1, Adh2 and Adh3 [11]. The Adh1 enzyme is crucial during fermentation, since it replenishes the NAD + assembly in order to produce ethanol; In addition, glucose manages to suppress the enzyme Adh2, thus allowing the oxidation of ethanol, this happens only under certain conditions of need. And last but not least, Adh3 is expressed in the mitochondria allowing to carry out its main function which is to maintain the redox balance of the process [1, 25, 29].

1.4 PHYSIOLOGICAL GROWTH AND SUBSTRATE UTILIZATION

Remarkably, under aerobic conditions and with the correct supplies of glucose, ammonium salts, and inorganic ions, most of the yeast species studied could survive successfully, being organisms with relatively simple nutritional needs. Where for its adequate growth it would be necessary to involve macronutrients and micronutrients, administered in millimolar and micromolar concentrations, respectively. We commonly call macronutrients sources of carbon, nitrogen, among others; and micronutrients include trace elements, such as calcium [30].

In consequence, nutritional requirements and specific characteristics of yeasts vary considerably according to the substrates used for their growth and/or fermentation, plus the environmental conditions in which it is found, in addition to the variety of the strain in question [30–32]. Some simple sugars as glucose or fructose can be easily harnessed by yeasts since these are assimilated right after being hydrolyzed either inside or outside the cell [33].

For the growth process of yeasts not all sugars are used simultaneously, the process is started with the fermentation of sucrose with the help of the enzyme invertase, they cause a hydrolysis allowing the generation of glucose and fructose for the use of yeast, during the initiation phase within fermentation, glucose is used in a greater proportion than fructose [34], due to the glycolytic property of yeast. When the values of these saccharides decrease under among concentration threshold maltose consumption increases if its available, which is variable due to its entry into the cell thanks to the action of maltose-permease enzyme; when it is present, it proceeds to the gradual use of maltotriose [34, 35].

To mention some of the yeasts capable of fermenting maltotriose, maltose, sucrose, glucose, and even fructose, they are *Saccharomyces cerevisiae and Saccharomyces uvarum*. The importance of fermentable sugars derives from their usefulness for the production of ethanol, CO_2, and some representative amino acids [36, 37].

The presence of carbon sources in a critical level turns cells metabolism from respiratory growth into fermentative growth. Among of fermentable sugars, glucose is the one who suppresses the enzyme that intervenes in the metabolism of other sugars and in respiratory growth. Then when glucose finds at critic concentration is converted into ethanol and CO_2, this in spite of a surplus of air; the optimal range for this to occur would be from 35 to 280 mg/L, which is equivalent to 5% of concentration [35, 38, 39].

Over the course of fermentation, a small amount of oxygen is present in the broth, which goes hand in hand with the tolerance to ethanol and the viability of the cell, a fact that allows the synthesis of sterols and is related to fatty acids unsaturated around the membrane [30, 35].

In a broader perspective on the carbon sources, there may be decomposed into four different sections; starting with the simple disaccharides also known as hexoses, continuing with the starches, mainly of vegetable origin, to this triad are also joined the lignocellulosic materials, and finally the agro-industrial waste [25, 33, 36]. It is important to mention that each one has its own pros and cons, for example, in terms of simple sugar-based materials, they can be easily and immediately used as the sole substrate by microbial strains capable of producing ethanol [25]. On the other hand, due to the excess of agricultural production, the use of starch for the production of ethanol has emerged, however, it is recognized that the levels of performance of this are too low to achieve the absolute replacement of gasoline [36]. After this is mandatory to mention the use of lignocellulosic materials as very rentable source for its low cost and easy obtention, without leaving aside the energy required to obtain the necessary sugars from LCB, it is important to mention that the main disadvantage is the availability of said substrates [25, 36].

Furthermore, the complexity of the substrate is what will dictate whether a hydrolysis is used, either enzymatic or acidic, or if it is worked in a crude way, the latter being applicable to agro-industrial waste. Therefore, for this to suck, it is necessary to provide a rich source of nitrogen, since not all yeasts are capable of using mineral nitrogen [33, 34]. Even within these there is a classification; first, we have the sources of mineral nitrogen where both ammonium salts and urea are located, as they have an assimilation capacity almost on par; second, there are the sources of organic nitrogen, where the presence of glutamic and aspartic acids are noted, as well as their amines (asparagine) in both D and L forms [35]. The aim of nitrogen present in nitrogenous compounds is that it dissociates in the form of ammonium ions and hydrogens that provide alkaline pH values to the growth medium, achieving this by adding such compounds.

It is important to mention that some yeasts and even filamentous fungi use nitrate as a source of nitrogen, whose assimilation happens thanks to the reduction to ammonia due to the action of the enzyme ammonium reductase [40].

1.5 YEAST PHYSIOLOGY AND ENVIRONMENT PARAMETERS

The environmental parameters, better known as extrinsic factors, are those that, despite not being part of the study subject, are still capable of causing changes in their properties. As for yeasts, the most notable could be said to be the temperature, pH, and surface tension of oxygen, in some cases the activity of water is considered. These parameters make the yeasts modify both their structure and some defense mechanisms, all in order to survive under the new conditions. In other words, these adaptations could be done by two ways, either by changes in their genetic makeup or by changes in their phenotype [35, 41]. In addition, the extreme exposure to these factors can have serious consequences for the ethanol production process, where the least of them would be a low yield in the recovery of ethanol and the greatest could be to kill the yeast [41].

Thus, at the time when the yeasts are subjected to certain light doses of tension, some of their cells express a characteristic called cross-protection, with which they become resistant to large and generally lethal doses of other tensions. This characteristic is developed due to the activation of a specific and general stress response program, known as the environmental stress response (ESR), which is in charge of regulating the genes necessary to survive in adverse conditions along with providing defense mechanisms against changing conditions. Furthermore, this mechanism plays a potential role in the induction of mutagenesis in the presence of stresses that can cause genetic instability in microorganisms, in other words, genetic mutation [42, 43]. Within the concept of ESR, it is important to mention cell viability and vitality, which are defined as the ability to reproduce and the metabolic activity of a culture, respectively. So, for these two conditions to occur during propagation, the participation of storage carbohydrates, such as glycogen and trehalose, is necessary. Glycogen serves as the main carbohydrate that stores energy in yeast allows the synthesis of sterols, trehalose, and fatty acids, throughout the delay stage. Due to this, it is desirable that the growth media contain high concentrations of said polysaccharide so that the synthesis of sterols and fatty acids is carried out, which represent a high expenditure of energy. Glycogen accumulation is the consequence of the limited availability of nitrogen or carbon. As for trehalose, it is considered to be a protective oxidative disaccharide in situations of cellular stress such as high levels of osmolarity, decreased nutrients, hunger, high, and low temperatures, and high concentration of ethanol, which in high levels intracellular concentration allows cell viability through the initial stages of fermentation and, therefore, increases the rates of carbohydrate utilization [42, 44].

1.5.1 TEMPERATURE AS AN EXTRINSIC FACTOR

Temperature is the main factor that affects the production of ethanol, due to the fermentation process is monitored based on the development of heat, that is, the increase in heat energy in response to the metabolic activities of microorganisms. One of the ways of inducing heat stress would be the absence of cooling, which would lead to reduced growth of microorganisms and fermentation defects. Therefore, the production of ethanol in tropical countries has been managed at high temperatures in order to achieve a reduction in costs in the cooling process. In addition, high temperature management during fermentation provides a number of advantages, such as efficient saccharification and fermentation, a constant change from fermentation to distillation and even minimal risk of contamination. To achieve this type of fermentation it is essential to use a competent yeast strain that can tolerate stress conditions, so it is essential to know the strain to work in order to make a correct adaptation [1, 42, 45].

Warm, sugary, acid, and aerobic is as it would be described as the ideal medium for growing most yeast species and some fungi. According to the literature, the optimal growth temperature of yeasts is between 5°C and 37°C in general, since this varies according to the physiology of each strain [11, 35, 41, 44, 47], these values cannot be literally applied in practice because there are other factors that exert changes in the times of generation or growth. As a result of the fact that this range is quite wide, we could shorten it by saying that in general most species grow very well around 25°C [30].

As mentioned in paragraphs preceding high temperature stress (or heat shock) in yeast cells, heat damage can alter hydrogen bonds and even hydrophobic interactions, leading to the unfolding of proteins and nucleic acids. Therefore, anaerobic fermentation, being an exothermic reaction, ends up increasing the temperature in some species, which are known as thermotolerant, thus reaching fermentation temperatures above 40°C. The term "thermotolerant" is used for those yeasts that have the transient ability in their cells to survive subsequent lethal exposures at elevated temperatures, that is, after a sudden thermal shock. Heat shock responses in yeast occur when cells move rapidly at elevated temperatures, and if this is near-fatal, it will produce "heat shock proteins (HSPs)" with a high level of conservation, in response to the induced synthesis of a specific set of proteins. HsP perform numerous physiological functions, including thermo-protection [11, 30, 41, 44].

1.5.2   INFLUENCE OF pH

Fermentative yeasts are acidophilic mesophiles; therefore, they manage to ferment under a pH of acid range (4 to 6). This characteristic provides us with a great advantage by acting as a growth inhibitor of some agene micro-organism in the medium, that is, it acts as a microbiological control agent [28, 30]. This is why yeast culture media acidified with organic acids (for example, acetic, lactic acid) are better at inhibiting growth compared to those acidified with mineral acids (for example, hydrochloric, phosphoric acids), since organic acids are responsible for lowering the intracellular pH just after its translocation through the fungal plasma membranes [28, 47]. Exposure to organic acids causes cells to deplete their energy (ATP) when they strive to maintain pH homeostasis through the activities of the proton-pumping ATPase complex in the plasma membrane. This forms the basis for the action of weak acidic preservatives to inhibit the growth of fungi and/or yeast from food breakdown [47]. *S. cerevisiae* is the most commonly used yeast in the industrial production of ethanol, since it tolerates a wide pH range, making the process less susceptible to infection [11].

1.5.3   OXYGEN SURFACE TENSION

A factor that influences fermentation and improves biochemical processes is the aeration of the must, since the presence of oxygen in the initial fermentation phase is essential for the rapid reproduction of yeast and the complete restoration of sugar. By maintaining the correct amount of glycogen and trehalose, the correct aeration improves the vitality of the biomass and the immunity of the yeast cells to different tensions; carrying out in this way a selective exchange of metabolites and a correct extraction of nutrients from the environment. When a low amount of oxygen is present in the medium, a delay in fermentation or a change in the sensory properties of the product is caused [48].

The aeration of a culture has the advantage of providing the necessary oxygen for respiration and the elimination of CO_2 produced by the metabolism of carbonate compounds. Therefore, the supply of oxygen through aeration must be sufficient to avoid the generation of alcohol (EtOH) throughout the fermentation process and it is recommended that the added air be rich in oxygen. Then, during growth and metabolism, the transfer of nutrients, and solid, solid, and gaseous metabolites between the environment and the cell is essential and continuous, this is where the solubility of the gas enters the liquid phase, so that it allows the exchange between phases. liquid and gaseous [35, 49, 52].

Yeasts require appropriate amounts of oxygen to carry out their oxidative metabolism and thus proceed with the optimal production of ethanol; by using this gas, they are able to synthesize unsaturated fatty acids and sterols, which are necessary for continuous anaerobic growth and cell division. Even *S. cerevisiae*, which is anaerobic in growth, requires small amounts of molecular oxygen for the synthesis of fatty acids and sterols. Furthermore, pentose fermentation yeasts require very small amounts of oxygen, which must be constantly monitored. So, since the fermentation rate is proportional to the number of metabolically active yeasts present, this means that the attenuation of the medium goes hand in hand with the availability of oxygen

[28, 48]. In consequence, insufficient aeration can lead to insufficient yeast revitalization, increasing deficiency, and low fermentation rates. In another hand, excessive aeration can lead to a large amount of biomass. Therefore, when yeasts are not aerated, they only grow during fermentation, and sugar consumption is also low [48].

Surprisingly, sterols and fatty acids become a limiting factor because industrial fermentation takes place without aeration. Whereas if the inhibitors present in the lignocellulosic hydrolysates are used in high concentrations of cell mass, this will affect the volumetric productivity of those fermentations inoculated with lower cell concentrations [44]. This is because optimization of aeration not only leads to an increase in the speed ratio of the fermentation process, but also improves the final characteristics of the product. Yeast with low oxygen requirements generally produces a slightly higher number of esters, especially the desired isoamyl acetate [48].

1.6 YEAST PHYSIOLOGY AND ETHANOL PRODUCTION PROCESSES

The process of producing ethanol through yeasts is dictated by the raw material to be used. Hence, this process is mainly divided into three phases:

- Obtaining fermentable sugars;
- Fermentation of sugars (production of ethanol); and
- Separation and purification of ethanol.

Raw materials are pre-treated in order to reduce their size and therefore facilitate subsequent processes. The hemicellulose and cellulose will then hydrolyze to fermentable sugars. The yeasts then begin to work to ferment these sugars into ethanol. Finally, the recovery of ethanol is carried out by means of various separation technologies, and thus it can be used as fuel [28].

Over time fermentation under high levels of stress has been improving in terms of efficiency and tolerance, so some strains of yeast have begun to gain strength within the industry. According to the literature, it is said that there is a close relationship between fermentation efficiency and resistance to stress, that is, the ability of a yeast strain to develop adaptability to inappropriate environments and growth conditions. Thus, industrial fermentation occasionally does not complete or progress at a slower rate. This is mainly due to slow and stagnant fermentations. High concentrations of ethanol, carbohydrate stress and/or heat are some of the factors that can cause abnormal and non-optimal fermentation [42].

A typical batch growth curve is composed of lag, exponential, and stationary phases, this is a response to the inoculation of fungal cells in media with the necessary nutrients and conditions [42]. First, the period of zero population growth is called the lag phase, where the inoculated cells begin a stage of adjustment to their new chemical and physical environment (synthesizing ribosomes and enzymes) [44]. Second, the exponential phase is a period of doubling of logarithmic cells (or mycelial biomass in the case of filamentous growth) and a constant maximum specific growth rate (μmax, in reciprocal dimensions of time, per hour), the value of which accurate depends on the prevalence of growing conditions [45]. During the exponential phase of balanced growth, cells carry out a rudimentary metabolism, that is, they begin to run those metabolic pathways essential for cell growth [42]. When the final objective of fermentation is both to achieve the highest levels of biomass production and the optimal extraction of primary metabolites, what is sought is to extend the growth phase, regularly by means of batch or continuous fed cultures. Consequently, they begin with the stationary phase, in which fungal biomass manages to remain constant and the growth rate returns to zero [41]. After prolonged periods in the stationary phase, individual cells can die and self-autolyze [48]. The stationary phase is characterized by being the stage in which the microorganism reaches long periods of survival without the need for the addition of nutrients. Not only the absence of nutrients can induce the stationary phase, there are other facilitating causes such as the presence of toxic metabolites (for example, ethanol in the case of yeast), low pH, high CO_2, variable O_2 and high temperature [30]. During the stationary phase of unbalanced growth, fungi can undergo secondary metabolism, specifically initiating metabolic pathways that are not essential for cell growth but are involved in the organism's survival. Therefore, the industrial production of secondary fungal metabolic compounds such as *penicillin* and *ergot alkaloids* essentially compromise the stability of the cell population during the stationary phase [30, 41].

There are three processes that are commonly used in the production of bioethanol which are separate hydrolysis and fermentation (SHF), simultaneous saccharification and fermentation (SSF), and simultaneous saccharification and co-fermentation (SSCF). In SHF, the hydrolysis of lignocellulosic materials is separated from ethanol fermentation, in which the separation of enzymatic hydrolysis and fermentation makes it easier for the enzyme to operate at high temperature for better performance, whereas fermentation organisms can be operated at a moderate temperature to optimize the use of sugar. In contrast, the other two methods (SSF and SSCF) have a short general process since the enzymatic hydrolysis and fermentation process occurs simultaneously to keep the glucose concentration low; that is, for SSF, the fermentation of glucose is separated from the pentose while the SSCF ferments the glucose and the pentose in the same reactor. In turn, these processes offer a series of benefits such as lower cost, higher ethanol yield and less processing time [49–52].

Bioethanol fermentation can be carried out in batches, batch feeds, repeat batches or in continuous mode. In batch process, the substrate is provided at the beginning of the process without adding or removing the medium; this system is known as simpler bioreactor with multi-vessel control process, flexible, and easy; where, the fermentation process is carried out in a closed-circuit system with a high concentration of sugars and inhibitors at the beginning and ends with a high concentration of product. Therefore, the batch system offers a number of benefits, which include complete sterilization, requires no job skills, is easy to handle raw materials, can be easily controlled, and flexible to various product specifications. However, productivity is low and requires high and intensive labor costs. The presence of a high concentration of sugar in the fermentation medium can lead to inhibition of the substrate and inhibit cell growth and the production of ethanol [1, 11, 41].

Cellular Recycling Batch Fermentation is a strategic method for effective ethanol production, as it reduces the time and cost of inoculum preparation; some of its advantages are easy cell harvesting, stable operation, and long-term productivity. Besides, sugar materials and immobilized yeast cells are used as facilitators in the cell separation necessary to carry out cell recycling [37, 48].

1.7   INDUSTRIAL IMPORTANCE OF ETHANOL

Ethanol is a chemical compound that has been produced for thousands of years for human consumption, the fermentation process of ethanol is obtained from raw materials. In recent years, the fermentation process for the production of ethanol has received great attention for its chemical and edible purposes. The best-known use of ethanol is as an alcoholic beverage [55], the increasing demand for various industrial applications that include its use to preserve biological samples, as a solvent in the manufacture of perfumes, varnishes, adhesives, pesticides, in the preparation of pharmaceutical products such as medicines and drugs, essences, deodorizers, and as a disinfectant, among many others, that has requited increasing production of ethanol [54]. In a relatively short period of time, technologies have been developed for the production of a wide range of bio-based products. Consequently, making the production of liquid fuels the focus of attention due to the great market potential for the replacement of oil in the fuel supply. Ethanol has the longest history and is the first commercially produced biomass-derived liquid fuel [51].

In Brazil the application of ethanol as fuel has obtained a great boom, since it is possible to manufacture at a low cost through the fermentation of sugar cane [56], and in the US, corn is the dominant biomass feedstock [55]. For its application fails to be more profitable various mixtures of ethanol are used with oil [57]. It can be implemented in neat form or in a mixture with gasoline. It can be used in percentages up to 10% with gasoline in common spark-ignition engines without further modification or can be operated at higher percentages in mixtures with gasoline in slightly modified engines which are called flexi-fuel vehicles. It is possible to use pure gasoline or ethanol blends up to 85% (even 100% ethanol in Brazil) in flex-fuel engines [58].

The manufacture of these biofuels has increased since 2000 [60]. This being the case, ethanol is more significant when it is considered according to the volume produced. Today, the United States and Brazil are the world's largest ethanol producers, able to generate ethanol production exceeding 94 billion liters per year, which represents around 85% of world production [61]. In particular, Brazil is the leading manufacturer of ethanol for automobiles, with an annual production rate of 4 billion gallons of ethanol from sugar [62].

In 2016, biofuels supplied around 4.5% of total fuel for road transportation worldwide and its planned share of world transport fuel by 2050 is estimated at 25%. [53]. As the lowest-cost manufacturer globally, the United States remains to retain its place as the most reliable and affordable source of ethanol internationally, reaching up to 57.8 billion liters of global manufacture in 2016. Brazil, which produced roughly 27.6 billion L, is accountable for about 27% of world production, while the EU follows with 5%. Other proper leaders to mention are China and Canada about ethanol manufacture [57].

The geographic distribution of the production and consumption of ethanol is related to many factors, such as manufacturing destinations, government policies, natural resources availability, and environmental regulations. Different world regions can be perceived as distinct markets with diverse demands and supply. Estimates denote that the USA is clearly the largest manufacturer and consumer of ethanol and it is followed by Brazil [63]. With regards to assessing these main producers, a difference from the perspective of the increase in production by 2020 is noted [59].

1.8 FINAL REMARKS

Finally, it is important to reaffirm the ability of yeast cells to grow, metabolize complex industrial raw materials and withstand the hostile environments of large-scale fermenters. Throughout this chapter, important considerations of yeast stress and nutritional physiology that influence yeast fermentative activities have been addressed. In addition to the yeast optimization and improvement strategies used in alcohol production, it is a priority to know the yeast strain to use in order to carry out the correct approach when carrying out the fermentation process. Even today, yeasts continue to be one of the least understood organisms, and in the same way they are among the most important as input in ethanol production processes. Even so, under the correct conditions of both feeding and environment, it is currently possible to obtain volumes of more than 20% of ethanol production, when these conditions are not met, leading to stress in the lead, stuck, slow, and inefficient fermentations are obtained. Therefore, it is essential to optimize alcoholic fermentations to understand aspects of yeast cell physiology, particularly when lignocellulosic substrates are used for the production of second-generation bioethanol.

KEYWORDS

- alcohol dehydrogenase
- bioethanol production
- biorefinery
- growth conditions
- lignocellulosic biomass
- yeast physiology

REFERENCES

1. Dzialo, M. C., Park, R., Steensels, J., Lievens, B., & Verstrepen, K. J., (2017). Physiology, ecology, and industrial applications of aroma formation in yeast. *FEMS Microbiol. Rev., 41*, S95–S128. https://doi.org/10.1093/femsre/fux031.
2. Buzzini, P., (2006). Yeast biodiversity and biotechnology. *Yeast Handbook; Biodivers. Ecophysiol. Yeasts*, pp. 533–559. https://doi.org/10.1007/3-540-30985-3_22.

3. Van, D. S. P., (2015). Approaches to production of natural flavors. In: Parker, J. K., Elmore, J. S., & Methven, L. B., (eds.), *Flavor Development, Analysis and Perception in Food and Beverages* (pp. 235–248). Woodhead Publishing. https://doi.org/https://doi.org/10.1016/B978-1-78242-103-0.00011-4.
4. Araújo, W. A., (2016). Ethanol industry: Surpassing uncertainties and looking forward. In: *Global Bioethanol.* (pp. 1–33). Academic Press.
5. Bhatia, L., Johri, S., & Ahmad, R., (2012). An economic and ecological perspective of ethanol production from renewable agro waste: A review. *AMB Express, 2*(1), 1–19.
6. Câmara, M. M., Soares, R. M., Feital, T., Naomi, P., Oki, S., Thevelein, J. M., & Pinto, J. C., (2017). On-line identification of fermentation processes for ethanol production. *Bioprocess and Biosystems Engineering, 40*(7), 989–1006.
7. Lopes, M. L., De Lima, P. S. C., Godoy, A., Cherubin, R. A., Lorenzi, M. S., Giometti, F. H. C., & De Amorim, H. V., (2016). Ethanol production in Brazil: A bridge between science and industry. *Brazilian Journal of Microbiology,* (47), 64–76.
8. Serafim, F. A. T., & Lanças, F. M., (2019). Sugarcane spirits (cachaça) quality assurance and traceability: An analytical perspective. In: Grumezescu, A. M., & Holban, A. M., (eds.), *Production and Management of Beverages* (pp. 335–359). Elsevier Inc. https://doi.org/10.1016/B978-0-12-815260-7.00011-0.
9. Vasconcelos De, J. N., (2015). Ethanol fermentation. In: Santos, F., Caldas, C., & Borém, A., (eds.), *Sugarcane: Agricultural Production, Bioenergy, and Ethanol* (pp. 311–340). Elsevier Inc. All. https://doi.org/10.1016/B978-0-12-802239-9.00015-3.
10. Dombek, K. M., & Ingram, L., (1987). Ethanol production during batch fermentation with *saccharomyces cerevisiae*: Changes in glycolytic enzymes and internal PH. *Appl. Environ. Microbiol., 53*, 1286–1291.
11. Sarris, D., & Papanikolaou, S., (2016). Biotechnological production of ethanol: Biochemistry, processes and technologies. *Eng. Life Sci., 16*(4), 307–329. https://doi.org/10.1002/elsc.201400199.
12. Zhang, B., Li, L., Zhang, J., & Gao, X., (2013). Improving ethanol and xylitol fermentation at elevated temperature through substitution of xylose reductase in *Kluyveromyces marxianus*. *J. Ind. Microbiol. Biotechnol., 40*, 305–316. https://doi.org/10.1007/s10295-013-1230-5.
13. Arshadi, M., & Grundberg, H., (2011). Biochemical production of bioethanol. In: Luque, R., Campelo, J., & Clark, J., (eds.), *Handbook of Biofuels Production* (pp. 199–220). Woodhead Publishing Limited. https://doi.org/10.1533/9780857090492.2.199.
14. Alperstein, L., Gardner, J. M., Sundstrom, J. F., Sumby, K. M., & Jiranek, V., (2020). Yeast bioprospecting versus synthetic biology — which is better for innovative beverage fermentation? *Appl. Microbiol. Biotechnol., 104*, 1939–1953. https://doi.org/10.1007/s00253-020-10364-x.
15. Akhtar, N., Karnwal, A., Upadhyay, A. K., Paul, S., & Mannan, M. A., (2018). *Saccharomyces cerevisiae* bio-ethanol production, a sustainable energy alternative. *Article Asian J. Microbiol. Biotechnol. Environ. Sci., 20*, 1–5.
16. Hong, K. K., & Nielsen, J., (2012). Metabolic engineering of *Saccharomyces cerevisiae*: A key cell factory platform for future biorefineries. *Cell. Mol. Life Sci., 69*, 2671–2690. https://doi.org/10.1007/s00018-012-0945-1.
17. Techaparin, A., Thanonkeo, P., & Klanrit, P., (2017). Biotechnology and industrial microbiology high-temperature ethanol production using thermotolerant yeast newly isolated from greater Mekong subregion. *Brazilian J. Microbiol., 48*, 461–475. https://doi.org/10.1016/j.bjm.2017.01.006.

18. Joshi, J., Dhungana, P., Prajapati, B., Maharjan, R., Poudyal, P., Yadav, M., Mainali, M., et al., (2019). Enhancement of ethanol production in electrochemical cell by *saccharomyces cerevisiae* (CDBT2) and wickerhamomyces anomalus. *Front. Energy Res., 7*, 1–11. https://doi.org/10.3389/fenrg.2019.00070.

19. Sukwong, P., Sunwoo, I. Y., Lee, M. J., & Ra, C. H., (2018). Application of the severity factor and HMF removal of red macroalgae *Gracilaria* verrucosa to production of bioethanol by *Pichia stipitis* and *Kluyveromyces marxianus* with adaptive evolution. *Appl. Biochem. Biotechnol.*

20. Gargalo, C. L., Chairakwongsa, S., Quaglia, A., Sin, G., & Gani, R., (2015). Methods and tools for sustainable chemical process design. In: Klemeš, J. J., (ed.), *Assessing and Measuring Environmental Impact and Sustainability* (pp. 277–321). Elsevier Inc.

21. Quintero, J. A., Rincón, L. E., & Cardona, C. A., (2011). Production of bioethanol from agro-industrial residues as feedstocks. In: Pandey, A., Ricke, S. C., Gnansounou, E., Larroche, C., & Dussap, C. G., (eds.), *Biofuels: Alternative Feedstocks and Conversion Processes* (pp. 251–285). Elsevier Inc. https://doi.org/10.1016/C2010-0-65927-X.

22. Thammasittirong, S. N., Thirasaktana, T., Thammasittirong, A., & Srisodsuk, M., (2013). Improvement of ethanol production by ethanol-tolerant *Saccharomyces cerevisiae* UVNR56. *SpringerPlus, 2*, 1–5. https://doi.org/10.1186/2193-1801-2-583.

23. Watanabe, T., Watanabe, I., Yamamoto, M., Ando, A., & Nakamura, T., (2011). Bioresource technology a UV-Induced mutant of *Pichia stipitis* with increased ethanol production from xylose and selection of a spontaneous mutant with increased ethanol tolerance. *Bioresour. Technol., 102*(2), 1844–1848. https://doi.org/10.1016/j.biortech.2010.09.087.

24. Arshad, M., Hussain, T., Iqbal, M., & Abbas, M., (2017). Enhanced ethanol production at commercial scale from molasses using high gravity technology by mutant *S. cerevisiae*. *Brazilian J. Microbiol., 48*, 403–409. https://doi.org/10.1016/j.bjm.2017.02.003.

25. Chen, H., & Fu, X., (2016). industrial technologies for bioethanol production from lignocellulosic biomass. *Renew. Sustain. Energy Rev., 57*, 468–478. https://doi.org/10.1016/j.rser.2015.12.069.

26. OECD/FAO, (2017). "Biofuels," in OECD-FAO Agricultural Outlook, 2017–2026, OECD Publishing, París. https://doi.org/10.1787/agr_outlook-2017-en.

27. Mohd, A. S. H., Abdulla, R., Jambo, S. A., Marbawi, H., Gansau, J. A., Mohd, F. A. A., & Rodrigues, K. F., (2017). Yeasts in sustainable bioethanol production: A review. *Biochem. Biophys. Reports, 10*, 52–61. https://doi.org/10.1016/j.bbrep.2017.03.003.

28. Yang, B., Dai, Z., Ding, S. Y., & Wyman, C. E., (2011). Enzymatic hydrolysis of cellulosic biomass. *Biofuels, 2*(4), 421–449. https://doi.org/10.4155/bfs.11.116.

29. Olsson, L., & Hahn-Hägerdal, B., (1996). Fermentation of lignocellulosic hydrolysates for ethanol production. *Enzyme Microb. Technol., 18*(5), 312–331. https://doi.org/10.1016/0141-0229(95)00157-3.

30. Rehman, A., Tong, Q., Jafari, S. M., Assadpour, E., Shehzad, Q., Aadil, R. M., Iqbal, M. W., et al., (2020). Carotenoid-loaded nanocarriers: A comprehensive review. *Adv. Colloid Interface Sci., 275*, 102048. https://doi.org/10.1016/j.cis.2019.102048.

31. Walker, G. M., & White, N. A., (2005). Introduction to fungal physiology. *Fungi Biol. Appl.*, 1–34. https://doi.org/10.1002/0470015330.ch1.

32. Sun, Y., & Cheng, J., (2002). Hydrolysis of lignocellulosic materials for ethanol production: A review. *Bioresour. Technol., 83*(1), 1–11. https://doi.org/10.1016/S0960-8524(01) 00212-7.

33. Jin, H., Liu, R., & He, Y., (2012). Kinetics of batch fermentations for ethanol production with immobilized *Saccharomyces cerevisiae* growing on sweet sorghum stalk juice. *Procedia Environ. Sci., 12*, 137–145. https://doi.org/10.1016/j.proenv.2012.01.258.

34. Rubio-Arroyo, M. F., Vivanco-Loyo, P., Juárez, M., Poisot, M., & Ramírez-Galicia, G., (2011). Bioethanol obtained by fermentation process with continuous feeding of yeast. *J. Mex. Chem. Soc., 55*(4), 242–245.

35. Oura, E., & Suomalainen, H., (1982). Biotin-active compounds, their existence in nature and the biotin. *The S. J. Inst. Brew, 88*, 299–308.

36. Arévalo, S., (1998). In: Trevan, M. D., & Goulding, K. H., (eds.), *Biotecnology – Biological Principles* (pp. 1–18). Editorial ACRIBIA, S.A Zaragoza Spain-1990. University of Lleida.

37. Byadgi, S. A., & Kalburgi, P. B., (2016). Production of bioethanol from waste newspaper. *Procedia Environ. Sci., 35*, 555–562. https://doi.org/10.1016/j.proenv.2016.07.040.

38. Walker, G. M., & Walker, R. S. K., (2018). *Enhancing Yeast Alcoholic Fermentations* (Vol. 105). Elsevier Ltd. https://doi.org/10.1016/bs.aambs.2018.05.003.

39. Castaño, H., & Mejia, C., (2008). Production of ethanol from cassava starch using the simultaneous saccharification-fermentation process strategy (SSF). *Vitae, Rev. La Fac. Química Farm., 15*(2), 251–258.

40. Kostas, E. T., White, D. A., Du, C., & Cook, D. J., (2016). Selection of yeast strains for bioethanol production from UK seaweeds. *J. Appl. Phycol., 28*(2), 1427–1441. https://doi.org/10.1007/s10811-015-0633-2.

41. Aksu, Z., & Eren, A. T., (2007). Production of carotenoids by the isolated yeast of *Rhodotorula glutinis. Biochem. Eng. J., 35*(2), 107–113. https://doi.org/10.1016/j.bej.2007.01.004.

42. Walker, G. M., & Basso, T. O., (2020). Mitigating stress in industrial yeasts. *Fungal Biol., 124*(5), 387–397. https://doi.org/10.1016/j.funbio.2019.10.010.

43. Saini, P., Beniwal, A., Kokkiligadda, A., & Vij, S., (2018). Response and tolerance of yeast to changing environmental stress during ethanol fermentation. *Process Biochem., 72*, 1–12. https://doi.org/10.1016/j.procbio.2018.07.001.

44. Święciło, A., (2016). Cross-stress resistance in *Saccharomyces cerevisiae* yeast—new insight into an old phenomenon. *Cell Stress Chaperones, 21*(2), 187–200. https://doi.org/10.1007/s12192-016-0667-7.

45. Van, D. M., Erdei, B., Galbe, M., Nygård, Y., & Olsson, L., (2019). Strain-dependent variance in short-term adaptation effects of two xylose-fermenting strains of *Saccharomyces cerevisiae. Bioresour. Technol., 292*, 121922. https://doi.org/10.1016/j.biortech.2019.121922.

46. Du Preez, J., (1994). Process parameters and environmental factors affecting D-xylose fermentation by yeasts. *Enzyme Microb. Technol., 16*, 944–956.

47. Timoumi, A., Guillouet, S. E., Molina-Jouve, C., Fillaudeau, L., & Gorret, N., (2018). Impacts of environmental conditions on product formation and morphology of yarrowia lipolytica. *Appl. Microbiol. Biotechnol., 102*(9), 3831–3848. https://doi.org/10.1007/s00253-018-8870-3.

48. Zemančíková, J., Kodedová, M., Papoušková, K., & Sychrová, H., (2018). Four saccharomyces species differ in their tolerance to various stresses though they have similar basic physiological parameters. *Folia Microbiol. (Praha), 63*(2), 217–227. https://doi.org/10.1007/s12223-017-0559-y.

49. Kucharczyk, K., & Tuszyński, T., (2017). The effect of wort aeration on fermentation, maturation and volatile components of beer produced on an industrial scale. *J. Inst. Brew., 123*(1), 31–38. https://doi.org/10.1002/jib.392.

50. Ishizaki, H., & Hasumi, K., (2013). *Ethanol Production from Biomass*. Elsevier, https://doi.org/10.1016/B978-0-12-404609-2.00010-6.
51. Ravanal, M. C., Camus, C., Buschmann, A. H., Gimpel, J., Olivera-Nappa, Á., Salazar, O., & Lienqueo, M. E., (2019). Production of bioethanol from brown algae. *Adv. Feed. Convers. Technol. Altern. Fuels Bioprod. New Technol. Challenges Oppor.*, 69–88. https://doi.org/10.1016/B978-0-12-817937-6.00004-7.
52. Teter, S. A., Sutton, K. B., & Emme, B., (2014). *Enzymatic Processes and Enzyme Development in Biorefining*. https://doi.org/10.1533/9780857097385.1.199.
53. Silveira, M. H. L., Vanelli, B. A., & Chandel, A. K., (2017). *Second Generation Ethanol Production: Potential Biomass Feedstock, Biomass Deconstruction, and Chemical Platforms for Process Valorization*. Potential biomass feedstock, biomass deconstruction, and chemical platforms for process valorization. Elsevier Inc. https://doi.org/10.1016/B978-0-12-804534-3.00006-9.
54. International Energy Agency (IEA), (2011). *Technology Roadmap: Biofuels for Transport*. Organization for Economic Cooperation and Development, IEA, Paris. http://www.iea.org/publications/freepublications/publication/biofuels_roadmap.pdf (accessed on 28 October 2021).
55. Joshi, V. K., Walia, A., & Rana, N. S., (2012). Production of bioethanol from food industry waste: Microbiology, biochemistry, and technology. In: *Biomass Conversion* (pp. 251–311). Springer, Berlin, Heidelberg.
56. Ghosh, T. K., & Prelas, M. A., (2011). Ethanol. In: *Energy Resources and Systems* (pp. 419–493). Springer, Dordrecht.
57. Matsuoka, S., Ferro, J., & Arruda, P., (2009). The Brazilian experience of sugarcane ethanol industry. *In Vitro Cellular & Developmental Biology-Plant, 45*(3), 372–381.
58. Kohler, M., (2019). Economic assessment of ethanol production. In: *Ethanol*, (pp. 505–521) Elsevier.
59. Amiri, T. Y., & Ghasemzadeh, K., (2019). Ethanol economy: Environment, demand, and marketing. In: *Ethanol*. (451–504). Elsevier.
60. Kolling, D. F., Dalla, C. V. F., & Oliveira, C. A. O., (2014). Global market issues in the liquid biofuels industry. In: *Liquid Biofuels: Emergence, Development and Prospects* (pp. 55–72). Springer, London.
61. Lopes, M. L., Cristina, S., Paulillo, D. L., Godoy, A., Cherubin, R. A., Lorenzi, M. S., Henrique, F., et al., (2016). Ethanol production in Brazil: A bridge between science and industry. *Brazilian J. Microbiol.*, 1–13. https://doi.org/10.1016/j.bjm.2016.10.003.
62. Anđelković, D., Antić, B., Vujanić, M., Subotić, M., & Radovanović, L., (2017). The perspectives of applying ethanol as an alternate fuel. Energy Sources, *Part B Econ. Planning, Policy, 12*(9), 1–10. https://doi.org/10.1080/15567249.2012.683930.
63. Janda, K., Kristoufek, L., & Zilberman, D., (2012). Biofuels: Policies and impacts. *Agricultural Economics, 58*(8), 372–386.

CHAPTER 2

Physiology of Ethanol Production by *Zymomonas mobilis*

LAURA ANDREA PÉREZ-GARCÍA, CINDY NATALY DEL RIO-ARELLANO, DAVID FRANCISCO LAFUENTE RINCÓN, and NORMA M. DE LA FUENTE-SALCIDO

Faculty of Biological Sciences, Autonomous University of Coahuila, Torreón, Coahuila, México, E-mail: normapbr322@gmail.com (N. M. D. L. Fuente-Salcido)

ABSTRACT

The bacteria *Zymomonas mobilis* is a natural ethanologenic with important desirable physiological features suitable for industrial production of bioethanol. This anaerobic bacterium is considered a powerful biocatalyst system for biofuel production and here will be compared with typical microorganisms involved for ethanol production. The main comparison focuses on the typical model ethanologenic, *Saccharomyces cerevisiae*, which uses the Embden-Meyerhof-Parnas (EMP) pathway for glucose fermentation because *Z. mobilis* uses the Entner-Doudoroff (ED) pathway, and also, the energy production in both will be described. Also, the utilization of carbohydrates, fermentative process conditions, yield, and perspectives will review, as well as some genetic modifications to optimize the production of bioethanol by harnessing the *Zymomonas* metabolism. Finally, diverse industrial perspectives for *Z. mobilis* application will be mentioned briefly.

2.1 INTRODUCTION

Since ancient civilization, the use of alcoholic beverages has taken part of the universal culture, traditions, and the economic strength of nations. However, that is why they thought that initially fermentations were considered a spontaneous process, normally depending on yeasts and bacteria that were already present in the raw material, the equipment, or introduced by insects. Nowadays, the production of ethanol is one of the significant biotechnological processes in the world in many areas like production of beverages, economic importance, and research. For this process, there are different sources of sugars such as cereals, fruits, and other carbon sources that can be used by various microorganisms to obtain products through the fermentation process such as wine, beer, or even biofuel that are economically essential products (Figure 2.1).

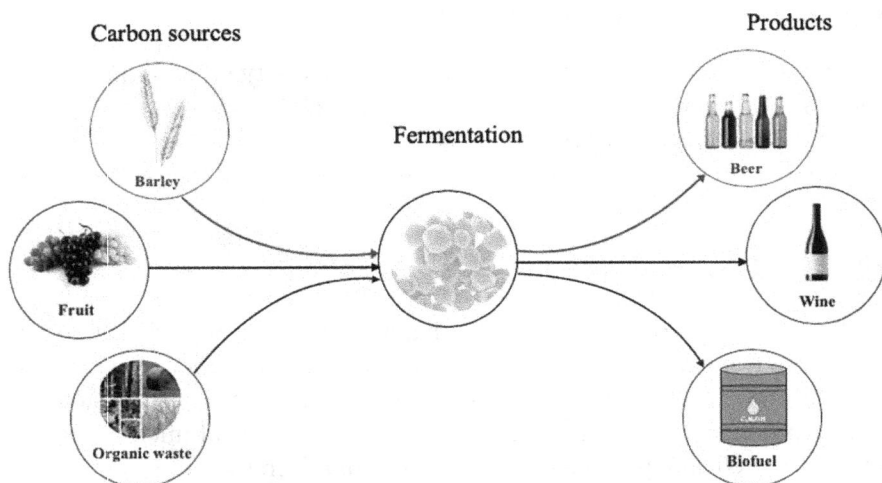

FIGURE 2.1   The fermentation process for the elaboration of alcoholic beverages and biofuel with different carbon sources.

Bioethanol production uses different substrates. The products derived from corn or based on starch, are produced primarily by the United States to be used in the transport sector. Another substrate for the production of ethanol derived from sugar cane or based on sucrose is produced mainly in Brazil. These biofuels are of the first generation because they are derived from the fermentation of hydrolyzed edible starch materials. Nowadays, there are other generations of bioethanol production that can be used without involving the socio-economic troubles, and especially second-generation bioethanol derived from the recovery-conversion of various waste streams are the most promising substrate applied around the world. These kinds of processes also contribute to enhancing the economy of many countries through renewable energy production [1–2].

The simple or complex substrates have the required quality to be used as an excellent source of carbon for microbial growth and to be converted into ethanol. Both substrates result in the generation of various carbohydrates as hexoses, pentoses, or glycerol that will be fermented by microorganisms to be converted into ethanol. The most recognized and important ethanol producers are the yeast *S. cerevisiae* and the bacterium *Z. mobilis*. The utilization of both microorganisms for the production of alcoholic beverages or biofuel for the industry is exposed to severe environmental changes and fermentation processes. There exist different factors that are influenced the process like temperature shock, osmotic stress caused by the high concentration of carbon sources and ethanol tolerance [4]. In relation to the yield, the factors mentioned previously can change depending on the microorganism used, as well as the type of fermentation that is carried out, and the parameters that are used to obtain the highest production of ethanol per gram of carbon source. In this chapter aims to integrate the current knowledge about the differences in the yeast *S. cerevisiae* and the bacterium *Z. mobilis* regarding to ethanolic fermentation. Also, the central role, the potential use of biological features of *Z. mobilis* and the common biotechnological mechanisms and tools for the production of bioethanol will be discussed.

2.2 FERMENTATIVE BIOPROCESS TO PRODUCE ETHANOL

Ancient civilizations brewed beer around the year 6000 A.C. through the fermentation process. For 2017, the average ethanol production is more than 1 million barrels (159 million liters) per day, with an annual rate of 16 million gallons (60 million liters). There are many microorganisms that produce ethanol but in each one has certain limitations due to economic problems of the production of bioethanol, such as the use of different substrates and yield. Within the microorganisms, yeasts are leaders in the production of bioethanol, however, bacteria such as *Z. mobilis*, *Escherichia coli*, or *Clostridium sp.*, are being developed to increase the production of bioethanol and eliminate the limitations such as the kind of substrate or tolerance to different factors that can change with the help of genetic engineering [5].

In addition, ethanol derived from yeast fermentation represents more than 80% of the world's renewable fuels and more than 85% of the alcohol produced is provided by the United States and Brazil. The *S. cerevisiae* produces more than 100 billion liters of ethanol per year [6, 7].

One of the best prospects to the production of ethanol is *Z. mobilis* because have several important and distinctive physiological features that give it advantages for bioethanol production. Among these characteristics is growth at high concentrations of six-carbon sugar in the medium, tolerance to high concentrations of ethanol, biological activity at a wide pH range for the production of bioethanol. In the fermentation process the cells of some strains of *Zymomona sp.* can grow up on glucose concentrations above to 400 g of glucose L^{-1} [5, 8]. On the other hand, the yeast cells undergo osmotic stress when inoculated into the grape must (generally above 200 g of sugar L^{-1}). This induces the yeast to synthesize glycerol, in this way it reduces the permeability of the glycerol and consequently, buffer the osmotic gradient between the inside and the outside of the cell [9].

Both *Z. mobilis* and *S. cerevisiae* have a different pathway to use a carbon source during the bioethanol production. Alcoholic fermentation commonly includes diverse biochemical degradation pathways as glycolysis, alcoholic fermentation, glycerol-pyruvic fermentation. Also, respiration processes for utilization of hexoses, xylose catabolic pathways for utilization of pentoses and glycerol assimilation, and glycolysis for glycerol-converting microorganisms, and regulation between fermentation and respiration. In the particular case of *Z. mobilis* uses ED pathway to anaerobically ferment glucose for ethanol production. But *S. cerevisiae* uses EMP pathway for glycolysis [2]. Fermentation process consists for *Z. mobilis* 1 mole of ATP is yielded per mole of glucose through the ED pathway required less enzymatic protein because together with *Pdc* and two alcohol dehydrogenases (*Adh*) "form the backbone" of this glycolysis metabolic pathway. On the other hand, *S. cerevisiae* one molecule of glucose ($C_6H_{12}O_6$) is converted into two molecules of pyruvic acid ($C_3H_4O_3$) during the process of glycolysis. Pyruvic acid is further decarboxylated to generate two molecules of acetaldehyde (CH_3CHO), which is reduced to ethanol (C_2H_5OH). The process described produce two molecules of ATP as gain, two molecules of ethanol and two molecules of CO_2 [10]. In the conversion of glucose to ethanol and CO_2 during fermentation, a total of 12 yeast enzymes are involved, 10 to degrade glucose to pyruvate with the generation of ATP. After that, pyruvic acid is converted to acetaldehyde through PDC and the final stage ADH converts acetaldehyde to ethanol in this way NAD^+ is regenerated to allow glycolysis to continue (Figure 2.2) [7]. This has a big importance because *Zymomonas* use less energy than *Saccharomyces* to make the fermentation process and these can be traduced to "higher production of ethanol, using less energy to make the process" depending on the strains.

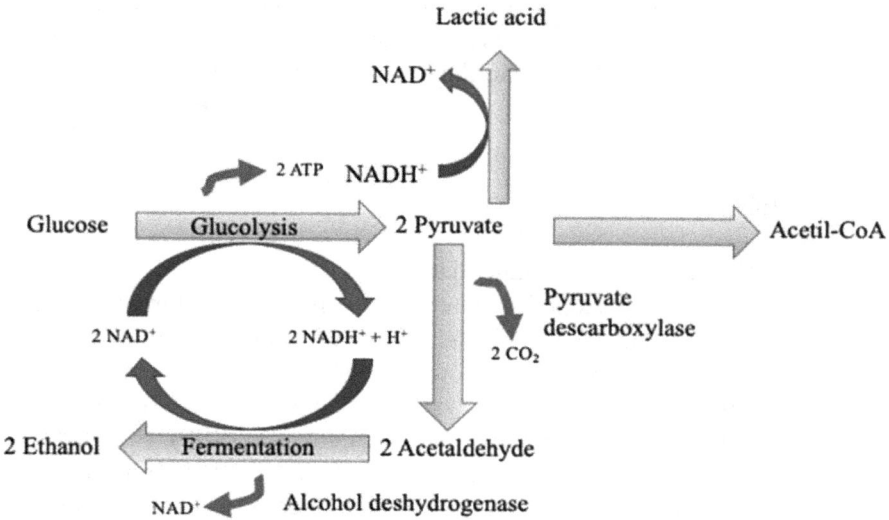

FIGURE 2.2 Glycolysis and fermentation in yeast.

2.3 BIOETHANOL PRODUCTION (SUBSTRATES, MICROORGANISMS, PROCESS, YIELD)

2.3.1 SUBSTRATES

The utilization of biomass such as cellulosic agricultural residues are valuable, economical, profitable, and sustainable renewable natural sources that are used for the production of bioenergy such as biodiesel, biohydrogen, and bioethanol [11]. The carbon source influences the quality of the ethanol produced from the fermentation, since the ethanol yield will depend on the type of sugar obtained [12]. The use of different substrates such as lignocellulosic residues are considered a viable alternative for the production of biodiesel [13, 14].

2.3.2 MICROORGANISMS

A wide variety of microorganisms' ferment hexose and pentose sugars such as; *Bacillus, Klebsiella, Escherichia coli, Thermoanerobacter, Aeromonas, S. cerevisiae, Candida shehatae, Pichia stipitis, C. brassicae, Pachysolen tannophilus, Fusarium, Mucor indicus, Neurospora, Monilia, Z. mobilis, and Rhizopus mobilis* are capable to produce ethanol [11, 15]. However, not all microorganisms can produce the same amount of ethanol, because each one has their own limitations, like the type of fermentation or the different tolerance like high sugar concentration or ethanol tolerance.

2.3.3   PROCESS

A fermentation process is sought that is as efficient as possible. Several methods have been developed to integrate hydrolysis and fermentation to improve cellulose availability [15]. There are new technologies that have been found that make it possible to carry out the viable hydrolysis and fermentation process, the most common include separate hydrolysis and fermentation (SHF), simultaneous saccharification and fermentation (SSF), simultaneous saccharification and fermentation (SSCF), consolidated bioprocessing (CBP), pre-saccharification followed by simultaneous saccharification and fermentation (PSSF), separate hydrolysis and co-fermentation (SHCF), and finally, pre-saccharification followed by simultaneous saccharification and fermentation (PSSF) [16, 17].

2.3.4   YIELD

The yield depends on the amount of metabolite required, the amount produced by the native organism and the biological activity of the compound [18].

2.4   GENERAL OVERVIEW OF *Zymomonas mobilis*

Biotechnology has allowed the increase of innovative fermentation processes to develop clean, almost inexhaustible, and highly competitive energies such as renewable energy. One of the most outstanding renewable energies is the production of ethanol by fermentation, a fuel product for motor vehicles, which contributes to mitigating the planet's climate change.

During the last decades, alcoholic fermentation has been extensively studied to improve and optimize the production of bioethanol, an economically profitable and eco-friendly biofuel. In this sense, the unit operations, culture media (substrates, organic waste) and microbial growth conditions involved in the fermentation process have been improved to increase bioethanol yield. However, the success of the synthesis of bioproducts (such as bioethanol) focuses mainly on the physiological characteristics and fermentative capacities of the producing microorganism. The microorganisms commonly used to produce ethanol were yeasts, mainly *S. cerevisiae*, however, successful processes have been achieved with extraordinary facultative bacteria such as *Zymomonas mobilis* [5]. This bacterium has been attracting significant attention for usage in large-scale biofuels bioprocessing.

The *Zymomonas* is a bacterial genus isolated frequently from some fermented beverages or plants surface, including the succulent Maguey cactus indigenous from México, African palm wine and also is considered as spoilage organism in beer or cider ("cider sickness") from Europe [19, 20]. It was first found in 1924 and reported in 1928 by Linder in pulque isolates, a popular Mexican fermented pre-Hispanic beverage and reported as "*Thermobacterium mobile.*" However, until 1936 Taxon *Z. mobilis* subsp. *mobilis* according to the proposal of Kluyver and van Niel, and formalized in 1976 by the classification proposal and reported by De Ley, J., and J. Swings.

The commonly known taxonomic hierarchy of *Zymomonas* is described as follows:

Phylum	Proteobacteria
Class	Alphaproteobacteria
Order	*Sphingomonadales*
Family	*Sphingomonadaceae*
Genus	*Zymomonas*
Species	*Zymomonas mobilis*
Subspecies	*Zymomonas mobilis francensis*
	Zymomonas mobilis mobilis
	Zymomonas mobilis pomaceae

The bacteria description included in the Bergey's Manual of Systematic Bacteriology listed the characteristics of the bacterium [21].

This facultative aerobic *Z. mobilis* is, gram-negative bacteria, a non-spore-forming, rod-shaped grouped in pairs and size of 2–6 × 1.0–1.4 μm. Commonly non-motile but some exceptions may be possessed one to four polar flagella. Growth of *Z. mobilis* on standard medium agar [D-glucose (20 g L^{-1}), yeast extract (5 g L^{-1})], grows forming glistening colonies, regularly edged, white to cream-colored, 1–2 mm in diameter after 48 h at 30°C incubation. Growth optimal conditions includes 30°C and at pH 3.5–7.5 and vitamins (biotin, pantothenate) as micronutrients [70]. Their nutrition includes fermentable sugars (glucose, fructose, sucrose), is chemoorganotrophic and highly efficient ethanol producer through the ED pathway [22]. Shows constitutive ethanol-tolerance (5%) and also is acid tolerance, but is sensible to novobiocin. The distinctive features of *Z. mobilis* are the genotypically their DNA content mol %G+C is 47.5–49.5, and phenotypically, the cell membrane contains pentacyclic triterpenoids of the hopane series and lack of fatty acid C14:0 2OH (nonhydroxy myristic acid), commonly present in all others α-*Sphingomonas* species.

Physiologically *Zymomonas* is facultatively anaerobic and has a strictly fermentative metabolism, with multiple applications in biotechnology [23–27].

2.5   AMAZING FEATURES OF *Zymomonas mobilis* FOR BIOTECHNOLOGICAL APPLICATIONS

The physiology of a microorganism is decisive key to ensure the synthesis of several high added value products by advanced fermentation technologies. *Z. mobilis* physiology is particularly fascinating, mostly by their capacity to synthesize economically and sustainably important products such as bioethanol among others biochemical as bionic acid, levan, sorbitol, among others [5, 28].

In this sense, the ED metabolic pathway of *Zymomonas mobilis* strains is considered a unique ethanologenic pathway due to its efficiency, characterized by a decrease in cellular ATP consumption, which promotes an increase in the transformation of sugars into biomass, and even with better yields and productivity of the bioethanol found in *S. cerevisiae* [23]. The ethanologenic *Zymomonas* strains are recognized as the most efficient bacterial ethanol producer, mainly due to that employ the ED pathway as a strictly step during a fermentative process, in addition to having a physiology perfectly adapted to industrial-scale bioethanol production. Sequential steps to glucose transformation by the ED pathway, the glucose phosphate is catabolized into 2-keto-3-deoxy-6-phosphogluconic acid (KDPG) and then is cleaved by KDPG aldolase to pyruvate and glyceraldehyde-3-phosphate (GAP). The GAP is oxidized to pyruvate by glycolytic enzymes and ATP produced by substrate-level phosphorylation. The next reaction is reduction of pyruvic acid to ethanol and CO_2. The overall reaction is:

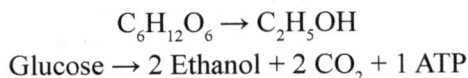

$$C_6H_{12}O_6 \rightarrow C_2H_5OH$$
$$\text{Glucose} \rightarrow 2 \text{ Ethanol} + 2 \text{ } CO_2 + 1 \text{ ATP}$$

As a catabolic main route exclusive of prokaryotes, the enzymatic divergence of the ED pathway is it uses 6-phosphogluconate dehydratase and 2-keto-3 deoxyphosphogluconate aldolase to synthesize pyruvate from glucose. Therefore, the pathway rises a yield of 1 ATP for each glucose molecule catabolized, as well as 1 NADH and 1 NADPH. The obvious comparison with the EMP pathway, indicates that glycolysis rises a yield of 2 ATP and 2 NADH for each glucose molecule processed [71]. This physiological feature distinguishes *Z. mobilis* as a profitable strain for industrial applications.

Metabolic efficiency of *Z. mobilis* as ethanol producer has broad application in industrial-scale ethanol production. Bioethanol is considered an alternative renewable energy source, indispensable, and urgently necessary in the very short term [29, 30]. However, it is very important to consider the advantages and disadvantages of using outstanding ethanologenic bacteria.

The main advantage of using *Zymomonas* (and yeast) is the clean production of ethanol during the fermentative process, while the most microorganisms generate mixtures of metabolic products in addition to alcohol. The disadvantage of the use of *Zymomonas* lies in its specificity for glucose as the only fermentable sugar and its lacks the enzymes to break down another carbon source, including the enzymes to degrade carbohydrate polymers such as starch and cellulose [31], and this would limit its biotechnological application.

Different strategies, including kinetic modeling and rational metabolic engineering to understand how central metabolism of *Zymomonas* work, and novel genome sequence are widely studied and applied to improve ethanol synthesis. In addition, the techniques promoted by the application of synthetic biology, metabolic engineering, and the ongoing world trend of industrial bioethanol production from the use of agro-industrial waste by the metabolism of *Zymomonas mobilis* has facilitated the development of genetically modified bacteria for co-culturing in cellulose-rich media [14, 32, 72–75].

2.6 FERMENTATIVE PATHWAYS OF *Zymomonas mobilis* AND *Saccharomyces cerevisiae*

Particularly, *Z. mobilis* uses the metabolic pathway of ED, which is a modified form of the glycolysis pathway that produces only one molecule of ATP per metabolized glucose unlike the EMP pathway that produces two glycolytic ATP molecules used by yeast. The bacterial pathway of ED also generates reducing equivalents as NADH and NADPH (Figure 2.3) [7].

Alcohols are derived from the catabolism of amino acids through a metabolic pathway. The amino acids, such as valine, leucine, isoleucine, methionine, and phenylalanine, are absorbed slowly throughout the fermentation process by the Ehrlich pathway. After a transamination reaction, the keto acids produced are converted into alcohols or acids by this pathway [33].

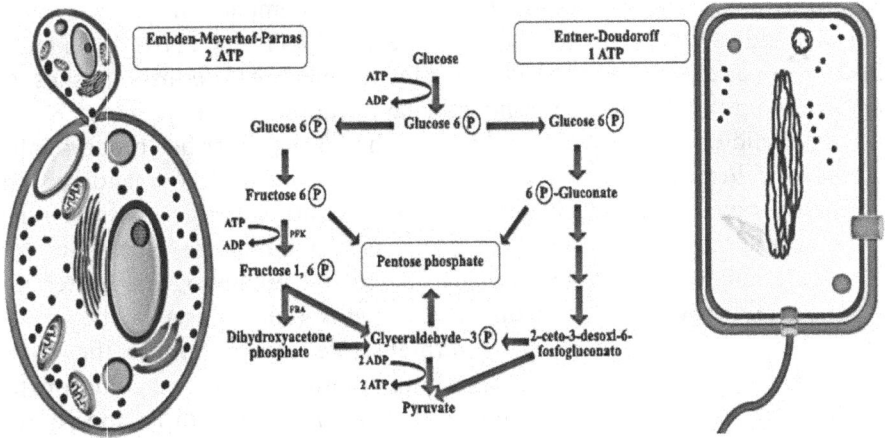

FIGURE 2.3   Metabolic pathway of *Saccharomyces cerevisiae* and *Zymomonas mobilis.*

The metabolism of aromatic amino acids (AAA) in yeast directly influences a more significant alcohol formation. The repression of nitrogen metabolites (NCR) can control the amino acid catabolism, which functions like a complex regulatory system that confers to S. cerevisiae to use sources with a high concentration of nitrogen to the yeast. Studies showed that the enzymes involved in the three principal stages of the Ehrlich pathway, which are transamination, decarboxylation, and reduction, are encoded by the *Bat2, Pdc1* and *Adh1* genes. These genes in the process of alcoholic fermentation, present similar expression profiles. Overexpression of these two genes generates significant increases in two different alcohols isobutanol and isoamyl alcohol [33].

2.7   PHYSIOLOGICAL ADVANTAGES IN BIOETHANOL PRODUCTION OF *Zymomonas mobilis*

Knowing the differences between the different metabolic routes that exist between the bacteria Z. mobilis and the yeast S. cerevisiae. Lack of different enzymes in the *Zymomonas* metabolic pathway is found as phosphofructokinase (*Pfk*) in the EMP pathway, phosphogluconate dehydrogenase (*Pgd*) and transaldolase *(Tal)* in the PPP pathway, as well as 2-oxoglutarate dehydrogenase complex (sucABCD) and malate dehydrogenase (*Mdh*) in the TCA cycle and this enzyme deficiency carry more carbon into the ethanol production pathway and highly efficient glycolysis resulting in the theoretical maximum ethanol production [34]. *Zymomonas sp.* strains have a temperature range of 25 to 31°C but there are strains reported that can grow at temperatures of 40°C. It has a wide range of pH 3.7–7.5 so it has a high tolerance to acidic pH′s. However, most strains of *Zymomonas sp.* grow up at an optimal pH of 5–7. Some strains of *Zymomonas* also can grow up at high concentration of glucose medium more than 200 gL⁻¹ but there is other that can grow up in 400 gL⁻¹ glucose medium [8]. Another important characteristic is that the ZM4 strain is facultative aerobic, so this bacterium can perform the metabolism for the production of bioethanol under aerobic conditions, however it is reported that the production is lower than in the anaerobic condition [35]. Also, *Z. mobilis* possess a high tolerance of ethanol like 11–16% v/v [29].

2.8 GENETIC FEATURE OF *Zymomonas mobilis* AND *Saccharomyces cerevisiae*

According to previous reports about the complete genome of *Z. mobilis* ZM4 corresponds to a circular chromosome with 2,056,416 bp and harbor five circular plasmids [34]. The chromosome and five plasmids have GC contents of 46.22%, 43.46%, 45.41%, 43.23%, 41.79%, 37.63% and 41.31% respectively. The genetic material includes 1,875 protein-encoding genes, 48 tRNA and 6 rRNA genes [19]. The most established product by *Z. mobilis* recombinant strains is ethanol, which has been extensively investigated. The most important detected genes are *pdc* and *adh* for ethanol production by *Zymomonas* and these genes have been inserted in diverse microorganisms modified by molecular biology. For ED pathway genes (*glk, zwf, pgl, pgk,* and *eno*) and the PDC-encoding *pdc* gene were observed to be more abundant under anaerobic conditions than aerobic conditions depending on the strain [5]. Otherwise, *S. cerevisiae* has a genome size of 12 Mb distributed among 16 chromosomes. The entire genome encodes 6,000 genes, of which 5,000 are individually nonessential [36]. The sequence of the *S. cerevisiae* has about 12,068 Kb of which defines 5,885 potential protein-encoding genes, 140 genes specifying ribosomal RNA, 40 genes for small nuclear RNA molecules, and 275 transfer RNA genes. Also, the advantage to have the complete sequence *S. cerevisiae* provides information about the higher-order organization of yeast's 16 chromosomes and allows some insight into their evolutionary history [37].

Among the most important genes of *Z mobilis* are *Pdc*, this gen synthesizes the protein that transforms pyruvate into acetaldehyde and the gen

Adh synthesizes the protein that transforms acetaldehyde into ethanol in the metabolic pathway ED, however, genetic engineering is looking for different ways to increase the production of ethanol and inhibit acetate production. On the other hand, there are reports of strains that are already modified with the inclusion of different gene such as *XylA, XylB, talB, and tktA* that give them the ability to use 5 C sugars and thus, be able to use expanding the range of raw material used as a carbon source [38].

Regarding to *S. cerevisiae* have only 5,538 genes encoding 100 amino acids and contains 18 genes for which orthologs are not identified. These may be species-specific genes in *S. cerevisiae*, but alternatively, they could reflect the gaps in the draft genomic sequences available [39, 40]. Yeast cells exposed to ethanol synthesize a range of *Hsps*, including *Hsp104, Hsp82, Hsp70, Hsp26, Hsp30,* and *Hsp12*. The *Hsp104* and *Hsp12* physiologically influence the ethanol tolerance in yeast [40]. Cell wall from *S. cerevisiae* contains 85% polysaccharides and 15% proteins approximately. If there is an increase in the level of ethanol, the stability of the membrane will be affected, which will cause damage to the proteins and, therefore, endocytosis is inhibited through the membrane [10].

In Figure 2.4, the yeast the *Hog*1 gene is responsible for encoding the high osmolarity glycerol pathway (HOG), this pathway mainly encodes two enzymes GPD1 and GPD2 that catalyze the conversion of dihydroxyacetone phosphate (DHAP) through glycerol-3-phosphate (G3P) to glycerol. The Fps1 channel helps to glycerol accumulation and contributes to raising the osmotic pressure within the cell. Within the stress response are the proteins trehalose-6-phosphate synthase (TPS) and trehalose-6-phosphate phosphatase (TPP), enzymes responsible for the synthesis of trehalose. In the same way, heat shock transcription factor (*Hsf*1) induces the production of HSPs [40]. Also, in Figure 2.4 in *Z. mobilis* the *Zms*4 and *Zms*6 genes are involved in direct sRNA and target mRNA interactions. The accumulation of ethanol within cells under stress increases the expressions of these genes. Increasing the level of *Zms*4 accelerates ethanol catabolism through upregulation of the aldehyde dehydrogenase gene (*SsdA*/*Zms*ZMO1754) directly and the alcohol dehydrogenase 1 gene (*Adh*A) indirectly to mask ethanol to other carboxylic acids. On the other hand, *Zms*6 upregulated under ethanol stress regulates the expression of the lysine export *Zms*1437 gene and negatively regulates the expression of the methylase gene of *Zms*1934 N-6 DNA to improve ethanol tolerance and avoid import of methylated DNA created by ethanol damage, respectively. However, several regulatory functions of this *Zms*6 gene are still unknown. Otherwise, *Zms*16 gene interacts directly as a target only with the gen *Zms*6 [41].

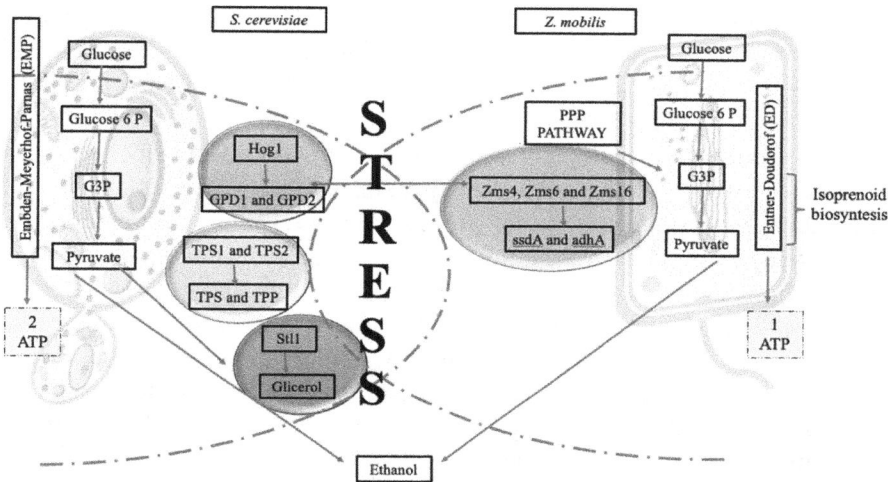

FIGURE 2.4 Metabolic pathway of *Saccharomyces cerevisiae* and *Zymomonas mobilis* with principal genes.

2.9 BIOETHANOL YIELD PRODUCTION OF *Zymomonas mobilis* AND *Saccharomyces cerevisiae*

Currently ethanol production depends on the fermentation of various carbon sources, which is why continuous improvement in industrial production is necessary [42]. Usually, ethanol production is using strains that have specific characteristics, such as tolerant to high temperatures, acidity, and ethanol [43]. To produce ethanol with *Zymomonas* strains, various culture media and fermentation processes, both batch and continuous, have been used. Different fermentation strategies have been used, as well as various culture practices for the evaluation of the alcohol fermentation by *Z. mobilis* the most common is batch fermentation [5]. The *Zymomonas* is a highly biotechnologically relevant ethanogenic bacterium, because it has a larger cell surface than *S. cerevisiae*, consumes glucose faster, leading to a higher production of ethanol, being up to 12% *v/v* and the tolerance of ethanol is 16% *v/v*. While *Z. mobilis* uses the ED pathway that results in 1 ATP per mole of glucose, *S. cerevisiae* obtains 2 ATP per mole of glucose. Therefore, *Z. mobilis* work with 50% less ATP, which leads to a better ethanol yield [5, 41, 44].

The TJ14 hybrid used by Benjaphokee in 2012 [43] produce 46.6 gL^{-1} of ethanol with 10% glucose medium at 41°C and pH 3, a considerable yield, considering tolerance to high temperatures. Most yeast prefers ferment glucose than fructose when both are in the culture medium [5, 42, 43, 45, 46, 77].

The aforementioned yield is similar to reported by Shihui Yang [78] with production of 0.50 g g^{-1} at pH 4.5 and 37C and 100 gL^{-1} of glucose with 12 hours of fermentation, using *Z. mobilis* (ATCC 10988) strain, which uses the ED fermentation pathway, and has a larger surface area than *Saccharomyces*. Fatty acids are produced by yeast and bacteria in response to stress. In this study, Wang et al. in 2016 [45] modified an operon (CP4) that induced their overproduction and high ethanol yield of 78 gL^{-1} at 32°C in 60 h. Davis in 2006 [42] evaluated the ethanol production by *Z. mobilis* ZM4 and, the comparison of *S. cerevisiae* in the same conditions, using the same substrate. Ethanol production was 0.49 gg^{-1} with 100 g of glucose at 30°C with *Z. mobilis* ZM4 strain and 0.46 gg^{-1} with *S. cerevisiae*, this can be attributed to the medium supplementing with KH_2PO_4 and $(NH_4)\ 2SO_4$ among other substances [5, 42]. Mazaheri and Pirouzi in 2020 [46], evaluate *Z. mobilis* for bioethanol production from potato peel waste, obtaining 23.3 gL^{-1} with initial sugar concentration of 61.3 gL^{-1}. The combination of enzymes for the hydrolysis process could have effectively released the fermentable sugars from the solids of the potato peel, however, the final concentration of ethanol obtained is not high enough for industrial scales as it causes high energy consumption in the purification stage [46].

2.10   STRESS CONDITIONS IN FERMENTATION IN *Zymomonas mobilis* AND *Saccharomyces cerevisiae*

The four major stress responses studied during fermentation are temperature stress, osmotic stress, lack of nutrients and ethanol produced during fermentation [40].

2.10.1   TOLERANCE TO HIGH CONCENTRATIONS OF GLUCOSE

There are different strains to *Zymomonas* that can support high concentration of glucose to grow up in 400 gL^{-1} medium, but the optimal glucose concentration is 200 gL^{-1} to obtain the maximum concentration of ethanol, 96 gL^{-1}. This bacterium has a higher glucose tolerance than *S. cerevisiae* and, therefore, highly concentrated glucose media can potentially be used. The bacterium metabolizes glucose, sucrose, and fructose. However, there are some strains that are modified genetically to use other carbon sources like xylose [47].

The yeasts are classified according to the respiratory-fermentative metabolism regulation. The Crabtree-positive species ferment under aerobic conditions, and contrary, the extent of fermentative metabolism for Crabtree-negative species is very limited under sufficient oxygen conditions [48]. Hypothetically, in high sugar levels, Crabtree-positive yeast can adapt and exploit sugars faster than Crabtree-negative competitors [49].

2.10.2   ETHANOL TOLERANCE

Osmotic pressure and temperature are environmental factors that influence ethanol tolerance in yeast, as well as available nutrients and growth substrates [50]. High ethanol tolerance is a hallmark of *Saccharomyces cerevisiae* and a wide variety of genes are responsible for a moderate ethanol tolerance ranging from 6 to 12% in laboratory strains and 16 to 20% in industrial strains, respectively [79]. According to Ming [80], wild-type *Z. mobilis* has an adaptive mechanism in response to ethanol stress and could tolerate up to 13%. Therefore, with these results, we can see that *Zymomonas* has a great advantage compared to yeast for tolerance to ethanol in the medium, this being one of the most important factors to consider choosing a microorganism to carry out the fermentation process in this bioprocess for ethanol production [81].

2.10.3   TEMPERATURE TOLERANCE

Temperature is one of the most important factors that affect the production of ethanol by different microorganisms. The fermentation without cooling, produces microbial heat shock stress, the microbial growth is reduced, and ethanol production yield is affected. There are some *Zymomonas* strains thermotolerant, can growth even at 39°C, that is, 5–10°C higher than the optimum temperature. This Z. mobilis harbor *deg*P gene from heat shock proteins (HSP) family [51, 52]. The HSF family of proteins (heat shock transcription factor) are primary modulators of heat shock response (HSR), as well as the MSN2 and MSN4 genes of significant use in the expression of the heat shock gene. Heat shock factor in *S. cerevisiae* (Hsf1) is an essential protein that assists HSR in adapting to oxidative stress and glucose deficiency [40].

It is noted that despite the fact that some *Saccharomyces* strains can grow at high temperatures, ethanol production is lower than *Zymomonas* strains that grow at the same range of temperatures [43, 52–56].

2.11 CONCLUSION

Currently, innovative, and more efficient microbial alternatives for ligno-cellulosic biorefineries are needed to improve the biofuels (bioethanol) production worldwide. Physiologically, *Z. mobilis* is an excellent alternative by both ethanologenic nature and highly efficient capacity to biodegrade several lignocellulosic residues in bioethanol. In order to apply *Zymomonas* in industrial fermentations, multiple genetic tools, and metabolic engineering methods, as well as biofilm reactor implementation to enhance and increase the bioethanol production sustainably have been investigated.

Finally, the results of decades of multidisciplinary scientific research strongly suggest that the unique and amazing physiology of *Z. mobilis* provides efficient biosynthetic machinery for industrial-scale biotechnological production of bioethanol.

KEYWORDS

- **bioethanol**
- **biofuels**
- **heat shock proteins**
- **Saccharomyces cerevisiae**
- **trehalose-6-phosphate phosphatase**
- **Zymomona mobilis**

REFERENCES

1. Varela, C., (2016). The impact of non-*Saccharomyces* yeasts in the production of alcoholic beverages. *Applied Microbiology and Biotechnology, 100*(23), 9861–9874. https://doi.org/10.1007/s00253-016-7941-6.

2. Sarris, D., & Papanikolaou, S., (2016). Biotechnological production of ethanol: Biochemistry, processes and technologies. *Engineering in Life Sciences, 16*(4), 307–329. https://doi.org/10.1002/elsc.201400199.

3. Mohd, A. S. H., Abdulla, R., Jambo, S. A., Marbawi, H., Gansau, J. A., Mohd, F. A. A., & Rodrigues, K. F., (2017). Yeasts in sustainable bioethanol production: A review. *Biochemistry and Biophysics Reports, 10*, 52–61. https://doi.org/10.1016/j.bbrep.2017.03.003.

4. Bleoanca, I., & Bahrim, G., (2013). Overview on brewing yeast stress factors. *Romanian Biotechnological Letters, 18*(5), 8559–8572.

5. Yang, S., Fei, Q., Zhang, Y., Contreras, L. M., Utturkar, S. M., Brown, S. D., Himmel, M. E., & Zhang, M., (2016). *Zymomonas mobilis* as a model system for production of biofuels and biochemicals. *Microb. Biotechnol., 9*, 699–717.

6. Shaw, A. J., Lam, F. H., Hamilton, M., Consiglio, A., MacEwen, K., Brevnova, E. E., & Stephanopoulos, G., (2016). Metabolic engineering of microbial competitive advantage for industrial fermentation processes. *Science, 353*(6299), 583–586. https://doi.org/10.1126/science.aaf6159.

7. Walker, G. M., & Walker, R. S. K., (2018). Enhancing yeast alcoholic fermentation. In: *Advances in Applied Microbiology* (Vol. 105). https://doi.org/10.1016/bs.aambs.2018.05.003.

8. Rogers, P. L., Lee, K. J., Skotnicki, M. L., & Tribe, D. E., (1982). Ethanol production by *Zymomonas mobilis. Biotechnology and Bioengineering, 26*(3), 247–251. https://doi.org/10.1002/bit.260260308.

9. Zhang, J., Zhang, W., & Gu, Y., (2018). Enzyme-free isothermal target-recycled amplification combined with page for direct detection of microRNA-21. *Analytical Biochemistry, 550*, 117–122. https://doi.org/10.1016/j.ab.2018.04.024.

10. Mannan, M. A., Upadhyay, A. K., & Karnwal, A., (2017). *Saccharomyces cerevisiae* bio-ethanol production a sustainable energy. *Asian Journal of Microbiology, Biotechnology and Environmental Sciences*, 20.

11. Gupta, A., & Prakash, J., (2015). Sustainable bio-ethanol production from agro-residues: A review. *Renewable and Sustainable Energy Reviews, 41*, 550–567. https://doi.org/10.1016/j.rser.2014.08.032.

12. Aditiya, H. B., Mahlia, T. M. I., Chong, W. T., Nur, H., & Sebayang, A. H., (2016). Second-generation bioethanol production: A critical review. *Renewable and Sustainable Energy Reviews, 66*, 631–653. https://doi.org/10.1016/j.rser.2016.07.015.

13. Souza, C. J. A. D., Costa, D. A., Rodrigues, M. Q. R. B., Ancély, F., Lopes, M. R., Abrantes, A. B. P., & Fietto, L. G., (2012). Bioresource technology the influence of presaccharification, fermentation temperature and yeast strain on ethanol production from sugarcane bagasse. *Bioresource Technology, 109*, 63–69. https://doi.org/10.1016/j.biortech.2012.01.024.

14. Carrillo-Nieves, D., Rostro, A. M. J., De La Cruz, Q. R., Ruiz, H. A., Iqbal, H. M. N., & Parra-Saldívar, R., (2019). Current status and future trends of bioethanol production from agro-industrial wastes in Mexico. *Renewable and Sustainable Energy Reviews, 102*, 63–74. https://doi.org/10.1016/j.rser.2018.11.031.

15. Rastogi, M., & Shrivastava, S., (2017). Recent advances in second-generation bioethanol production: An insight to pretreatment, saccharification and fermentation processes. *Renewable and Sustainable Energy Reviews, 80*, 330–340. https://doi.org/10.1016/j.rser.2017.05.225.

16. Carrillo-nieves, D., Rostro, M. J., De, R., & Quiroz, C., (2019). *Current Status and Future Trends of Bioethanol Production from Agro-Industrial Wastes in Mexico, 102*, 63–74. https://doi.org/10.1016/j.rser.2018.11.031.

17. Jahnavi, G., Prashanthi, G. S., Sravanthi, K., & Rao, L. V., (2017). Status of availability of lignocellulosic feedstocks in India: Biotechnological strategies involved in the production of bioethanol. *Renewable and Sustainable Energy Reviews, 73*, 798–820. https://doi.org/10.1016/j.rser.2017.02.018.

18. Connor, S. E. O., (2015). *Engineering of Secondary Metabolism*, 1–24. https://doi.org/10.1146/annurev-genet-120213-092053.

19. Pappas, K. M., Kouvelis, V. N., Saunders, E., Brettin, T. S., Bruce, D., Detter, C., Balakireva, M., et al., (2011). Genome sequence of the ethanol-producing *Zymomonas mobilis* subsp. *mobilis* lectotype strain ATCC 10988. *Journal of Bacteriology, 193*(18), 5051, 5052. https://doi.org/10.1128/JB.05395-11.
20. Ojeda-Linares, C. I., Vallejo, M., Lappe-Oliveras, P., & Casas, A., (2020). Traditional management of microorganisms in fermented beverages from cactus fruits in Mexico: An ethnobiological approach. *J. Ethnobiol. Ethnomed., 16*(1), 1. doi: 10.1186/s13002-019-0351-y.
21. Bergey's Manual of Systematics of Archaea and Bacteria, Online © 2015 Bergey's Manual Trust. This article is © 2005 Bergey's Manual Trust. doi: 10.1002/9781118960608. gbm00925. Published by John Wiley & Sons, Inc., in association with Bergey's Manual Trust.
22. Chacon-Vargas, K., Chirino, A. A., Davis, M. M., Debler, S. A., Haimer, W. R., Wilbur, J. J., Mo, X., et al., (2017). Genome sequence of *Zymomonas mobilis* subsp. *mobilis* NRRL B-1960. *Genome Announc., 5*(30), e00562–17. https://doi.org/10.1128/genomeA.00562-17.
23. Geng, B., Cao, L., Li, F., Song, H., Liu, C. H., Zhao, X. Q., & Bai, F. W., (2020). Potential of *Zymomonas mobilis* as an electricity producer in ethanol production. *Biotechnol Biofuels, 13*, 36. https://doi.org/10.1186/s13068-020-01672-5.
24. Musatti, A., Cappa, C., Mapelli, C., Alamprese, C., & Rollini, M., (2020). *Zymomonas mobilis* in bread dough: Characterization of dough leavening performance in presence of sucrose. *Foods, 9*: 89. doi: 10.3390/foods9010089.
25. Wang, W., Dai, L., Wu, B., Bu-Fan, Q., Tian-Fang, H., Guo-Quan, H., & Ming-Xiong, H., (2020). Biochar-mediated enhanced ethanol fermentation (BMEEF) in *Zymomonas mobilis* under furfural and acetic acid stress. *Biotechnol Biofuels, 13*, 28. https://doi.org/10.1186/s13068-020-1666-6.
26. Musatti, A., Mapelli, C., Rollini, M., Foschino, R., & Picozzi, C., (2018). Can *Zymomonas mobilis* substitute *Saccharomyces cerevisiae* in cereal dough leavening? *Foods, 7*, 61.
27. Rogers, P. L., Jeon, Y. J., Lee, K. J., & Lawford, H. G., (2007). *Zymomonas mobilis* for fuel ethanol and higher-value products. *Adv Biochem Eng Biotechnol., 108*, 263–288. doi: 10.1007/10_2007_060.
28. Zhang, K., et al., (2019). New technologies provide more metabolic engineering strategies for bioethanol production in *Zymomonas mobilis*. *Applied Microbiology and Biotechnology, 103*, 2087–2099.
29. Fuchino, K., Kalnenieks, U., Rutkis, R., Grube, M., & Bruheim, P., (2020). *Metabolic Profiling of Glucose-Fed Metabolically Active Resting Zymomonas Mobilis Strains Metabolites, 10*, 81. doi: 10.3390/metabo10030081.
30. Carreón-Rodríguez, O. E., Gutiérrez-Ríos, R. M., Acosta, J. L., Martinez, A., & Cevallos, M. A., (2019). Phenotypic and genomic analysis of *Zymomonas mobilis* ZM4 mutants with enhanced ethanol tolerance. *Biotechnology Reports (Amsterdam, Netherlands), 23*, e00328. https://doi.org/10.1016/j.btre.2019.e00328.
31. Clark, & Pazdernik, (2016). Synthetic biology. In: *Biotechnology*. Elsevier. http://dx.doi.org/10.1016/B978-0-12-385015-7.00013-2.
32. Fuchino, K., Kalnenieks, U., Rutkis, R., Grube, M., & Bruheim, P., (2020). Metabolic profiling of glucose-fed metabolically active resting *zymomonas mobilis* strains. *Metabolites, 10*(3), 6–8. https://doi.org/10.3390/metabo10030081.
33. Belda, I., Ruiz, J., Esteban-Fernández, A., Navascués, E., Marquina, D., Santos, A., & Moreno-Arribas, M. V., (2017). Microbial contribution to wine aroma and its intended

use for wine quality improvement. *Molecules, 22*(2), 1–29. https://doi.org/10.3390/molecules22020189.

34. Wang, X., He, Q., Yang, Y., Wang, J., Haning, K., Hu, Y., & Wu, B., (2018). *Advances and Prospects in Metabolic Engineering of Zymomonas Mobilis. 50*, 57–73. https://doi.org/10.1016/j.ymben.2018.04.001.
35. Lee, K. Y., Park, J. M., Kim, T. Y., Yun, H., & Lee, S. Y., (2010). The genome-scale metabolic network analysis of *Zymomonas mobilis* ZM4 explains physiological features and suggests ethanol and succinic acid production strategies. *Microbial Cell Factories, 9*, 1–12. https://doi.org/10.1186/1475-2859-9-94.
36. Annaluru, N., Annaluru, N., Muller, H., Mitchell, L. A., Ramalingam, S., Stracquadanio, G., & Han, J. S., (2014). *Total Synthesis of a Functional Designer Eukaryotic Chromosome, 55*. https://doi.org/10.1126/science.1249252.
37. Goffeau, A., Barrell, B. G., Bussey, H., Davis, R. W., Dujon, B., Feldmann, H., & Oliver, S. G., (1996). Life with 6000 genes conveniently among the different interna- old questions and new answers the genome. At the beginning of the se- of its more complex relatives in the eukary- cerevisiae has been completely sequenced *Schizosaccharomyces pombe* indicate. *Science, 274*, 546–567. https://doi.org/jyu.
38. He, M. X., Wu, B., Qin, H., Ruan, Z. Y., Tan, F. R., Wang, J. L., & Hu, Q. C., (2014). *Zymomonas mobilis*: A novel platform for future biorefineries. *Biotechnology for Biofuels, 7*(1), 1–15. https://doi.org/10.1186/1754-6834-7-101.
39. Kellis, M., Patterson, N., Endrizzi, M., Birren, B., & Lander, E. S., (2003). Sequencing and comparison of yeast species to identify genes and regulatory elements. *Nature, 423*(6937), 241–254. https://doi.org/10.1038/nature01644.
40. Saini, P., Beniwal, A., Kokkiligadda, A., & Vij, S., (2018). Response and tolerance of yeast to changing environmental stress during ethanol fermentation. *Process Biochemistry, 72*, 1–12. https://doi.org/10.1016/j.procbio.2018.07.001.
41. Han, R., Haning, K., Gonzalez-rivera, J. C., Yang, Y., & Li, R., (2020). *Multiple Small RNAs Interact to Co-regulate Ethanol Tolerance in Zymomonas Mobilis, 8*, 1–19. https://doi.org/10.3389/fbioe.2020.00155.
42. Davis, L., Rogers, P., Pearce, J., & Peiris, P., (2006). *Evaluation of Zymomonas -based Ethanol Production from a Hydrolyzed Waste Starch Stream, 30*, 809–814. https://doi.org/10.1016/j.biombioe.2005.05.003.
43. Benjaphokee, S., Hasegawa, D., Yokota, D., Asvarak, T., Auesukaree, C., Sugiyama, M., & Harashima, S., (2012). Highly efficient bioethanol production by a *Saccharomyces cerevisiae* strain with multiple stress tolerance to high temperature acid and ethanol. *New Biotechnology, 29*(3), 379–386. https://doi.org/10.1016/j.nbt.2011.07.002.
44. Panesar, P. S., Marwaha, S. S., & Kennedy, J. F., (2006). *Zymomonas Mobilis: An Alternative Ethanol Producer, 635*, 623–635. https://doi.org/10.1002/jctb.1448.
45. Wang, H., Cao, S., Tianshuo, W., Kaven, W., & Wang, T., (2016). Very high gravity ethanol and fatty acid production of *Zymomonas mobilis* without amino acid and vitamin. *Journal of Industrial Microbiology & Biotechnology.* https://doi.org/10.1007/s10295-016-1761-7.
46. Mazaheri, D., & Pirouzi, A., (2020). Valorization of *Zymomonas mobilis* for bioethanol production from potato peel: Fermentation process optimization. *Biomass Conv. Bioref.* https://doi.org/10.1007/s13399-020-00834-7.
47. Ajit, A., Sulaiman, A. Z., & Chisti, Y., (2017). Production of bioethanol by *Zymomonas mobilis* in high-gravity extractive fermentations. *Food and Bioproducts Processing, 102*, 123–135. https://doi.org/10.1016/j.fbp.2016.12.006.

48. Gonzalez, R., Quirós, M., & Morales, P. (2013). Yeast respiration of sugars by non-Saccharomyces yeast species: a promising and barely explored approach to lowering alcohol content of wines. *Trends in Food Science & Technology*, *29*(1), 55–61. https://doi.org/10.1016/j.tifs.2012.06.015.
49. Boynton, P. J., & Greig, D., (2014). The Ecology and Evolution of Non-domesticated *Saccharomyces Species*, 449–462. https://doi.org/10.1002/yea.
50. Riles, L., & Fay, J. C., (2019). Genetic basis of variation in heat and ethanol tolerance in *Saccharomyces cerevisiae*. *G3 (Bethesda, Md.)*, *9*(1), 179–188. https://doi.org/10.1534/g3.118.200566.
51. Anggarini, S., Murata, M., Kido, K., Kosaka, T., Sootsuwan, K., Thanonkeo, P., & Yamada, M., (2020). Improvement of thermotolerance of *Zymomonas mobilis* by genes for reactive oxygen species-scavenging enzymes and heat shock proteins. *Frontiers in Microbiology*, *10*, 1–14. https://doi.org/10.3389/fmicb.2019.03073.
52. Charoensuk, K., Sakurada, T., Tokiyama, A., Murata, M., Kosaka, T., Thanonkeo, P., & Yamada, M., (2017). Thermotolerant genes essential for survival at a critical high temperature in thermotolerant ethanologenic *Zymomonas mobilis* TISTR 548. *Biotechnology for Biofuels*, *10*(1), 1–11. https://doi.org/10.1186/s13068-017-0891-0.
53. D'Amore, T., Celotto, G., Russell, I., & Stewart, G. G., (1989). Selection and optimization of yeast suitable for ethanol production at 40 C. *Enzyme and Microbial Technology*, *11*(7), 411–416. https://doi.org/10.1016/0141-0229(89)90135-X.
54. Sree, N. K., Sridhar, M., Suresh, K., Banat, I. M., & Rao, L. V., (2000). Isolation of thermotolerant, osmotolerant, flocculating *Saccharomyces cerevisiae* for ethanol production. *Bioresource Technology*, *72*(1), 43–46. https://doi.org/10.1016/S0960-8524(99)90097-4.
55. De Souza, E. A., Gomes, D. S., Panek, A. D., & Eleutherio, E. C. A., (2003). The role of glutathione in yeast dehydration tolerance. *Cryobiology*, *47*(3), 236–241. https://doi.org/10.1016/j.cryobiol.2003.10.003.
56. Costa, D. A., De Souza, C. J., Costa, P. S., Rodrigues, M. Q., Dos, S. A. F., Lopes, M. R., & Fietto, L. G., (2014). Physiological characterization of thermotolerant yeast for cellulosic ethanol production. *Applied Microbiology and Biotechnology*, *98*(8), 3829–3840. https://doi.org/10.1007/s00253-014-5580-3.
57. Choudhary, J., Singh, S., & Nain, L., (2016). Thermotolerant fermenting yeasts for simultaneous saccharification fermentation of lignocellulosic biomass. *Electronic Journal of Biotechnology*, *21*, 82–92. https://doi.org/10.1016/j.ejbt.2016.02.007.
58. Chen, C., Wu, L., Cao, Q., Shao, H., Li, X., Zhang, Y., et al., (2018). Genome comparison of different *Zymomonas mobilis* strains provides insights on conservation of the evolution. *PLoS One*, *13*(4), e0195994. https://doi.org/10.1371/journal.pone.0195994.
59. Bergey's Manual of Systematics of Archaea and Bacteria First published: 17 April 2015. Online ISBN: 9781118960608| doi: 10.1002/9781118960608.
60. Kluyver, A. J., & Van, N. K., (1936). Prospects for a natural system of classification of bacteria. *Zentralb. Bakteriol. Parasitenkd. Infektionskr. Hyg. Abt.*, *2*(94), 369–403.
61. De Ley, J., & Swings, J., (1976). Phenotypic description, numerical analysis and a proposal for an improved taxonomy and nomenclature of the genus zymomonas kluyver and Van Niel 1936. *Int. J. Syst. Bacteriol.*, *26*, 146–157.
62. Escalante, A., Giles-Gómez, M., Hernández, G., Córdova-Aguilar, M. S., López-Munguía, A., Gosset, G., & Bolívar, F., (2008). Analysis of bacterial community during the fermentation of pulque, a traditional Mexican alcoholic beverage, using a polyphasic

approach. *International Journal of Food Microbiology, 124*(2), 126–134. https://doi.org/10.1016/j.ijfoodmicro.2008.03.003.

63. Lindner, P. (1928). Fermentation studies on pulque in Mexico. *Report of the West Prussian Botanical-Zoological Association 50*, 253–255.

64. Swingsm, J., & De Ley, J., (1977). The biology of zymomonas. *Bacteriol. Revs., 41*, 1–46.

65. Jeong-Sun, S., Chong, H., Park, H. S., Kyoung-Oh, Y., Jung, C., Kim, J. J., Hong, J. H., et al., (2005). The genome sequence of the ethanologenic bacterium *Zymomonas mobilis* ZM4. *Nat. Biotechnol., 23*(1), 63–8. doi: 10.1038/nbt1045.

66. Jiang, X., Dong, D., Bian, L., Zou, D., He, X., & Ao, D., (2016). *Rapid Detection of Candida Albicans by Polymerase Spiral Reaction Assay in Clinical Blood Samples, 7*, 1–6. https://doi.org/10.3389/fmicb.2016.00916.

67. Li, L., Wang, X., Jiao, X., & Qin, S., (2017). Differences between flocculating yeast and regular industrial yeast in transcription and metabolite profiling during ethanol fermentation. *Saudi Journal of Biological Sciences, 24*(3), 459–465. https://doi.org/10.1016/j.sjbs.2017.01.013.

68. Swinnen, S., Goovaerts, A., Schaerlaekens, K., Dumortier, F., Verdyck, P., Souvereyns, K., Van, Z. G., et al., (2015). Auxotrophic mutations reduce tolerance of *Saccharomyces cerevisiae* to very high levels of ethanol stress. *Eukaryotic Cell, 14*(9), 884–897. https://doi.org/10.1128/EC.00053-15.

69. Randez-Gil, F., Córcoles-Sáez, I., & Prieto, J. A., (2013). Genetic and phenotypic characteristics of baker's yeast: Relevance to baking. *Annual Review of Food Science and Technology, 4*(1), 191–214. https://doi.org/10.1146/annurev-food-030212-182609.

70. Bergey's Manual of Systematics of Archaea and Bacteria, Online © 2015 Bergey's Manual Trust. sThis article is © 2005 Bergey's Manual Trust. Published by John Wiley & Sons, Inc., in association with Bergey's Manual Trust. doi: 10.1002/9781118960608.gbm00925.

71. Schatschneider, S., Huber, C., Neuweger, H., Watt, T. F., Pühler, A., Eisenreich, W., & Vorhölter, F. J. (2014). Metabolic flux pattern of glucose utilization by Xanthomonas campestris pv. campestris: prevalent role of the Entner–Doudoroff pathway and minor fluxes through the pentose phosphate pathway and glycolysis. *Molecular BioSystems, 10*(10), 2663–2676. https://doi.org/10.1039/C4MB00198B.

72. Díaz, V. H. G., & Willis, M. J. (2019). Ethanol production using *Zymomonas mobilis*: Development of a kinetic model describing glucose and xylose co-fermentation. *Biomass and Bioenergy, 123*, 41–50. https://doi.org/10.1016/j.biombioe.2019.02.004.

73. Fuchino, K., Kalnenieks, U., Rutkis, R., Grube, M., & Bruheim, P. (2020). Metabolic profiling of glucose-fed metabolically active resting *Zymomonas mobilis* strains. *Metabolites, 10*(3), 81. https://doi.org/10.3390/metabo10030081.

74. Seo, J. S., Chong, H., Park, H. S., Yoon, K. O., Jung, C., Kim, J. J., ... & Kang, H. S. (2005). The genome sequence of the ethanologenic bacterium *Zymomonas mobilis* ZM4. *Nature Biotechnology, 23*(1), 63-68. https://doi.org/10.1038/nbt1045.

75. Kalnenieks, U., Balodite, E., Strähler, S., Strazdina, I., Rex, J., Pentjuss, A., ... & Bettenbrock, K. (2019). Improvement of acetaldehyde production in *Zymomonas mobilis* by engineering of its aerobic metabolism. *Frontiers in Microbiology, 10*, 2533. https://doi.org/10.3389/fmicb.2019.02533.

76. Zhang, K., Lu, X., Li, Y., Jiang, X., Liu, L., & Wang, H. (2019). New technologies provide more metabolic engineering strategies for bioethanol production in *Zymomonas mobilis*. *Applied Microbiology and Biotechnology, 103*(5), 2087–2099. https://doi.org/10.1007/s00253-019-09620-6.

77. Magyar, I., & Tóth, T. (2011). Comparative evaluation of some oenological properties in wine strains of *Candida stellata, Candida zemplinina, Saccharomyces uvarum* and S*accharomyces cerevisiae. Food Microbiology, 28*(1), 94–100. https://doi.org/10.1016/j. fm.2010.08.011.

78. Yang, S., Fei, Q., Zhang, Y., Contreras, L. M., Utturkar, S. M., Brown, S. D., ... & Zhang, M. (2016). *Zymomonas mobilis* as a model system for production of biofuels and biochemicals. *Microbial Biotechnology, 9*(6), 699–717. https://doi.org/10.1111/1751-7915.12408.

79. Swinnen, S., Goovaerts, A., Schaerlaekens, K., Dumortier, F., Verdyck, P., Souvereyns, K., Van Zeebroeck, G., Foulquié-Moreno, M. R., & Thevelein, J. M. (2015). Auxotrophic Mutations Reduce Tolerance of *Saccharomyces cerevisiae* to Very High Levels of Ethanol Stress. *Eukaryotic Cell, 14*(9), 884–897. https://doi.org/10.1128/EC.00053-15.

80. He, M. X., Wu, B., Shui, Z. X., Hu, Q. C., Wang, W. G., Tan, F. R., Tang, X. Y., Zhu, Q. L., Pan, K., Li, Q., & Su, X. H. (2012). Transcriptome profiling of *Zymomonas mobilis* under ethanol stress. *Biotechnology for Biofuels, 5*(1), 75. https://doi.org/10.1186/1754-6834-5-75.

81. Hacking, A. J., Taylor, I. W. F., & Hanas, C. M. (1984). Selection of yeast able to produce ethanol from glucose at 40 C. *Applied Microbiology and Biotechnology, 19*(5), 361–363. https://doi.org/10.1007/BF00253786.

CHAPTER 3

Physiology of Ethanol Production by *Clostridium thermocellum*

D. B. ARYA,[1] SALOM GNANA THANGA VINCENT,[1] and
NAGAMANI BALAGURUSAMY[2]

[1]*Department of Environmental Sciences, University of Kerala,
Thiruvananthapuram, Kerala, India*

[2]*Bioremediation Laboratory, Autonomous University of Coahuila, México*

ABSTRACT

Clostridium thermocellum has been extensively studied as a model organism for microbial cellulose degradation due to its ability to rapidly solubilize biomass and use cellulose as a carbon and energy source, which is attributed to the presence of cellulosome, an extracellular multienzyme complex. *C. thermocellum* is a thermophilic, rod-shaped anaerobe and is capable of producing ethanol directly from a wide range of substrates. It was first isolated in 1926, followed by detailed characterization studies in the subsequent years. Recently this organism has received increased attention as a potential candidate for the consolidated bioprocessing (CBP) of lignocellulosic biomass (LCB) into biofuels such as ethanol, where CBP is a single step deconstruction and conversion of lignocellulose into useful products, like acetate, lactate, hydrogen, and ethanol. Although engineered *C. thermocellum* can tolerate ethanol concentrations up to 80 g/L, the ethanol production under normal conditions is <30 g/L. This is because of the inefficiency of the ethanol production pathway of *C. thermocellum*, wherein, the key enzyme for ethanol production, ferredoxin oxidoreductase has lower affinity than lactate dehydrogenase (LDH) and phosphotransacetylase (PTA) resulting in the diversion of carbon flux towards production of more undesirable by-products like acetic and lactic acids. Nevertheless, several reports describe the successful attempts to increase ethanol production involving pathway modifications, strain improvement, or genetic engineering to achieve a significant increase in ethanol yield up to 75% of the theoretical maximum.

3.1 INTRODUCTION

Renewable energy sources receive much attraction in the present scenario due to the reduced supply and high cost of available nonrenewable resources. The alternative fuel for nonrenewable resources must contain some essential characteristics: they have to be sufficient to meet the worlds energy demand, should be environment friendly and cost-effective. Biofuels are one of the promising alternatives for nonrenewable resources and hence, significant technologies should be explored to meet world energy demand for low carbon fuels [1]. Cellulosic biofuels can make a significant contribution to rural economy [2]. Ethanol from lignocellulosic material is a very cost-effective alternative. Due to the complex structure and recalcitrant nature of LCB, the production of bioethanol from lignocellulosic material needs expensive chemical or physical pretreatments along with enzyme treatment that has a negative impact on its industrial large-scale production [3, 4].

Pretreatment is necessary to convert cellulosic material to fermentable form, such treatment enable the feedstock readily available for the enzyme action [5]. Acid or hydrolytic enzymes can be used as pretreatment, in which acid or enzyme degrade complex cellulose to glucose monomers by disrupting the hydrolytic linkage [6]. The main problem associated with the production of microbial ethanol from lignocellulosic material is the conversion of cellulosic material to fermentable form [7]. It is important to overcome these impacts for an economically feasible sustainable production by integrating all processes into CBP and genetic manipulation of ethanol producers [8]. This review consolidates the characteristic features and the physiology of *C. thermocellum* that makes it an attractive and potential candidate for ethanol production.

3.2 WHY *Clostridium thermocellum?*

For the first-generation bioethanol production, mesophilic organisms such as *Saccharomyces cerevisiae* and *Zymomonas mobilis* were used. The growing interest of second-generation ethanol due to the obstacles of first-generation ethanol led to the genetic modification of *S. cerevisiae* and *Z. mobilis* and the use of other microbe which can utilize the feedstock. In 1980, the idea of using thermophiles for second generation ethanol production was put forth [8]. The main advantage of thermophiles over mesophiles is that the thermophiles can utilize a wide range of substrate and allow direct-ethanol recovery through *in situ* vacuum distillation. Moreover, they have the advantages of tolerating extreme pH and salt concentration and low nutritional requirement [4].

Cellulolytic bacteria are efficient among the microbes for the direct conversion of LCB into ethanol. The important ethanologenic thermophilic anaerobe include members of the genera Clostridium, Caldanaerobacter, Thermoanaerobacter, Bacillus, Geobacillus, Pencibacillus, and Caloramator [4, 9–11, 42]. Among these, Clostridium is widely studied due to its strain diversity. They are strict anaerobes and capable of utilizing a wide spectrum of substrates and thrive under mesophilic or thermophilic temperature conditions [12]. The widely explored Clostridium species is *C. thermocellum*, which could grow at a temperature range of 50 and 68°C and are able to degrade complex cellulosic structures with the presence of cellulosome, a complex hydrolytic enzyme [13] which has demonstrated remarkable hydrolysis efficiency of cellulose [14]. Moreover, thermophilic cellulolytic microorganisms like *C. thermocellum* are regarded as potential candidates because their growth at high temperatures reduces the risk of contamination and also increases the substrate solubility [15].

Clostridium thermocellum is an anaerobic, rod-shaped, Gram-positive thermophilic microbe which is able to produce ethanol directly from cellulose. C. *thermocellum* is an ideal bacterium for synthesizing various alcohols including ethanol and isobutanol [16, 17]. Thermophilic cellulolytic microorganisms are well suited for this process because thermophilic condition reduces the risk of contamination and increases the solubility of feedstock [13]. C. *thermocellum* is an ideal candidate for CBP which includes a single step which includes the degradation of lignocellulosic material and their conversion to useful products [16]. In 1926, Viljoen et al. [18] attempted to identify the organism capable of degrading cellulose, which lead to the basic identification of C. *thermocellum*. Further, McBee [19] characterized C. *thermocellum* and observed that they can grow in between 50 to 60°C in substrates such as cellulose, cellobiose, and hemicellulose. The various byproducts were CO_2, hydrogen gas, acetic acid, succinic acid, and ethanol. After these studies, Freier et al. [20] revealed that the optimum pH and temperature for these organisms are 6 to 7 and 55°C respectively. *C. thermocellum* can be cultured in both batch and continuous culture, with growth rates of 0.10/h and 0.16/h, respectively [21]. The attractive feature of *C. thermocellum* is the presence of cellulosome, which is an extracellular multienzyme complex containing around 20 enzymes able to reduce the resistance of lignocellulosic material from degradation [22]. Also, they possess multiple carbohydrate esterases (CEs) which help in the reduction of LCB in a concentrated enzyme reaction [23–25].

3.3   PHYSIOLOGY OF *C. thermocellum*

The central metabolism of thermophilic anaerobic bacteria like *C. thermocellum* is the bioconversion of sugars. Through modified Embden-Meyerhof-parnas (EMP) glycolytic pathway *C. thermocellum* converts glucose and cellodexins to pyruvate by using mixed acid fermentation [26] (Figure 3.1). The entire pathway is controlled by a set of cofactors such as NADH, ATP, etc., through glycolysis and hexose is converted to pyruvate. After that pyruvate is reduced to acetyl coA by pyruvate ferredoxin oxidoreductase, that leads to the further reduction of acetyl coA to ethanol by alcohol dehydrogenase (ADH). Final end products are mixed acids. The formation of these byproducts is the main obstacle for high ethanol yield. Understanding the electron transfer mechanism is very important for the metabolic engineering of desired organism [27].

3.4   STRAIN IMPROVEMENT FOR ENHANCED ETHANOL PRODUCTION

One of the important by-products of cellulose fermentation by *C. thermocellum* is ethanol, which attracts attention as a renewable energy resource. The formation of other by-products like acetate and lactate and their inability to utilize xylose is a major drawback. In order to avoid this, Shaw et al. [28] attempted to delete the genes encoding lactate dehydrogenase (LDH) and phosphotransacetylase (PTA) to increase ethanol production. Wild type *C. thermocellum* can only tolerate ethanol up to 5 g/L, above which, the organism is significantly inhibited [29]. Several studies were carried out to overcome this sensitivity problem. The sensitivity occurs due to the presence of specialized lipids in the membrane structure. So, the wild type strain must be genetically modified to tolerate large amounts of ethanol. Ethanol production improved after the adaptation of strain for improved growth and high yield [30]. Studies were also focused to increase ethanol production by disrupting the normal pathways [31]. Recent studies showed that thermophilic anaerobes, especially *C. thermocellum* is more effective for the degradation of LCB than industrial standard fungal cellulases in a wide range of conditions [1]. In 2015, Biswas et al. [31] and Rydzak et al. [32] attempted to increase the ethanol production by disrupting competitive pathways for the electric flux by hydrogenase deletion to increase the yield to 64% of theoretical maximum. Papanek et al. [33] attempted to increase ethanol production by deleting genes encoding for lactate, formate, and other compounds and showed similar output of hydrogenase deletion. Other attempts were carried out to introduce n-butanol/isobutanol synthesis pathway to native species [34], which achieved a production rate of 5.5 g/L [35].

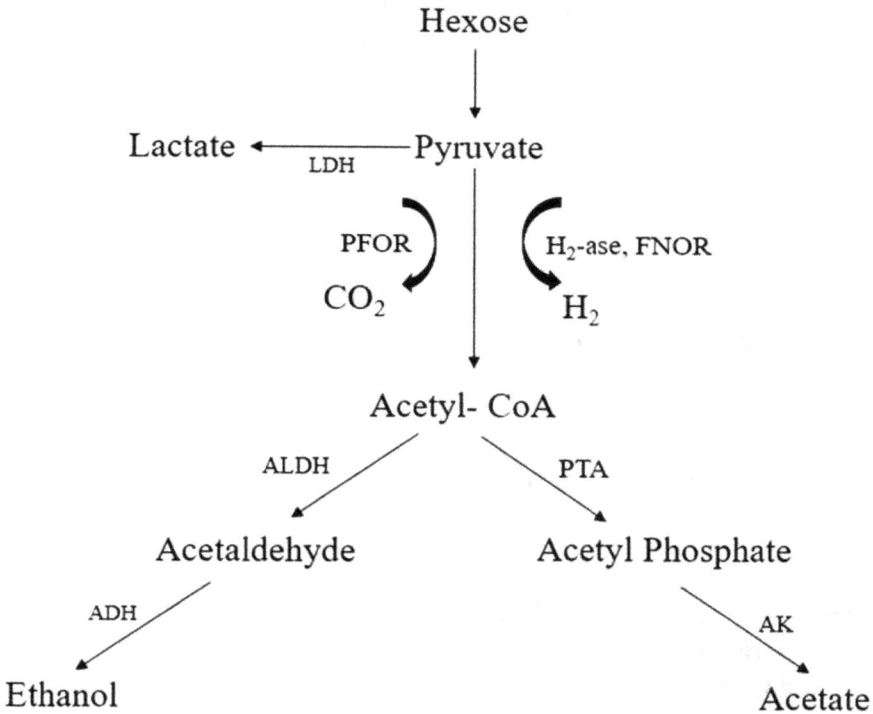

FIGURE 3.1 Simplified scheme of glucose degradation to various end products by strictly anaerobic bacteria. [*Enzyme abbreviations*: ALDH: acetaldehyde dehydrogenase; ADH: alcohol dehydrogenase; AK: acetate kinase; FNOR: ferredoxin oxidoreductase; H_2-ase: hydrogenase; LDH: lactate dehydrogenase; PFOR: pyruvate ferredoxin oxidoreductase; and PTA: phosphotransacetylase].

Genetically engineered *C. thermocellum* exhibit high ethanol tolerance as compared to wild strains of *C. thermocellum* and other thermophiles [1]. Recently developed engineered strain of *C. thermocellum* produced ethanol at 75% of theoretical yield and titer of 25 g/L. Since microbial tolerance to ethanol is complex, related to membrane alterations and limited by electron flux, further studies are required using approaches to genetically modify *C. thermocellum*. A recombinant strain of *C. thermocellum* by introduction of PDC gene from *Zymomonas mobilis* enhanced ethanol production ability which can be further increased by improvements involving bioprocess optimizations [36]. Genes encoding NfnAB and Rnf are studied widely for biofuel production due to their presence in several anaerobic microorganisms and their importance in bioethanol production [27]. Introducing four genes: adhE, nfnA, nfnB, and adhA involved in ethanol production in *Thermoanaerobacterium saccharolyticum* were insufficient to achieve expected ethanol yields in *C. thermocellum* [37], which suggests the presence of missing components and requires further investigations.

3.5   CONSOLIDATED BIOPROCESSING (CBP)

Consolidated bioprocessing (CBP) technologies combine the enzyme production, hydrolysis, and fermentation stages into a single step, which helps to increase processing efficiencies, eliminating the need for added exogenous hydrolytic enzymes, and reducing the sugar inhibition of cellulases [38]. CBP, formerly known as direct microbial conversion (DMC), is identified as an economic process for the production of second-generation biofuel based on a candidate CBP microorganism or group of microorganisms having the ability for combined hydrolysis and fermentation. The process is economic due to simpler feedstock processing, lower energy inputs and higher conversion efficiencies. Relevant CBP microorganism is required for an economically feasible production and *C. thermocellum* is identified as a potential candidate for CBP by enhancing pathway thermodynamic function and feasibility due to the presence of complex cellulosome [39], its fast rate of cellulose digestion and also the ability to hydrolyze both hemicellulose and cellulose in a broad range of substrates. Moreover, owing to the higher fermentation temperature, the fermentation process is not easily polluted and ethanol is easily separated. CBP is cost-effective because the CBP microorganisms do not require exogenous saccharifying enzymes and they produce their own cellulolytic and hemicellulolytic enzymes for the degradation of LCB [14]. The cost-effectiveness and industrial relevance of CBP is determined based on the target ethanol yield of >90%. For CBP to be economically feasible, the CBP microorganisms should have an enzyme system capable of fermenting sugars to ethanol at a rate of more than 1 g/L/h [43]. As no microorganisms with such efficiency is discovered, genetic engineering technologies can be employed for creating efficient CBP microorganisms. The fermentation process can also be improved through mutagenesis as well as improvement of technological and culture conditions.

The ideal microorganisms for CBP should have several essential characteristics including simultaneous utilization and conversion of multiple sugars like cellobiose, glucose, and xylose and also the ability to tolerate toxic by-products. Most of the studies involving CBP is based on ethanol fermentation using model substrates like cellulose and xylan and studies pertaining to real substrate-based ethanol fermentation is limited. Hence, enhanced ethanol production from native LCB like rice straw biomass is the goal of sustainable CBP. Co-culture of different cellulolytic and sugar-fermenting thermophilic anaerobic bacteria has been widely studied to achieve improved ethanol production using CBP. Singh et al. [40] demonstrated the direct fermentation ability of a thermophilic anaerobic cellulolytic bacteria, *C. thermocellum* DSM 1313 isolated from Himalayan hot spring, to convert various cellulosic and hemicellulosic substrates into ethanol using a CBP based approach without addition of any exogenous enzymes. This strain showed good ethanol production on pretreated rice straw. Use of CBP in butanol production from LCB is a good option due to the high energy density of butanol. A co-culture of *C. thermocellum* and *C. saccharoperbutylacetonicum* has been reported to produce a significant amount of butanol using crystalline cellulose as substrate [41], and further research is also focused on developing new pathways of butanol production by genetic engineering of *C. thermocellum*.

3.6 CONCLUSION

The use of biofuels is a suitable alternative to cope up with the increasing energy demand. Second-generation biofuel production technologies make use of lignocellulosic feedstock to produce ethanol. Microorganisms can degrade these feedstocks in several ways. Thermophilic cellulolytic bacteria are well-known candidates for ethanol yield, among which *C. thermocellum* is well known for the degradation of lignocellulosic materials as well as ethanol production and thus can be used to solve the dual problem of waste disposal and energy production. The genetic alteration of the native strains improves ethanol production and opens a way towards a sustainable world.

KEYWORDS

- *C. thermocellum*
- consolidated bioprocessing
- ethanol production
- lactate dehydrogenase
- phosphotransacetylase
- strain improvement

REFERENCES

1. Lynd, L. R., Liang, X., Biddy, M. J., Allee, A., Cai, H., Foust, T., Himmel, M. E., Lser, M. S., et al., (2017). Cellulosic ethanol: status and innovation. *Current Opinion in Biotechnology, 45*, 202–211.

2. Jordan, N., Boody, G., Broussard, W., Glover, J. D., Keeney, D., McCown, B. H., McIsaac, G., et al., (2007). *Sustainable Development of the Agricultural Bio-Economy, 316*(5831), 1570. Science-New York then Washington.

3. Sanchez, O. J., & Cardona, C. A., (2008). Trends in biotechnological production of fuel ethanol from different feedstocks. *Bioresource Technology, 99*(13), 5270–5295.

4. Taylor, M. P., Eley, K. L., Martin, S., Tuffin, M. I., Burton, S. G., & Cowan, D. A., (2009). Thermophilic ethanologenesis: Future prospects for second-generation bioethanol production. *Trends in Biotechnology, 27*(7), 398–405.

5. Gupta, R. B., & Demirbas, A., (2010). *Gasoline, Diesel, and Ethanol Biofuels from Grasses and Plants.* Cambridge University Press.

6. Orozco, A., Ahmad, M., Rooney, D., & Walker, G., (2007). Dilute acid hydrolysis of cellulose and cellulosic bio-waste using a microwave reactor system. *Process Safety and Environmental Protection, 85*(5), 446–449.

7. Viikari, L., Vehmaanperä, J., & Koivula, A., (2012). Lignocellulosic ethanol: From science to industry. *Biomass and Bioenergy, 46*, 13–24.

8. Scully, S. M., & Orlygsson, J., (2015). Recent advances in second generation ethanol production by thermophilic bacteria. *Energies, 8*(1), 1–30.

9. Sveinsdottir, M., Baldursson, S. R. B., & Örlygsson, J., (2009). Ethanol production from monosugars and lignocellulosic biomass by thermophilic bacteria isolated from Icelandic hot springs. *Icel. Agric. Sci., 22*, 45–48.

10. Fong, J. C., Svenson, C. J., Nakasugi, K., Leong, C. T., Bowman, J. P., Chen, B., Glenn, D. R., et al., (2006). Isolation and characterization of two novel ethanol-tolerant facultative-anaerobic thermophilic bacteria strains from waste compost. *Extremophiles, 10*(5), 363–372.

11. Crespo, C., Pozzo, T., Karlsson, E. N., Alvarez, M. T., & Mattiasson, B., (2012). *Caloramator boliviensis* sp. nov., a thermophilic, ethanol-producing bacterium isolated from a hot spring. *International Journal of Systematic and Evolutionary Microbiology, 62*(7), 1679–1686.

12. Wiegel, J., Tanner, R., & Rainey, F. A., (2006). An introduction to the family *Clostridiaceae. The Prokaryotes, 4,* 654–678.

13. Demain, A. L., Newcomb, M., & Wu, J. D., (2005). Cellulase, clostridia, and ethanol. *Microbiology and Molecular Biology Reviews, 69*(1), 124–154.

14. Lu, Y., Zhang, Y. H. P., & Lynd, L. R., (2006). Enzyme - microbe synergy during cellulose hydrolysis by *Clostridium thermocellum. Proc. Natl. Acad. Sci. U.S.A. 103,* 16165–16169. doi: 10.1073/pnas.0605381103.

15. Blumer-Schuette, S. E., Brown, S. D., Sander, K. B., Bayer, E. A., Kataeva, I., Zurawski, J. V., et al., (2013). Thermophilic lignocellulose deconstruction. *FEMS Microbiol. Rev., 38,* 393–448. doi: 10.1111/1574-6976.12044.

16. Akinosho, H., Yee, K., Close, D., & Ragauskas, A., (2014). The emergence of *Clostridium thermocellum* as a high utility candidate for consolidated bioprocessing applications. *Frontiers in Chemistry, 2,* 66.

17. Holwerda, E. K., Thorne, P. G., Olson, D. G., Amador-Noguez, D., Engle, N. L., Tschaplinski, T. J., Van, D. J. P., & Lynd, L. R., (2014). The exometabolome of *Clostridium thermocellum* reveals overflow metabolism at high cellulose loading. *Biotechnology for Biofuels, 7*(1), 155.

18. Viljoen, J. A., Fred, E. B., & Peterson, W. H., (1926). The fermentation of cellulose by thermophilic bacteria. *The Journal of Agricultural Science, 16*(1), 1–17.

19. McBee, R. H., (1954). The characteristics of *Clostridium thermocellum. Journal of Bacteriology, 67*(4), 505.

20. Freier, D., Mothershed, C. P., & Wiegel, J., (1988). Characterization of *Clostridium thermocellum* JW20. *Applied and Environmental Microbiology, 54*(1), 204–211.

21. Lynd, L. R., Grethlein, H. E., & Wolkin, R. H., (1989). Fermentation of cellulosic substrates in batch and continuous culture by *Clostridium thermocellum. Applied and Environmental Microbiology, 55*(12), 3131–3139.

22. Bayer, E. A., Henrissat, B., & Lamed, R., (2009). The cellulosome: A natural bacterial strategy to combat biomass recalcitrance. *Biomass Recalcitrance: Deconstructing the Plant Cell Wall for Bioenergy,* 407–435.

23. Bomble, Y. J., Lin, C. Y., Amore, A., Wei, H., Holwerda, E. K., Ciesielski, P. N., Donohoe, B. S., et al., (2017). Lignocellulose deconstruction in the biosphere. *Current Opinion in Chemical Biology, 41,* 61–70.

24. Himmel, M. E., Xu, Q., Luo, Y., Ding, S. Y., Lamed, R., & Bayer, E. A., (2010). Microbial enzyme systems for biomass conversion: Emerging paradigms. *Biofuels, 1*(2), 323–341.

25. Pawar, P. M. A., Koutaniemi, S., Tenkanen, M., & Mellerowicz, E. J., (2013). Acetylation of woody lignocellulose: Significance and regulation. *Frontiers in Plant Science, 4,* 118.

26. Rydzak, T., McQueen, P. D., Krokhin, O. V., Spicer, V., Ezzati, P., Dwivedi, R. C., Shamshurin, D., et al., (2012). Proteomic analysis of *Clostridium thermocellum* core metabolism: Relative protein expression profiles and growth phase-dependent changes in protein expression. *BMC Microbiology, 12*(1), 214.

27. Lo, J., Olson, D. G., Murphy, S. J. L., Tian, L., Hon, S., Lanahan, A., Guss, A. M., & Lynd, L. R., (2017). Engineering electron metabolism to increase ethanol production in *Clostridium thermocellum. Metabolic Engineering, 39,* 71–79.

28. Shaw, A. J., Podkaminer, K. K., Desai, S. G., Bardsley, J. S., Rogers, S. R., Thorne, P. G., & Lynd, L. R., (2008). Metabolic engineering of a thermophilic bacterium to produce ethanol at high yield. *Proceedings of the National Academy of Sciences, 105*(37), 13769–13774.

29. Herrero, A. A., & Gomez, R. F., (1980). Development of ethanol tolerance in *Clostridium thermocellum:* Effect of growth temperature. *Applied and Environmental Microbiology, 40*(3), 571–577.

30. Argyros, D. A., Tripathi, S. A., Barrett, T. F., Rogers, S. R., Feinberg, L. F., Olson, D. G., Foden, J. M., et al., (2011). High ethanol titers from cellulose by using metabolically engineered thermophilic, anaerobic microbes. *Applied and Environmental Microbiology, 77*(23), 8288–8294.

31. Biswas, R., Zheng, T., Olson, D. G., Lynd, L. R., & Guss, A. M., (2015). Elimination of hydrogenase active site assembly blocks H$_2$ production and increases ethanol yield in *Clostridium thermocellum. Biotechnology for Biofuels, 8*(1), 1–8.

32. Rydzak, T., Lynd, L. R., & Guss, A. M., (2015). Elimination of formate production in *Clostridium thermocellum. Journal of Industrial Microbiology & Biotechnology, 42*(9), 1263–1272.

33. Papanek, B., Biswas, R., Rydzak, T., & Guss, A. M., (2015). Elimination of metabolic pathways to all traditional fermentation products increases ethanol yields in *Clostridium thermocellum. Metabolic Engineering, 32*, 49–54.

34. Gaida, S. M., Liedtke, A., Jentges, A. H. W., Engels, B., & Jennewein, S., (2016). Metabolic engineering of *Clostridium cellulolyticum* for the production of n-butanol from crystalline cellulose. *Microbial. Cell Factories, 15*(1), 1–11.

35. Lin, P. P., Mi, L., Morioka, A. H., Yoshino, K. M., Konishi, S., Xu, S. C., Papanek, B. A., et al., (2015). Consolidated bioprocessing of cellulose to isobutanol using *Clostridium thermocellum. Metabolic Engineering, 31*, 44–52.

36. Kannuchamy, S., Mukund, N., & Saleena, L. M., (2016). Genetic engineering of *Clostridium thermocellum* DSM1313 for enhanced ethanol production. *BMC Biotechnology, 16*(Suppl 1), 34.

37. Hon, S., Olson, D. G., Holwerda, E. K., Lanahan, A. A., Murphy, S. J. L., Maloney, M. I., Zheng, T., et al., (2017). The ethanol pathway from *Thermoanaerobacterium saccharolyticum* improves ethanol production in *C. thermocellum. Metabolic Engineering, 42*, 175–184.

38. Xu, Q., Singh, A., & Himmel, M. E., (2009). Perspectives and new directions for the production of bioethanol using consolidated bioprocessing of lignocellulose. *Current Opinion in Biotechnology, 20*(3), 364–371.

39. Dash, S., Olson, D. G., Chan, S. H. J., Amador-Noguez, D., Lynd, L. R., & Maranas, C. D., (2019). Thermodynamic analysis of the pathway for ethanol production from cellobiose in *Clostridium thermocellum. Metabolic Engineering, 55*, 161–169.

40. Singh, N., Mathur, A. S., Tuli, D. K., Gupta, R. P., Barrow, C. J., & Munish, P., (2017). Cellulosic ethanol production via consolidated bioprocessing by a novel thermophilic anaerobic bacterium isolated from a Himalayan hot spring. *Biotechnol. Biofuels, 10*, 73.

41. Nakayama, S., Kiyoshi, K., Kadokura, T., & Nakazato, A., (2011). Butanol production from crystalline cellulose by cocultured *Clostridium thermocellum* and *Clostridium saccharoperbutylacetonicum* N1-4. *Appl. Environ. Microbiol., 77*, 6470–6475. doi: 10.1128/AEM.00706-11.

42. Tomás, A. F. (2013). *Optimization of Bioethanol Production from Carbohydrate Rich Wastes by Extreme Thermophilic Microorganisms* (Doctoral dissertation, PhD Thesis, Technical University of Denmark, Copenhagen, Denmark).

43. Dien, B. S., Cotta, M. A., & Jeffries, T. W. (2003). Bacteria engineered for fuel ethanol production: current status. *Applied microbiology and biotechnology, 63*(3), 258–266.

CHAPTER 4

Genetic Regulation of Principal Microorganisms (Yeast, *Zymomonas mobilis,* and *Clostridium thermocellum*) Producing Bioethanol/Biofuel

DANIA SANDOVAL-NUÑEZ,[1] TERESA ROMERO-GUTIÉRREZ,[2] MELCHOR ARELLANO-PLAZA,[1] ANNE GSCHAEDLER,[1] and LORENA AMAYA-DELGADO[1]

[1]*Industrial Biotechnology Unit, Center for Research and Assistance in Technology and Design of the State of Jalisco A.C., Camino Arenero–1227, El Bajio del Arenal, C.P.–45019, Zapopan, Jalisco, Mexico*

[2]*Computer Sciences Department, Exact Sciences and Engineering University Centre, Universidad de Guadalajara, Blvd. Gral. Marcelino Garcia Barragan–1421, Olimpica. Guadalajara, Mexico*

ABSTRACT

Interest in ethanol production has been increasing since the 1980s because it is considered an alternative source of clean energy. The search for new and cheaper raw materials drove the development of new processes to obtain ethanol, such as the fermentation of lignocellulosic hydrolysates. In lignocellulosic ethanol production, microbial cells are under a stress-inducing environment, and their physiological behavior is altered significantly, indicating that different gene regulatory mechanisms are activated from those in non-stress-inducing conditions. Ethanologenic yeasts and bacteria tune gene expression to regulate response mechanisms and quickly adapt their global metabolic networks to grow and produce ethanol under unfavorable circumstances. Understanding the regulatory mechanism of gene expression during ethanol production has been a main topic in molecular and cellular biology with the goal of developing more efficient and resistant ethanol-producing microorganisms.

In this chapter, we present the state of knowledge in the field of genetic regulatory mechanisms in the most important microorganisms used for ethanol production, the yeasts *Saccharomyces cerevisiae* and *Kluyveromyces marxianus*, and the bacteria *Zymomonas mobilis*, and *Clostridium thermocellum*.

4.1 INTRODUCTION

Ethanol is becoming increasingly important in energy supply and economic development. Industrial ethanol production is commonly carried out by the yeast *Saccharomyces cerevisiae*, which is capable of fermenting C6 sugars only [1]. However, the degradation of other materials such as lignocellulosic biomass (LCB) also produces C5 sugars; this problem has been mitigated by using recombinant strains of *S. cerevisiae* and bacteria such as *Z. mobilis* and some thermophiles such as *Clostridium thermocellum*. Interestingly, microorganisms such as *C. thermocellum* can not only efficiently degrade cellulose and hemicellulose but also ferment hexose sugars to ethanol. Regardless of the microorganism used for ethanol production, fermentation efficiency depends on cellular metabolism, which is directly related to gene regulation.

Genetic regulation is essential for microorganisms in several biological processes, including the generation of biomolecules such as ethanol. During ethanol production, gene regulation plays a critical role in the cellular capacity to adapt quickly to the physicochemical conditions of the process, the ability to utilize carbon sources, and the ability to use alternative metabolic pathways to overcome nutrient-limiting conditions or respond to stressors [2]. Genetic regulation can be exerted on different levels in the cell; the most basic level is transcription. Regulation in eukaryotic cells requires the coordination of a whole set of genes that are scattered over the genome. This mechanism is different in bacteria, where genes, which are regulated in the same way, are often organized into operons, and transcribed from one regulatory sequence. When ethanol is produced, two types of regulation can be present: stimulation of the expression of specific genes (positive regulation) or inhibition of their expression (negative regulation) [1, 2]. Several studies have demonstrated that some restrictions are present in bioethanol production when wild-type microorganisms are used. These restrictions involve different biotic (formation of unwanted products) and abiotic effects (source of carbon, nitrogen, nutrients, temperature, oxygen, and the presence of stressful molecules) on the microorganisms. These restrictions may lead to a low conversion rate and low yield in industrial ethanol production. Thus, it is critical to know and control genetic regulation in yeast and bacteria to improve ethanol production.

The regulation of genes involved in the production of ethanol has been extensively studied. For example, alcohol dehydrogenase II (ADH2) is encoded by an *ADH* gene, and its primary function is to catalyze acetaldehyde conversion to ethanol. Therefore, it may be possible to improve ethanol production by modifying the function of ADH2. To date, many genetic engineering methods have been applied to disrupt the expression of the *ADH2* gene in *S. cerevisiae*, *Z. mobilis*, and *C. thermocellum* [2–4]. Therefore, knowing the regulatory mechanisms and generation of intermediary molecules during ethanol production allows the establishment of strategies for the positive or negative regulation of specific genes that increase ethanol yields. This chapter presents the state of knowledge in the field of genetic regulatory mechanisms in the most important microorganisms used to produce ethanol. It is based on experimental evidence of how microorganisms combine different regulatory mechanisms to coordinate multiple metabolic pathways during ethanol production.

4.2 GENE REGULATION IN ETHANOLOGENIC YEAST

Gene regulation is a cellular process consisting of activating or deactivating genes, which can occur at any point in the transcription-translation process. Gene regulation occurs most frequently at the transcriptional level. The regulation of gene expression in yeast can take place in different stages (Figure 4.1). In the nucleus, the chromatin remodeling process regulates the availability of a gene for transcription. Once transcribed, the primary mRNA transcript, or pre-mRNA, undergoes RNA processing, which involves splicing and adding a 5' cap and 3' poly (A) tail to produce a mature mRNA in the nucleus. Mature mRNA is exported from the nucleus to the cytoplasm, where its lifespan varies. Outside the nucleus, localization factors can direct mature mRNAs to specific regions of the cytoplasm where they are translated into polypeptides. The resulting polypeptides can undergo posttranslational modifications, which can regulate protein folding, glycosylation, intracellular transport, and protein activation and degradation.

Gene regulation is crucial for all eukaryotic organisms, such as yeast, and it is mainly required to adapt rapidly to environmental changes and conditions. This regulation may involve adaptation to different carbon sources, the ability to use alternative metabolic pathways to overcome limiting nutrient or environmental conditions, and responses to stress factors such as temperature, cold, chemical agents, ethanol, and the generation of metabolites [1]. Gene regulation in yeast is crucial during the generation of products or molecules of industrial interest, such as ethanol. Because ethanol production requires a series of biochemical stages, it is imperative to know the regulatory mechanisms carried out in yeasts during the fermentation process. This knowledge allows control of the fermentation processes to achieve higher yields and productivities. Some of the primary yeast taxa most studied in ethanol production processes are *S. cerevisiae* and *K. marxianus*.

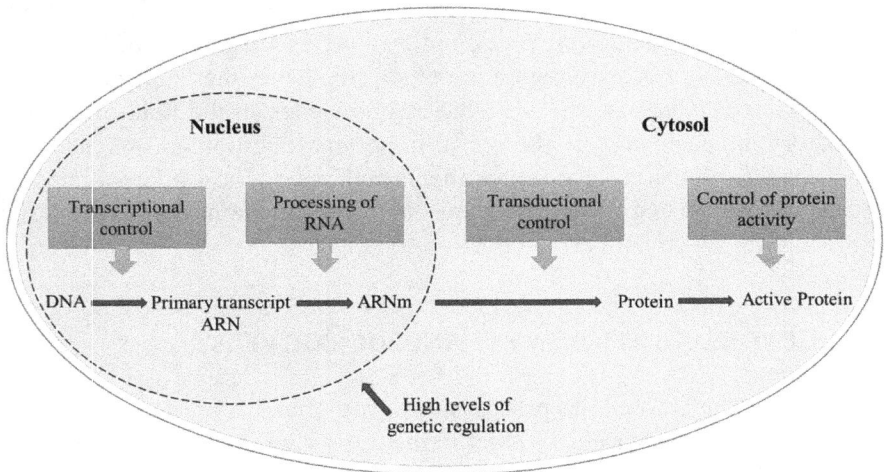

FIGURE 4.1   Representation of levels of genetic regulation in yeasts.

4.2.1 *SACCHAROMYCES CEREVISIAE AS A MODEL STRAIN IN ETHANOL PRODUCTION*

Saccharomyces cerevisiae is an undisputed model yeast in ethanol production at the industrial level. The production of first-generation ethanol (1G ethanol) is carried out mainly by various strains of *S. cerevisiae*, which are used by large ethanol-producing companies such as Lallemand Biofuels and Distilled Spirits, ABMauri Biotek, and Lesaffre Advanced Fermentation. First-generation ethanol is produced mainly from sucrose from sugar cane and starches from cereal grains such as corn [5]. The integration of consolidated processes is being sought where the use of other types of carbon sources is required, such as residues from LCB for the production of second-generation ethanol (2G ethanol).

One of the main challenges faced in the fermentation process for the production of 2G ethanol is the presence of various toxic compounds that affect yeast physiology and metabolism, inhibiting or reducing ethanol yields. *S. cerevisiae* has been studied in several fermentation processes where stress factors such as high temperatures, the absence of nutrients, and the presence of cell growth inhibitors such as organic acids, furans, phenols, and aldehydes intervene [6–9]. The presence of inhibitory compounds causes changes in yeast gene regulation, mainly during the transcription process, where the overexpression or inactivation of genes is involved in the consumption of the carbon source and therefore in ethanol production. However, the glycolysis pathway is not the only pathway affected by the presence of inhibitors, since for the carbon source to be consumed and subsequently converted to ethanol, the inhibitors present in the medium must be metabolized by yeast [10, 11]. Once the inhibitors are metabolized, the yeast is able to consume the sugars quickly.

Nevertheless, inhibitors activate a series of defense mechanisms and homeostatic balance in yeasts, achieving an ideal intracellular balance for ethanol production [9]. The detoxification process in *S. cerevisiae* employs a series of oxidation-reduction reactions to remove the high concentration of reactive oxygen species (ROS) in the mitochondria. On the other hand, *S. cerevisiae* uses a series of metabolic pathways to synthesize amino acids, transporters, and membrane lipids, produce secondary metabolites, and overexpress enzymes that facilitate the detoxification process and ethanol generation (Figure 4.2).

4.2.1.1 TRANSCRIPTION FACTOR GENES AND INTERMEDIATE ENZYMES IN ETHANOL PRODUCTION

Alcohol dehydrogenase (ADH) enzymes are the most studied enzymes in ethanol-producing yeasts since they catalyze acetaldehyde reduction to ethanol. In *S. cerevisiae*, seven ADH isozymes (ADH1 to ADH7) have been identified that have different physiological roles and functions. The ADH1 enzyme catalyzes acetaldehyde reduction to ethanol during glucose fermentation; it can also catalyze the reverse reaction. ADH1 is constitutively expressed, while the reduction of intracellular glucose concentration induces the expression of ADH2. The main function of ADH2 is to oxidize ethanol to acetaldehyde. On the other hand, ADH3 is a mitochondrial isozyme with an important role under anaerobic conditions: ADH3 forms part of the ethanol acetaldehyde shuttle, helping to shuttle mitochondrial NADH to the cytosol for NAD+ regeneration. ADH3 is repressed by glucose, and its expression is derepressed after glucose depletion. ADH4 and ADH5 are associated with ethanol production, and ADH4 expression is upregulated under low zinc concentrations or zinc starvation. The ADH5 isozyme might be expressed under conditions in which none of the other four ADHs (ADH1 to ADH4) are functional. In fermentation processes in the presence of furan aldehydes (furfural and HMF), the overproduction of ATP and NADPH has been observed since the responses to stress by furan aldehydes generate oxidative stress, which provokes a considerable increase in the net production of NADPH in strains of *S. cerevisiae* (> 4-fold). In this condition, the additional NADPH could be used by ADHs (ADH6 and ADH7) to convert furan aldehydes into less toxic compounds, lowering the toxicity of the medium for the yeast [12].

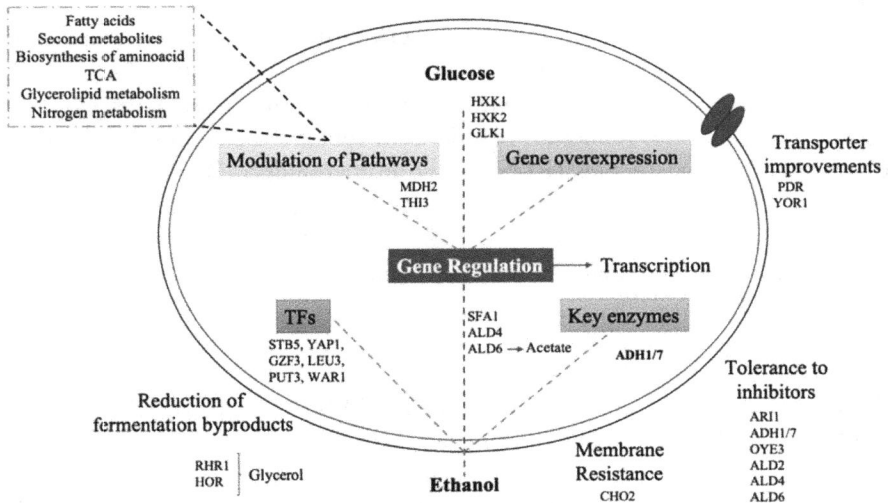

FIGURE 4.2 Schematic representation of the modulation pathways and genes involved in ethanol production under stress conditions. [Abbreviations: HXK: hexokinase; GLK: glucokinase; MDH: malate dehydrogenase; THI: thiamine metabolism regulatory protein; PDR: transcription factor; YOR: oligomycin resistance ATP-dependent permease; STB: protein STB, DNA-binding transcription factor; YAP: AP-1-like transcription factor; GZF: DNA-binding transcription factor; LEU: regulatory protein; PUT: proline utilization trans-activator; WAR: weak acid resistance protein; SFA: S-(hydroxymethyl)glutathione dehydrogenase; ALD4: potassium-activated aldehyde dehydrogenase; ALD6: magnesium-activated aldehyde dehydrogenase; ADH: alcohol dehydrogenase; RHR: glycerol 3-phosphate phosphohydrolase; HOR: glycerol-1-phosphate phosphohydrolase; CHO: phosphatidylethanolamine N-methyltransferase; ARI: NADPH-dependent aldehyde reductase; OYE: NADPH dehydrogenase; ALD: aldehyde dehydrogenase; TFs: transcription factors].

Multiple sequence alignment analysis of the ADHs of *S. cerevisiae* shows the similarities and differences between the seven ADHs (Figure 4.3). ADH1 and ADH2 present the highest amino acid sequence similarity of 93%. ADH3, ADH5, ADH6 and ADH7 maintain 79, 76, 29, and 27% similarity with ADH1, respectively. Amino acid sequence analysis of the ADH4 isozyme revealed a very low similarity (18%) with ADH1.

Advances in synthetic biology have focused on reengineering the ADH gene to achieve higher substrate specificity and improved catalytic activity. In addition, genome engineering methods allow us to obtain yeasts that overexpress genes encoding enzymes associated with the ethanol production process, improve ethanol tolerance, and assimilate a wide range of carbon sources [13, 14].

In addition to the study of the expression of various critical genes involved in ethanol production, transcription factors (TFs) play an essential role in the production of 2G ethanol, participating in processes of adaptation and yeast resistance to stressful conditions. TFs such as YAP1, STB5, DAL81, GZF3, LEU3, PUT3, and WAR1 have been widely studied because they are associated with various stress response mechanisms. These TFs confer appropriate biological characteristics on yeast to carry out ethanol production [15]. For example, YAP1 is associated with the glutathione pathway and is essential in the intracellular detoxification of ROS. On the other hand, some TFs, such as LEU3 and DAL81, are involved in carbon metabolism, amino acid biosynthesis, and nitrogen catabolism, allowing yeasts to use the available energy in the form of ATP and NADPH to carry out detoxification processes [7, 15].

4.2.1.2 GENETIC REGULATION OF SACCHAROMYCES CEREVISIAE DURING ETHANOL PRODUCTION UNDER STRESS CONDITIONS

During the saccharification process of LCB used as raw material for 2G ethanol production, several inhibitors are generated, limiting the metabolic activity of yeasts in the fermentation process. Among the main inhibitory compounds produced are furfural and 5-hydroxymethylfurfural (HMF). Even today, the genetic mechanisms involved in tolerance, adaptation, and resistance to stress caused by these compounds are not fully understood. The search for strains with higher tolerance to inhibitors promises to guarantee the desired concentrations and productivities of the processes. However, finding or generating resistant strains is not easy, not only because all the mechanisms for tolerance to furan-type inhibitors are unknown but also because the inhibitors do not act independently but synergistically, even with ethanol [9, 16]. The TFs involved in different stress responses are

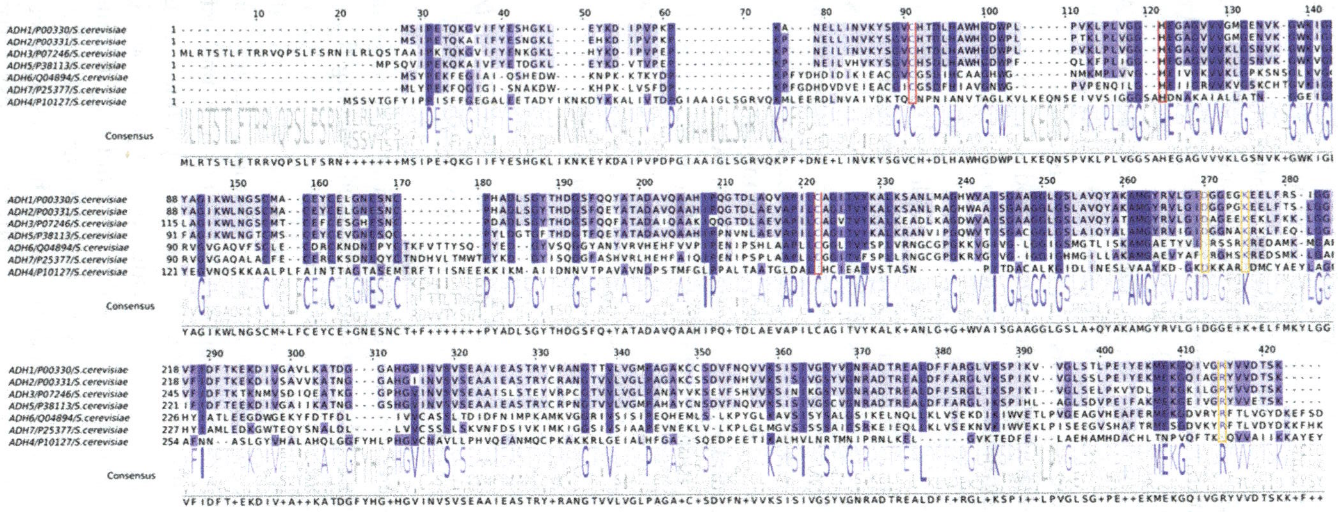

FIGURE 4.3 Multiple sequence alignment analysis of the alcohol dehydrogenases (ADH1 to ADH7) identified in *Saccharomyces cerevisiae*. The conserved residues are highlighted in purple. Residues boxed in red correspond to the catalytic zinc-binding motif. Residues boxed in orange belong to the binding site with NADH/NADPH. The multiple sequence alignment was obtained with the MUSCLE algorithm (https://www.ebi.ac.uk/Tools/msa/muscle/), using the ADHs sequences of *Saccharomyces cerevisiae* (strain ATCC 204508/S288c). Alignments were edited with Jalview software (https://www.jalview.org/).

activated or repressed by other TFs, forming a complex regulatory network (Figure 4.4). This complexity can be exemplified by the function of the YAP1 transcription factor. This transcription activator is related to the oxidative stress response and redox homeostasis. Its function is the regulation of genes encoding antioxidant enzymes, including the thioredoxin system (TRX2, TRR1), the glutaredoxin system (GSH1, GLR1), superoxide dismutase (SOD1, SOD2), glutathione peroxidase (GPX2), and thiol-specific peroxidases (TSA1, AHP1), which are components of the cellular thiol-redox pathways. YAP1 acts in conjunction with the transcription factor SKN7 to induce the expression of antioxidant enzymes. YAP1 is induced by oxidative and carbon stress but is not activated by high temperature, acidic pH, or ionic stress. On the other hand, SKN7 forms part of the SLN1-YPD1-SNN7 two-component regulatory system, which is related to the expression control of genes involved in the response to changes in extracellular osmolarity.

As mentioned above, gene regulation occurs most frequently at the transcriptional level. The regulation of gene expression in *S. cerevisiae* under stress-inducing conditions can affect several biological processes and trigger a series of regulatory events mediated by numerous TFs (Figure 4.4). Genes related to the processes of detoxification, adaptation, and tolerance in *S. cerevisiae* are regulated by key TFs, such as the YAP gene family, PDR gene family, RPN4, and HSF1 [7, 9, 15].

The primary genes activated during 2G ethanol production in the presence of inhibitors such as furfural and HMF are described below.

4.2.1.3 ADAPTATION AND TOLERANCE OF SACCHAROMYCES CEREVISIAE TO ACETIC ACID, FURFURAL, AND HMF

In recent years, various studies have been carried out to find strategies to understand the molecular mechanisms involved in the adaptation and tolerance of *S. cerevisiae* with respect to the main groups of growth and fermentation inhibitors, such as aldehydes, phenols, ketones, and weak organic acids. Furfural, HMF, and acetic acid are the three main compounds that participate in the inhibition of the 2G ethanol production process [17].

For ethanol production, fermentation is carried out at pH values between 4 and 5, causing the dissociation of acetic acid (pKa = 4.76), which diffuses through the plasma membrane of yeast [18]. Inside the cell, the acid dissociates and releases a proton, making it necessary to use ATP to reach the cell equilibrium required for essential biological functions and prevent cytosolic acidification. The use of ATP causes a decrease in yeast growth during the first

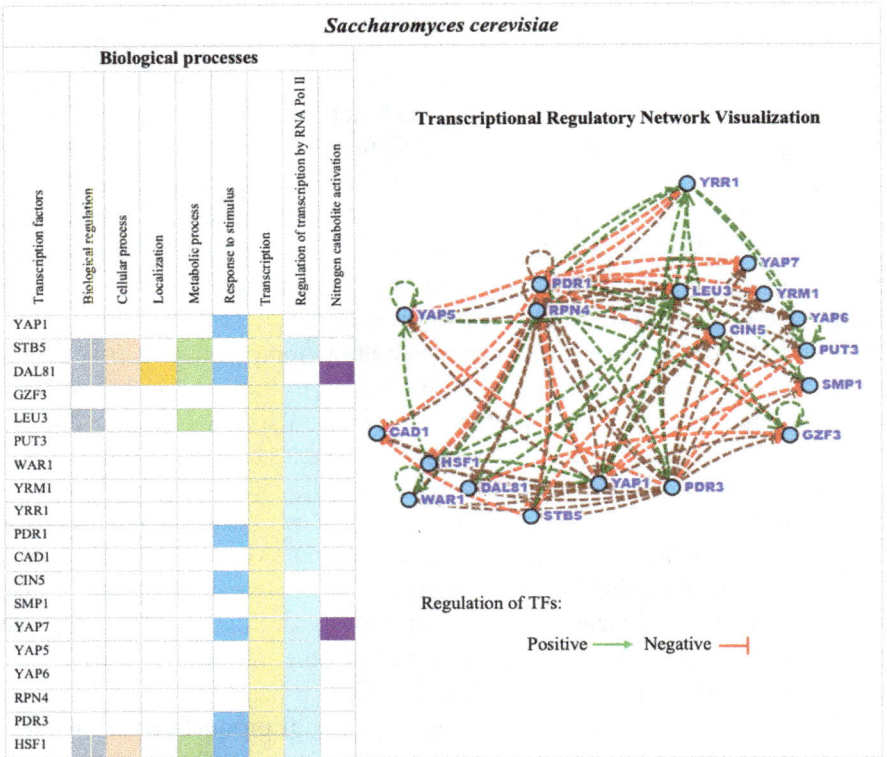

FIGURE 4.4   Transcriptional regulatory network visualization of *Saccharomyces cerevisiae*. [*Abbreviations:* YAP: AP-1-like transcription factor; STB: protein STB, DNA-binding transcription factor; DAL: transcriptional activator protein; GZF3: transcriptional regulator, DNA-binding transcription factor; LEU: regulatory protein; PUT: proline utilization transactivator; WAR: weak acid resistance protein; YRM: zing finger transcription factor; YRR: zing finger transcription factor; PDR1: transcription factor, positive regulator of proteins involved in permeability; CAD: YAP2, AP-1-like transcription factor; CIN: YAP4, AP-1-like transcription factor; SMP: transcription factor, involved in osmostress response; RPN: transcriptional activator protein; PDR: transcription factor; HSF: heat shock factor protein]. The classification of transcription factors in biological processes was carried out utilizing PANTHER and the networks using YEASTRACT (http://yeastract.com/). Visualization of the transcriptional regulatory network makes it possible to correlate transcriptional associations such as gene activation or deactivation of transcription factors, green (positive), red (negative), positive, and negative (brown).

hours of contact; this condition remains until the yeast manages its genetic machinery to adapt to this adverse condition. The acetic acid concentration can be higher than 5 g/l in lignocellulosic residue hydrolysates, increasing the inhibition problem caused by acetic acid. At a high concentration of acetic acid, the physiological ability of yeasts to carry out fermentation is decreased compared to that in normal conditions. Therefore, tolerance to acetic acid at low culture pH is a key goal in yeast strain development for 2G ethanol production. Meijnen et al. discovered that tolerance to acetic acid is the result of a polygenic response of yeast, making it difficult for a targeted genetic change to generate resistant yeast [19]. One of the most important factors in resistance to acetic acid stress is Haa1p [20]. Haa1p is related directly or indirectly to the regulation of the transcription of approximately 80% of the genes encoding protein kinases, multidrug resistance transporters, proteins involved in lipid metabolism and in the processing of nucleic acids, and proteins of unknown function, suggesting that this factor is one of the most important in the response of yeast to acetic acid. Other important factors in resistance to acetic acid stress are Rim101p and Msn2p, which are also displayed under other stress types, such as ethanol, high concentrations of glucose and oxidative stress [21–23].

Furfural is well known to be transformed into furfuryl alcohol by nicotinamide adenine dinucleotide-dependent dehydrogenases (NADH) under anaerobic conditions. Therefore, enzymes that participate in the reduction of aldehyde groups are essential to reduce the concentration of inhibitors in lignocellulosic residues. The genes reported to participate in the detoxification of furfural and HMF are ADH6, ADH7, ALD4, GRE3 [9]; ADH1 [24]; ARI1 [25]; and GRE2 [26]; in addition, the Y62 and Y76 genes participate only in the detoxification of furfural, but no evidence has been found that they also participate in HMF. Studies developed to evaluate the capacity of ADH6 and ADH7 in the reduction of furfural and HMF show that these enzymes have a high capacity to reduce aldehydes, as much as 100 times greater than their oxidation capacity [27]. The synergistic activity between the aldehyde dehydrogenase enzymes Ald5, Pad1 decarboxylase, and the alcohol acetyltransferases Atf1 and Atf2 has been determined to provide resistance to phenolic inhibitors [28]; thus, it is likely that they participate actively in the transformation of furans present in lignocellulosic residues. Ma and Liu demonstrated that transcriptional genes such as YAP1, PDR1, PDR3, RPN4 and HSF1 actively participate in the adaptation of *S. cerevisiae* to HMF [29].

In recent years, the existence of short-chain aldehyde reductase enzymes known as SDR has been evidenced; SDR enzymes participate in the transformation of aldehydes within yeast cells, including Ykl107wp. This enzyme can reduce acetaldehyde, glycolaldehyde, furfural, formaldehyde, HMF, and propionaldehyde, but it was not observed to reduce the six ketones corresponding to the same compounds [30]. Aldehyde reductase enzymes are highly diverse. SDRs have particular substrate preferences and are found in different places within the cell. In addition, considering that aldehydes can cross mitochondrial, nuclear, and cytoplasmic membranes, the presence of specific aldehyde reductases in each zone guarantees a decrease in cellular damage [31, 32].

Although it is possible to direct the genetic modification of yeasts to increase the production of TFs and increase resistance to acetic acid, furfural, and HMF, it should be considered that several transcriptional factors are activated during detoxification, so it is possible that affecting only one of these factors will not have an overall impact. Likewise, the great diversity of enzymes necessary to reduce cellular damage in each organelle should be considered. Therefore, evolutionary adaptation is an interesting strategy because, during evolution, strains can undergo the necessary changes to allow higher tolerance to the inhibitors present in the lignocellulosic hydrolysates and even increase ethanol production and productivity [33–35].

4.2.1.4   METABOLOMICS ANALYSIS TO IMPROVE SACCHAROMYCES CEREVISIAE STRESS TOLERANCE

The production of low-molecular-weight molecules or metabolites during ethanol production is one of the main results of gene regulation in yeasts. The complete collection of these metabolites integrates the metabolome and is closely related to phenotype. The molecules generated in the cell under specific culture conditions are directly associated with cellular maintenance, growth, and function; this association has been used to analyze the contribution of metabolic pathways to the cellular lifespan. However, multiple metabolic pathways are affected by external stimuli, and it is necessary to characterize the whole metabolome through metabolic analysis. In recent years, metabolomic analysis has been utilized as a promising tool to understand the molecular mechanism that drives the stress response and to improve stress tolerance in yeasts [36–38].

Metabolomics analyzes have demonstrated the association between ethanol stress tolerance and the production of specific metabolites in *S. cerevisiae*. Li et al. demonstrated that the levels of metabolites produced from intermediary molecules of the Embden-Meyerhof-Parnas (EMP) pathway, glycine, serine, threonine, glycerol, alanine, isoleucine, and valine, increase under ethanol stress, which indicates the inhibition of glycolysis. Similarly, the levels of fatty acids are altered by chemical stress; lower levels of palmitelaidic acid but higher levels of hexadecenoic and octadecanoic acids were detected in *S. cerevisiae* under ethanol stress [36]. The evaluation of membrane phospholipids in *S. cerevisiae* under furan aldehyde stress revealed an increase in phosphatidylethanolamine and its association with the loss of membrane stability [11].

Regarding metabolites related to ethanol tolerance, metabolomic studies have shown that intracellular accumulation of trehalose, proline, acetyl-CoA, fumaric acid, aspartic acid, and TCA cycle-related metabolites leads to improved ethanol tolerance [39]. These results have made it possible to establish strategies to improve tolerance in *S. cerevisiae* by deleting selected genes to increase (simultaneous deletion of *LEU4* and *LEU9*) or reduce (simultaneous deletion of *INM1* and *INM2*) the concentration of metabolites such as valine and inositol, which significantly increase ethanol tolerance [39].

Despite all the successful results obtained through the years in understanding stress response mechanisms in *S. cerevisiae*, it is evident that a rational strategy is necessary to improve yeast tolerance of ethanol or inhibitory compounds. Such a strategy should combine omic analyzes (genomics, transcriptomics, proteomics, metabolomics, and fluxomics) with predictive computational modeling or simulation to design tolerant yeasts with better metabolic properties for industrial ethanol production.

4.2.2 KLUYVEROMYCES MARXIANUS AS A POTENTIAL YEAST IN ETHANOL PRODUCTION

Kluyveromyces marxianus is an unconventional yeast of great interest for use in biotechnological processes aimed at the production of various metabolites, such as aromatic molecules, higher alcohols, and lytic enzymes [10, 40–42]. Wild-type strains of *K. marxianus* have been widely studied, demonstrating their ability and high potential to consume different carbon sources [43]. *K. marxianus* has also been used as a model in various systems and strategies of alcoholic fermentation to evaluate factors such as temperature, carbon source aeration, enzyme expression, and the detoxification processes of compounds present in hydrolysates. These tests showed ethanol yields from agro-industrial residues and defined culture media (Table 4.1).

The use of *K. marxianus* in ethanol production has increased in the last decade because it has several biological characteristics that give it interesting properties for use in numerous fermentation processes. Thermotolerance characteristics and robustness to chemical agents (organic acids, phenols, furans, and aldehydes) present in lignocellulosic hydrolysates make *K. marxianus* a model yeast suitable for use in improving ethanol yields in the presence of toxic compounds [56, 57].

TABLE 4.1 Ethanol Production by *Kluyveromyces marxianus* Using Different Carbon Sources

Raw Material	Strain	Fermentation Process	Ethanol (g/l h)	References
Barley straw	K. marxianus CECT 10875	SSF	0.40	[44]
Sugar cane juice	K. marxianus DMKU 3-1042	Batch	1.30	[45]
Wheat straw slurry	K. marxianus CECT 10875	Batch	1.69	[46]
Wheat straw slurry	K. marxianus CECT 10875	Batch	0.36	[47]
YPD	K. marxianus BUNL-21	Batch	0.06	[48]
Sweet sorghum juice	K. marxianus DBKKUY-103	Batch	1.42	[49]
Wheat straw	K. marxianus DBTIOC-35	SSF	0.86	[50]
YPDX	K. marxianus CBS712	Batch	1.18	[51]
Kanlow switchgrass	K. marxianus IMB4	SSF	0.23	[52]
Wheat straw	K. marxianus CECT 10875	SSF	0.50	[53]
Corncob residue	K. marxianus NBRC1777	Batch	1.05	[54]
Sugar cane bagasse	K. marxianus NRRL Y-50883 (SLP1)	Batch	0.29	[55]
Mineral medium	K. marxianus NRRL Y-50883 (SLP1)	Batch	0.30	[10]

4.2.2.1 GENETIC REGULATION OF KLUYVEROMYCES MARXIANUS DURING ETHANOL PRODUCTION UNDER STRESS CONDITIONS

Genetic regulation is one of the main biological mechanisms employed by yeast at all stages of the cell cycle. However, several extracellular factors contribute to the use of regulatory mechanisms associated with stress responses and cell repair. The participation of some TFs, the overexpression and inactivation of genes, and the synthesis and degradation of some metabolites have been demonstrated when *K. marxianus* is employed in ethanol production processes under certain fermentation conditions in the presence of different carbon sources, nutrients, and toxic molecules [10, 42, 56].

Ethanol production by *K. marxianus* has been evaluated mainly with the use of agro-industrial residues and in defined culture media. Therefore, understanding some of the molecular regulatory mechanisms employed by *K. marxianus* during ethanol production is of great interest. In addition to the carbon source used and its subsequent conversion to ethanol, several biochemical and metabolic biological events disrupt ethanol conversion pathways and mechanisms. Environmental or chemical effects provoke biological responses by yeasts and consequently intervene in gene regulation during the ethanol production process, inhibiting or decreasing the generation of ethanol [56, 58].

The presence of toxic molecules such as furfural and organic acids plays an important role in gene regulation, biochemically affecting ethanol production due to the overexpression of proteins and genes, affecting metabolic pathways such as glycolysis, the pentose phosphate pathway (PPP), glutathione, and butanoate metabolism, fatty acid metabolism, and the biosynthesis of many amino acids such as Ala, Arg, Asp, Glu, His, Ile, Leu, Pro, Trp, Tyr, and Val, in addition to MAPK signaling pathways. Some metabolic pathways, such as those involving secondary metabolites and the synthesis, and degradation of amino acids, TCA, and glyoxylate, act as intermediaries in regulatory processes related to defense mechanisms and homeostasis, maintaining the ideal intracellular balance for ethanol production [56, 59].

4.2.2.2 MOLECULAR STRATEGIES FOR THE GENETIC IMPROVEMENT OF KLUYVEROMYCES MARXIANUS

The molecular techniques mostly used to improve *K. marxianus* for in ethanol production include homologous recombination (HR) and CRISPR-Cas.

These techniques have allowed the integration or overexpression of genes that potentiate biological processes and reactions that participate as intermediaries in ethanol production [60].

Schabort et al. performed a metabolic regulation analysis using a transcriptomics study on *K. marxianus* UFS-Y-2791 during glucose and xylose consumption [61]. They concluded that gene regulation levels play an important role in the regulation of metabolic fluxes in central carbon metabolism, glycolysis reactions, and xylose adaptation. Gao et al. overexpressed thioredoxin TPX1 in *K. marxianus* Y179, achieving an increase in the maximum rate of glucose consumption and the rate of ethanol generation by the strain, a significant improvement compared to the control [62].

Studies focused on ethanol production from xylose as a carbon source have also achieved improvement by overexpressing enzymes through HR. For example, the overexpression of enzymes such as xylose reductase heterologous (XYL1) and xylitol dehydrogenase (XYL2) have increased the participation of the phosphate pentose pathway and the flow into ethanol production [63]. The increase in ethanol yields has been achieved through the overexpression of native xylokinase (XYL3), L-ribulose-5-phosphate 4-epimerase (RPE1), ribose-5-phosphate isomerase (RKI1), transcetolase (TKL1), and transaldolase (TAL1) genes as well as pyruvate decarboxylase (PDC1) and ADH2. The XYL1 and XYL2 heterologous genes were selected because of their preference for NADP(H) over NAD(H), thus helping to rectify an imbalance in cofactors during growth on xylose [64].

4.2.2.3 KEY GENES ASSOCIATED WITH ETHANOL PRODUCTION IN KLUYVEROMYCES MARXIANUS

Several genes and their association with the ethanol production process have been reported to be involved in posttranslational regulation. In ethanol production under certain stress conditions (temperature, ethanol, and chemical agents), protein folding and the expression or repression of genes coding for carbon source carriers and defense mechanisms have been found to affect oxidation-reduction processes and intracellular transport, in addition to genes involved in the inhibition of routes and signaling pathways associated with regulatory processes in ethanol production (Figure 4.5). Assays performed in *K. marxianus* demonstrated the biological importance of genes that code for intermediary enzymes in the ethanol production process, mainly enzymes involved in carbon metabolism for both xylose and glucose, enzymes involved in transport processes, mitochondrial enzymes, and mediators of cellular homeostasis processes under stress conditions during ethanol production, such as MDH2, TDHs, and ADHs (Figure 4.5).

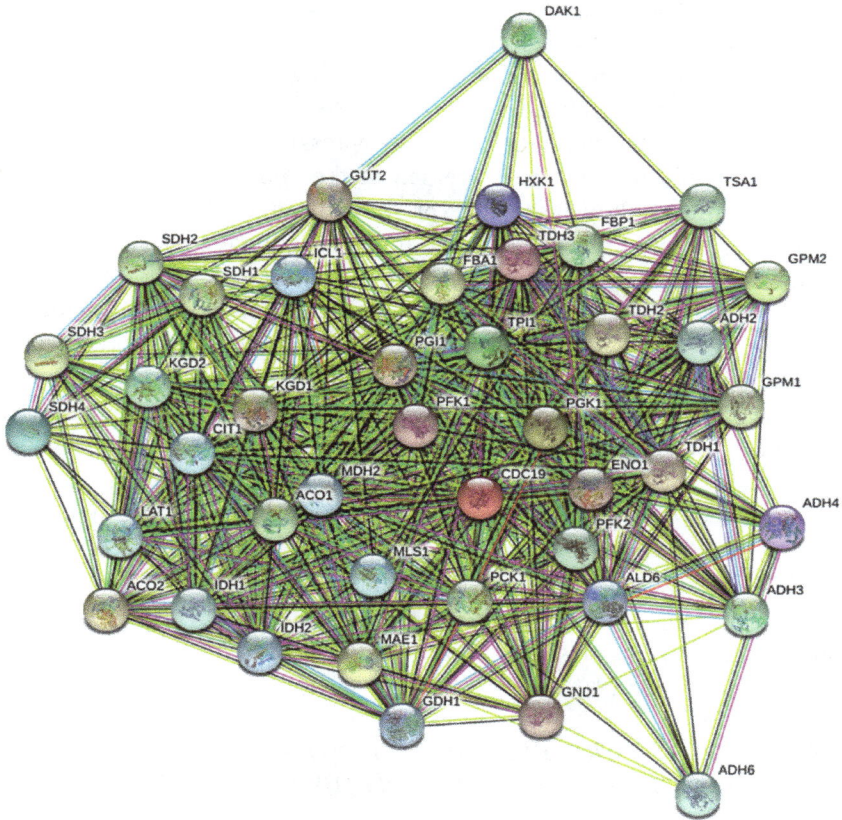

FIGURE 4.5 Interactome of genes involved in ethanol production in *Kluyveromyces marxianus*. [*Abbreviations*: DAK: dihydroxyacetone kinase; GUT: glycerol-3 phosphate dehydrogenase; HXK: hexokinase; TSA: peroxiredoxin; SDH: succinate dehydrogenase; ICL: isocitrate lyase; FBA: fructose-biphosphate aldolase; TDH: glyceraldehyde-3-phosphate dehydrogenase; FBP: fructose-1,6-biphosphatase; KGD: 2-oxoglutarate dehydrogenase complex; PGI: glucose-6-phosphate isomerase; TPI: triosephosphate isomerase; ADH: alcohol dehydrogenase; GPM: phosphoglycerate mutase; CIT: citrate synthase; PFK: ATP-dependent 6-phosphofructokinase; PGK: phosphoglycerate kinase; MDH: malate dehydrogenase; CDC: pyruvate kinase; ENO: enolase; LAT: dihydrolipoyllysine-resiude acetyltransferase component of pyruvate dehydrogenase complex; ACO: aconitate hydratase; IDH: isocitrate dehydrogenase; MLS: malate synthase; PCK: phosphoenolpyruvate carboxykinase; ALD: aldehyde dehydrogenase; MAE: NAD-dependent malic enzyme; GDH: glutamate dehydrogenase; GND: 6-phosphogluconate dehydrogenase]. The interactome was based on *S. cerevisiae* using STRING: functional protein association networks (https://string-db.org/)].

Seven ADHs (KmADH1 to KmADH7) have been identified in *K. marxianus*; however, the function of KmADH5 to KmADH7 has been little studied. Similar to *S. cerevisiae*, KmADH1 and KmADH2 are associated with ethanol production, and both are expressed at high levels in aerobic and anaerobic conditions. The function of KmADH3 is to use nonfermentable carbon sources, and it is upregulated during the stationary phase. The KmADH4 isozyme is related to ethanol detoxification, and its expression also increases in the stationary phase. Expression analysis of KmADH5 suggests that this isozyme is expressed at the end of fermentation under aerobic conditions; however, its expression in anaerobic conditions was constant throughout fermentation. The function of the KmADH6 isozyme is associated with cellular processes carried out in the stationary phase under anaerobic conditions. The metabolic function of KmADH7 is associated with the oxidation of hemiacetal as an alternative route for the synthesis of ethyl acetate, and its expression increases in the stationary phase [65].

Multiple sequence alignment analysis of the ADHs of *K. marxianus* shows the similarities and differences between the main four KmADHs. The similarity percentages of KmADH2 to KmADH4 with respect to KmADH1 are 82, 79, and 78% (Figure 4.6). Similar to that in *S. cerevisiae*, the amino acid sequence reveals the metal binding site (zinc catalytic site) and the NAD-binding site (according to NADH or NADPH dependence). Expression analysis demonstrated that the expression of KmADHs depends on the growth phase and carbon source [65].

4.2.2.4   *TRANSCRIPTION FACTORS (TFS), INTENSIFIERS, AND SILENCERS INVOLVED IN ETHANOL PRODUCTION IN KLUYVEROMYCES MARXIANUS*

In the process of regulating gene expression, the participation of expression modulating sequences plays a very important role. These sequences can intensify or silence the transcription process. Enhancers are sequences that stimulate transcription and whose location can be thousands of nucleotides away from the promoter. Silencers are sequences that inhibit transcription and can also be located far away from the promoter. For *K. marxianus*, few activating or repressor proteins have been reported that interact with enhancer sequences or silencers. For example, Spt15 is a protein that intensifies ethanol yields [66]. In contrast, the participation of TFs such as HSF1 and MSN2 confers resistance to chemicals such as furfural, phenol, and acetic acid and increases lignocellulosic ethanol production [60].

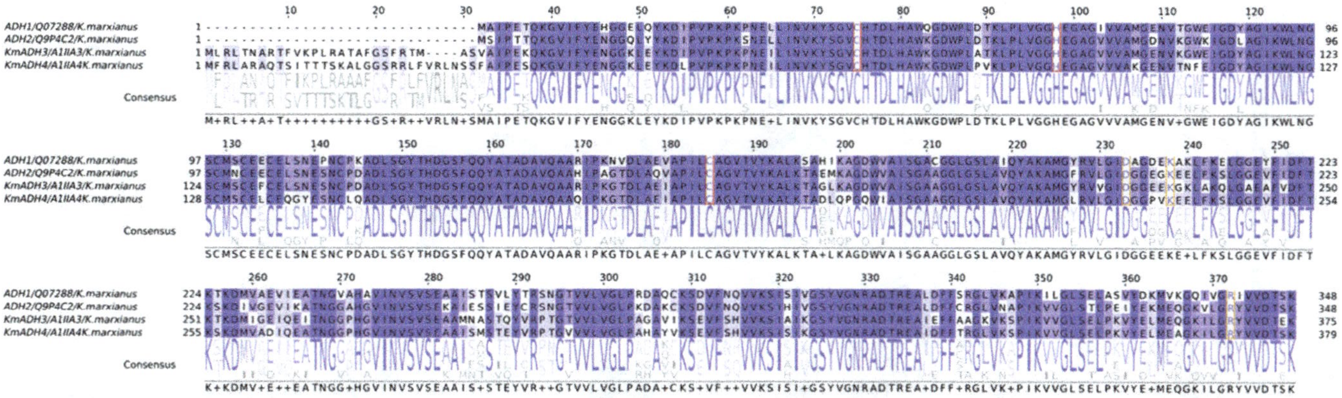

FIGURE 4.6 Multiple sequence alignment analysis of the alcohol dehydrogenases (KmADH1 to KmADH4) identified in *Kluyveromyces marxianus*. The conserved residues are highlighted in purple. Residues boxed in red correspond to the catalytic zinc-binding motif. Residues boxed in orange belong to the binding site with NADH/NADPH. The multiple sequence alignment was obtained with the MUSCLE algorithm (https://www.ebi.ac.uk/Tools/msa/muscle/), using the ADHs sequences of *Kluyveromyces marxianus* (strain ATCC 12424 for KmADH1-2, and strain DMKU 3-1042 for KmADH3-4). Alignments were edited with Jalview software (https://www.jalview.org/).

Similar to *S. cerevisiae* and other yeasts, gene regulation in *K. marxianus* is a complex regulatory network that affects different biological processes (Figure 4.7). Although *K. marxianus* is a very interesting yeast for producing 2G ethanol, information about the activation of TFs during its stress response is limited. However, the expression of several TFs by *K. marxianus* has been demonstrated during heat stress (HSF1, MSN2, SFP1, and RPN4), oxidative stress (MSN2, SNF2, GCNA), osmotic stress (MSN2), and stress by nutrient starvation (SFP1) [56, 60].

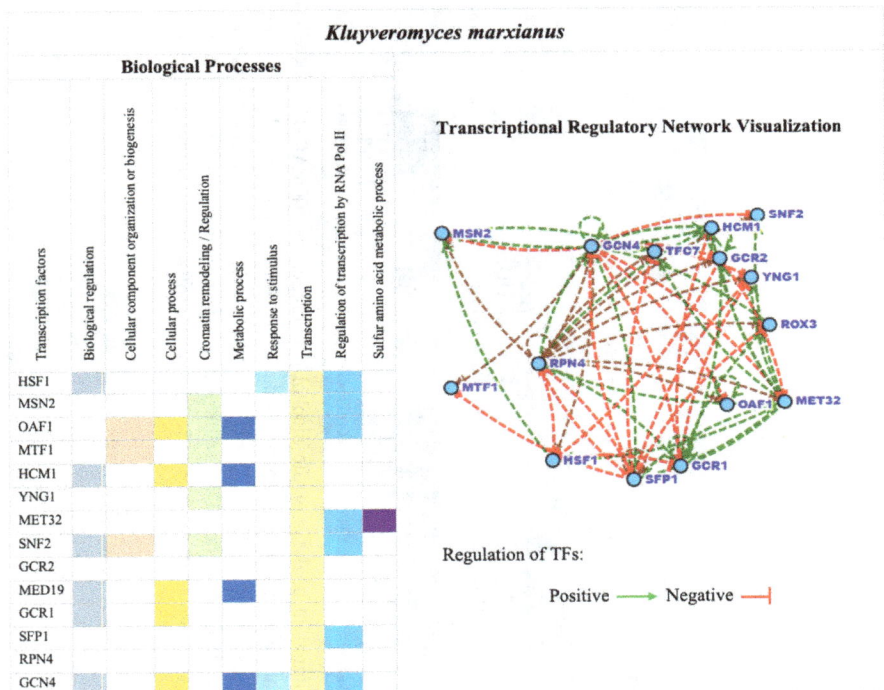

FIGURE 4.7 Transcriptional regulatory network visualization of *Kluyveromyces marxianus*. [*Abbreviation*: HFS: heat shock factor protein; MSN: zinc finger protein; MSN: zinc finger protein; OAF: oleate-activated transcription factor; MTF: mitochondrial transcription factor; HCM: forkhead transcription factor; YNG: chromatin modification-related protein; MET: transcriptional regulator; SNF: transcriptional regulatory protein; GCR: glycolytic genes transcriptional activator; MED: mediator of RNA polymerase II; SFP: transcription factor; RPN: transcriptional activator protein RPN; GCN: general control protein]. The visualization of the transcriptional regulatory network for *K. marxianus* represents an approximation of the transcriptional association between transcription factors, green (positive), red (negative), positive, and negative (brown). For its construction, *S. cerevisiae* was used as a reference.

4.2.2.5   *METABOLOMIC ANALYSIS IN KLUYVEROMYCES MARXIANUS*

Metabolomic analysis is a useful tool to increase knowledge about the stress response mechanisms in *K. marxianus*. Similar to *S. cerevisiae*, stress provoked by inhibitory compounds (ethanol, furan aldehydes, weak acids, among others) inhibited the glycolysis pathway and, consequently, the growth of *K. marxianus*. Several metabolites, such as amino acids, sugars, and lipids, increased in concentration when *K. marxianus* was under ethanol stress. Arginine, phenylalanine, glutamate, proline, and tryptophan are related to the ethanol adaptation process, as the concentrations of these amino acids increase when the yeast is exposed to ethanol stress. Trehalose is another metabolite that is recognized as a protein and plasma membrane protector [67]. Flores-Cosío et al. observed an increase in the concentration of phosphatidylethanolamine and a decrease in phosphatidylcholine when *K. marxianus* was stressed by furan aldehydes, illuminating the effect of these inhibitors on the plasma membrane and the capacity of *K. marxianus* to reorganize its membrane [11]. Despite the potential of metabolomic analyzes as tools to study the effect of inhibitory compounds on cellular metabolism and physiology, this omic technique has been rarely used in *K. marxianus*, so there are few studies that help to understand the stress response in this yeast through its metabolome.

Recent studies have demonstrated that *K. marxianus* is a promising yeast for ethanol production and other interesting metabolites, such as enzymes, flavor, and fragrance compounds, and xylitol. *K. marxianus* has advantages over other yeasts because it can assimilate several carbon sources (glucose, xylose, arabinose, sucrose, inulin, fructans, among others). In addition, *K. marxianus* is a thermotolerant yeast with a high growth rate and high resistance to inhibitory compounds, and it can quickly adapt its metabolic machinery to survive toxic environments. The main obstacle to its development at an industrial level has been the limited knowledge of its genetics and physiology, but this is rapidly changing thanks to the new omic technologies, making *K. marxianus* a promising yeast for ethanol production.

4.3   GENETIC REGULATION OF ETHANOLOGENIC BACTERIA

Bacteria, similar to yeasts, are exposed to changing environments in which biotic and abiotic factors can radically alter their metabolism. Bacteria respond to such variations by regulating their gene expression; thus, they can adjust the metabolic pathways depending on the carbon source, pH, temperature, toxic compounds, and other nutrients available. Different mechanisms regulate genes in bacteria and yeasts; bacterial genes are organized into operons or clusters of coregulated genes. In addition to being physically close in the genome, these genes are regulated to turn on or off together. This characteristic of grouping related genes under a common control mechanism allows bacteria to adapt rapidly to changes in the environment.

Transcription has three steps in bacteria: first, RNA polymerase binds to a promoter site on DNA to form a closed complex; then, RNA polymerase starts transcription by opening the DNA duplex to form a transcription bubble. In the second stage, termed elongation, the transcription bubble moves along DNA, and the RNA chain is extended by adding nucleotides in the 5' to 3' direction. Finally, transcription stops, and the DNA duplex reforms when RNA polymerase dissociates at a terminator site. In bacteria, each gene or operon is flanked by a promoter and a terminator. The promoter is a specific nucleotide sequence site where the RNA polymerase binds to DNA and starts making RNA (mRNA). The terminator is a similar instruction in the DNA where the RNA polymerase stops transcribing mRNA and dissociates from the DNA. This mechanism is the purest form of gene expression regulation in bacteria. Essential components of transcription are sigma factors (σ), which are subunits of all bacterial RNA polymerases. They are responsible for determining the specificity of promoter DNA binding and efficiently control transcription initiation. In conclusion, the first step in bacterial gene expression and the step most often controlled is transcription. Regulatory factors usually determine whether a specific gene is transcribed by RNA polymerase or not under specific environmental conditions.

4.3.1   *ZYMOMONAS MOBILIS*

Zymomonas mobilis is a gram-negative anaerobic bacterium that was first isolated from palm wine and pulque in Mexico. This microorganism has many desirable industrial characteristics due to its greater suitability than yeasts in ethanol production [68]. Such attributes include a faster production rate per cell, a lack of requirement for controlled oxygen during fermentation, a higher sugar absorption rate, a wide tolerance to ethanol, and an efficient growth rate across a broad pH range (3.5–7.5) [69, 70]. One of the most interesting physiological characteristics of *Z. mobilis* is its carbohydrate catabolism via the Entner-Doudoroff (ED) pathway [71]. The ED pathway is almost entirely restricted to aerobic gram-negative bacteria due to low ATP production compared with the EMP pathway for glycolysis, but it is also present in anaerobic bacteria such as *E. coli* and *Z. mobilis*, and it produces one ATP molecule per glucose consumed [69]. In this pathway, two key enzymes participate, PDC and two ADHs, which convert pyruvate to acetaldehyde and then to ethanol [70]. Lignocellulose-derived inhibitors have negative effects on the ethanol fermentation capacity of *Z. mobilis* and other microorganisms. Such inhibitors include acetic and formic acid, furfural, HMF, and phenols [72]. High temperatures, oxygen, and the ethanol produced in fermentation itself can also act as growth inhibitors.

Different metabolic engineering experiments were performed to improve bioethanol production through the use of genetic tools such as heterologous expression systems, the silencing of specific genes, and mutagenesis, among others [73]. One example is a strain of *Klebsiella oxytoca* that was modified by metabolic engineering by inserting the PDC gene from *Z. mobilis* to increase its capacity for pyruvate decarboxylation, a key enzyme in the homoethanol pathway of *Z. mobilis* [74]. In recent years, next-generation sequencing (NGS) technologies have allowed the characterization of complete genomes from different microorganisms of industrial interest, including *Z. mobilis*. Knowledge of the genetic structure of these bacteria has allowed us to deeply study the production of metabolites with biotechnological applications and the adaptation mechanisms to environmental factors. The first genome of *Z. mobilis* was sequenced in 2005, with a circular chromosome with a length of 2,056,416 bp and 5 plasmids, and it was further annotated in 2009 [75, 76]. Subsequently, other genomes were also sequenced with an approximate length of 2.01 to 2.22 Mb including 2 to 6 plasmids [73].

4.3.1.1 GENETIC REGULATION OF ZYMOMONAS MOBILIS UNDER TEMPERATURE AND OXYGEN STRESS

To date, several differential gene expression studies have been performed to understand the cellular dynamics of *Z. mobilis* under heat stress, under aerobic and anaerobic conditions, and in the presence of chemical inhibitors, including phenolic aldehydes, furfural, and ethanol. The objective of these studies is to characterize the metabolic phenomena that occur during cellular adaptation to biotic and abiotic agents and how this affects the expression dynamics and regulation of key genes, leading to a decrease or increase ethanol production.

A study of the heat stress adaptation of *Z. mobilis* ZM4 by RT-qPCR allowed the identification of overexpressed genes that are related to oxidative stress and heat shock proteins (HSPs), which increase cell viability

(Table 4.2) [77, 78]. The gene expression dynamics in *Z. mobilis* ZM4 were also evaluated under aerobic and anaerobic conditions. In the absence of oxygen, several overexpressed genes participating in the ED pathway that optimize glucose metabolism were identified. Additionally, overexpressed genes involved in ribosome-mediated polypeptide synthesis and amino acid and cofactor biosynthetic genes were identified (Table 4.2). The growth rate of *Z. mobilis* under aerobic conditions negatively influences fermentation performance, since it increases the concentration of toxic compounds such as aldehydes and decreases ethanol production due to the overexpression of genes involved in stress response, transcriptional regulation, the metabolism of sulfur compounds, and apoptosis and chemotaxis [79].

TABLE 4.2 Genes of *Zymomonas mobilis* Z4 Differentially Expressed Under Different Stress Conditions (Temperature and Aeration)[#]

Heat Stress		
Gene ID	**Cellular Process**	**References**
sod, cat, ZMO1573, ZZ6-0186, ahpc	Oxidative stress	[77]*
dnaKJ, hsp20, clpB, clpA, clpS	HSP genes	[78]*
		[75]*
Aeration conditions		
Anaerobic conditions		
glk, zwf, pgl, pgk, eno, pdc, adhB	ED pathway	[79]
rars1	Ribosome-mediated polypeptide synthesis	
leuC, trpB, argC, ilvI, ilvC, thrC, thiC, and ribC	Amino acid and co-factor biosynthetic genes	
Aerobic conditions		
ZMO0084, ZMO0641, ZMO0651	Chemotaxis	[79]
ZMO1022, ZMO1460, NT01ZM1467	Metabolism of sulfur compounds	
ZMO1121, ZMO1216, ZMO1387, ZMO1063	Transcriptional regulators	
nadE	Nicotinamide adenine dinucleotide de novo biosynthesis	
atpA, atpB	ATPase alpha/beta chains family	
tdsD, nifF	Flavoprotein transcripts	
ZMO1097, ZMO1830, ZMO1732, ZMO0279, ZMO1118, ZMO0749	Stress response	

[#]Genes highlighted in green are up regulated and those selected in red are down regulated. The experiments highlighted with (*) were performed by RT-qPCR.

4.3.1.2 GENETIC REGULATION OF ZYMOMONAS MOBILIS UNDER INHIBITORY COMPOUND STRESS

Chemical inhibitors are commonly present during *Z. mobilis* fermentation because these products may be present in LCB residues or may be derived from the fermentation itself [80]. The study of the expression profile of *Z. mobilis* with diverse inhibitors helps to characterize the metabolic pathways involved in cell detoxification processes. Changes in the cell growth and ethanol yield of *Z. mobilis* ZM4 in the presence of inhibitors such as phenolic aldehydes, furfural, and ethanol have been evaluated. Phenolic aldehydes are formed in the pretreatment of LCB used as raw material for the production of biofuels and have been reported as toxic agents that can affect cell growth and fermentation [81]. Yi et al. evaluated the genomic response of *Z. mobilis* ZM4 in the presence of the inhibitors 4-hydroxybenzaldehyde, syringaldehyde, and vanillin, identifying overexpressed genes from the respiratory chain and transporter genes (Table 4.3) that help reduce inhibitors to their corresponding phenolic alcohols and maintain ethanol production [82].

TABLE 4.3 Genes of *Zymomonas mobilis* Z4 Differentially Expressed Under Inhibitory Compounds Stress*

Phenolic Aldehydes		
Gene ID	**Cellular Process**	**Reference**
ZMO116, ZMO1696, **ZMO1885**	Respiratory chain	[82]
ZMO0282, ZMO0283, ZMO0799, ZMO0800	Transporter genes	
Furfural		
flhA, fliE, fliG, flgH, flgL, ZMO0619, ZMO0285, ZMO0780, ZMO1525, *oprM*, ZMO0307, ZMO0779, ZMO0064, ZMO0835, ZMO0197, ZMO01331	Cell motility and cell wall membrane biogenesis	[83]
ZMO0629, ZMO0356, ZMO0996, ZMO0216, ZMO1174, **ZMO1311**, ZMO0291		
rpsD, rpsF, rplI, rbsR, frr, rbfA, proS, alaS, leuS, glyS, pheT, valS, gnlA, trpA, trpB, argG, gltB, ilvE, glnB, serA, serC	Protein synthesis	
leuC		
ZMO0351, *addA, addB, rnhB, ung*, ZMO1185, ZMO1652	DNA replication, recombination, and repair	
ZMO1930, **ZMO1356, ZMO1426**, ZMO1588, ZMO0362, ZMO1231, **ZMO1584**, ZMO0354, **ZMO1193**		
ZMO1336, ZMO0050, **ZMO0774**, ZMO0471, ZMO1738, ZMO1944, **ZMO0281**, ZMO1283, **ZMO1623**	Transcriptional regulation	

*Genes highlighted in green are up regulated and those selected in red are down regulated.

Another inhibitor derived from LCB is furfural. This toxic compound is widely studied in yeasts due to growth inhibition and low ethanol production. Microarray analysis was conducted to evaluate the transcriptional response of *Z. mobilis* ZM4 under furfural stress [83]. The genes involved in furfural tolerance were identified as both up- and downregulated and are related to cell motility and cell wall membrane biogenesis: Several downregulated genes are implicated in protein synthesis and DNA replication, recombination, and repair, and upregulated genes are related to transcriptional regulation in response to DNA damage, which has been extensively studied in other bacteria, e.g., *Escherichia coli* and fermentative yeasts [84, 85].

4.3.1.3   GENETIC REGULATION OF ZYMOMONAS MOBILIS UNDER ETHANOL STRESS

Similar to other inhibitors, the intracellular and extracellular accumulation of ethanol is generally toxic to *Z. mobilis*. To evaluate the gene expression dynamics related to ethanol tolerance, a microarray analysis of *Z. mobilis* ZM4 was performed, identifying low expression levels in different genes involved in carbohydrate metabolism by the ED pathway, in addition to the *ldhA* gene, which is involved in ethanol formation [86]. Up- and downregulated genes related to cell motility and cell wall membrane biogenesis were also identified. Remarkably, the *fliE*, *fliG*, and ZMO01311 genes were also found to be downregulated in the presence of furfural [83].

Genes involved in the respiratory chain are also up- and downregulated: in particular, ZMO1885 is upregulated in the presence of ethanol and downregulated in the presence of phenolic aldehydes. Transcriptional regulation is a cellular process involved in the adaptation process of *Z. mobilis*, with the genes ZMO281 and ZMO0774 upregulated in the presence of ethanol and phenolic aldehydes. Upregulation of genes implicated in DNA replication, recombination, and repair also plays an important role in the adaptation process, and the genes ZMO1193, ZMO1356, ZMO1426 and ZMO1584 have high expression levels in both ethanol and phenolic aldehyde adaptation. Many genes in *Z. mobilis* are upregulated in relation to stress tolerance, and the ZMO1623 gene is also present in phenolic aldehyde stress (Table 4.4).

The study of the biology of *Z. mobilis* for more than 6 decades has revealed its exceptional metabolic characteristics and its great potential as an ethanol producer, combined with few growth cultivation restrictions. To date, we have information about its metabolism and physiology, in addition to genomic, transcriptomic, and metabolomic data. This bacterium has recently been studied through systems biology, which is a multidisciplinary approach involving biological sciences, mathematical models and computer science that aims to study the relationships that connect the components of a network and the components themselves. With this approach, we can develop predictive mathematical models of the biological processes of *Z. mobilis.*

TABLE 4.4 Genes of *Zymomonas mobilis* Z4 Differentially Expressed Under Ethanol Stress[*]

Ethanol		
Gene ID	**Cellular Process**	**References**
gnl	ED pathway	[86]
gntk	Pyruvate biosynthesis	
ldhA	Ethanol formation	
flhAB, fliDEFGHIKLMNPQRS, ZMO0613, ZMO0614, ZMO0604, ZMO605, ZMO607, ZMO608, ZMO609, ZMO610, ZMO611, ZMO612, ZMO619, ZMO624, ZMO632, ZMO634, ZMO635, ZMO642, ZMO643, ZMO648, ZMO649, ZMO651, ZMO652	Cell motility and cell wall membrane biogenesis	
ZMO1311		
ZMO0022, ZMO1571, ZMO1572, ZMO1032, ZMO1255, ZMO1256, ZMO1189, ZMO1669, ZMO0678, ZMO1812, ZMO1813, ZMO1814, *rnfAB*	Respiratory chain	
ZMO1844, ZMO0957, ZMO0958, ZMO1252, ZMO1253, ZMO1254, ZMO1479, ZMO1480, ZMO1113, **ZMO1885**, ZMO1809, ZMO1810, ZMO1811, ZMO0569		
ZMO1404, **ZMO1623**, ZMO0274, ZMO0626	Stress response	
ZMO1356, ZMO1426, ZMO1484, ZMO1417, ZMO1648, **ZMO1193**, ZMO1401, ZMO086, ZMO0598, ZMO1907, ZMO1187, ZMO1054, **ZMO1584**, ZMO812	DNA replication, recombination, and repair	
ZMO0054, ZMO2033, ZMO1697, **ZMO0281**, ZMO1547, **ZMO0774**, ZMO0190	Transcriptional regulation	
ZMO1107, ZMO0347		
ZMO1180, ZMO2018, ZMO1395, ZMO1804, ZMO1025, ZMO1855, ZMO1522, ZMO1425, ZMO1647, ZMO1262, ZMO0546	Transport systems	
ZMO1649, ZMO1757, ZMO0899		

[*]Genes highlighted in green are up regulated and those selected in red are down regulated.

4.3.2 CLOSTRIDIUM THERMOCELLUM

Clostridium thermocellum is an anaerobic thermophilic bacterium with a high growth rate on cellulose; this characteristic is due to its highly efficient extracellular free and multienzyme complex termed the cellulosome and its accessory enzymes. A model of the multienzymatic systems is represented in Figure 4.8; some proteins were renamed as follows: CipA (ScaA), OlpB (ScaB), Orf2p (ScaC), OlpA (ScaD), SdbA (ScaF), and OlpC (ScaG) [87]. This enzymatic complex gives *C. thermocellum* the ability to solubilize the cellulose contained in LCB and rapidly ferment it to produce ethanol. The metabolic ability of *C. thermocellum* to produce ethanol directly from LCB makes it the main candidate microorganism for bioethanol production via consolidated bioprocessing (CBP). Despite the biotechnological potential of *C. thermocellum*, the industrial application of this bacterium is relegated because of its disadvantages compared to other microorganisms, such as mixed acid fermentation, low ethanol productivity and titer, low ethanol tolerance, and low hemicellulose utilization [88–90]. To solve these biotechnological limitations, several strategies have been carried out to improve ethanol production by *C. thermocellum*, from technological strategies such as cocultivation with other bacteria to genomic strategies such as metabolic engineering. In parallel, several genetic and transcriptomic studies have been performed to understand the gene regulation involved in biomass degradation, as well as the ethanol tolerance mechanism used by *C. thermocellum* [91].

4.3.2.1 GENETIC REGULATION OF THE CELLULOSOME

The main attribute that makes *C. thermocellum* a promising bacterium for the production lignocellulosic ethanol is the ability to produce cellulosomes; as a consequence, the regulation of cellulosome expression has been widely studied. *C. thermocellum*, similar to other microorganisms, regulates gene expression in response to the environment to adapt and synchronize its metabolic reactions to new conditions. Genetic and metabolic analysis of *C. thermocellum* showed that this bacterium employs more than 100 genes for biomass degradation, including more than 70 genes that encode various cellulosomal enzymes.

One of the most studied regulation systems in *C. thermocellum* is the homologous LacI transcriptional regulatory network for the *celC* operon [93–96]. The *celC* operon is formed by a noncellulosomal GH5 endoglucanase gene (*celC*), the *glyR3* gene (GlyR3 protein) and the endo-1,3-β-d-glucosidase gene (*licA*);

this operon is negatively autoregulated by the binding of the GlyR3 protein to the *celC* promotor region [93]. The repression of the operon *celC* is relieved by laminaribiose, which impedes the binding of GlyR3 to the *celC* promoter. An extended model for the regulation of the six-gene cluster *celC-glyR3-licA-orf4-manB-celT* was proposed by Choi et al.; in this model, the protein GlyR3 coregulates the expression of the *celC* and *manB* genes [96]. In the extended model, the expression of the cellulosomal family 26 glycoside hydrolase ManB is repressed by a high concentration of the protein GlyR3 in the presence of laminaribiose. In contrast, at a basal concentration of GlyR3 (in the absence of laminaribiose), the *manB* gene is expressed (Figure 4.9). The mechanism of regulation of the *celT* gene that encodes a cellulosomal family 9 endoglucanase is unknown. In the same way, the mechanism regulating the *orf4* gene and the function of the protein it encodes are unknown.

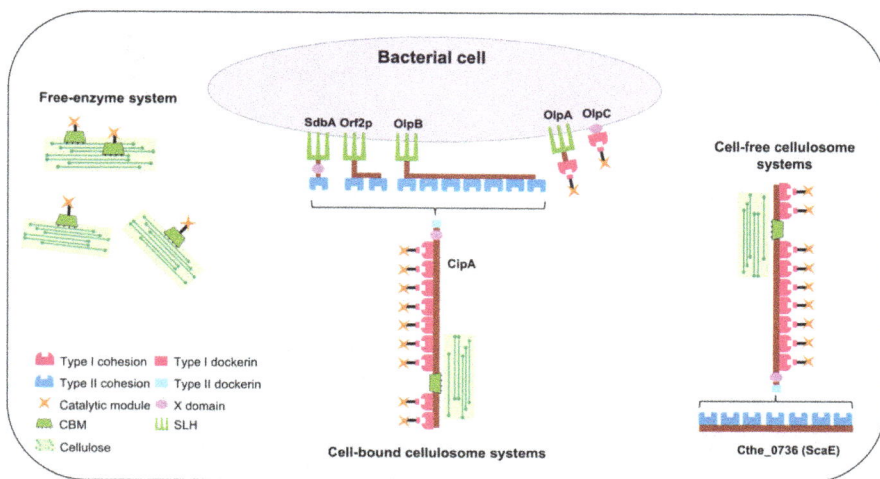

FIGURE 4.8 Model of *Clostridium thermocellum* cellulosome systems.
Source: Adapted from: Ref. [92].

On the other hand, several regulatory mechanisms have been proposed to be involved in the genetic regulation of the expression of the cellulosome, including carbon catabolite repression and alternative σ factors. It is well known that extracellular polysaccharides affect cellulosome composition by regulating which enzymes and structural components are expressed [97]. The expression of the individual components of the cellulosome in *C. thermocellum* is probably regulated by a cluster of at least 6 paralogous alternative

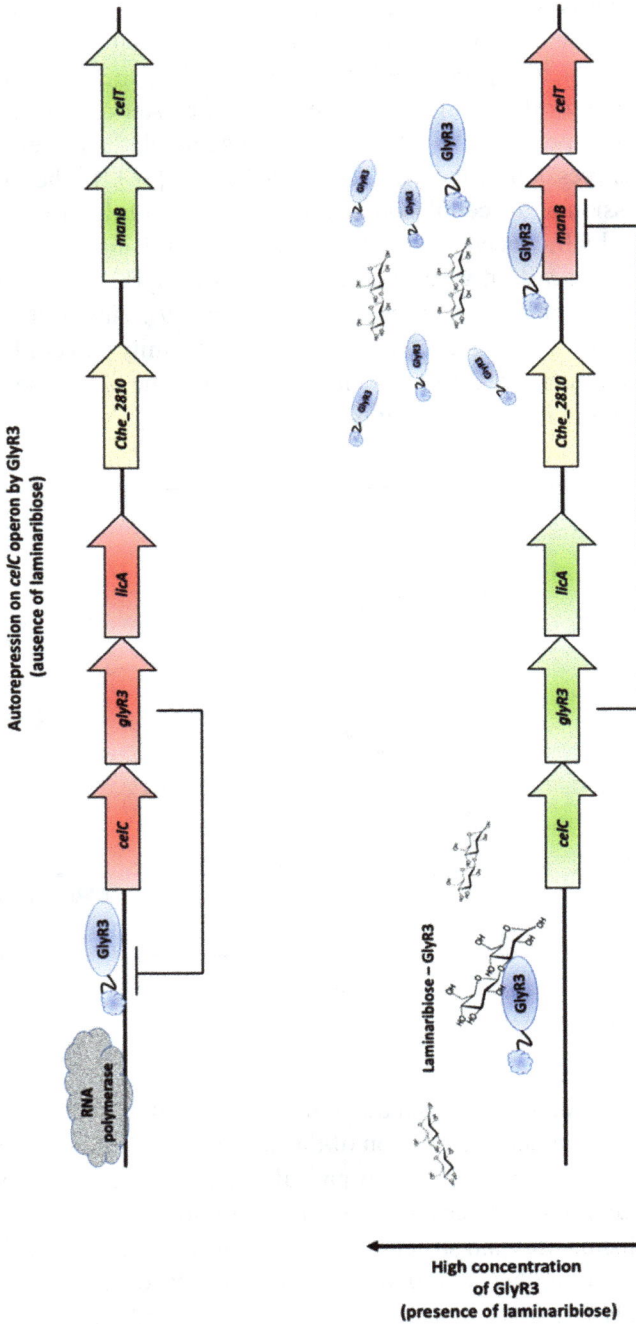

FIGURE 4.9 Expanded model of GlyR3 regulation in *Clostridium thermocellum*. Red and green indicate genetic repression and expression, respectively.

σ^I factors and their cognate membrane-associated anti-σ [97, 98]. Several investigations have revealed that the main structural component of the cellulosome in C. *thermocellum* is CipA, a protein scaffold that includes nine type I cohesin modules, a type II dockerin module, and a family III carbohydrate binding module known for a strong affinity for cellulose (Figure 4.8).

Ortiz de Ora et al. demonstrated that 5 σ^I factors (identified as σ^{11}, σ^{12}, σ^{13}, σ^{14} and σ^{16}) regulate the expression of 17 genes encoding different cellulosomal proteins [98]. They could relate the σ^{11}–σ^{16} factors to the genetic regulation of two of the most important components of the cellulosome, the primary scaffoldin (CipA) and the most abundant enzyme (Cel48S). In the same way, the regulons of the σ^I alternative factors were identified by bioinformatic analysis in conjunction with classical microbiology genetic tools and the application of the heterologous B. *subtilis* host system [98]. All σ^I factor genes (*sigI1-sigI6* and *sig24C*) are upregulated in the presence of extracellular polysaccharides. In this context, the proposed mechanism to activate alternate σ^I implied an extracellular carbohydrate-active module and an intracellular anti-σ peptide domain [97] (Table 4.5).

Studies performed to understand the mechanisms of genetic regulation of cellulosome production in C. *thermocellum* demonstrated that growth conditions regulate the expression of cellulosomal components and that this regulation is a response to the polysaccharides present in the culture media. Nataf et al. identified five sugar ABC transport systems: four are specific for β-1,4-linked glucose oligomers (cellodextrins), and one is specific for β-1,3-linked glucose dimer (laminaribiose) [99]. The sugar transporters and their substrate specificities demonstrated that C. *thermocellum* prefers to assimilate cellodextrins rather than cellobiose or glucose. Genome analysis also suggests that the bacterium lacks any other sugar ABC transporters, in agreement that this strain can grow only on β-glucans. Consequently, the sugars present in the culture media determine the composition of the cellulosome, which that subsequently influences the overall ability of the bacteria to degrade lignocellulosic substrates and produce ethanol.

4.3.2.2 GENETIC REGULATION OF CLOSTRIDIUM THERMOCELLUM DURING ETHANOL PRODUCTION

The regulation of genes involved in ethanol production is affected by the substrate and growth conditions, similar to the regulation of cellulosome expression. Although C. *thermocellum* is a promising bacterium for ethanol production, mixed acid fermentation and low ethanol tolerance are the major

TABLE 4.5 Regulatory Networks of σ^I Factors in *Clostridium thermocellum*

σ^I – anti-σ Factors	Regulon	Gen Product	C-Terminal Sensing of Anti-σ Factor/Activator Polysaccharides
σ^{I1}-RsgI1	*sigI1, cel8A, cel48S, sdbA, xgh74A*	σ^{I1}, GH8-DocI, GH48-DocI, CohlI-X-3(SLH), GH74-DocI	CBM3/cellulose
σ^{I2}-RsgI2	*sigI2*, Clo 1313_0420, Clo 1313_2216	σ^{I2}, DocI-UNK, Abf-DocI-GH43	CBM3/cellulose
σ^{I3}-RsgI3	*sigI3, cel48S, cipA, rga12A, rgI11A*	σ^{I3}, GH48-DocI, 2(CohI)-CBM3-6(CohI)-X-DocII, Rga-DocI-CMB35-Rga, Doc-CBM35-Rgl	PA14 dyad/pectin
σ^{I4}-RsgI4	*sigI4, cel8A, cel48S, cipA,* Clo1313_1436	σ^{I4}, GH8-DocI, GH48-DocI, 2(CohI)-CBM3-6(CohI)-X-DocII, HP	CBM3/cellulose
σ^{I5}-RsgI5	*sigI5*	–	CBM42/arabinoxylan
σ^{I6}-RsgI6	*sigI1, sigI6, cel9V, cel48S, cipA, cseP, rsgI5, xyn11B, xyn10D, xyn10Y, xyn10Z*	σ^{I1}, σ^{I6}, GH9-2-(CBM3)-DocI, GH48-DocI, 2(CohI)-CBM3-6(CohI)-X-DocII, CotH-DocI, anti-σ^{I5} (or RsgI5), GH11-CBM6-DocI, CBM22-GH10-DocI, CBM22-GH10-CBM22-DocI-CE1, CE1-CBM6-DocI-GH10	GH10/xylans, cellulose
σ^{24C}-Rsi24C	*sig24C*	–	GH5/cellulose

Abbreviations: GH: glycoside hydrolase; Doc: dockerin; CBM: carbohydrate-binding module; Coh: cohesin; X, X-module (module of unknown function); CotH: spore coat protein H; UNK: unknown sequence; HP: hypothetical protein; Abf: Alpha-L-arabinofuranosidase; Rga: rhamnogalacturan acetylesterase; Rgl: rhamnogalacturonan lyase; SLH: S-layer homology domain; CE: carbohydrate esterase.

Source: Adapted from: Refs. [97, 98].

obstacles to its commercial application. However, to overcome the limitations that slow down the use of *C. thermocellum* to produce ethanol industrially, some approaches have been pursued to understand the mechanism of inhibition by ethanol, as well as approaches focused on increasing ethanol yield, including metabolic engineering efforts to delete pathways for carbon flux to lactate and acetate [4, 91, 100, 101]. Wild-type *C. thermocellum* is inhibited by low ethanol concentrations (approximately 10 g/l) and is completely unable to grow at 20 g/l ethanol. However, wild-type *C. thermocellum* had to be adapted to tolerate 50–70 g/l ethanol [102, 103]. *C. thermocellum* uses the EMP glycolysis pathway to generate pyruvate from sugars (cellobiose or glucose); however, some differences from the traditional pathway are present [104]. The final steps of the EMP pathway involve the successive reduction of acetyl-CoA and acetaldehyde with electrons provided by NADH (i.e., the ALDH and ADH reactions); these two reactions are both catalyzed by the bifunctional alcohol dehydrogenase (ADHE) enzyme. This enzyme is related to the ethanol tolerance of *C. thermocellum*, as mutations in the bifunctional ADHE result in increased ethanol tolerance [101]. Previous work demonstrated that the expression of *ADH* genes, pyruvate ferredoxin oxidoreductase genes (Cthe_2390, Cthe_2391, Cthe_2392 and Cthe_0340) and other genes related to ethanol production were affected by the carbon source (cellulose or cellobiose) and the growth rate [105, 106].

Recently, global transcriptomic analysis of *C. thermocellum* ATCC 27405 growing in dilute acid-pretreated Populus and switchgrass showed overexpression in genes related to nitrogen uptake and metabolism at the end of fermentation when populus was used; this overexpression coincided with increases in ethanol concentration [107]. Similar results were obtained when *C. thermocellum* was grown on crystalline cellulose [108]. These results demonstrated that the expression of genes related to ethanol production is influenced by the carbon source and the fermentation time, as well as the growth rate.

Although several studies have been carried out to understand the mechanism of genetic regulation in one of the most promising bacteria to produce ethanol directly to LCB, *C. thermocellum*, substantial further effort is required to fully elucidate all the genetic regulatory mechanisms that control the expression of proteins involved in biomass degradation and ethanol production. The application of new omics technologies (genomics, transcriptomics, proteomics, and metabolomics) allowed us to reach a complete understanding of metabolism and its connection with the genetic regulation of *C. thermocellum*. An understanding of the genetic regulatory mechanisms in *C. thermocellum* can help improve this industrially relevant strain and promote the use of cellulosome-producing bacteria in the CBP of biomass.

4.3.3 OTHER ETHANOLOGENIC BACTERIA

Fermentation is a crucial process in bioethanol production, where ethanol is produced from the metabolic activity of a microorganism, either bacteria or yeast. We previously reviewed the characteristics of several bacteria, such as _Z. mobilis_ and cellulosic thermophiles, including _C. thermocellum_. Another interesting alternative is extreme thermophilic anaerobic bacteria, which are good candidates for bioethanol production due to their ability to ferment an extensive variety of substrates that include hexoses, pentoses, and disaccharides to produce ethanol, in addition to the relatively low contamination risk by other microorganisms due to high fermentation temperatures [109].

Extreme thermophilic anaerobic bacteria have been isolated from different environments, including geothermal areas, volcanic mud, and canned products [110–112]. These bacteria are facultatively anaerobic and tolerate extreme pH and salt concentrations during fermentation with minimal nutrient supplementation [113]. Despite the numerous advantages of employing thermophilic bacteria in fermentation for bioethanol production, it is known that yeasts and other bacteria such as _Z. mobilis_ can tolerate a higher ethanol concentration than extremophiles due to the fatty acid composition of the cell membrane [114].

Several studies have been performed with hemicellulolytic thermophiles of the genera _Thermoanaerobacter_ and _Thermoanaerobacterium_ to increase the ethanol yield and ethanol tolerance. These strategies include the suppression of other fermentation products through lactate and acetate metabolic pathway knockout in _T. saccharolyticum_ JW/SL-YS485 and the overexpression of enzymes such as NAD(P)H-dependent ADH in _T. mathranii_, which is directly related to increased ethanol production [109, 115]. The expression dynamics of the ADH enzymes ADHA, ADHB, and ADHE, which have a key role in ethanol formation in _T. ethanolicus_, were evaluated, and the expression of these enzymes was observed to be affected at high ethanol concentrations [116].

In conclusion, the search for more efficient and resistant microorganisms to produce ethanol continues. Microorganisms such as _S. cerevisiae_ and _Z. mobilis_ have been extensively studied due to their ability to produce ethanol; knowledge of their genomes and the application of omic techniques have considerably illuminated their mechanisms of gene regulation during ethanol production. Emerging microorganisms such as _K. marxianus_ and _C. thermocellum_ have also aroused interest for use in ethanol production due to their outstanding metabolic and physiological characteristics. However, knowledge about their gene regulation during ethanol production is still limited, so at present, several studies are focused on understanding why these microorganisms are promising for the industrial production of ethanol. The development of new omics techniques will allow a complete understanding of the regulatory mechanisms that govern the genetic expression of these bacteria and yeasts in the short and medium term. Consequently, we will be able to develop new strains capable of producing ethanol efficiently in highly stressful conditions.

KEYWORDS

- *Clostridium thermocellum*
- **ethanol production**
- **genetic regulation**
- *Kluyveromyces marxianus*
- *Saccharomyces cerevisiae*
- *Zymomonas mobilis*

REFERENCES

1. Prinz, B., & Lang, C., (2004). Gene regulation in yeast. *Genet. Biotechnol.*, 129–145. https://doi.org/10.1007/978-3-662-07426-8_8.
2. Barbosa De, M. F. S., & Ingram, L. O., (1994). Expression of the *Zymomonas mobilis* alcohol dehydrogenase II (AdhB) and pyruvate decarboxylase (Pdc) genes in *Bacillus*. *Curr. Microbiol., 28*(5), 279–282. https://doi.org/10.1007/BF01573206.
3. Xue, T., Liu, K., Chen, D., Yuan, X., Fang, J., Yan, H., Huang, L., Chen, Y., & He, W., (2018). Improved bioethanol production using CRISPR/Cas9 to disrupt the ADH2 gene in *Saccharomyces cerevisiae. World J. Microbiol. Biotechnol., 34*(10), *154*, 1–12. https://doi.org/10.1007/s11274-018-2518-4
4. Lo, J., Zheng, T., Hon, S., Olson, D. G., & Lynd, L. R., (2015). The bifunctional alcohol and aldehyde dehydrogenase gene, AdhE, Is necessary for ethanol production in *Clostridium thermocellum* and *Thermoanaerobacterium saccharolyticum. J. Bacteriol., 197*(8), 1386–1393. https://doi.org/10.1128/JB.02450-14.
5. Favaro, L., Jansen, T., & Van, Z. W. H., (2019). Exploring industrial and natural *Saccharomyces cerevisiae* strains for the bio-based economy from biomass: The case of bioethanol. *Crit. Rev. Biotechnol., 39*(6), 800–816. https://doi.org/10.1080/0738855 1.2019.1619157.

6. Yang, J., Ding, M. Z., Li, B. Z., Liu, Z. L., Wang, X., & Yuan, Y. J., (2012). Integrated phospholipidomics and transcriptomics analysis of *Saccharomyces cerevisiae* with enhanced tolerance to a mixture of acetic acid, furfural, and phenol. *Omi. A J. Integr. Biol., 16*(7,8), 374–386. https://doi.org/10.1089/omi.2011.0127.

7. Kim, D., & Hahn, J. S., (2013). Roles of the Yap1 transcription factor and antioxidants in *Saccharomyces cerevisiae*'s tolerance to furfural and 5-hydroxymethylfurfural, which function as thiol-reactive electrophiles generating oxidative stress. *Appl. Environ. Microbiol., 79*(16), 5069–5077. https://doi.org/10.1128/AEM.00643-13.

8. Ishida, Y., Nguyen, T. T. M., & Izawa, S., (2017). The yeast ADH7 promoter enables gene expression under pronounced translation repression caused by the combined stress of vanillin, furfural, and 5-hydroxymethylfurfural. *J. Biotechnol., 252*, 65–72. https://doi.org/10.1016/j.jbiotec.2017.04.024.

9. Liu, Z. L. L., (2018). Understanding the tolerance of the industrial yeast *Saccharomyces cerevisiae* against a major class of toxic aldehyde compounds. *Appl. Microbiol. Biotechnol., 102*(13), 5369–5390. https://doi.org/10.1007/s00253-018-8993-6.

10. Flores-Cosio, G., Arellano-Plaza, M., Gschaedler, A., & Amaya-Delgado, L., (2018). Physiological response to furan derivatives stress by *Kluyveromyces marxianus* SLP1 in ethanol production. *Rev. Mex. Ing. Quim., 17*(1), 189–202. https://doi.org/10.24275/uam/izt/dcbi/revmexingquim/2018v17n1/Flores.

11. Flores-Cosío, G., Herrera-López, E. J., Arellano-Plaza, M., Gschaedler-Mathis, A., Sanchez, A., & Amaya-Delgado, L., (2019). Dielectric property measurements as a method to determine the physiological state of *Kluyveromyces marxianus* and *Saccharomyces cerevisiae* stressed with furan aldehydes. *Appl. Microbiol. Biotechnol., 103*(23, 24), 9633–9642. https://doi.org/10.1007/s00253-019-10152-2.

12. Guo, W., Chen, Y., Wei, N., & Feng, X., (2016). Investigate the metabolic reprogramming of *Saccharomyces cerevisiae* for enhanced resistance to mixed fermentation inhibitors via 13C metabolic flux analysis. *PLoS One, 11*(8), 1–15. https://doi.org/10.1371/journal.pone.0161448.

13. Dellomonaco, C., Fava, F., & Gonzalez, R., (2010). The path to next generation biofuels: Successes and challenges in the era of synthetic biology. *Microb. Cell Fact., 9*, 1–15. https://doi.org/10.1186/1475-2859-9-1.

14. Matsushika, A., Inoue, H., Kodaki, T., & Sawayama, S., (2009). Ethanol production from xylose in engineered *Saccharomyces cerevisiae* strains: Current state and perspectives. *Appl. Microbiol. Biotechnol., 84*(1), 37–53. https://doi.org/10.1007/s00253-009-2101-x.

15. Wu, G., Xu, Z., & Jönsson, L. J., (2017). Profiling of *Saccharomyces cerevisiae* transcription factors for engineering the resistance of yeast to lignocellulose-derived inhibitors in biomass conversion. *Microb. Cell Fact., 16*(1), 1–15. https://doi.org/10.1186/s12934-017-0811-9.

16. Greetham, D., Wimalasena, T. T., Leung, K., Marvin, M. E., Chandelia, Y., Hart, A. J., Phister, T. G., et al., (2014). The genetic basis of variation in clean lineages of *Saccharomyces cerevisiae* in response to stresses encountered during bioethanol fermentations. *PLoS One, 9*(8), 1–14. https://doi.org/10.1371/journal.pone.0103233.

17. Cunha, J. T., Romaní, A., Costa, C. E., Sá-Correia, I., & Domingues, L., (2019). Molecular and physiological basis of *Saccharomyces cerevisiae* tolerance to adverse lignocellulose-based process conditions. *Appl. Microbiol. Biotechnol., 103*(1), 159–175. https://doi.org/10.1007/s00253-018-9478-3.

18. Beckner, M., Ivey, M. L., & Phister, T. G., (2011). Microbial contamination of fuel ethanol fermentations. *Lett. Appl. Microbiol., 53*(4), 387–394. https://doi.org/10.1111/j.1472-765X.2011.03124.x.

19. Meijnen, J. P., Randazzo, P., Foulquié-Moreno, M. R., Van, D. B. J., Vandecruys, P., Stojiljkovic, M., Dumortier, F., et al., (2016). Polygenic analysis and targeted improvement of the complex trait of high acetic acid tolerance in the yeast *Saccharomyces cerevisiae*. *Biotechnol. Biofuels, 9*(1), 1–18. https://doi.org/10.1186/s13068-015-0421-x.

20. Mira, N. P., Becker, J. D., & Sá-Correia, I., (2010). Genomic expression program involving the Haa1p-regulon in *Saccharomyces cerevisiae* response to acetic acid. *Omi. A J. Integr. Biol., 14*(5), 587–601. https://doi.org/10.1089/omi.2010.0048.

21. Mira, N. P., Teixeira, M. C., & Sá-Correia, I., (2010). Adaptive response and tolerance to weak acids in *Saccharomyces cerevisiae*: A genome-wide view. *Omi. A J. Integr. Biol., 14*(5), 525–540. https://doi.org/10.1089/omi.2010.0072.

22. Teixeira, M. C., Mira, N. P., & Sá-Correia, I., (2011). A genome-wide perspective on the response and tolerance to food-relevant stresses in *Saccharomyces cerevisiae*. *Curr. Opin. Biotechnol., 22*(2), 150–156. https://doi.org/10.1016/j.copbio.2010.10.011.

23. Saini, P., Beniwal, A., Kokkiligadda, A., & Vij, S., (2018). Response and tolerance of yeast to changing environmental stress during ethanol fermentation. *Process Biochem., 72*, 1–12. https://doi.org/10.1016/j.procbio.2018.07.001.

24. Almeida, J. R. M., Röder, A., Modig, T., Laadan, B., Lidén, G., & Gorwa-Grauslund, M. F., (2008). NADH- vs NADPH-coupled reduction of 5-hydroxymethyl furfural (HMF) and its implications on product distribution in *Saccharomyces cerevisiae*. *Appl. Microbiol. Biotechnol., 78*(6), 939–945. https://doi.org/10.1007/s00253-008-1364-y.

25. Liu, Z. L., & Moon, J., (2009). A novel NADPH-dependent aldehyde reductase gene from *Saccharomyces cerevisiae* NRRL Y-12632 involved in the detoxification of aldehyde inhibitors derived from lignocellulosic biomass conversion. *Gene, 446*(1), 1–10. https://doi.org/10.1016/j.gene.2009.06.018.

26. Moon, J., & Liu, Z. L., (2012). Engineered NADH-dependent GRE2 from *Saccharomyces cerevisiae* by directed enzyme evolution enhances HMF reduction using additional cofactor NADPH. *Enzyme Microb. Technol., 50*(2), 115–120. https://doi.org/10.1016/j.enzmictec.2011.10.007.

27. Larroy, C., Fernández, M. R., González, E., Parés, X., & Biosca, J. A., (2003). Properties and functional significance of *Saccharomyces cerevisiae* ADHVI. *Chem. Biol. Interact., 143, 144*, 229–238. https://doi.org/10.1016/S0009-2797(02)00166-7.

28. Adeboye, P. T., Bettiga, M., & Olsson, L., (2017). ALD5, PAD1, ATF1 and ATF2 facilitate the catabolism of coniferyl aldehyde, ferulic acid and p-coumaric acid in *Saccharomyces cerevisiae*. *Sci. Rep., 7*, 1–13. https://doi.org/10.1038/srep42635.

29. Ma, M., & Liu, Z. L., (2010). Comparative transcriptome profiling analyses during the Lag phase uncover YAP1, PDR1, PDR3, RPN4, and HSF1 as key regulatory genes in genomic adaptation to the lignocellulose derived inhibitor HMF for *Saccharomyces cerevisiae*. *BMC Genomics, 11*(660), 1–19. http://www.biomedcentral.com/1471-2164/11/660.

30. Wang, H., Li, Q., Zhang, Z., Zhou, C., Ayepa, E., Abrha, G. T., Han, X., et al., (2019). YKL107W from *Saccharomyces cerevisiae* encodes a novel aldehyde reductase for detoxification of acetaldehyde, glycolaldehyde, and furfural. *Appl. Microbiol. Biotechnol., 103*(14), 5699–5713. https://doi.org/10.1007/s00253-019-09885-x.

31. Allen, S. A., Clark, W., McCaffery, J. M., Cai, Z., Lanctot, A., Slininger, P. J., Liu, Z. L., & Gorsich, S. W., (2010). Furfural induces reactive oxygen species accumulation and cellular damage in *Saccharomyces cerevisiae*. *Biotechnol. Biofuels, 3*, 2.

32. Voulgaridou, G. P., Anestopoulos, I., Franco, R., Panayiotidis, M. I., & Pappa, A., (2011). DNA damage induced by endogenous aldehydes: Current state of knowledge. *Mutat. Res. - Fundam. Mol. Mech. Mutagen., 711*(1, 2), 13–27. https://doi.org/10.1016/j.mrfmmm.2011.03.006.

33. Bajwa, P. K., Ho, C. Y., Chan, C. K., Martin, V. J. J., Trevors, J. T., & Lee, H., (2013). Transcriptional profiling of *Saccharomyces cerevisiae* T2 cells upon exposure to hardwood spent sulphite liquor: Comparison to acetic acid, furfural and hydroxymethylfurfural. *Antonie Van Leeuwenhoek, Int. J. Gen. Mol. Microbiol., 103*(6), 1281–1295. https://doi.org/10.1007/s10482-013-9909-1.

34. Gu, H., Zhang, J., & Bao, J., (2014). Inhibitor analysis and adaptive evolution of *Saccharomyces cerevisiae* for simultaneous saccharification and ethanol fermentation from industrial waste corncob residues. *Bioresour. Technol., 157*, 6–13. https://doi.org/10.1016/j.biortech.2014.01.060.

35. Wright, J., Bellissimi, E., De Hulster, E., Wagner, A., Pronk, J. T., & Van, M. A. J. A., (2011). Batch and Continuous culture-based selection strategies for acetic acid tolerance in xylose-fermenting *Saccharomyces cerevisiae*. *FEMS Yeast Res., 11*(3), 299–306. https://doi.org/10.1111/j.1567-1364.2011.00719.x.

36. Li, H., Ma, M. L., Luo, S., Zhang, R. M., Han, P., & Hu, W., (2012). Metabolic responses to ethanol in *Saccharomyces cerevisiae* using a gas chromatography tandem mass spectrometry-based metabolomics approach. *Int. J. Biochem. Cell Biol., 44*(7), 1087–1096. https://doi.org/10.1016/j.biocel.2012.03.017.

37. Nugroho, R. H., Yoshikawa, K., & Shimizu, H., (2015). Metabolomic analysis of acid stress response in *Saccharomyces cerevisiae*. *J. Biosci. Bioeng., 120*(4), 396–404. https://doi.org/10.1016/j.jbiosc.2015.02.011.

38. Leupold, S., Hubmann, G., Litsios, A., Meinema, A. C., Takhaveev, V., Papagiannakis, A., Niebel, B., et al., (2019). *Saccharomyces cerevisiae* goes through distinct metabolic phases during its replicative lifespan. *Elife, 8*, 1–19. https://doi.org/10.7554/eLife.41046.

39. Ohta, E., Nakayama, Y., Mukai, Y., Bamba, T., & Fukusaki, E., (2016). Metabolomic approach for improving ethanol stress tolerance in *Saccharomyces cerevisiae*. *J. Biosci. Bioeng., 121*(4), 399–405. https://doi.org/10.1016/j.jbiosc.2015.08.006.

40. Raimondi, S., Zanni, E., Amaretti, A., Palleschi, C., Uccelletti, D., & Rossi, M., (2013). Thermal adaptability of *Kluyveromyces marxianus* in recombinant protein production. *Microb. Cell Fact., 12*(1), 1–7. https://doi.org/10.1186/1475-2859-12-34.

41. Morrissey, J. P., Etschmann, M. M. W., Schrader, J., & De Billerbeck, G. M., (2015). Cell factory applications of the yeast *Kluyveromyces marxianus* for the biotechnological production of natural flavor and fragrance molecules. *Yeast,* (32), 3–16. https://doi.org/10.1002/yea.

42. Iñiguez, M. L. E., Arellano-Plaza, M., Prado-Montes De, O. E., Kirchmayr, M. R., Segura-García, L. E., Amaya-Degado, L., & Gschaedler, M., (2019). The production of esters and gene expression by *Saccharomyces cerevisiae* during fermentation on *Agave tequilana* juice in continuous cultures. *Rev. Mex. Ing. Química, 18*(2), 451–462. https://doi.org/10.24275/uam/izt/dcbi/revmexingquim/2019v18n2/iniguez.

43. Rodrussamee, N., Lertwattanasakul, N., Hirata, K., Suprayogi, Limtong, S., Kosaka, T., & Yamada, M., (2011). Growth and ethanol fermentation ability on hexose and

pentose sugars and glucose effect under various conditions in thermotolerant yeast *Kluyveromyces marxianus*. *Appl. Microbiol. Biotechnol.*, *90*(4), 1573–1586. https://doi. org/10.1007/s00253-011-3218-2.

44. García-Aparicio, M. P., Oliva, J. M., Manzanares, P., Ballesteros, M., Ballesteros, I., González, A., & Negro, M. J., (2011). Second-generation ethanol production from steam exploded barley straw by *Kluyveromyces marxianus* CECT 10875. *Fuel, 90*(4), 1624–1630. https://doi.org/10.1016/j.fuel.2010.10.052.
45. Limtong, S., Sringiew, C., & Yongmanitchai, W., (2007). Production of fuel ethanol at high temperature from sugar cane juice by a newly isolated *Kluyveromyces marxianus*. *Bioresour. Technol.*, *98*(17), 3367–3374. https://doi.org/10.1016/j.biortech.2006.10.044.
46. Moreno, A. D., Ibarra, D., Fernández, J. L., & Ballesteros, M., (2012). Different laccase detoxification strategies for ethanol production from lignocellulosic biomass by the thermotolerant yeast *Kluyveromyces marxianus* CECT 10875. *Bioresour. Technol.*, *106*, 101–109. https://doi.org/10.1016/j.biortech.2011.11.108.
47. Moreno, A. D., Ibarra, D., Ballesteros, I., González, A., & Ballesteros, M., (2013). Comparing cell viability and ethanol fermentation of the thermotolerant yeast *Kluyveromyces marxianus* and *Saccharomyces cerevisiae* on steam-exploded biomass treated with laccase. *Bioresour. Technol.*, *135*, 239–245. https://doi.org/10.1016/j.biortech.2012.11.095.
48. Nitiyon, S., Keo-oudone, C., Murata, M., Lertwattanasakul, N., Limtong, S., Kosaka, T., & Yamada, M., (2016). Efficient conversion of xylose to ethanol by stress-tolerant *Kluyveromyces marxianus* BUNL-21. *Springerplus*, *5*(1), 1–12. https://doi.org/10.1186/s40064-016-1881-6.
49. Pilap, W., Thanonkeo, S., Klanrit, P., & Thanonkeo, P., (2018). The potential of the newly isolated thermotolerant *Kluyveromyces marxianus* for high-temperature ethanol production using sweet sorghum juice. *3 Biotech, 8*(2). https://doi.org/10.1007/s13205-018-1161-y.
50. Saini, J. K., Agrawal, R., Satlewal, A., Saini, R., Gupta, R., Mathur, A., & Tuli, D., (2015). Second generation bioethanol production at high gravity of pilot-scale pretreated wheat straw employing newly isolated thermotolerant yeast *Kluyveromyces marxianus* DBTIOC-35. *RSC Adv.*, *5*(47), 37485–37494. https://doi.org/10.1039/c5ra05792b.
51. Signori, L., Passolunghi, S., Ruohonen, L., Porro, D., & Branduardi, P., (2014). Effect of oxygenation and temperature on glucose-xylose fermentation in *Kluyveromyces marxianus* CBS712 strain. *Microb. Cell Fact.*, *13*(1), 1–13. https://doi.org/10.1186/1475-2859-13-51.
52. Suryawati, L., Wilkins, M. R., Bellmer, D. D., Huhnke, R. L., Maness, N. O., & Banat, I. M., (2008). Simultaneous Saccharification and fermentation of Kanlow switchgrass pretreated by hydrothermolysis using *Kluyveromyces marxianus* IMB4. *Biotechnol. Bioeng.*, *101*(5), 894–902. https://doi.org/10.1002/bit.21965.
53. Tomás-Pejó, E., Oliva, J. M., González, A., Ballesteros, I., & Ballesteros, M., (2009). Bioethanol production from wheat straw by the thermotolerant yeast *Kluyveromyces marxianus* CECT 10875 in a simultaneous saccharification and fermentation fed-batch process. *Fuel, 88*(11), 2142–2147. https://doi.org/10.1016/j.fuel.2009.01.014.
54. Zhang, B., Zhang, J., Wang, D., Han, R., Ding, R., Gao, X., Sun, L., & Hong, J., (2016). Simultaneous fermentation of glucose and xylose at elevated temperatures co-produces ethanol and xylitol through overexpression of a xylose-specific transporter in engineered *Kluyveromyces marxianus*. *Bioresour. Technol.*, *216*, 227–237. https://doi.org/10.1016/j. biortech.2016.05.068.
55. Sandoval-Nuñez, D., Arellano-Plaza, M., Gschaedler, A., Arrizon, J., & Amaya-Delgado, L., (2018). A comparative study of lignocellulosic ethanol productivities by

Kluyveromyces marxianus and *Saccharomyces cerevisiae. Clean Technol. Environ. Policy, 20*(7), 1491–1499. https://doi.org/10.1007/s10098-017-1470-6.

56. Wang, D., Wu, D., Yang, X., & Hong, J., (2018). Transcriptomic analysis of thermotolerant yeast: *Kluyveromyces marxianus* in multiple inhibitors tolerance. *RSC Adv., 8*(26), 14177–14192. https://doi.org/10.1039/c8ra00335a.

57. Matsumoto, I., Arai, T., Nishimoto, Y., Leelavatcharamas, V., Furuta, M., & Kishida, M., (2018). Thermotolerant yeast *Kluyveromyces marxianus* reveals more tolerance to heat shock than the brewery yeast *Saccharomyces cerevisiae. Biocontrol Sci., 23*(3), 133–138. https://doi.org/10.4265/bio.23.133.

58. Mo, W., Wang, M., Zhan, R., Yu, Y., He, Y., & Lu, H., (2019). *Kluyveromyces marxianus* developing ethanol tolerance during adaptive evolution with significant improvements of multiple pathways. *Biotechnol. Biofuels, 12*(1), 1–15. https://doi.org/10.1186/s13068-019-1393-z.

59. Lertwattanasakul, N., Kosaka, T., Hosoyama, A., Suzuki, Y., Rodrussamee, N., Matsutani, M., Murata, M., et al., (2015). Genetic basis of the highly efficient yeast *Kluyveromyces marxianus*: Complete genome sequence and transcriptome analyses. *Biotechnol. Biofuels, 8*(1), 1–14. https://doi.org/10.1186/s13068-015-0227-x.

60. Li, P., Fu, X., Zhang, L., Zhang, Z., Li, J., & Li, S., (2017). The transcription factors Hsf1 and Msn2 of thermotolerant *Kluyveromyces marxianus* promote cell growth and ethanol fermentation of *Saccharomyces cerevisiae* at high temperatures. *Biotechnol. Biofuels, 10*(1), 1–13. https://doi.org/10.1186/s13068-017-0984-9.

61. Schabort, D. T. W. P., Kilian, S. G., & Du Preez, J. C., (2018). Gene regulation in *Kluyveromyces marxianus* in the context of chromosomes. *PLoS One, 13*(1), 1–16. https://doi.org/10.1371/journal.pone.0190913.

62. Gao, J., Feng, H., Yuan, W., Li, Y., Hou, S., Zhong, S., & Bai, F., (2017). Enhanced fermentative performance under stresses of multiple lignocellulose-derived inhibitors by overexpression of a typical 2-Cys peroxiredoxin from *Kluyveromyces marxianus. Biotechnol. Biofuels, 10*(1), 1–13. https://doi.org/10.1186/s13068-017-0766-4.

63. Kwon, D. H., Park, J. B., Hong, E., & Ha, S. J., (2019). Ethanol production from xylose is highly increased by the *Kluyveromyces marxianus* mutant 17694-DH1. *Bioprocess Biosyst. Eng., 42*(1), 63–70. https://doi.org/10.1007/s00449-018-2014-0.

64. Zhang, J., Zhang, B., Wang, D., Gao, X., Sun, L., & Hong, J., (2015). Rapid ethanol production at elevated temperatures by engineered thermotolerant *Kluyveromyces marxianus* via the NADP(H)-preferring xylose reductase-xylitol dehydrogenase pathway. *Metab. Eng., 31*, 140–152. https://doi.org/10.1016/j.ymben.2015.07.008.

65. Löbs, A. K., Engel, R., Schwartz, C., Flores, A., & Wheeldon, I., (2017). CRISPR-Cas9-enabled genetic disruptions for understanding ethanol and ethyl acetate biosynthesis in *Kluyveromyces marxianus. Biotechnol. Biofuels, 10*(1), 1–14. https://doi.org/10.1186/s13068-017-0854-5.

66. Li, P., Fu, X., Li, S., & Zhang, L., (2018). Engineering TATA-binding protein Spt15 to improve ethanol tolerance and production in *Kluyveromyces marxianus. Biotechnol. Biofuels, 11*(1), 1–13. https://doi.org/10.1186/s13068-018-1206-9.

67. Alvim, M. C. T., Vital, C. E., Barros, E., Vieira, N. M., Da Silveira, F. A., Balbino, T. R., Diniz, R. H. S., et al., (2019). Ethanol stress responses of *Kluyveromyces marxianus* CCT 7735 revealed by proteomic and metabolomic analyses. *Antonie Van Leeuwenhoek, Int. J. Gen. Mol. Microbiol., 112*(6), 827–845. https://doi.org/10.1007/s10482-018-01214-y.

68. Lee, K. J., & Rogers, P. L., (1983). The fermentation kinetics of ethanol production by *Zymomonas mobilis. Chem. Eng. J., 27*(2). https://doi.org/10.1016/0300-9467(83)80067-7.

69. Kremer, T. A., LaSarre, B., Posto, A. L., McKinlay, J. B., & Ingram, L. O., (2015). N$_2$ gas is an effective fertilizer for bioethanol production by *Zymomonas mobilis*. *Proc. Natl. Acad. Sci. U. S. A., 112*(7), 2222–2226. https://doi.org/10.1073/pnas.1420663112.

70. Wang, X., He, Q., Yang, Y., Wang, J., Haning, K., Hu, Y., Wu, B., et al., (2018). Advances and prospects in metabolic engineering of *Zymomonas mobilis*. *Metab. Eng., 50*, 57–73. https://doi.org/10.1016/j.ymben.2018.04.001.

71. Yang, S., Fei, Q., Zhang, Y., Contreras, L. M., Utturkar, S. M., Brown, S. D., Himmel, M. E., & Zhang, M., (2016). *Zymomonas mobilis* as a model system for production of biofuels and biochemicals. *Microb. Biotechnol., 9*(6), 699–717. https://doi.org/10.1111/1751-7915.12408.

72. Todhanakasem, T., Yodsanga, S., Sowatad, A., Kanokratana, P., Thanonkeo, P., & Champreda, V., (2018). Inhibition analysis of inhibitors derived from lignocellulose pretreatment on the metabolic activity of *Zymomonas mobilis* biofilm and planktonic cells and the proteomic responses. *Biotechnol. Bioeng., 115*(1), 70–81. https://doi.org/10.1002/bit.26449.

73. He, M. X., Wu, B., Qin, H., Ruan, Z. Y., Tan, F. R., Wang, J. L., Shui, Z. X., et al., (2014). *Zymomonas mobilis*: A novel platform for future biorefineries. *Biotechnol. Biofuels, 7*(1), 1–15. https://doi.org/10.1186/1754-6834-7-101.

74. Kannuchamy, S., Mukund, N., & Saleena, L. M., (2016). Genetic engineering of *Clostridium thermocellum* DSM1313 for enhanced ethanol production. *BMC Biotechnol., 16*(1), 1–6. https://doi.org/10.1186/s12896-016-0260-2.

75. Seo, J. S., Chong, H., Park, H. S., Yoon, K. O., Jung, C., Kim, J. J., Hong, J. H., et al., (2005). The genome sequence of the ethanologenic bacterium *Zymomonas mobilis* ZM4. *Nat. Biotechnol., 23*(1), 63–68. https://doi.org/10.1038/nbt1045.

76. Yang, S., Pappas, K. M., Hauser, L. J., Land, M. L., Chen, G. L., Hurst, G. B., Pan, C., et al., (2009). Improved genome annotation for *Zymomonas mobilis*. *Nat. Biotechnol., 27*(10), 893–894. https://doi.org/10.1038/nbt1009-893.

77. Anggarini, S., Murata, M., Kido, K., Kosaka, T., Sootsuwan, K., Thanonkeo, P., & Yamada, M., (2020). Improvement of thermotolerance of *Zymomonas mobilis* by genes for reactive oxygen species-scavenging enzymes and heat shock proteins. *Front. Microbiol., 10*, 1–14. https://doi.org/10.3389/fmicb.2019.03073.

78. Charoensuk, K., Irie, A., Lertwattanasakul, N., Sootsuwan, K., Thanonkeo, P., & Yamada, M., (2011). Physiological Importance of cytochrome c peroxidase in ethanologenic thermotolerant *Zymomonas mobilis*. *J. Mol. Microbiol. Biotechnol., 20*(2), 70–82. https://doi.org/10.1159/000324675.

79. Yang, S., Tschaplinski, T. J., Engle, N. L., Carroll, S. L., Martin, S. L., Davison, B. H., Palumbo, A. V., et al., (2009). Transcriptomic and metabolomic profiling of *Zymomonas mobilis* during aerobic and anaerobic fermentations. *BMC Genomics, 10*. https://doi.org/10.1186/1471-2164-10-34.

80. Jönsson, L. J., & Martín, C., (2016). Pretreatment of lignocellulose: Formation of inhibitory by-products and strategies for minimizing their effects. *Bioresour. Technol., 199*, 103–112. https://doi.org/10.1016/j.biortech.2015.10.009.

81. Yang, S., Vera, J. M., Grass, J., Savvakis, G., Moskvin, O. V., Yang, Y., McIlwain, S. J., et al., (2018). Complete genome sequence and the expression pattern of plasmids of the model ethanologen *Zymomonas mobilis* ZM4 and its xylose-utilizing derivatives 8b and 2032. *Biotechnol. Biofuels, 11*(1), 1–20. https://doi.org/10.1186/s13068-018-1116-x.

82. Yi, X., Gu, H., Gao, Q., Liu, Z. L., & Bao, J., (2015). Transcriptome analysis of *Zymomonas mobilis* ZM4 reveals mechanisms of tolerance and detoxification of phenolic aldehyde

inhibitors from lignocellulose pretreatment. *Biotechnol. Biofuels, 8*(1), 1–15. https://doi.org/10.1186/s13068-015-0333-9.

83. He, M. X., Wu, B., Shui, Z. X., Hu, Q. C., Wang, W. G., Tan, F. R., Tang, X. Y., et al., (2012). Transcriptome profiling of *Zymomonas mobilis* under furfural stress. *Appl. Microbiol. Biotechnol., 95*(1), 189–199. https://doi.org/10.1007/s00253-012-4155-4.

84. Glebes, T. Y., Sandoval, N. R., Reeder, P. J., Schilling, K. D., Zhang, M., & Gill, R. T., (2014). Genome-wide mapping of furfural tolerance genes in *Escherichia coli*. *PLoS One, 9*(1). https://doi.org/10.1371/journal.pone.0087540.

85. Qi, L., Zhang, K., Wang, Y. T., Wu, J. K., Sui, Y., Liang, X. Z., Yu, L. Z., et al., (2019). Global analysis of furfural induced genomic instability using a yeast model. *Appl. Environ. Microbiol., 85*(18), 1–15. https://doi.org/10.1128/AEM.01237-19.

86. He, M. X., Wu, B., Shui, Z. X., Hu, Q. C., Wang, W. G., Tan, F. R., Tang, X. Y., et al., (2012). Transcriptome profiling of *Zymomonas mobilis* under ethanol stress. *Biotechnol. Biofuels, 5*, 1–10. https://doi.org/10.1186/1754-6834-5-75.

87. Brás, J. L. A., Pinheiro, B. A., Cameron, K., Cuskin, F., Viegas, A., Najmudin, S., Bule, P., et al., (2016). Diverse specificity of cellulosome attachment to the bacterial cell surface. *Sci. Rep., 6*, 1–12. https://doi.org/10.1038/srep38292.

88. Singh, N., Mathur, A. S., Gupta, R. P., Barrow, C. J., Tuli, D., & Puri, M., (2018). Enhanced cellulosic ethanol production via consolidated bioprocessing by *Clostridium thermocellum* ATCC 31924. *Bioresour. Technol., 250*, 860–867. https://doi.org/10.1016/j.biortech.2017.11.048.

89. Zhu, X., Cui, J., Feng, Y., Fa, Y., Zhang, J., & Cui, Q., (2013). Metabolic adaption of ethanol-tolerant *Clostridium thermocellum*. *PLoS One, 8*(7), 1–9. https://doi.org/10.1371/journal.pone.0070631.

90. Tian, L., Perot, S. J., Hon, S., Zhou, J., Liang, X., Bouvier, J. T., Guss, A. Met al., (2017). Enhanced ethanol formation by *Clostridium thermocellum* via pyruvate decarboxylase. *Microb. Cell Fact., 16*(1), 1–10. https://doi.org/10.1186/s12934-017-0783-9.

91. Holwerda, E. K., Olson, D. G., Ruppertsberger, N. M., Stevenson, D. M., Murphy, S. J. L., Maloney, M. I., Lanahan, A. A., et al., (2020). Metabolic and evolutionary responses of *Clostridium thermocellum* to genetic interventions aimed at improving ethanol production. *Biotechnol. Biofuels, 13*(1), 1–20. https://doi.org/10.1186/s13068-020-01680-5.

92. Xu, Q., Resch, M. G., Podkaminer, K., Yang, S., Baker, J. O., Donohoe, B. S., Wilson, C., et al., (2016). Cell biology: Dramatic performance of *Clostridium thermocellum* explained by its wide range of cellulase modalities. *Sci. Adv., 2*(2). https://doi.org/10.1126/sciadv.1501254.

93. Newcomb, M., Chen, C. Y., & Wu, J. H. D., (2007). Induction of the CelC operon of *Clostridium thermocellum* by laminaribiose. *Proc. Natl. Acad. Sci. U. S. A, 104*(10), 3747–3752. https://doi.org/10.1073/pnas.0700087104.

94. Newcomb, M., Millen, J., Chen, C. Y., & Wu, J. H. D., (2011). Co-transcription of the CelC gene cluster in *Clostridium thermocellum*. *Appl. Microbiol. Biotechnol., 90*(2), 625–634. https://doi.org/10.1007/s00253–011–3121-x.

95. Wilson, C. M., Klingeman, D. M, Schlachter, C., Syed, M. H., Wu, C., Guss, A. M., Brown, S. D., (2017). LacI transcriptional regulatory networks in *Clostridium thermocellum* DSM1313. *Appl Environ Microbiol., 83*, e02751–16. https://doi.org/10.1128/AEM.02751-16.

96. Choi, J., Klingeman, D. M., Brown, S. D., & Cox, C. D., (2017). The LacI family protein GlyR3 co-regulates the CelC operon and ManB in *Clostridium thermocellum*. *Biotechnol. Biofuels, 10*(1), 1–11. https://doi.org/10.1186/s13068-017-0849-2.

97. Nataf, Y., Bahari, L., Kahel-Raifer, H., Borovok, I., Lamed, R., Bayer, E. A., Sonenshein, A. L., & Shoham, Y., (2010). *Clostridium thermocellum* cellulosomal genes are regulated by extracytoplasmic polysaccharides via alternative sigma factors. *Proc. Natl. Acad. Sci. U. S. A., 107*(43), 18646–18651. https://doi.org/10.1073/pnas.1012175107.

98. Ortiz De, O. L., Lamed, R., Liu, Y. J., Xu, J., Cui, Q., Feng, Y., Shoham, Y., et al., (2018). Regulation of biomass degradation by alternative σ factors in cellulolytic clostridia. *Sci. Rep., 8*(1), 1–11. https://doi.org/10.1038/s41598-018-29245-5.

99. Nataf, Y., Yaron, S., Stahl, F., Lamed, R., Bayer, E. A., Scheper, T. H., Sonenshein, A. L., & Shoham, Y., (2009). Cellodextrin and laminaribiose ABC transporters in *Clostridium thermocellum. J. Bacteriol., 91*(1), 203–209. https://doi.org/10.1128/JB.01190-08.

100. Olson, D. G., Maloney, M., Lanahan, A. A., Hon, S., Hauser, L. J., & Lynd, L. R., (2015). Identifying promoters for gene expression in *Clostridium thermocellum. Metab. Eng. Commun., 2*, 23–29. https://doi.org/10.1016/j.meteno.2015.03.002.

101. Tian, L., Cervenka, N. D., Low, A. M., Olson, D. G., & Lynd, L. R., (2019). A mutation in the AdhE alcohol dehydrogenase of *Clostridium thermocellum* Increases tolerance to several primary alcohols, including isobutanol, n-butanol and ethanol. *Sci. Rep., 9*(1), 1–7. https://doi.org/10.1038/s41598-018-37979-5.

102. Williams, T. I., Combs, J. C., Lynn, B. C., & Strobel, H. J., (2007). Proteomic profile changes in membranes of ethanol-tolerant *Clostridium thermocellum. Appl. Microbiol. Biotechnol., 74*(2), 422–432. https://doi.org/10.1007/s00253-006-0689-7.

103. Shao, X., Raman, B., Zhu, M., Mielenz, J. R., Brown, S. D., Guss, A. M., & Lynd, L. R., (2011). Mutant selection and phenotypic and genetic characterization of ethanol-tolerant strains of *Clostridium thermocellum. Appl. Microbiol. Biotechnol., 92*(3), 641–652. https://doi.org/10.1007/s00253-011-3492-z.

104. Cui, J., Stevenson, D., Korosh, T., Amador-Noguez, D., Olson, D. G., & Lynd, L. R., (2020). Developing a cell-free extract reaction (CFER) system in *Clostridium thermocellum* to identify metabolic limitations to ethanol production. *Front. Energy Res., 8*. https://doi.org/10.3389/fenrg.2020.00072.

105. Stevenson, D. M., & Weimer, P. J., (2005). Expression of 17 genes in. *Society, 71*(8), 4672–4678. https://doi.org/10.1128/AEM.71.8.4672.

106. Riederer, A., Takasuka, T. E., Makino, S. I., Stevenson, D. M., Bukhman, Y. V., Elsen, N. L., & Fox, B. G., (2011). Global gene expression patterns in *Clostridium thermocellum* as determined by microarray analysis of chemostat cultures on cellulose or cellobiose. *Appl. Environ. Microbiol., 77*(4), 1243–1253. https://doi.org/10.1128/AEM.02008-10.

107. Wilson, C. M., Rodriguez, M., Johnson, C. M., Martin, S. L., Chu, T. M., Wolfinger, R. D., Hauser, L. J., et al., (2013). Global transcriptome analysis of *Clostridium thermocellum* ATCC 27405 during growth on dilute acid pretreated Populus and switchgrass. *Biotechnol. Biofuels, 6*(1), 1–18. https://doi.org/10.1186/1754-6834-6-179.

108. Raman, B., McKeown, C. K., Rodriguez, M., Brown, S. D., & Mielenz, J. R., (2011). Transcriptomic analysis of *Clostridium thermocellum* ATCC 27405 cellulose fermentation. *BMC Microbiol., 11*. https://doi.org/10.1186/1471-2180-11-134.

109. Yao, S., & Mikkelsen, M. J., (2010). Metabolic engineering to improve ethanol production in *Thermoanaerobacter mathranii. Appl. Microbiol. Biotechnol., 88*(1), 199–208. https://doi.org/10.1007/s00253-010-2703-3.

110. Xue, Y., Xu, Y., Liu, Y., Ma, Y., & Zhou, P., (2001). *Thermoanaerobacter tengcongensis* Sp. Nov., a novel anaerobic, saccharolytic, thermophilic bacterium isolated from a hot

spring in tengcong, China. *Int. J. Syst. Evol. Microbiol., 51*(4), 1335–1341. https://doi.org/10.1099/00207713-51-4-1335.

111. Slobodkin, A. I., Tourova, T. P., Kuznetsov, B. B., Kostrikina, N. A., Chernyh, N. A., & Bonch-Osmolovskaya, E. A., (1999). *Thermoanaerobacter siderophilus* sp. Nov., a novel dissimilatory Fe(III)- reducing, anaerobic, thermophilic bacterium. *Int. J. Syst. Bacteriol., 49*(4), 1471–1478. https://doi.org/10.1099/00207713-49-4-1471.

112. Carlier, J. P., Bonne, I., & Bedora-Faure, M., (2006). Isolation from canned foods of a novel thermoanaerobacter species phylogenetically related to *Thermoanaerobacter mathranii* (Larsen 1997): Emendation of the species description and proposal of *Thermoanaerobacter mathranii* subsp. *Alimentarius* subsp. Nov. *Anaerobe, 12*(3), 153–159. https://doi.org/10.1016/j.anaerobe.2006.03.003.

113. Sittijunda, S., Tomás, A. F., Reungsang, A., O-thong, S., & Angelidaki, I., (2013). Ethanol Production from glucose and xylose by immobilized *Thermoanaerobacter pentosaceus* at 70°C in an up-flow anaerobic sludge blanket (UASB) reactor. *Bioresour. Technol., 143*, 598–607. https://doi.org/10.1016/j.biortech.2013.06.056.

114. Michael, S. S., & Orlygsson, J., (2019). Progress in second generation ethanol production with thermophilic bacteria. *Fuel Ethanol Prod. from Sugarcane.* https://doi.org/10.5772/intechopen.78020.

115. Desai, S. G., Guerinot, M. L., & Lynd, L. R., (2004). Cloning of L-lactate dehydrogenase and elimination of lactic acid production via gene knockout in *Thermoanaerobacterium saccharolyticum* JW/SL-YS485. *Appl. Microbiol. Biotechnol., 65*(5), 600–605. https://doi.org/10.1007/s00253-004-1575-9.

116. Pei, J., Zhou, Q., Jiang, Y., Le, Y., Li, H., Shao, W., & Wiegel, J., (2010). *Thermoanaerobacter* Spp. control ethanol pathway via transcriptional regulation and versatility of key enzymes. *Metab. Eng., 12*(5), 420–428. https://doi.org/10.1016/j.ymben.2010.06.001.

CHAPTER 5

Metabolic Engineering of Yeast, *Zymomonas mobilis*, and *Clostridium thermocellum* to Increase Yield of Bioethanol

S. SÁNCHEZ-MUÑOZ,[1] M. J. CASTRO-ALONSO,[1] F. G. BARBOSA,[1]
E. MIER-ALBA,[1] T. R. BALBINO,[1] D. RUBIO-RIBEAUX,[1]
I. O. HERNÁNDEZ-DE LIRA,[2] J. C. SANTOS,[3] C. N. AGUILAR,[4] and
S. S. DA SILVA[1]

[1]*Bioprocesses and Sustainable Products Laboratory,
Department of Biotechnology, Engineering School of Lorena,
University of São Paulo (EEL-USP), Lorena–12.602.810, SP, Brazil,
E-mail: silviosilverio@gmail.com (S. S. Da Silva)*

[2]*Bioremediation Laboratory, Biological Science Faculty,
Autonomous University of Coahuila (UAdeC), Torreón Campus,
Coahuila–27276, México*

[3]*Bioprocesses, Biopolymers, Simulation, and Modeling Laboratory,
Department of Biotechnology, Engineering School of Lorena,
University of São Paulo (EEL-USP), Lorena–12.602.810, SP, Brazil*

[4]*Bioprocesses and Bioproducts Group, Food Research Department,
School of Chemistry, Autonomous University of Coahuila (UAdeC),
Saltillo Campus, Coahuila–25280, Mexico*

ABSTRACT

The price fluctuation of petroleum-based fuels makes biofuels one of the most promising alternative energies for global economies. Since the biological production of fuels, from vegetal biomass, offers sustainable and economically attractive options compared to petroleum-based production. However, the complexity of vegetal biomass does not allow the process to be simple. Furthermore, there are still many important biological and technological barriers for the processing of biofuels being competitive. Besides, bioprocesses normally required multiple steps of feedstock pretreatment and subsequent conversion to fuel. Those steps are being consolidated into single microbial processes using metabolically engineered species (e.g., *Saccharomyces cerevisiae*, *Zimmomona mobilis*, and *Clostridium thermocellum*). This chapter will review the advance in metabolic engineering and synthetic biology on the main ethanol producers' microorganisms, which allow the development of new engineered systems aiming for a better transition from fossil fuels to biofuels.

5.1 INTRODUCTION

Economical fluctuations and CO_2 emissions of oil-based fuels are increasing day by day. These factors control the equilibrium of developing economies and produce asymmetric responses in social and economic growth, also harming the environment [1, 2]. Thus, biofuels production as alternative energy gained attention and potentialized efforts in many research areas [3]. However, the possibility of a sustainable power source has been around for a long time but did not receive considerable governmental attention mainly because of low-cost oil since last 50 years. Furthermore, industrial 2G bioprocesses are not well integrated to achieve cost-effective bioproducts in established biorefineries [4, 5].

To compete with oil-based energies, the biofuels process can take advantage of 2G biomass and must be enhanced in relevant steps, mainly in the synthesis of biomolecules by microorganisms. Although microbes have the inherent metabolic pathways for generation of these valuable molecules, the natural biofuel synthesis is significantly low and limits its production and commercialization at the industrial level [6]. Thus, for achieving those goals, different strategies in synthetic biology and metabolic engineering fields have been studied. Also, those areas could offer the sustainability factor and the possibility of producing new molecules to become a profitable bioprocess [7, 8]. These disciplines have been applied to improve the microbial production level of advanced biofuels through the over-expression of specific regulators and target enzymes, heterologous gene expression, orthogonal pathway construction, protein engineering, co-factor balancing, blocking of competitor pathways and down-regulating genes, among others [6, 8–10].

Established microbial industrial hosts like the bacterium *Escherichia coli* and the yeast *Saccharomyces cerevisiae* are preferred for metabolic engineering systems, because of their well-known genetic handling with available 'omics' (genomic, proteomics, transcriptomics, or metabolomics) databases, growth speed, low incubation costs, and their now established use in industrial bioprocesses [6, 11, 12]. However, other microorganisms such as the bacterium *Zymomonas mobilis* and the fungus *Clostridium thermocellum* are of great interest in research field, because of their versatility to produce several biomolecules under industrial conditions, that made them excellent alternatives for biotechnology [13, 14].

In this chapter, we have addressed the current state of metabolic engineering in biofuel production and described the recent progress made in producing bioethanol from the main genetically modified microorganisms.

5.2 METABOLIC ENGINEERING TO ENHANCE BIOPROCESSES

Biological organisms, enzymes, and their genetic modification formed the basis of a growing collection of techniques grounded in molecular biology and cell biology. As a result, biotechnology, and bioprocesses are knowledge areas that industries and researchers have worked on, looking up to develop and commercialize the building blocks of life to offer products and services [15, 16]. For this purpose, metabolic engineering is the science that makes possible the correct performance of the biotechnological field, rewriting the metabolism of cells for different aims in each bioprocess [7].

Bioprocesses are used to produce pharmaceuticals, enzymes, vaccines, foods, flavors, pigments, polymers, amino acids, fuels, and other important chemical molecules [17, 18]. Briefly, a bioprocess usually consists of feedstock preparation and pretreatment (upstream process), fermentation or biocatalysis (core process), and separation for product recovery and purification (downstream process) [19, 20]. Also, it could involve genetic engineering for the manipulation of plants, animals, and microorganisms such as yeasts, bacteria, and fungi, since those microorganisms may not be able or well-performed for the production of our desired product [19, 21, 22]. Furthermore, the metabolic characteristics of living organisms often impose challenges on bioprocessing, thus a wide study of cells is always an important prerequisite for successful engineering design [17].

Due to diverse research developments, several synthetic gene constructs and circuits have been designed within a wide number of host microorganisms leading to a positive impact on different areas [23]. Nevertheless, metabolic engineered tools applied to industrial strains to generate high yield in bioprocess still have important challenges to consider them as promising alternatives in most research areas [24].

There are some examples that can be listed to show the potential of metabolic engineering in different steps of bioprocesses. One of them is the conversion of lignocellulosic biomass (LCB) (upstream process) into monosaccharides using heterologous cellulases, a main critical step in biorefineries, as its enhancement can help in cost saving achievements [25, 26]. Another important step is fermentation (core process), and at this process level, several strategies could be applied. For example, deletion of genes or sequences that interrupt strategic cell responses, such as, autophagy, that is a recycling process of cellular components; in some cases, its interruption enhance the biomass growth rate (ex. yeast) and consequently, a positive influence in the production of some products like ethanol could occur [27–29]. Fermentation has been well-studied for the application of several molecular tools. The most used one could be the overexpression of specific enzymes. For example, in citric acid production, pyruvate carboxylase is a key enzyme in the reduction of the tricarboxylic acid cycle, and it is closely related to the formation of this important biomolecule, thus the overexpression of this enzyme could lead to a higher production of this organic acid [30, 31]. Other examples of metabolic engineering tools applied to microorganisms for enhancing the production of biomolecules are shown in Table 5.1.

5.3 ROLE OF GENETIC ENGINEERING IN SUSTAINABLE BIOFUELS/ETHANOL PRODUCTION: BOTTLENECKS AND MODEL STRAINS

The progress in metabolic engineering allows the reconstruction and development of "microbial cell factories" as a promising strategy for large scale production of ecofriendly fuels [38]. The fermentation processes through wild-type microorganisms produce low titer of biofuels since the characteristics of end-product and conditions of processes generate several stresses, making the microbial adaptability to large-scale industrial processes very difficult [39]. Thereby, industrial biofuel production requires microorganisms with ability to tolerate several stressing conditions during bioprocesses, such as osmotolerance, shearing forces, organic acids, and high temperature. The physiological responses related to those stress conditions can be improved by genetic manipulation of a single or a few genes [8]. Hence, several conventional genetic engineering tools and synthetic biology techniques, such as genetic transformation, gene targeting, use

TABLE 5.1 Metabolic Engineering Strategies Applied to Several Microorganisms for Enhancing the Production of Trending Industrial Compounds (B: bacteria; Y: yeast; F: fungus)

Product	Strain	Genetic Modification	Enhancement*	Application	References
Clavulanic acid	*Streptomyces clavuligerus* F613-1 (B)	Co-transcription of the late-stage gene *claR* with the reporter gene *neo*	33%	Antibiotic	[24]
Coenzyme B12	*Pseudomonas denitrificans* ATCC 13867 (B)	Removal of riboswitches and the replacement of promoters in operons of cluster I	2-fold	Essential cofactor in many biological rearrangement reactions	[32]
Glargine (insulin analog)	*Escherichia coli* JM109 (B)	Heterologous expression of the prepeptide glargine insulin	–	Diabetes mellites	[12]
Ethanol	*Saccharomyces cerevisiae* (Y)	Replacement of the regulatory gene *PHO4*	5.3%	Fuel	[33]
Citric acid	*Yarrowia lipolytica* (Y)	Over expression of pyruvate carboxylase gene	98.5%	Additive for several products	[34]
Ethanol	*Saccharomyces cerevisiae* (Y)	Disruption of ATG32 region	2.76%	Beverages (sake)	[29]
Cellulases/ Xylanases	*Trichoderma reesei* RUT C30 (F)	Construction of transcriptional activation linked to a domain of herpes simplex virus protein VP16	50–80%	Saccharification of lignocellulosic biomass	[35]
Pullulan	*Aureobasidium pullulans* NRRL Y-12974 (F)	Homologous expression of UGPase	1.7-fold	Blood plasma substitutes, food preservation, adhesive, and others	[36]
Gibberellin (GA$_3$)	*Fusarium fujikuroi* (F)	^{60}Co γ-ray radiation combined with lithium chloride treatment to generate mutants	21-fold	Growth hormones for plants	[37]

*Enhancement of production of modified strain compared with wild strain.

of different promoters, clustered regularly interspaced short palindromic repeats (CRISPR) and RNA-interference (RNAi) among others, had been widely used to address bottlenecks to achieve large-scale production of biofuels [40–42]. The major advances of conventional and modern genetic tools to improve industrial biofuels production are discussed below.

5.3.1   GENETIC ENGINEERING TO ENHANCING SOLVENT TOLERANCE OF BIOFUELS

One of the major challenges for achieve high levels of biofuels production is that fermentation performance is affected by toxicity of end products [43]. The toxicity effects of biofuels, particularly organic solvents, are related to the logarithm of its octanol-water partition coefficient (*logP*). Organic solvents with *logP* values between a range of 0.8 and 5 cause disruption of cellular membranes and interferes with several vital metabolic functions, such as membrane transport and energy generation. Moreover, solvents generate damage to biological macromolecules, such as DNA and RNA [44]. Several genetic engineering strategies had been employed to attenuate the toxicity of organic solvents and biofuels, and to elucidate stress responses of microorganism through the overexpression of heat shock protein genes [45], redirection of carbon metabolic flux [46], regulation extrusion of solvents efflux pumps genes [47] among others.

HSP are involved in the synthesis, transport, and folding of proteins closely related to metabolic activities [43, 48]. The role of HSP in enhancement of biofuel tolerance was showed for the first time in *Clostridrium acetobutylicum*. In that study, the overexpression of chaperone *GroESL* improved tolerance to n-butanol (85%) [45]. Another work from this research group demonstrated that the overexpression of *GroESL*, also increased the expression of other HSP involved in solvent tolerance, including *dnaKJ, hsp18, hsp90* [49]. In other studies, with isobutanol, the microbial transcriptional profiles revealed main genes related to heat shock stress and protein misfolding, including *rpoH, dnaJ, htpG*, and *ibpAB* [50, 51]. The same strategy was applied in *Escherichia coli, Lactobacillus plantarum* and *Zymomonas mobilis* and showed significant improvement in survival and viability of cells in presence of ethanol, 1-butanol, and 2,3-butanediol by the overexpression of *GroESL, ClpB* chaperones and *SecB* multi-tasking chaperone [44, 52–55]. On the other hand, Zhang et al. [56] showed that genes involved in central carbon flux and TCA cycle (*gapA* and *sdhB*) are closed related to end product tolerance. Genes *gapA* (glyceraldehyde-3-phosphate dehydrogenase A) and *sdhB* (FeS subunit of succinate dehydrogenase) provides high energy storage to regulate the solvent stress.

Redirection of carbon metabolic flux can be performed by CRISPR systems to improve the tolerance and production of biofuels. Wang et al. [57] applied CRISPR interference in *Klebsiella pneumoniae* and found that amino acid synthesis pathways interfere with n-butanol synthesis. The repression of the main genes that encode valine, leucine, isoleucine, threonine, and alanine pathways increased n-butanol tolerance and led to higher production (154%) of this fuel. Similarly, Otoupal and Chatterjee [58], using the CRISPR perturbations system in *E. coli*, identified the unknown genes *yjjZ* and *yehS* with strong potential to improve tolerance to n-butanol and n-hexane. At the moment, it is known that *yjjZ* gene plays a regulatory role in cellular metabolism as small RNA (sRNA) [58–60]. In another work, Huang and Geng [61] developed a novel replicative CRISPR/Cas9 plasmids system (pdC99 and pPC9) to integrate the 2,3-butanediol pathway in *Saccharomyces cerevisiae* that redirected carbon metabolic flux to improve tolerance and lead to higher production of 2,3 butanediol (50.5 g/L).

Efflux pumps are membrane transporters responsible for recognizing and exporting toxic compounds from the cytoplasm to out of cell. Those transporters are comprised by three types of proteins: inner membrane proteins, periplasmic linkers, and outer membrane channels [43]. Several studies observed that overexpression of exogenous efflux pumps and the cloning of their operons are effective strategies to enhance the tolerance to biofuels. Early studies reported that genes encoding the efflux pump SrpABC of *Pseudomonas putida* from the RND family were expressed in the presence of n-butanol [62, 63]. Since the elucidation of this mechanism, other studies revealed that the overexpression of efflux pump srpABC or the srpB subunit from *P. putida* in *E. coli* enhances n-butanol tolerance (20–40%) [64, 65]. Similarly, the co-overexpression of the membrane ATP binding cassette ABC and the efflux pumps *Snp2* and *Pdr5* in *S. cerevisiae* increased the tolerance and cellular growth when it was underexposed to exogenous n-decane [66]. Reyes et al. [47], found that overexpression of another membrane transporter associated to the genes *ygfO, setA, mdtA,* and *pgsA* improve n-butanol tolerance in *E. coli.* Furthermore, this study also exhibited that underexposure to n-butanol can lead to adaptive structural changes of the membrane lipid structure, since *pgsA* gene is involved in cardiolipin biosynthesis. In a recent work, a negative feedback network was introduced to *E. coli* strain using promoter *PgntK,* which regulated the expression of the butanol efflux pump AcrB and controls membrane protein expression to optimize growth. That study reported an increase in tolerance

(40%) and production (35%) of n-butanol [67]. Other studies introduced efflux pump *AcrB* in *E. coli* through directed evolution strategy and showed increase significant tolerance to n-butanol (>25%), α-pinene (400%) and n-octane (47%) [68, 69]. In contrast, He et al. [70] found that deletion of *acrB* gene in *E. coli* also raised tolerance of n-butanol in terms of cell density, which was improved by 82.8%.

5.3.2 GENETIC ENGINEERING FOR IMPROVED INHIBITOR TOLERANCE

Inhibitors derived from fermentation process due to the pre-processing of lignocellulosic material limits the metabolism of strains in terms of microbial growth and biofuel production [71]. These inhibitors can be categorized into three major types: (i) furan derivatives, (ii) short-chain aliphatic acids and (iii) phenolic compounds, as reviewed by Palmqvist and Hahn-Hägerdal [72]. Among these compounds, furan derivatives such as furfural and 5-hydroxymethylfurfural (HMF) are some of the most toxic substances since they induced serious damages to cellular metabolism. Furfural and HMF inhibits glycolytic and fermentative enzymes (pyruvate, acetaldehyde, and alcohol dehydrogenases (ADHs)) and cause breaks in double-stranded DNA [73]. Thus, furan derivates also generate reduction in concentration of intracellular ATP and NAD(P)H because the energy flux is redirected to pentose phosphate pathway (PPP) for repairing the induced damages [74, 75]. Moreover, furfural also affect mitochondria and vacuole membranes, as it induces the formation of intracellular ROS [76]. All the cellular damages mentioned above result in prolongation of lag phase of microbial growth and decrease specific growth rate, which lead to low yield of biofuel production [46]. In addition, furfural also increase the toxicity of acetic acid and phenolic compounds [77, 78].

Extensive studies have been performed to improved microbial tolerance to these inhibitors through classical and novel mutagenesis strategies. Miller et al. [75] reported that silencing of NADPH-dependent oxidoreductases (*yqhD* and *dkgA*) increased furfural tolerance in *E. coli*. In contrast, several recent studies showed that overexpression of oxidoreductases and proteins involved in oxidative stresses improved inhibitor tolerance. Co-overexpression of oxidoreductases (yqhD and FucO) showed increased tolerance to furfural, improved glucose utilization, and enhance the production of isobutanol (110%) [79]. Similarly, Song et al. [80] employed the strategy of increasing NAD(P)H pool through overexpression of *pncB* and *nadE* genes. These genes are involved in nicotine amide salvage pathway that led to reduce furfural toxicity and improved the production of isobutanol (2.5-fold higher). Oh et al. [81] found that overexpression *RCK1* gene encoding for a protein kinase involved in oxidative stress improved acetic acid tolerance under glucose and xylose conditions and reduce 40% of intracellular ROS levels. More recently, Liu et al. [82] showed that the manipulation of intracellular redox potential of genes encoding cofactors related oxidoreductases in *Z. mobilis* is a promising novel strategy to improve tolerance to acetic acid, furfural, and phenolic compounds in industrial biofuels production. In other studies, glycerol supplementation results in other promising strategy to furfural detoxification [83, 84]. Agu et al. [84] demonstrated that overexpression of two glycerol dehydrogenases dhaD1 and gldA1 reduced 68% of furfural toxicity in *Clostridium beijerinckii*.

5.3.3   GENETIC ENGINEERING TO DEVELOPMENT OF THERMOSTABILITY

The temperature of industrial fermentation process should be up to 40°C, which prevents contamination and reduces cooling costs by decreasing water and energy consumption. Moreover, high operating temperature improves productivities in simultaneous saccharification and fermentation (SSF) process because of the optimal temperature for enzymes that catalyze the saccharification of biomass at $\geq 50°C$ [85–87]. However, this temperature is higher in comparison with optimal growth temperatures of mesophilic microorganisms, typically used in industrial processes of biofuels production, e.g., *E. coli*.

Higher temperature causes serious cellular damages such as degradation of cytoskeleton proteins, morphological abnormalities, and inhibition of cell division and growth [86–88]. Thus, development of thermotolerant microbial strains is other major challenge for the first and second-generation biofuels production [8]. Adaptation laboratory evolution, random mutagenesis, and CRISPR had been employed to elucidate mechanisms involved in microbial thermostability. Several early genetic studies revealed that microorganism acquired thermotolerance to $\geq 50°C$ through accumulation of trehalose and overexpression of heat stress proteins such as chaperones, ubiquitin, among others [89–91]. In other studies, single amino acid modification in the pyruvate kinase, C-5 sterol desaturase and NADH dehydrogenase led to increase the thermotolerance and biofuel production in *S. cerevisiae* and *Z. mobilis* [85, 92]. Likewise, Nasution et al. [93] showed that the deletion of *Dfg5* gene encoding glycosyl phosphatidylinositol-anchored membrane protein improved thermotolerance and decreased level of ROS.

In last years, researches had been developed artificial systems using synthetic biology to improve the thermostability of microorganisms through the incorporation of systems constituted by heat shock and antioxidant proteins. Jia et al. [94] developed an intelligent microbial heat-regulating engine (IMHeRE) to improve the thermo-robustness of *E. coli* through the integration of a thermotolerant system and a quorum-regulating system. At cellular level, the thermotolerant system composed of different HSP and RNA thermometers hierarchically led to increase the optimum temperature by sensing heat changes. At community level, the quorum-regulating system dynamically regulates the altruistic sacrifice of individual cells to reduce metabolic heat release by detection of temperature and cell density. This study showed that the synthetic IMHeRE system improves cell growth (10%) at >40°C. Xu et al. [95] engineered an artificial antioxidant defense system to improve thermo-tolerance of yeast. They introduced several antioxidant genes from *S. cerevisiae* and *Thermus thermophiles* HB8 to construct "Angel yeast." This synthetic yeast showed an increase of thermotolerance in terms of cell density (65.2%) at >40°C. Li et al. [87] used CRISPR/Cas-based gene activation screening in *S. cerevisiae* and identified the essential role of delta-9 desaturase gene *OLE1* in increasing fatty acid unsaturation and reduction of lipid peroxidation caused by heat stress at 42°C. Similarly, Li et al. [96] elucidated the role of 4-Hydroxy-2,2,6,6-tetramethylpiperidin-1-oxyl (TEMPOL) as a stress-alleviating agent in *S. cerevisiae*. TEMPOL is a redox-cycling nitroxide and membrane-permeable antioxidant escorted by dual redox potential forms. This system enhanced expression of important genes and the activity of enzymatic antioxidant defense system, altering the ratio of cofactors to improve the non-enzymatic antioxidant defense system, or by directly degrading ROS with the help of NADH. Recently, Tao et al. [97] used 'Cas9 nickase-based genome editing' and found that the inactivation of the gene *MspI* improved the thermotolerance of *C. cellulolyticum*. Despite the *MspI* gene belongs to the restriction-modification system, this study revealed that the disruption of *MspI* gene can influence the expression of other thermotolerant genes.

5.3.4 MOLECULAR TOOLS APPLIED TO VEGETABLE BIOMASS

Agricultural wastes and byproducts are complex substrates mainly composed of cellulose, hemicellulose, and lignin, which proportions depend on the source and species of plants [98]. Cellulose and hemicellulose are polymers composed of hexoses (glucose, mannose, and galactose), pentoses (arabinose and xylose), and sugar acids (uronic acids). Lignin is an amorphous macromolecule, composed by phenylpropanoids (guacyl and siringyl units). Lignin together with cellulose and hemicellulose forms a closed structure that confers biological, chemical, and mechanical resistance to cell walls [99]. These complex structures are one of the main bottlenecks in biorefinery, which affect the production cost of biofuels. Lignin and cellulose also limit the accessibility of enzymes to sugar polymers, thus restricting their hydrolysis. Hence, lignocellulosic pretreatments are a key step in bioprocessing.

Hydrolysates obtained from sugar fractions of biomass are a mixture of hexoses, pentoses, and products derived from lignin. The products derived from lignin generate seriously microbial damages that inhibit fermentation performance. The artificial modification and genetic engineering of plants that result in low content of lignin are potential strategies to enhance plant biomass hydrolysis to obtain maximum sugar releases [100]. Furthermore, genetic strategies have more advantages than physical and chemical pretreatments considering that they do not require additional energy or chemicals input and result in lower environmental pollution [101].

The sense, antisense, or RNAi approaches has been used to upregulation or downregulation of genes involved in lignin biosynthesis pathway to alter their content, location, and type (Syringyl: Guaiacyl ratio) [102, 103]. Early studies revealed that the expression of *p-coumaroyl-CoA 3-hydroxylase (C3'H)* gene is related to lignin synthesis [104–107]. Subsequently, Coleman et al. [108] reported that RNAi suppression of C3'H expression in hybrid poplar led to significant reduction of lignin content (56–59%). Similarly, Chen and Dixon [109] reported that downregulation of the *hydroxycinnamoyl transferase (HCT)* and *caffeic acid 3-O-methyltransferase (COMT)* genes, reduced lignin content (40%) and increase sugar releases (166%) in alfalfa. Shafrin et al. [110] introduced hpRNA-based vectors for downregulation of *Cinnamate 4-hydroxylase (C4H)* and *COMT* genes in jute, which showed significant reduction of soluble lignin (13–23%) and fiber lignin (13–17%). RNAi downregulation of *COMT* gene also reduces the Syringyl:Guaiacyl (S:G) ratio, improved the fermentable sugars yield (>34%) and ethanol production yield (≥ 38%) in switchgrass and sugarcane bagasse, respectively [111, 228]. In contrast, other studies showed that reduction of S:G lignin ratio had no correlation with the reduction of biomass recalcitrance [112, 113]. More recently, Zhang et al. [114] demonstrated that the overexpression of *PdPFD2.2* gene encoding a prefoldin (PFD) protein, resulted in the increase of lignin S:G ratio and sugar releases (7.6%) in transgenic poplar. Likewise, Fan et al. [225] identify a novel miRNA (miR6443) that modules syringyl lignin biosynthesis by specially regulating *F5H2* in *Paulownia tomentosa*. This strategy showed that the improve of saccharification efficiency (24.5%) was positively correlated with the increase of S:G ratio. Also, Sakamoto et al. [115] discover two bacterial genes encoding coniferaldehyde dehydrogenase *calB* and *couA* can function as novel genetic tools for lignin manipulation. In this study, glucose releases increased significantly after modifications of *CalB*, *CouA*, and *F5H*, by 21%, 55%, and 31%, respectively.

5.3.5 GENETIC ENGINEERING STRATEGIES TO IMPROVE BIOETHANOL PRODUCTION

Among biofuels, bioethanol is the most produced and commercialized in the word. Industrial bioethanol plants depend on sucrose and starch-based raw materials such as sugarcane in Brazil, corn in USA, and wheat, sugar beet, and barley in Europe. Bioethanol generated from these feedstocks is known as first-generation. First-generation bioethanol has several advantages like low production cost and energy efficient production methods that result in lower fossil fuel consumption [229]. However, first-generation biofuels production generates substantial economic and environmental concerns related to the increasing demand for crops competing for cultivable land. Thus, researches have been interested in the development of commercial production of second-generation bioethanol, which is produced from non-food lignocellulosic materials such as agricultural wastes and industrial byproducts as feedstocks. However, bioethanol generation from lignocellulosic materials still confronts challenges in its commercial production, due to the complexity of biomass chemical structure, metabolic characteristics of microorganisms, and fermentation conditions [229, 230].

At present, the most widely studied microorganisms for the first- and second-generation bioethanol production include yeast, bacteria, and fungi, such as *Saccharomyces cerevisiae*, *Zymomonas mobilis*, and *Clostridium thermocellum* (Figure 5.1) [230, 231]. The metabolic performance of these microorganisms has been improved by genetic strategies in several aspects, such as substrate conversion, development of substrate assimilation, and improved inhibitor tolerance, as discussed in the following sections.

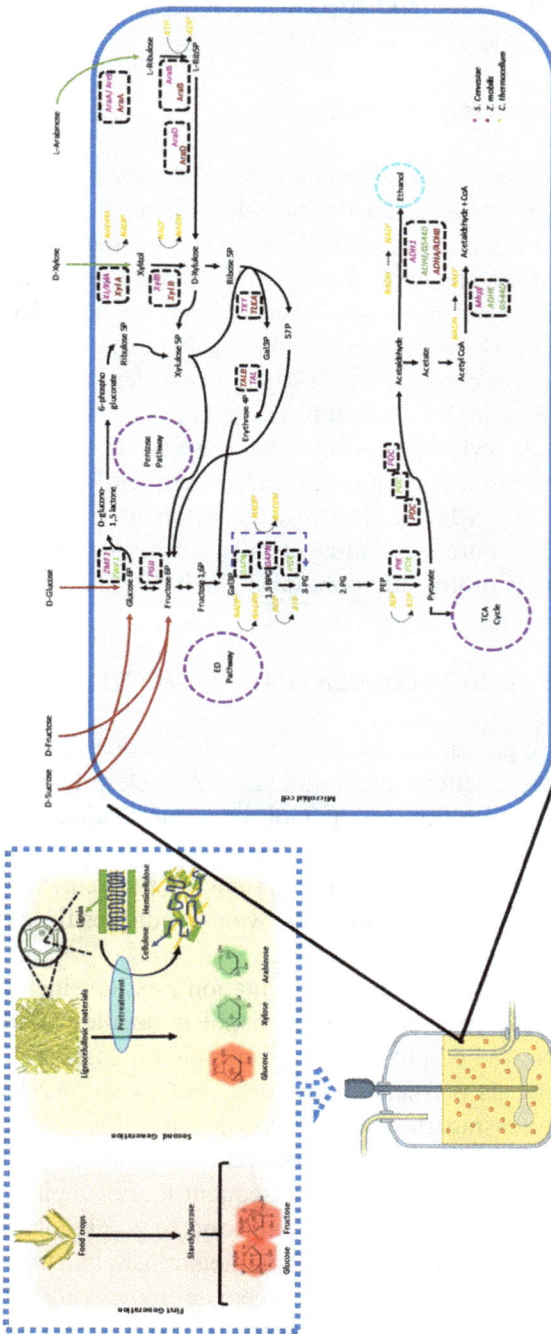

FIGURE 5.1 Bioethanol production from 1st and 2nd generation biomass: metabolic routes integration. Genes in the dotted box are the most manipulated to improve metabolic performance for bioethanol production in *Saccharomyces cerevisiae* (•), *Zymomonas mobilis* (•) and *Clostridium thermocellum* (•).

Source: Own authorship.

5.4   MAIN METABOLIC ENGINEERED MICROORGANISMS FOR ETHANOL PRODUCTION

5.4.1   YEAST: SACCHAROMYCES CEREVISIAE

Yeasts, especially *Saccharomyces cerevisiae*, when compared to other microorganisms, have advantageous characteristics that make them suitable for bioethanol production. Features such as the ability to ferment high concentration of sugars, high productivity of end products, and high tolerance to ethanol and inhibitor compounds, are desirable aspects observed in this yeast [116–118].

These microorganisms are already commonly used for industrial bioethanol production refineries [119, 120]. During fermentation, some stress conditions, such as high temperatures, ethanol concentration and the presence of sugars (e.g., xylose), can limit the process. These factors lead to the inhibition of the production process and consequently, the reduction of the ethanol yield [117, 121, 122]. Thus, recombinant DNA techniques and genetic modifications are constantly carried out in yeasts, in order to improve the performance of this microorganism [120, 122].

5.4.1.1   STRESS DUE TO HIGH CONCENTRATION OF ETHANOL

In industrial process (e.g., Sake), the ethanol concentration reaches values approximately 20% in the culture medium [123, 124]. Despite some yeasts are tolerant to high concentrations of ethanol, its accumulation inhibits cell growth and metabolic activities that limits the production yield [125–127]. In addition, negative tolerance to ethanol promotes the reduction of the synthesis and activity of RNA and enzymes, which could result in punctual mutations [120, 128].

As the concentration of ethanol in fermentation medium increases, the permeability of the membrane is altered. Ethanol molecules interact with the hydrophilic portion of the lipid bilayer, promoting the loss of membrane integrity and increasing its permeability, which induces an increase in the influx of protons, decreasing the pH to toxic levels [126, 129, 130]. In addition to membrane alterations, high concentrations of ethanol promote the denaturation of cellular proteins and the consequent loss of their functions, for example, enzymes involved in specific pathways (e.g., glycolysis) could be affected, which influences in central carbon metabolism [129, 131, 132].

Ethanol-tolerant strains can be obtained by genetic engineering techniques. Studies have carried out analysis of microarrays to determine the transcriptional response to ethanol to generate mutations that increase end product tolerance, from identified sensitive genes [133, 134]. Hirasawa et al. [133] also performed microarray analysis followed by gene knockout and showed that the overexpression of genes related to tryptophan biosynthesis can confer tolerance to ethanol stress for yeast cells. Techniques such as global transcription machinery engineering (gTME), are also tools for obtaining ethanol tolerant strains. In this case, genes of mutant transcription factors (TFs) that control the genes of cellular metabolism are introduced. Studies have shown that a recombinant *S. cerevisiae* with a mutation in the *SPT15* gene (gene encoding the TATA binding factor) can exhibit a new pattern of gene expression for several genes. This recombinant isolate, with a completely different gene expression pattern, was able to tolerate high levels of ethanol [134].

5.4.1.2 STRESS DUE TO TEMPERATURE

As well as the high concentration of ethanol, high temperatures influence yeast and other microorganisms, causing an imbalance of the plasma membrane, increasing the permeability of the membrane and the flux of protons, making it difficult to regulate the intracellular pH [117, 123, 135].

Tolerance to temperature stress is controlled by many genes involved in protecting cells. Overexpression of genes involved in tolerance to high temperatures is a commonly used tool. Studies found that the RSP5 gene in *Saccharomyces cerevisiae* is responsible for the thermotolerant phenotype, when it undergoes a mutation process in the promoter region, the RSP5-C allele causes an increase in the transcription of the RSP5 gene, making it thermotolerant [136, 137].

High temperature conditions, as well as stress due to ethanol tolerance, are answered by two main cellular mechanisms: the accumulation of HSP (described in Section 5.3.3) and carbohydrates (trehalose) [129, 132, 138]. In these stress conditions, the HSP and trehalose are responsible for folding proteins and reducing membrane permeability [139–141]. Additionally, the accumulation of trehalose makes cells able to tolerate high levels of ethanol and high temperatures, exchanging it with water and stabilizing the membrane and proteins [142]. Some genetic engineering tools aimed to reduce trehalose degradation. One of the enzymes responsible for trehalose degradation is vacuolar acid trehalase (ATH1). Performing a Null mutation of this gene caused high survival rates under various stress conditions, and the inhibition of *ATH1* gene activity by antisense RNA decreased trehalose degradation and increased ethanol tolerance [143, 144].

5.4.1.3   *FERMENTATION OF 5-CARBON SUGARS*

Currently, many feedstocks used in the production of bioethanol have 5-carbon carbohydrates, such as xylose, in their composition [145, 146]. However, the presence of these carbohydrates is often a challenge in bioethanol fermentation because of some yeasts, such as *S. cerevisiae*, do not have the ability to assimilate pentose sugars [147–149]. Some species such as *Pichia, Schizosaccharomyces*, and *Pachysolen* have been described as good producers of ethanol from pentose sugars [150]. Furthermore, other yeasts (e.g., *Kluyveromyces marxianus*) have the ability to co-ferment hexose and pentose sugars [151].

Pentose fermentation difficulties have been solved using genetically modified organisms or co-culture of two yeast strains [152, 153]. To metabolize xylose, the microorganism must be able to incorporate pentoses into the cell via membrane transporters. Subsequently, two different pathways for xylose isomerization in xylulose can be followed: the balanced redox oxidoreductase and the isomerase pathway. The first occurs when the enzyme xylose reductase (XR) converts D-xylose to xylitol, then, the enzyme xylitol dehydrogenase (XD) transforms xylitol into D-xylulose. Later, xylose is isomerized to xylulose by the enzyme Xylose Isomerase (XI) without the need for a cofactor [149, 154]. Finally, this D-xylulose is further phosphorylated by a xylulokinase in D-xylulose-5-P, which can enter the PPP [232].

Studies have been successful in the production of ethanol from xylose using recombinant *S. cerevisiae*. These strains are constructed by inserting genes that encode the heterologous metabolic enzymes XR and xylitol dehydrogenase (XDH) from *Scheffersomyces stipitis, Candida intermedia* or other strains [155, 227, 233]. In addition to the construction of recombinant yeasts, studies carried out genetic modifications to increase xylose consumption. For example, the overexpression of the non-oxidative enzymes of the PPP, endogenous *XKS1* gene (encoding xylulokinase) and the deletion of *GRE3* (encoding reductase enzyme capable of converting xylose to xylitol using NADPH) in *S. cerevisiae*, resulted in higher growth rate during xylose fermentation [156]. Examples of other genetic strategies of yeasts to improve bioethanol production are shown in Table 5.2.

5.4.2   *ZYMOMONAS MOBILIS*

In addition to yeasts, other microorganisms have been gaining attention for ethanol production, like the gram-negative bacteria *Zymomonas mobilis*. The

TABLE 5.2 Metabolic Engineering Strategies in *S. cerevisiae* to Enhance Bioethanol Production

Challenge	Strategy	Genetic Alteration	Enhancement in Mutants	Ethanol Yield (-fold)	References
Tolerance to ethanol	Deletion of non-essential genes	Deletion of *ACE2* encoding a transcription factor.	Higher tolerance between 5–15% (v/v)	1.37	[157]
Xylose assimilation	Transformation	Insertion of an endogenous gene that assimilates xylose with the overexpression of the ADH1 (alcohol dehydrogenase) gene	Ability to consume xylose	2.0	[158]
Xylose assimilation	Transformation and overexpression	Overexpression of *GLN1* with knockout of *FPS1* and *GPD2*	Increased consumption of xylose and reduction in glycerol production.	1.2	[159]
Xylose assimilation	Construction of recombinant yeast strain	Construction of a recombinant strain for co-fermentation of glucose and xylose. With insertion of the *PirXylA* and *OrpXylA* genes	Co-assimilation of glucose and xylose.	1.86	[160]
Cellulose assimilation	Transformation	Insertion of gene encoding heterogeneous cellulase system (sestc) in the *S. cerevisiae* genome	Assimilation of cellulose	57.86	[161]
Xylose assimilation	Transformation, superexpression	Overexpression of the xylulose kinase gene	Increased capacity to co-ferment glucose and xylose from lignocellulosic biomass	1.48	[162]
Tolerance to stress	Transcriptomic analysis and transformation	Expression of the genes involved in the degradation of ROS and acetic acid species	Enhanced yeast tolerance to stress due to elevated reactive oxygen species (ROS) and acetic acid	1.17	[163]
Xylose assimilation	Gene deletion	Deletion of *PDE1* and *PDE2* genes involved in the PKA activity	The recombinant strain showed increased consumption of xylose by 50% in relation to the mutant strain (*pde1Δ pde2Δ*)	2.0	[33]

TABLE 5.2 *(Continued)*

Challenge	Strategy	Genetic Alteration	Enhancement in Mutants	Ethanol Yield (-fold)	References
Tolerance to ethanol	Phenotypic analysis and knockout	Simultaneous knockout of *LEU4* and *LEU9* (leading to the accumulation of valine) or *INM1* and *INM2* (leading to the reduction of inositol)	Increased ethanol tolerance	–	[164]
Xylose assimilation	overexpression, evolutionary engineering	Insertion of genes involved in the consumption of xylose (*XYL1; XYL2; XKS1; TAL1; PYK1; MGT05196*)	Increased capacity to co-ferment glucose and xylose	~2.0	[165]

Z. mobilis has desirable characteristics such as a generally recognized as safe (GRAS) status, high glucose uptake and tolerance (up to 400 g/L), ethanol tolerance (16% v/v), and grows in a broad pH range (3.5 to 7.5) [13, 166]. To produce ethanol, this bacterium uses only Entner-Doudoroff (ED) pathway, in consequence, low ATP and cellular biomass is produced, resulting in high conversion efficiency (0.51 g ethanol/g glucose) [167, 168]. Besides ED pathway, the high expression of pyruvate decarboxylase (PDC) and ethanol dehydrogenase also makes *Z. mobilis* a good ethanol-producing microorganism [169, 170]. Despite of these advantages, there are some difficulties to use *Z. mobilis* in biotechnological processes, for example, the use of pentose sugars (e.g., xylose, and arabinose) [171], the toxicity of ethanol, and inhibitors from lignocellulosic hydrolysates [13, 172]. In face of these limitations, researchers have been focused on the development of genetic strategies in *Z. mobilis* for improving ethanol production, as shown in Table 5.3.

As described before, some of the problems with a strong impact in ethanol production by *Z. mobilis* are the inability of these bacteria to grow in presence of inhibitors formed during the biomass pretreatment, such as acetic acid and furfural [173]. The acetic acid, generated by hemicellulose and lignin deacetylation, can cross the cytoplasmic membrane of microorganisms by diffusion when it is present in its undissociated form [174]. Inside the cell, acetic acid is dissociated into proton and corresponding ion, resulting in a decrease of intracellular pH, higher necessity of ATP to pumping protons out of the cells, and growth inhibition [175, 176]. In an attempt to solve the problem caused by the presence of acetic acid, Liu et al. [177] obtained acid-tolerant mutants through chemical mutation and adaptive evolution. Similarly, Wu et al. [178] studied acetic acid-tolerant mutant strains obtained via a multi-round atmospheric and room temperature plasma (mARTP) mutagenesis.

On the other hand, furfural is formed during lignocellulose pretreatment due to the dehydration of pentose sugars [166, 179]. Even in small concentrations, this inhibitor can demonstrate a negative effect on terpenoid biosynthesis and mRNAs related to the ED pathway, alterations of the central carbon metabolism, membrane perturbation and decrease of NADH and ATP concentrations [180, 181]. Regards to furfural inhibition, *Z. mobilis* mutants were constructed by the error-prone PCR-based whole genome shuffling to improve the tolerance to furfural [168]. The authors hypothesized that the enhanced NADH-dependent furfural reductase activity detected during the early log phase may be related to an increase of furfural tolerance of the mutants, allowing an accelerated furfural detoxification process. Interestingly, Wang et al. [182] considered both acetic acid and furfural inhibition

TABLE 5.3 Metabolic Engineering Strategies in *Z. mobilis* to Enhanced Bioethanol Production

Challenge	Strategy	Genetic Alteration	Enhancement in Mutants	Ethanol Yield (-fold)	References
Sodium ion inhibition	Transformation	Overexpression of *ZMO0119* gene (Na+/H+ antiporter)	Higher glucose uptake, cell growth and Na$^+$ tolerance with 150 mM Na$^+$ or hydrolysate added 90 mM Na+	2.09	[193]
Salt inhibition	EZ-Tn5-based transposon insertion mutagenesis system	Gene *ZMO1122* (*himA*) fractured, higher expression of *pdc* and *adh* genes under 2% NaCl	Higher tolerance under 1.5 and 2% NaCl exposure	1.2	[194]
Ethanol inhibition	Directed adaptative evolution	Alterations in *clpP* (ATP-dependent protease), *spoT/relA*, ((p)ppGpp synthetase/hydrolase), and *clpB* (ATP-dependent chaperone)	Higher tolerance in different ethanol concentration, grow in up to 100 g/L of ethanol, improvement in cell viability	–	[185]
Ethanol inhibition	Random mutagenesis	Global transcription factor RpoD protein (σ70), higher *pdc* gene expression and lower *adh* gene expression under 9% (v/v) ethanol exposure	Higher ethanol tolerance (10% v/v) and glucose consumption	0.78	[186]
Acetic acid inhibition	Chemical mutagenesis and adaptative evolution	Deletion in terminator region of *ZMO0117* (hydroxylamine reductase) and upstream of *ZMO0119*	Higher tolerance to acetate (244 mM), Higher Y$_{P/S}$ and increase in the conversion rate	19.26	[177]
Acetic acid and furfural inhibition	Genome shuffling mediated by protoplast electrofusion	Single nucleotides inserted between ZMO_RS04290, (monofunctional biosynthetic peptidoglycan transglucosylase) and *ZMO04295* (cytochrome c)	Higher tolerance and ethanol productivity under 7 g/L acetic acid and 3 g/L furfural exposure	2.32	[182]

TABLE 5.3 *(Continued)*

Challenge	Strategy	Genetic Alteration	Enhancement in Mutants	Ethanol Yield (-fold)	References
Acetic acid inhibition	Multiplex atmospheric and room temperature plasma mutagenesis	Single nucleotides variations inserted in regions between genes *ZMO0952* (tRNA methyltransferase)-*ZMO0956* (ubiquinol-cytochrome C reductase); *ZMO0152* (pyruvate kinase)-*ZMO0153* (DNA binding transcriptional factor); *ZMO0373* (hypothetical protein)-*ZMO0374* (levansucrase)	Higher acetic acid tolerance (5–8 g/L)	44.7	[178]
Furfural inhibition	Error-prone PCR-based whole genome shuffling	Higher NADH-dependent furfural reductase activity	Higher tolerance to furfural (3 g/L) and higher concentration of glucose uptake	1.13	[168]
Phenolic aldehydes bioconversion	Transformation	Expression of NAD+-dependent aldehyde dehydratase	Increase of the conversion rate of furfural, hydroxybenzaldehyde, and vanillin to less-toxic compounds	1.77	[189]
Use of biogas slurry such as substrate	Atmospheric and room temperature plasma mutagenesis combined with adaptive laboratory evolution	Nucleotides deletion in CDS of gene *ZMO_RS07255* (carbamoyl-phosphate synthase large subunit) and between gene *ZMO_RS06410* (FUSC family protein) and *ZMO_RS06415* (DNA polymerase III subunit delta)	Ethanol was produced from biogas slurry to replace water and nutrients, under sterilized and unsterilized condition.	1.62	[195]

to obtain *Z. mobilis* tolerant mutants through genome shuffling mediated by protoplast electrofusion.

As well documented, high concentration of ethanol can also influence the growth rate and cause inhibition of intracellular metabolites, damage of peptidoglycan cell wall, and consequently reducing the potential of pumping protons across membrane [183, 184]. Regarding this problem, Carreón-Rodríguez et al. [185] obtained and characterized two ethanol-tolerant mutant strains, through the cultivation of *Z. mobilis* ZM4 in medium by increasing ethanol concentrations in consecutive steps. They observed that overexpression of the *spoT/relA* gene increased the synthesis of (p) ppGpp alarmone, that influenced the increase expression of genes related to amino acid synthesis mutation. Despite the improvement in ethanol tolerance, most of the ethanol-tolerant mutant strains did not increase ethanol production. However, Tan et al. [186] obtains ethanol-tolerant mutants and enhance ethanol production yield by utilizing random mutagenesis of the sigma factor RpoD protein (σ70).

Phenolic aldehydes are also produced during lignocellulose pretreatment by the degradation of lignin components and can pass across cell membranes due to its low molecular weight [166]. These inhibitors can cause damage to internal structures and DNA, leading to the inhibition of RNA and protein synthesis. Besides, it causes an increase of the cell membrane fluidity, decrease of the intracellular potassium levels and cell growth, and cell morphology abnormalities [187]. An alternative to reduce toxic effects of these inhibitors is the direct conversion of phenolic aldehydes to less-toxic compounds. However, the phenolic aldehydes are recalcitrant molecules with low water solubility, which disturbs their bioconversion [188]. In this context, Yi et al. [189] constructed an intracellular oxidative pathway to enable simultaneous biodetoxification of phenolic aldehydes and fermentation in *Z. mobilis*. Experiments in vivo demonstrated a strong oxidative capacity of aldehyde dehydrogenase and upregulation of key genes of the ED pathway and the oxidative phosphorylation.

Z. mobilis is a promising candidate to be used in the industrial process with economically viable production [173]. Comparing to the model yeast *S. cerevisiae*, this bacterium presents higher ethanol productivity (about four-fold faster) [190, 191]. Moreover, the possibility to use low pH and non-sterile conditions to grow *Z. mobilis* reduces the cost process and the chance of contamination [192]. Therefore, the characteristics of *Z. mobilis* added to studies of metabolic engineering that promote enhancement in bioproduction, allow large-scale commercial production of bioethanol with a positive economic balance for industries.

5.4.3   CLOSTRIDIUM THERMOCELLUM

In addition to the powerful metabolic machinery available in *Saccharomyces cerevisiae* and *Zimomonas mobilis*, other microorganisms are gaining attention in the industrial field due to their potential for bioethanol production from a wide range of carbohydrates (e.g., lignocellulosic biomass (LCB)) [196–199].

On this matter, *Clostridium thermocellum* is a gram-positive anaerobic bacterium that grows at temperatures above 50°C and can metabolize different substrates such as lignocellulose, cellobiose, xylose, and hemicelluloses [14, 200]. In the case of the lignocellulosic material, the hydrolysis occurs through the cellulosome, which is a hydrolytic enzyme complex with a synergistic action bound to a backbone scaffold protein and attached on the surface of the cell wall of the microorganism [14]. Owing to *C. thermocellum* can solubilize lignocellulose and produce ethanol as an additional fermentation product, its application is considered as a low-cost consolidated bioprocess with many advantages in one single step [201]. However, *C. thermocellum* is inhibited in high concentrations of ethanol. This fact is related to alterations in the membrane composition observed through the accumulation of long-chain branched fatty acids during its culture [202]. Consequently, this ethanol inhibition difficulties its widespread industrial adoption. In this sense, metabolic engineering has been crucial to enhance the bioethanol production capacities of *C. thermocellum* (Table 5.4). Hence, to turn this microorganism a better ethanol producer, different protocols of transformation and deletion have been widely tested by researchers [203–205]. For instance, deletion experiments are routinely performed in strain DSM 1313 due to high transformation efficiency [206, 207]. On the other hand, the development of counter selectable markers such as tdk, hpt, and pyrF make possible an easier manipulation of the *C. thermocellum* chromosome [203, 208, 209]. Furthermore, genetic tools such as gene deletion and targeted mutagenesis have contributed with additional modifications as an attempt to meet the industry demands [210, 211].

Among genetic tools, gene overexpression has provided substantial information about the complexity of the clostridial metabolism and the possible ways for increasing bioethanol production [212]. Similarly, heterologous expression of bioethanol production pathways has been performed in native strains as a potential solution to the commercialization of this biofuel [213, 214]. In addition, the carbon flux and metabolic pathways affected in *C. thermocellum* by the inactivation or complete removal of specific genes have been another alternative selected by researchers [215]. Since bioethanol

TABLE 5.4 Metabolic Engineering Strategies in *Clostridium thermocellum* to Improve Bioethanol Production

Challenge	Strategy	Genetic Alteration	Enhancement in Mutants	Ethanol Yield (-fold)	References
Increasing of ethanol yield	Growth-based evolutionary engineering	Deletion of the genes responsible for organic acid formation *ldh* and *pta*	Growth in batch culture for 2,000 h, without pyruvate production and increased the ethanol titer 4-fold relative to the wild type	4.2	[216]
Increasing of ethanol yield	Gene deletions from markers inserted and removed in the host chromosome	Deletion of the genes responsible for organic acid formation *ldh* and *pta*	Ethanol selectivity of 40:1 relative to organic acids	1.5	
Conversion of phosphoenol-pyruvate (PEP) to pyruvate	Transformation	Heterologous expression of pyruvate kinase (Tsac_1363) from *T. saccharolyticum*, with modification of the star codon GTG of phosphoenolpyruvate carboxykinase	H$_2$ production reduced by 55%, carbon recovery of 97%, grow in up to 35 g/L of ethanol	3.25	[215]
Conversion of phosphoenol-pyruvate (PEP) to pyruvate	Transformation	Gene for malic enzyme and part of malate dehydrogenase deleted	H$_2$ production reduced by 65%, carbon recovery of 94.2%, grow in up to 35 g/L of ethanol, no NADH-dependent activity detected	3	
Increasing of ethanol yield	Transformation	Deletion of deletion of the gene initiation regulator, spo0A (Cthe_0812) and phosphotransacetylase (pta, Cthe_1029) by removing base pairs +174 to +1,074 of the 1,077 bp gene	No phosphate acetyltransferase specific activity detected; pyruvate concentrations two folds higher	1.7	[220]

TABLE 5.4 *(Continued)*

Challenge	Strategy	Genetic Alteration	Enhancement in Mutants	Ethanol Yield (-fold)	References
Increasing the flux to ethanol by removing side product formation	Transformation	Deletion of hydrogenase maturase (*hydG*), lactate dehydrogenase (Clo1313_1160; *ldh*), pyruvate-formate lyase (Clo1313_1717; *pflB*), pfl-activating enzyme (Clo1313_1716; *pflA*), phosphotrans-acetylase (Clo1313_1185; *pta*) and acetate kinase (Clo1313_1186; *ack*)	Maximum ethanol titer of 73 mM at 20 g/L Avicel, growth in the absence of pH control	3.19	[221]
Redirection of carbon and electron flux	Cloning and transformation	Deletion of genes encoding pyruvate: formate lyase (*pflB*) and PFL-activating enzyme (*pflA*)	Formate production eliminated, acetate production decreased by 50% on both complex and defined medium	–	[217]
Increasing the affinity of the pyruvate ferredoxin oxidoreductase	Cloning and transformation	Heterologous expression of pyruvate decarboxylase gene (pNW33N-*pdc*) from *Zymomonas mobilis*	Two-fold increase in pyruvate carboxylase activity	2.03	[222]
Improving ethanol yield	Cloning and transformation	Heterologous expression of *T. saccharolyticum* ethanol production operon (*adhA, nfnAB,* and *adhE*^{G544D})	Hydrogen production decreased	1.5	[214]
Improving ethanol yield	Cloning and transformation	Deletion of hydrogen production	Maximum ethanol titer of 280 mM (12.90 g/L) within 75 h	1.2	

production demands the reduction of metabolic power for increasing its yield, this has been a key point to achieve large-scale processes. Thus, to enhance the NADPH pool, the most frequently engineered metabolic route is PPP [216, 217]. In this sense, due to the abundance of pentose sugars in LCB, and the lack of xylose consumption by *C. thermocellum,* studies have been developed to enhance the assimilation of this economic substrate [218, 219]. For instance, Verbeke et al. [218] reported that the deletion of the ATP-dependent transporter (CbpD) in *C. thermocellum* partially alleviated xylose inhibition. Moreover, the authors observed a decrease in the total and molar yields (mol xylitol: mol cellobiose consumed) of ~41% and ~46% respectively, when deleted a putative XD, encoded by Clo1313_0076. Banerjee et al. [122] also analyzed the heterologous expression in *C. thermocellum* of the genes *xylA* (xylose isomerase) and *xylB* (xylulokinase) from a thermophilic anaerobic bacterium *Thermoanaerobacter ethanolicus.* The results from this study suggested that the combined activity of these enzymes converted xylose to xylulose-5-phosphate, which was then converted to ethanol through the incorporation to PPP, glycolysis, and malate shunt pathway. Searching for new metabolic reactions through enzyme engineering is one of the expanding areas to improve actual scenario in the industrial sector [198].

5.5   CURRENT STATUS AND FUTURE PERSPECTIVES OF ENGINEERED MICROORGANISM FOR ETHANOL PRODUCTION: INDUSTRY

Fossil fuels can potentially be replaced by biofuels in our daily lives, albeit much work must develop to make the current biofuels production technology economically feasible. Several considerations have to take in count for cost reduction, such as maximum theorical yields in energy terms and cheaper biomass feedstock fermentation. The transition from conventional fuels to bioenergy systems requires the design and discovery of new metabolic pathways, as well of new enzymes, new approaches in the bioprocess engineering and development of microbial consortium capable of tolerate high ethanol concentrations [223].

As described in sections above, recent advances and future perspectives in microbial genetic engineering due to the fusion of biological approaches derived from data of transcriptomics, proteomics, and metabolomics have led to design novel synthetic circuits that might allow us to cross the gap between the laboratory scale to the industrial marketplace [8].

Additionally, to researches presented before, other novel genetic strategies (e.g., CRISPR/Cas) continue to be developed to simplify the use of model microorganisms (*S. cerevisiae, Z. mobilis* and *C. thermocellum*) in biorefineries. For example, a recent study utilized a novel strategy (SHPERM-bCGHR) to replace the *PHO4* gene in *S. cerevisiae*, without the introduction of antibiotic resistance genes and giving stability for long-term industrial ethanol production [33]. Furthermore, numerous efforts have been started to boost ethanol production by suppressing glycerol formation. Liu et al. [207] developed a strategy in *S. cerevisiae* based on CRISPR/Cas9 novel system, which optimize the ethanol metabolic pathway by the disruption and combinatorically of three genes (*ADH, GPD, and ALD*). Results showed an increase of ethanol production and yield by 40% and 22% respectively. In addition, Xue et al. [224] used CRISPR/Cas9 technology to disrupt the alcohol dehydrogenase 2 (*ADH2*) gene via complete deletion of the gen in *S. cerevisiae*. Their results demonstrated an improvement up to 75% in ethanol yield compared with the native strain.

5.6 CONCLUSIONS

Demanding on fuels cost reduction have occasioned in increased alternatives to replace fossil fuels. In addition, several novel strategies have been implemented to improve biofuel production by the microbial metabolism engineering in order to look forward for better yields and improvement in industrial biofuels production. Data recollected in this chapter indicates that the field of metabolic engineering and synthetic biology has been expanding rapidly in the case of yeast. This has made possible the construction of complex metabolic pathways, well-regulated metabolic networks, synthetic strains, useful recombinant vectors, specific genetic manipulation, and high-capacity systems. Well-established metabolic engineering and synthetic biology provides not only a novel biochemical approaches insight into the lignocellulosic biofuel industry, but also an additional insight into deciphering the pathways behind bioethanol production by different microorganisms. Thus, the main improvements to scale up bioethanol production process, must be related with the development of high-performance strains, such as high ethanol concentration tolerance, thermostability, and detoxification of inhibitors. Finally, economical scenarios must be assessed to determine the true commercial value of bioethanol production through metabolic engineered microorganisms.

KEYWORDS

- **bioethanol**
- **biofuels**
- **global transcription machinery engineering**
- **metabolic engineering**
- **prefoldin**
- **synthetic biology**

REFERENCES

1. Shahbaz, M., Tiwari, A. K., & Nasir, M., (2013). The effects of financial development, economic growth, coal consumption and trade openness on CO_2 emissions in South Africa. *Energy Policy, 61*, 1452–1459.
2. Awodumi, O. B., & Adewuyi, A. O., (2020). The role of non-renewable energy consumption in economic growth and carbon emission: Evidence from oil producing economies in Africa. *Energy Strategy Reviews, 27*, 100434.
3. Ching-Sung, T., Kwak, S., Turner, T. L., & Yong-Su, J., (2015). Yeast synthetic biology toolbox and applications for biofuel production. *FEMS Yeast Research, 15*(1), 1–15.
4. Abdulkareem-Alsultan, G., Asikin-Mijan, N., Lee, H. V., & Taufiq-Yap, Y. H., (2020). Biofuels: Past, present, future. In: *Innovations in Sustainable Energy and Cleaner Environment* (pp. 489–504). Springer, Singapore.
5. Vasconcelos, M. H., Mendes, F. M., Ramos, L., Dias, M. O. S., Bonomi, A., Jesus, C. D. F., & Dos, S. J. C., (2020). Techno-economic assessment of bioenergy and biofuel production in integrated sugarcane biorefinery: Identification of technological bottlenecks and economic feasibility of dilute acid pretreatment. *Energy*, 117422.
6. Das, M., Patra, P., & Ghosh, A., (2020). Metabolic engineering for enhancing microbial biosynthesis of advanced biofuels. *Renewable and Sustainable Energy Reviews, 119*, 109562.
7. Nielsen, J., & Keasling, J. D., (2016). Engineering cellular metabolism. *Cell, 164*(6), 1185–1197.
8. Madhavan, A., Jose, A. A., Binod, P., Sindhu, R., Sukumaran, R. K., Pandey, A., & Castro, G. E., (2017). Synthetic biology and metabolic engineering approaches and its impact on non-conventional yeast and biofuel production. *Frontiers in Energy Research, 5*, 8.
9. Adrio, J. L., & Demain, A. L., (2006). Genetic improvement of processes yielding microbial products. *FEMS Microbiology Reviews, 30*(2), 187–214.
10. Lian, J., Mishra, S., & Zhao, H., (2018). Recent advances in metabolic engineering of *Saccharomyces cerevisiae*: New tools and their applications. *Metabolic Engineering, 50*, 85–108.

11. Jiang, M., Stephanopoulos, G., & Pfeifer, B. A., (2012). Toward biosynthetic design and implementation of *Escherichia coli*-derived paclitaxel and other heterologous polyisoprene compounds. *Applied and Environmental Microbiology, 78*(8), 2497–2504.

12. Hwang, H. G., Kim, K. J., Lee, S. H., Kim, C. K., Min, C. K., Yun, J. M., & Son, Y. J., (2016). Recombinant glargine insulin production process using *Escherichia coli. J Microbiol. Biotechnol., 26*(10), 1781–9.

13. Zhang, K., Lu, X., Li, Y., Jiang, X., Liu, L., & Wang, H., (2019). New technologies provide more metabolic engineering strategies for bioethanol production in *Zymomonas mobilis*. In: *Applied Microbiology and Biotechnology* (Vol. 103, No. 5, pp. 2087–2099). https://doi.org/10.1007/s00253-019-09620-6.

14. Brown, S. D., Sander, K. B., Wu, C. W., & Guss, A. M., (2015). *Clostridium thermocellum*: Engineered for the production of bioethanol. In: *Direct Microbial Conversion of Biomass to Advanced Biofuels* (pp. 321–333). Elsevier.

15. Hopkins, M. M., Kraft, A., Martin, P. A., Nightingale, P., & Mahdi, S., (2007). *Is the Biotechnology Revolution a Myth? 1*(17), 591–613.

16. Papagianni, M., (2017). Microbial bioprocesses. In: *Current Developments in Biotechnology and Bioengineering* (pp. 45–72). Elsevier.

17. Doran, P. M., (2013). Bioprocess development: An interdisciplinary challenge. *Bioprocess Eng. Princ.*, 3–11.

18. Ko, Y. S., Kim, J. W., Lee, J. A., Han, T., Kim, G. B., Park, J. E., & Lee, S. Y., (2020). Tools and strategies of systems metabolic engineering for the development of microbial cell factories for chemical production. *Chemical Society Reviews*.

19. Singh, R., (2005). Hybrid membrane systems-applications and case studies. *Hybrid Membrane Systems for Water Purification, 131*. Elsevier Science, Amsterdam.

20. Yang, S. T., (2007). Bioprocessing-from biotechnology to biorefinery. In: *Bioprocessing for Value-Added Products from Renewable Resources* (pp. 1–24). Elsevier.

21. Stephanopoulos, G., Aristidou, A. A., & Nielsen, J., (1998). *Metabolic Engineering: Principles and Methodologies*. Elsevier.

22. Vigneswaran, C., Ananthasubramanian, M., & Kandhavadivu, P., (2014). *Enzymes Technology in Bioprocessing of Textiles.* WPI Publishing.

23. Ledesma-Amaro, R., Nikel, P. I., & Ceroni, F., (2020). Synthetic Biology-Guided Metabolic Engineering. *Frontiers in Bioengineering and Biotechnology, 8.*

24. Qin, R., Zhong, C., Zong, G., Fu, J., Pang, X., & Cao, G., (2017). Improvement of clavulanic acid production in *Streptomyces clavuligerus* F613-1 by using a claR-neo reporter strategy. *Electronic Journal of Biotechnology, 28*, 41–46.

25. Parisutham, V., Kim, T. H., & Lee, S. K., (2014). Feasibilities of consolidated bioprocessing microbes: From pretreatment to biofuel production. *Bioresource Technology, 161*, 431–440.

26. Biddy, M. J., Davis, R., Humbird, D., Tao, L., Dowe, N., Guarnieri, M. T., & Beckham, G. T., (2016). The techno-economic basis for coproduct manufacturing to enable hydrocarbon fuel production from lignocellulosic biomass. *ACS Sustainable Chemistry & Engineering, 4*(6), 3196–3211.

27. Okamoto, K., Kondo-Okamoto, N., & Ohsumi, Y., (2009). Mitochondria-anchored receptor Atg32 mediates degradation of mitochondria via selective autophagy. *Developmental Cell, 17*(1), 87–97.

28. Eiyama, A., Kondo-Okamoto, N., & Okamoto, K., (2013). Mitochondrial degradation during starvation is selective and temporally distinct from bulk autophagy in yeast. *FEBS Letters, 587*(12), 1787–1792.

29. Shiroma, S., Jayakody, L. N., Horie, K., Okamoto, K., & Kitagaki, H., (2014). Enhancement of ethanol fermentation in *Saccharomyces cerevisiae* sake yeast by disrupting mitophagy function. *Applied and Environmental Microbiology, 80*(3), 1002–1012.

30. Peksel, A., Torres, N., Liu, J., Juneau, G., & Kubicek, C., (2002). 13 C-NMR analysis of glucose metabolism during citric acid production by *Aspergillus Niger. Applied Microbiology and Biotechnology, 58*(2), 157–163.

31. Chi, Z., Wang, Z. P., Wang, G. Y., Khan, I., & Chi, Z. M., (2016). Microbial biosynthesis and secretion of l-malic acid and its applications. *Critical Reviews in Biotechnology, 36*(1), 99–107.

32. Nguyen-Vo, T. P., Ainala, S. K., Kim, J. R., & Park, S., (2018). Analysis and characterization of coenzyme B12 biosynthetic gene clusters and improvement of B12 biosynthesis in *Pseudomonas denitrificans* ATCC 13867. *FEMS Microbiology Letters, 365*(21), fny211.

33. Wu, R., Chen, D., Cao, S., Lu, Z., Huang, J., Lu, Q., & Huang, R., (2020). Enhanced ethanol production from sugarcane molasses by industrially engineered *Saccharomyces cerevisiae* via replacement of the PHO4 gene. *RSC Advances, 10*(4), 2267–2276.

34. Tan, M. J., Chen, X., Wang, Y. K., Liu, G. L., & Chi, Z. M., (2016a). Enhanced citric acid production by a yeast yarrowia lipolytica over-expressing a pyruvate carboxylase gene. *Bioprocess and Biosystems Engineering, 39*(8), 1289–1296.

35. Zhang, J., Wu, C., Wang, W., & Wei, D., (2018). Construction of enhanced transcriptional activators for improving cellulase production in *Trichoderma reesei* RUT C30. *Bioresources and Bioprocessing, 5*(1), 40.

36. Li, H., Zhang, Y., Gao, Y., Lan, Y., Yin, X., & Huang, L., (2016). Characterization of UGPase from *Aureobasidium pullulans* NRRL Y-12974 and application in enhanced pullulan production. *Applied Biochemistry and Biotechnology, 178*(6), 1141–1153.

37. Zhang, B., Lei, Z., Liu, Z. Q., & Zheng, Y. G., (2020a). Improvement of gibberellin production by a newly isolated *Fusarium fujikuroi* mutant. *Journal of Applied Microbiology.*

38. Paudel, S., & Menze, M. A., (2014). Genetic engineering, a hope for sustainable biofuel production. *International Journal of Environment,* 311.

39. Shanmugam, S., Ngo, H. H., & Wu, Y. R., (2019). Advanced CRISPR/Cas-based genome editing tools for microbial biofuels production: A review. *Renewable Energy.* doi: 10.1016/j.renene.2019.10.107.

40. Dai, Z., Zhang, S., Yang, Q., Zhang, W., Qian, X., Dong, W., & Xin, F., (2018). Genetic tool development and systemic regulation in biosynthetic technology. *Biotechnology for Biofuels, 11*(1), 152.

41. Jagadevan, S., Banerjee, A., Banerjee, C., Guria, C., Tiwari, R., Baweja, M., & Shukla, P., (2018). Recent developments in synthetic biology and metabolic engineering in microalgae towards biofuel production. *Biotechnology for Biofuels, 11*(1), 185.

42. Javed, M. R., Noman, M., Shahid, M., Ahmed, T., Khurshid, M., Rashid, M. H., & Khan, F., (2018). Current situation of biofuel production and its enhancement by CRISPR/Cas9-mediated genome engineering of microbial cells. *Microbiological Research.* doi: 10.1016/j.micres.2018.10.010.

43. Dunlop, M. J., (2011). Engineering microbes for tolerance to next-generation biofuels. *Biotechnology for Biofuels, 4*(1), 1–9.

44. Zingaro, K. A., & Terry, P. E., (2013). GroESL overexpression imparts *Escherichia coli* tolerance to i-, n-, and 2-butanol, 1,2,4-butanetriol and ethanol with complex and unpredictable patterns. *Metabolic Engineering, 15*, 196–205. doi: 10.1016/j.ymben.2012.07.009.

45. Tomas, C. A., Welker, N. E., & Papoutsakis, E. T., (2003). Overexpression of groESL in *Clostridium acetobutylicum* results in increased solvent production and tolerance, prolonged metabolism, and changes in the cell's transcriptional program. *Applied and Environmental Microbiology, 69*(8), 4951–4965.

46. Wang, S., He, Z., & Yuan, Q., (2017a). Xylose enhances furfural tolerance in candida tropicalis by improving NADH recycle. *Chemical Engineering Science, 158*, 37–40.

47. Reyes, L. H., Abdelaal, A. S., & Kao, K. C., (2013). Genetic determinants for n-butanol tolerance in evolved *Escherichia coli* mutants: Cross adaptation and antagonistic pleiotropy between n-butanol and other stressors. *Applied and Environmental Microbiology, 79*(17), 5313–5320.

48. Mukhopadhyay, A., (2015). Tolerance engineering in bacteria for the production of advanced biofuels and chemicals. *Trends in Microbiology, 23*(8), 498–508.

49. Tomas, C. A., Beamish, J., & Papoutsakis, E. T., (2004). Transcriptional analysis of butanol stress and tolerance in *Clostridium acetobutylicum*. *Journal of Bacteriology, 186*(7), 2006–2018.

50. Brynildsen, M. P., & Liao, J. C., (2009). An integrated network approach identifies the isobutanol response network of *Escherichia coli*. *Molecular Systems Biology, 5*(1), 277.

51. Rutherford, B. J., Dahl, R. H., Price, R. E., Szmidt, H. L., Benke, P. I., Mukhopadhyay, A., & Keasling, J. D., (2010). Functional genomic study of exogenous n-butanol stress in *Escherichia coli*. *Applied and Environmental Microbiology, 76*(6), 1935–1945.

52. Clark, D. S., Whitehead, T., Robb, F. T., Laksanalamai, P., & Jiemjit, A., (2014). *U.S. Patent No. 8,685,729*. Washington, DC: U.S. Patent and Trademark Office.

53. Luan, G., Dong, H., Zhang, T., Lin, Z., Zhang, Y., Li, Y., & Cai, Z., (2014). Engineering cellular robustness of microbes by introducing the GroESL chaperonins from extremophilic bacteria. *Journal of Biotechnology, 178*, 38–40.

54. Abdelaal, A. S., Ageez, A. M., Abd El, A. E. H. A., & Abdallah, N. A., (2015). Genetic improvement of n-butanol tolerance in *Escherichia coli* by heterologous overexpression of groESL operon from *Clostridium acetobutylicum*. *3 Biotech, 5*(4), 401–410.

55. Xu, G., Wu, A., Xiao, L., Han, R., & Ni, Y., (2019a). Enhancing butanol tolerance of *Escherichia coli* reveals hydrophobic interaction of multi-tasking chaperone SecB. *Biotechnology for Biofuels, 12*(1), 164.

56. Zhang, F., Qian, X., Si, H., Xu, G., Han, R., & Ni, Y., (2015). Significantly improved solvent tolerance of *Escherichia coli* by global transcription machinery engineering. *Microbial Cell Factories, 14*(1), 175.

57. Wang, M., Liu, L., Fan, L., & Tan, T., (2017b). CRISPRi based system for enhancing 1-butanol production in engineered *Klebsiella pneumoniae*. *Process Biochemistry, 56*, 139–146.

58. Otoupal, P. B., & Chatterjee, A., (2018). CRISPR gene perturbations provide insights for improving bacterial biofuel tolerance. *Frontiers in Bioengineering and Biotechnology, 6*, 122.

59. Chen, S., Lesnik, E. A., Hall, T. A., Sampath, R., Griffey, R. H., Ecker, D. J., & Blyn, L. B., (2002). A bioinformatics-based approach to discover small RNA genes in the *Escherichia coli* genome. *Biosystems, 65*(2, 3), 157–177.

60. Erickson, K. E., Otoupal, P. B., & Chatterjee, A., (2017). *Transcriptome-Level Signatures in Gene Expression and Gene Expression Variability During Bacterial Adaptive Evolution, 2*(1). MSphere.

61. Huang, S., & Geng, A., (2020). High-copy genome integration of 2, 3-butanediol biosynthesis pathway in *Saccharomyces cerevisiae* via in vivo DNA assembly and replicative CRISPR-Cas9 mediated delta integration. *Journal of Biotechnology, 310,* 13–20.

62. Kieboom, J., Dennis, J. J., De Bont, J. A., & Zylstra, G. J., (1998). Identification and molecular characterization of an efflux pump involved in Pseudomonas putida S12 solvent tolerance. *Journal of Biological Chemistry, 273*(1), 85–91.

63. Segura, A., Duque, E., Mosqueda, G., Ramos, J. L., & Junker, F., (1999). Multiple responses of gram-negative bacteria to organic solvents. *Environmental Microbiology,* 1(3), 191–198.

64. Lee, J. Y., Geraldi, A., Rahman, Z., Lee, J. H., & Kim, S. C., (2015). Improved n-butanol tolerance in *Escherichia coli* by controlling membrane related functions. *Journal of Biotechnology, 204,* 33–44.

65. Jiménez-Bonilla, P., Zhang, J., Wang, Y., Blersch, D., de-Bashan, L. E., Guo, L., & Wang, Y., (2020). Enhancing the tolerance of *Clostridium saccharoperbutylacetonicum* to lignocellulosic-biomass-derived inhibitors for efficient biobutanol production by overexpressing efflux pumps genes from Pseudomonas putida. *Bioresource Technology,* 123532.

66. Ling, H., Chen, B., Kang, A., Lee, J. M., & Chang, M. W., (2013). Transcriptome response to alkane biofuels in *Saccharomyces cerevisiae*: Identification of efflux pumps involved in alkane tolerance. *Biotechnology for Biofuels, 6*(1), 95.

67. Boyarskiy, S., Davis, L. S., Kong, N., & Tullman-Ercek, D., (2016). Transcriptional feedback regulation of efflux protein expression for increased tolerance to and production of n -butanol. *Metabolic Engineering, 33,* 130–137. doi: 10.1016/j.ymben.2015.11.005.

68. Fisher, M. A., Boyarskiy, S., Yamada, M. R., Kong, N., Bauer, S., & Tullman-Ercek, D., (2014). Enhancing tolerance to short-chain alcohols by engineering the *Escherichia coli* AcrB efflux pump to secrete the non-native substrate n-butanol. *ACS Synthetic Biology, 3*(1), 30–40.

69. Foo, J. L., & Leong, S. S. J., (2013). Directed evolution of an E. coli inner membrane transporter for improved efflux of biofuel molecules. *Biotechnology for Biofuels, 6*(1), 1–12.

70. He, X., Xue, T., Ma, Y., Zhang, J., Wang, Z., Hong, J., Hui, L., et al., (2019). Identification of functional butanol-tolerant genes from *Escherichia coli* mutants derived from error-prone PCR-based whole-genome shuffling. *Biotechnology for Biofuels, 12*(1), 73.

71. Bhagat, R., Panakkal, H., Gupta, I., & Ingle, A. P., (2020). Recent advances in the production of biodiesel using lignocellulosic biomass. *Lignocellulosic Biorefining Technologies,* 69–85. doi: 10.1002/9781119568858.ch5.

72. Palmqvist, E., & Hahn-Hägerdal, B., (2000). Fermentation of lignocellulosic hydrolysates. II: Inhibitors and mechanisms of inhibition. *Bioresource Technology,* 74(1), 25–33. doi: 10.1016/s0960-8524(99)00161-3.

73. Almeida, J. R., Modig, T., Petersson, A., Hähn-Hägerdal, B., Lidén, G., & Gorwa-Grauslund, M. F., (2007). Increased tolerance and conversion of inhibitors in lignocel-lulosic hydrolysates by *Saccharomyces cerevisiae*. *Journal of Chemical Technology & Biotechnology: International Research in Process, Environmental & Clean Technology,* 82(4), 340–349.

74. Gorsich, S. W., Dien, B. S., Nichols, N. N., Slininger, P. J., Liu, Z. L., & Skory, C. D., (2006). Tolerance to furfural-induced stress is associated with pentose phosphate

pathway genes ZWF1, GND1, RPE1, and TKL1 in *Saccharomyces cerevisiae. Applied Microbiology and Biotechnology, 71*(3), 339–349.

75. Miller, E. N., Jarboe, L. R., Turner, P. C., Pharkya, P., Yomano, L. P., York, S. W., & Ingram, L. O., (2009). Furfural inhibits growth by limiting sulfur assimilation in ethanologenic *Escherichia coli* strain LY180. *Applied and Environmental Microbiology, 75*(19), 6132–6141.

76. Allen, S. A., Clark, W., McCaffery, J. M., Cai, Z., Lanctot, A., Slininger, P. J., & Gorsich, S. W., (2010). Furfural induces reactive oxygen species accumulation and cellular damage in *Saccharomyces cerevisiae. Biotechnology for Biofuels, 3*(1), 1–10.

77. Zaldivar, J., Martinez, A., & Ingram, L. O., (1999). Effect of selected aldehydes on the growth and fermentation of ethanologenic *Escherichia coli. Biotechnology and Bioengineering, 65*(1), 24–33.

78. Almarsdottir, A. R., Sigurbjornsdottir, M. A., & Orlygsson, J., (2012). Effect of various factors on ethanol yields from lignocellulosic biomass by *Thermoanaerobacterium* AK17. *Biotechnology and Bioengineering, 109*(3), 686–694.

79. Seo, H. M., Jeon, J. M., Lee, J. H., Song, H. S., Joo, H. B., Park, S. H., & Lee, H., (2016). Combinatorial application of two aldehyde oxidoreductases on isobutanol production in the presence of furfural. *Journal of Industrial Microbiology & Biotechnology, 43*(1), 37–44.

80. Song, H. S., Jeon, J. M., Kim, H. J., Bhatia, S. K., Sathiyanarayanan, G., Kim, J., & Yang, Y. H., (2017). Increase in furfural tolerance by combinatorial overexpression of NAD salvage pathway enzymes in engineered isobutanol-producing *E. coli. Bioresource Technology, 245*, 1430–1435.

81. Oh, E. J., Wei, N., Kwak, S., Kim, H., & Jin, Y. S., (2019). Overexpression of RCK1 improves acetic acid tolerance in *Saccharomyces cerevisiae. Journal of Biotechnology.* doi: 10.1016/j.jbiotec.2018.12.013.

82. Liu, C. G., Cao, L. Y., Wen, Y., Li, K., Mehmood, M. A., Zhao, X. Q., & Bai, F. W., (2020a). Intracellular redox manipulation of *Zymomonas mobilis* for improving tolerance against lignocellulose hydrolysate-derived stress. *Chemical Engineering Science,* 115933.

83. Ujor, V., Agu, C. V., Gopalan, V., & Ezeji, T. C., (2014). Glycerol supplementation of the growth medium enhances in situ detoxification of furfural by *Clostridium beijerinckii* during butanol fermentation. *Applied Microbiology and Biotechnology, 98*(14), 6511–6521.

84. Agu, C. V., Ujor, V., & Ezeji, T. C., (2019). Metabolic engineering of *Clostridium beijerinckii* to improve glycerol metabolism and furfural tolerance. *Biotechnology for Biofuels, 12*(1), 50.

85. Caspeta, L., Chen, Y., Ghiaci, P., Feizi, A., Buskov, S., Hallström, B. M., & Nielsen, J., (2014). Altered sterol composition renders yeast thermotolerant. *Science, 346*(6205), 75–78.

86. Li, P., Fu, X., Zhang, L., Zhang, Z., Li, J., & Li, S., (2017). The transcription factors Hsf1 and Msn2 of thermotolerant *Kluyveromyces marxianus* promote cell growth and ethanol fermentation of *Saccharomyces cerevisiae* at high temperatures. *Biotechnology for Biofuels, 10*(1), 1–13.

87. Li, P., Fu, X., Zhang, L., & Li, S., (2019). CRISPR/Cas-based screening of a gene activation library in *Saccharomyces cerevisiae* identifies a crucial role of OLE 1 in thermotolerance. *Microbial. Biotechnology, 12*(6), 1154–1163.

88. Tessarz, P., Schwarz, M., Mogk, A., & Bukau, B., (2009). The yeast AAA+ chaperone Hsp104 is part of a network that links the actin cytoskeleton with the inheritance of damaged proteins. *Molecular and Cellular Biology, 29*(13), 3738–3745.

89. De Virgilio, C., Hottiger, T., Dominguez, J., Boller, T., & Wiemken, A., (1994). The role of trehalose synthesis for the acquisition of thermotolerance in yeast: I. Genetic evidence that trehalose is a thermoprotectant. *European Journal of Biochemistry, 219* (1-2), 179–186.

90. Lindquist, S., & Kim, G., (1996). Heat-shock protein 104 expression is sufficient for thermotolerance in yeast. *Proceedings of the National Academy of Sciences, 93*(11), 5301–5306.

91. Ribeiro, M. J., Reinders, A., Boller, T., Wiemken, A., & De Virgilio, C., (1997). Trehalose synthesis is important for the acquisition of thermotolerance in *Schizosaccharomyces pombe*. *Molecular Microbiology, 25*(3), 571–581.

92. Benjaphokee, S., Koedrith, P., Auesukaree, C., Asvarak, T., Sugiyama, M., Kaneko, Y., & Harashima, S., (2012a). CDC19 encoding pyruvate kinase is important for high-temperature tolerance in *Saccharomyces cerevisiae*. *New Biotechnology, 29*(2), 166–176.

93. Nasution, O., Lee, J., Srinivasa, K., Choi, I. G., Lee, Y. M., Kim, E., & Kim, W., (2015). Loss of D fg5 glycosylphosphatidylinositol-anchored membrane protein confers enhanced heat tolerance in *Saccharomyces cerevisiae*. *Environmental Microbiology, 17*(8), 2721–2734.

94. Jia, H., Sun, X., Sun, H., Li, C., Wang, Y., Feng, X., & Li, C., (2016). Intelligent microbial heat-regulating engine (IMHeRE) for improved thermo-robustness and efficiency of bioconversion. *ACS Synthetic Biology, 5*(4), 312–320. doi: 10.1021/acssynbio.5b00158.

95. Xu, K., Gao, L., Hassan, J. U., Zhao, Z., Li, C., Huo, Y. X., & Liu, G., (2018). Improving the thermo-tolerance of yeast base on the antioxidant defense system. *Chemical Engineering Science, 175*, 335–342. doi: 10.1016/j.ces.2017.10.016.

96. Li, K., Zhang, J. W., Liu, C. G., Mehmood, M. A., & Bai, F. W., (2020). Elucidating the molecular mechanism of TEMPOL-mediated improvement on tolerance under oxidative stress in *Saccharomyces cerevisiae*. *Chemical Engineering Science, 211*, 115306.

97. Tao, X., Xu, T., Kempher, M. L., Liu, J., & Zhou, J., (2020). Precise promoter integration improves cellulose bioconversion and thermotolerance in *Clostridium cellulolyticum*. *Metabolic Engineering*.

98. Zhao, X., Zhang, L., & Liu, D., (2012). Biomass recalcitrance. Part I: The chemical compositions and physical structures affecting the enzymatic hydrolysis of lignocellulose. *Biofuels, Bioproducts and Biorefining, 6*(4), 465–482.

99. Ballesteros, L. F., Michelin, M., Vicente, A. A., Teixeira, J. A., & Cerqueira, M. Â., (2018). Lignocellulosic materials: Sources and processing technologies. In: *Lignocellulosic Materials and Their Use in Bio-based Packaging* (pp. 13–33). Springer, Cham.

100. Bhatia, S. K., Jagtap, S. S., Bedekar, A. A., Bhatia, R. K., Patel, A. K., Pant, D., & Yang, Y. H., (2020). Recent developments in pretreatment technologies on lignocellulosic biomass: Effect of key parameters, technological improvements, and challenges. *Bioresource Technology, 300*, 122724.

101. Xu, N., Liu, S., Xin, F., Jia, H., Xu, J., Jiang, M., & Dong, W., (2019b). Biomethane production from lignocellulose: Biomass recalcitrance and its impacts on anaerobic digestion. *Frontiers in Bioengineering and Biotechnology, 7*, 191.

102. Lu, S., Li, L., & Zhou, G., (2010). Genetic modification of wood quality for second-generation biofuel production. *GM Crops, 1*(4), 230–236.

103. Sharma, R., Joshi, R., & Kumar, D., (2020a). Present status and future prospect of genetic and metabolic engineering for biofuels production from lignocellulosic biomass. In: *Genetic and Metabolic Engineering for Improved Biofuel Production from Lignocellulosic Biomass* (pp. 171–192). Elsevier.

104. Franke, R., McMichael, C. M., Meyer, K., Shirley, A. M., Cusumano, J. C., & Chapple, C., (2000). Modified lignin in tobacco and poplar plants over-expressing the *Arabidopsis* gene encoding ferulate 5-hydroxylase. *The Plant Journal, 22*(3), 223–234.

105. Abdulrazzak, N., Pollet, B., Ehlting, J., Larsen, K., Asnaghi, C., Ronseau, S., & Ullmann, P., (2006). A coumaroyl-ester-3-hydroxylase insertion mutant reveals the existence of nonredundant meta-hydroxylation pathways and essential roles for phenolic precursors in cell expansion and plant growth. *Plant Physiology, 140*(1), 30–48.

106. Ralph, J., (2006). What makes a good monolignol substitute. T*he Science and Lore of the Plant Cell Wall Biosynthesis, Structure and Function*, 285–293.

107. Besseau, S., Hoffmann, L., Geoffroy, P., Lapierre, C., Pollet, B., & Legrand, M., (2007). Flavonoid accumulation in Arabidopsis repressed in lignin synthesis affects auxin transport and plant growth. *The Plant Cell, 19*(1), 148–162.

108. Coleman, H. D., Park, J. Y., Nair, R., Chapple, C., & Mansfield, S. D., (2008). RNAi-mediated suppression of p-coumaroyl-CoA 3′-hydroxylase in hybrid poplar impacts lignin deposition and soluble secondary metabolism. *Proceedings of the National Academy of Sciences, 105*(11), 4501–4506.

109. Chen, F., & Dixon, R. A., (2007). Lignin modification improves fermentable sugar yields for biofuel production. *Nature Biotechnology, 25*(7), 759–761.

110. Shafrin, F., Ferdous, A. S., Sarkar, S. K., Ahmed, R., Hossain, K., Sarker, M., & Khan, H., (2017). Modification of monolignol biosynthetic pathway in jute: Different gene, different consequence. *Scientific Reports, 7*(1), 1–12.

111. Jung, J. H., Fouad, W. M., Vermerris, W., Gallo, M., & Altpeter, F., (2012). RNAi suppression of lignin biosynthesis in sugarcane reduces recalcitrance for biofuel production from lignocellulosic biomass. *Plant Biotechnology Journal, 10*(9), 1067–1076.

112. Ragauskas, A. J., Williams, C. K., Davison, B. H., Britovsek, G., Cairney, J., Eckert, C. A., & Mielenz, J. R., (2006). The path forward for biofuels and biomaterials. *Science, 311*(5760), 484–489.

113. Studer, M. H., DeMartini, J. D., Davis, M. F., Sykes, R. W., Davison, B., Keller, M., & Wyman, C. E., (2011). Lignin content in natural Populus variants affects sugar release. *Proceedings of the National Academy of Sciences, 108*(15), 6300–6305.

114. Zhang, J., Xie, M., Li, M., Ding, J., Pu, Y., Bryan, A. C., & Lindquist, E. A., (2020b). Overexpression of a prefoldin β subunit gene reduces biomass recalcitrance in the bioenergy crop Populus. *Plant Biotechnology Journal, 18*(3), 859–871.

115. Sakamoto, S., Kamimura, N., Tokue, Y., Nakata, M. T., Yamamoto, M., Hu, S., & Kajita, S., (2020). Identification of enzymatic genes with the potential to reduce biomass recalcitrance through lignin manipulation in *Arabidopsis. Biotechnology for Biofuels, 13*, 1–16.

116. Tesfaw, A., & Assefa, F., (2014). Current trends in bioethanol production by *Saccharomyces cerevisiae*: Substrate, inhibitor reduction, growth variables, coculture, and immobilization. *International Scholarly Research Notices, 2014*.

117. Azhar, S. H. M., Abdulla, R., Jambo, S. A., Marbawi, H., Gansau, J. A., Faik, A. A. M., & Rodrigues, K. F., (2017). Yeasts in sustainable bioethanol production: A review. *Biochemistry and Biophysics Reports, 10*, 52–61.

118. Deparis, Q., Claes, A., Foulquié-Moreno, M. R., & Thevelein, J. M., (2017). Engineering tolerance to industrially relevant stress factors in yeast cell factories. *FEMS Yeast Research, 17*(4).

119. Lau, M. W., Gunawan, C., Balan, V., & Dale, B. E., (2010). Comparing the fermentation performance of *Escherichia coli* KO11, *Saccharomyces cerevisiae* 424A (LNH-ST) and *Zymomonas mobilis* AX101 for cellulosic ethanol production. *Biotechnology for Biofuels, 3*(1), 11.

120. Doğan, A., Demirci, S., Aytekin, A. Ö., & Şahin, F., (2014). Improvements of tolerance to stress conditions by genetic engineering in *Saccharomyces cerevisiae* during ethanol production. *Applied Biochemistry and Biotechnology, 174*(1), 28–42.

121. Albergaria, H., & Arneborg, N., (2016). Dominance of *Saccharomyces cerevisiae* in alcoholic fermentation processes: Role of physiological fitness and microbial interactions. *Applied Microbiology and Biotechnology, 100*(5), 2035–2046.

122. Banerjee, S., Mishra, G., & Roy, A., (2019). Metabolic engineering of bacteria for renewable bioethanol production from cellulosic biomass. *Biotechnology and Bioprocess Engineering*, 1–21.

123. Banat, I. M., Nigam, P., Singh, D., Marchant, R., & McHale, A. P., (1998). Ethanol production at elevated temperatures and alcohol concentrations: Part I–yeasts in general. *world Journal of Microbiology and Biotechnology, 14*(6), 809–821.

124. Yamaoka, C., Kurita, O., & Kubo, T., (2014). Improved ethanol tolerance of *Saccharomyces cerevisiae* in mixed cultures with *Kluyveromyces* lactis on high-sugar fermentation. *Microbiological Research, 169*(12), 907–914.

125. Ansanay-Galeote, V., Blondin, B., Dequin, S., & Sablayrolles, J. M., (2001). Stress effect of ethanol on fermentation kinetics by stationary-phase cells of *Saccharomyces cerevisiae*. *Biotechnology Letters, 23*(9), 677–681.

126. Auesukaree, C., (2017). Molecular mechanisms of the yeast adaptive response and tolerance to stresses encountered during ethanol fermentation. *Journal of Bioscience and Bioengineering, 124*(2), 133–142.

127. Riles, L., & Fay, J. C., (2019). Genetic basis of variation in heat and ethanol tolerance in *Saccharomyces cerevisiae*. *G3: Genes, Genomes, Genetics, 9*(1), 179–188.

128. Hu, X. H., Wang, M. H., Tan, T., Li, J. R., Yang, H., Leach, L., & Luo, Z. W., (2007). Genetic dissection of ethanol tolerance in the budding yeast *Saccharomyces cerevisiae*. *Genetics, 175*(3), 1479–1487.

129. Stanley, D., Bandara, A., Fraser, S., Chambers, P. J., & Stanley, G. A., (2010). The ethanol stress response and ethanol tolerance of *Saccharomyces cerevisiae*. *J. Appl. Microbiol., 109*(1), 13–24.

130. Charoenbhakdi, S., Dokpikul, T., Burphan, T., Techo, T., & Auesukaree, C., (2016). Vacuolar H+-ATPase protects *Saccharomyces cerevisiae* cells against ethanol-induced oxidative and cell wall stresses. *Applied and Environmental Microbiology, 82*(10), 3121–3130.

131. Pascual, C., Alonso, A., Garcia, I., Romay, C., & Kotyk, A., (1988). Effect of ethanol on glucose transport, key glycolytic enzymes, and proton extrusion in *Saccharomyces cerevisiae*. *Biotechnology and Bioengineering, 32*(3), 374–378.

132. Ma, M., & Liu, Z. L., (2010). Mechanisms of ethanol tolerance in *Saccharomyces cerevisiae*. *Applied Microbiology and Biotechnology, 87*(3), 829–845.

133. Hirasawa, T., Yoshikawa, K., Nakakura, Y., Nagahisa, K., Furusawa, C., Katakura, Y., & Shioya, S., (2007). Identification of target genes conferring ethanol stress tolerance

to *Saccharomyces cerevisiae* based on DNA microarray data analysis. *Journal of Biotechnology, 131*(1), 34–44.

134. Zhao, X. Q., & Bai, F. W., (2009). Mechanisms of yeast stress tolerance and its manipulation for efficient fuel ethanol production. *Journal of Biotechnology, 144*(1), 23–30.

135. Nuwamanya, E., Chiwona-Karltun, L., Kawuki, R. S., & Baguma, Y., (2012). Bio-ethanol production from non-food parts of cassava (Manihot esculenta Crantz). *Ambio, 41*(3), 262–270.

136. Benjaphokee, S., Hasegawa, D., Yokota, D., Asvarak, T., Auesukaree, C., Sugiyama, M., & Harashima, S., (2012). Highly efficient bioethanol production by a *Saccharomyces cerevisiae* strain with multiple stress tolerance to high temperature, acid and ethanol. *New Biotechnology, 29*(3), 379–386.

137. Shahsavarani, H., Sugiyama, M., Kaneko, Y., Chuenchit, B., & Harashima, S., (2012). Superior thermotolerance of *Saccharomyces cerevisiae* for efficient bioethanol fermentation can be achieved by overexpression of RSP5 ubiquitin ligase. *Biotechnology Advances, 30*(6), 1289–1300.

138. Mahmud, S. A., Hirasawa, T., & Shimizu, H., (2010). Differential importance of trehalose accumulation in *Saccharomyces cerevisiae* in response to various environmental stresses. *Journal of Bioscience and Bioengineering, 109*(3), 262–266.

139. Wiemken, A., (1990). Trehalose in yeast, stress protectant rather than reserve carbohydrate. *Antonie van Leeuwenhoek, 58*(3), 209–217.

140. Singer, M. A., & Lindquist, S., (1998). Thermotolerance in *Saccharomyces cerevisiae*: The yin and yang of trehalose. *Trends in Biotechnology, 16*(11), 460–468.

141. Vianna, C. R., Silva, C. L., Neves, M. J., & Rosa, C. A., (2008). *Saccharomyces cerevisiae* strains from traditional fermentations of Brazilian cachaca: Trehalose metabolism, heat and ethanol resistance. *Antonie Van Leeuwenhoek, 93*(1, 2), 205–217.

142. Cray, J. A., Stevenson, A., Ball, P., Bankar, S. B., Eleutherio, E. C., Ezeji, T. C., & Hallsworth, J. E., (2015). Chaotropicity: A key factor in product tolerance of biofuel-producing microorganisms. *Current Opinion in Biotechnology, 33*, 228–259.

143. Nwaka, S., Mechler, B., & Holzer, H., (1996). Deletion of the ATH1 gene in *Saccharomyces cerevisiae* prevents growth on trehalose. *FEBS Letters, 386*(2, 3), 235–238.

144. Jung, Y. J., & Park, H. D., (2005). Antisense-mediated inhibition of acid trehalase (ATH1) gene expression promotes ethanol fermentation and tolerance in *Saccharomyces cerevisiae*. *Biotechnology Letters, 27*(23, 24), 1855–1859.

145. Caspeta, L., Buijs, N. A., & Nielsen, J., (2013). The role of biofuels in the future energy supply. *Energy & Environmental Science, 6*(4), 1077–1082.

146. Caspeta, L., Castillo, T., & Nielsen, J., (2015). Modifying yeast tolerance to inhibitory conditions of ethanol production processes. *Frontiers in Bioengineering and Biotechnology, 3*, 184.

147. Kumar, A., Singh, L. K., & Ghosh, S., (2009). Bioconversion of lignocellulosic fraction of water-hyacinth (Eichhornia crassipes) hemicellulose acid hydrolysate to ethanol by *Pichia stipitis*. *Bioresource technology, 100*(13), 3293–3297.

148. Laluce, C., Schenberg, A. C. G., Gallardo, J. C. M., Coradello, L. F. C., & Pombeiro-Sponchiado, S. R., (2012). Advances and developments in strategies to improve strains of *Saccharomyces cerevisiae* and processes to obtain the lignocellulosic ethanol: A review. *Applied Biochemistry and Biotechnology, 166*(8), 1908–1926.

149. Komesu, A., Oliveira, J., Neto, J. M., Penteado, E. D., Diniz, A. A. R., & Da Silva, M. L. H., (2020). Xylose fermentation to bioethanol production using genetic engineering

microorganisms. In: *Genetic and Metabolic Engineering for Improved Biofuel Production from Lignocellulosic Biomass* (pp. 143–154). Elsevier.

150. Mussatto, S. I., Machado, E. M., Carneiro, L. M., & Teixeira, J. A., (2012). Sugars metabolism and ethanol production by different yeast strains from coffee industry wastes hydrolysates. *Applied Energy, 92*, 763–768.

151. Yanase, S., Hasunuma, T., Yamada, R., Tanaka, T., Ogino, C., Fukuda, H., & Kondo, A., (2010). Direct ethanol production from cellulosic materials at high temperature using the thermotolerant yeast *Kluyveromyces marxianus* displaying cellulolytic enzymes. *Applied Microbiology and Biotechnology, 88*(1), 381–388.

152. Hahn-Hägerdal, B., Karhumaa, K., Jeppsson, M., & Gorwa-Grauslund, M. F., (2007). Metabolic engineering for pentose utilization in *Saccharomyces cerevisiae*. In: *Biofuels* (pp. 147–177). Springer, Berlin, Heidelberg.

153. Matsushika, A., Inoue, H., Kodaki, T., & Sawayama, S., (2009). Ethanol production from xylose in engineered *Saccharomyces cerevisiae* strains: Current state and perspectives. *Applied Microbiology and Biotechnology, 84*(1), 37–53.

154. Hector, R. E., Dien, B. S., Cotta, M. A., & Mertens, J. A., (2013). Growth and fermentation of D-xylose by *Saccharomyces cerevisiae* expressing a novel D-xylose isomerase originating from the bacterium *Prevotella* ruminicola TC2-24. *Biotechnology for Biofuels, 6*(1), 1–12.

155. Katahira, S., Mizuike, A., Fukuda, H., & Kondo, A., (2006). Ethanol fermentation from lignocellulosic hydrolysate by a recombinant xylose-and cellooligosaccharide-assimilating yeast strain. *Applied Microbiology and Biotechnology, 72*(6), 1136–1143.

156. Karhumaa, K., Hahn-Hägerdal, B., & Gorwa-Grauslund, M. F., (2005). Investigation of limiting metabolic steps in the utilization of xylose by recombinant *Saccharomyces cerevisiae* using metabolic engineering. *Yeast, 22*(5), 359–368.

157. Fang, T., Yan, H., Li, G., Chen, W., Liu, J., & Jiang, L., (2020). Chromatin remodeling complexes are involvesd in the regulation of ethanol production during static fermentation in budding yeast. *Genomics, 112*(2), 1674–1679.

158. Fukuda, A., Kuriya, Y., Konishi, J., Mutaguchi, K., Uemura, T., Miura, D., & Okamoto, M., (2019). Kinetic modeling and sensitivity analysis for higher ethanol production in self-cloning xylose-using *Saccharomyces cerevisiae*. *Journal of Bioscience and Bioengineering, 127*(5), 563–569.

159. Xiong, M., Chen, G., & Barford, J., (2014). Genetic engineering of yeasts to improve ethanol production from xylose. *Journal of the Taiwan Institute of Chemical Engineers, 45*(1), 32–39.

160. Liu, T., Huang, S., & Geng, A., (2018). Recombinant diploid *Saccharomyces cerevisiae* strain development for rapid glucose and xylose co-fermentation. *Fermentation, 4*(3), 59.

161. Yang, P., Zhang, H., & Jiang, S., (2016). Construction of recombinant *sestc Saccharomyces cerevisiae* for consolidated bioprocessing, cellulase characterization, and ethanol production by in situ fermentation. *3 Biotech, 6*(2), 192.

162. Ko, J. K., Jung, J. H., Altpeter, F., Kannan, B., Kim, H. E., Kim, K. H., & Lee, S. M., (2018). Largely enhanced bioethanol production through the combined use of lignin-modified sugarcane and xylose fermenting yeast strain. *Bioresource Technology, 256*, 312–320.

163. Qin, L., Dong, S., Yu, J., Ning, X., Xu, K., Zhang, S. J., & Li, C., (2020). Stress-driven dynamic regulation of multiple tolerance genes improves robustness and productive capacity of *Saccharomyces cerevisiae* in industrial lignocellulose fermentation. *Metabolic Engineering*.

164. Ohta, E., Nakayama, Y., Mukai, Y., Bamba, T., & Fukusaki, E., (2016). Metabolomic approach for improving ethanol stress tolerance in *Saccharomyces cerevisiae. Journal of Bioscience and Bioengineering, 121*(4), 399–405.

165. Sun, Y., Xue, Q., Hou, J., Kong, M., Li, X., He, B., & Cao, L., (2020). *Research Square.* (preprint version) PPR129700 DOI: 10.21203/rs.2.22006/v1.

166. Yang, Y., Rong, Z., Song, H., Yang, X., Li, M., & Yang, S., (2020). Identification and characterization of ethanol-inducible promoters of *Zymomonas mobilis* based on omics data and dual reporter-gene system. Biotechnology and *Applied Biochemistry, 67*(1), 158–165.

167. Charoensuk, K., Sakurada, T., Tokiyama, A., Murata, M., Kosaka, T., Thanonkeo, P., & Yamada, M., (2017). Thermotolerant genes essential for survival at a critical high temperature in thermotolerant ethanologenic *Zymomonas mobilis* TISTR 548. *Biotechnology for Biofuels, 10*(1), 1–11.

168. Huang, S., Xue, T., Wang, Z., Ma, Y., He, X., Hong, J., Zou, S., Song, H., & Zhang, M., (2018). Furfural-tolerant *Zymomonas mobilis* derived from error-prone PCR-based whole genome shuffling and their tolerant mechanism. *Applied Microbiology and Biotechnology, 102*(7), 3337–3347.

169. Fuchino, K., Kalnenieks, U., Rutkis, R., Grube, M., & Bruheim, P., (2020). Metabolic profiling of glucose-fed metabolically active resting *Zymomonas mobilis* strains. *Metabolites, 10*(3), 6–8. https://doi.org/10.3390/metabo10030081.

170. Liu, Y., Ghosh, I. N., Martien, J., Zhang, Y., Amador-Noguez, D., & Landick, R., (2020b). Regulated redirection of central carbon flux enhances anaerobic production of bioproducts in *Zymomonas mobilis. Metabolic Engineering, 61*, 261–274. https://doi.org/10.1016/j.ymben.2020.06.005.

171. Sarkar, S., Mukherjee, A., Das, S., Ghosh, B., Chaudhuri, S., Bhattacharya, D., Sarbajna, A., & Gachhui, R., (2019). Nitrogen deprivation elicits dimorphism, capsule biosynthesis and autophagy in *Papiliotrema laur*entii strain RY1. *Micron, 124*, 102708. https://doi.org/10.1016/j.micron.2019.102708.

172. Ma, K., Ruan, Z., Shui, Z., Wang, Y., Hu, G., & He, M., (2016). Open fermentative production of fuel ethanol from food waste by an acid-tolerant mutant strain of *Zymomonas mobilis. Bioresource Technology, 203*, 295–302. https://doi.org/10.1016/j.biortech.2015.12.054.

173. Todhanakasem, T., Wu, B., & Simeon, S., (2020). Perspectives and new directions for bioprocess optimization using *Zymomonas mobilis* in the ethanol production. *World Journal of Microbiology and Biotechnology, 36*(8), 1–16. https://doi.org/10.1007/s11274-020-02885-4.

174. Yin, H., Zhang, R., Xia, M., Bai, X., Mou, J., Zheng, Y., & Wang, M., (2017). Effect of aspartic acid and glutamate on metabolism and acid stress resistance of *Acetobacter pasteurianus. Microbial Cell Factories, 16*(1), 1–14. https://doi.org/10.1186/s12934-017-0717-6.

175. Guan, N., & Liu, L., (2020). Microbial response to acid stress: Mechanisms and applications. *Applied Microbiology and Biotechnology, 104*(1), 51–65. -https://doi.org/10.1007/s00253-019-10226-1.

176. Todhanakasem, T., Yodsanga, S., Sowatad, A., Kanokratana, P., Thanonkeo, P., & Champreda, V., (2018). Inhibition analysis of inhibitors derived from lignocellulose pretreatment on the metabolic activity of *Zymomonas mobilis* biofilm and planktonic cells and the proteomic responses. *Biotechnology and Bioengineering, 115*(1), 70–81. https://doi.org/10.1002/bit.26449.

177. Liu, Y. F., Hsieh, C. W., Chang, Y. S., & Wung, B. S., (2017). Effect of acetic acid on ethanol production by *Zymomonas mobilis* mutant strains through continuous adaptation. *BMC Biotechnology, 17*(1), 1–10. https://doi.org/10.1186/s12896-017-0385-y.

178. Wu, B., Qin, H., Yang, Y., Duan, G., Yang, S., Xin, F., Zhao, C., et al., (2019). Engineered *Zymomonas mobilis* tolerant to acetic acid and low pH via multiplex atmospheric and room temperature plasma mutagenesis. *Biotechnology for Biofuels, 12*(1), 1–13. https://doi.org/10.1186/s13068-018-1348-9.

179. Singh, B., Kumar, P., Yadav, A., & Datta, S., (2019). Degradation of fermentation inhibitors from lignocellulosic hydrolysate liquor using immobilized bacterium, *Bordetella* sp. BTIITR. *Chemical Engineering Journal, 361*, 1152–1160. https://doi.org/10.1016/j.cej.2018.12.168.

180. He, M. X., Wu, B., Qin, H., Ruan, Z. Y., Tan, F. R., Wang, J. L., Shui, Z. X., et al., (2014). *Zymomonas mobilis*: A novel platform for future biorefineries. In: *Biotechnology for Biofuels* (Vol. 7, No. 1, pp. 1–15).

181. Kurgan, G., Panyon, L. A., Rodriguez-Sanchez, Y., Pacheco, E., Nieves, L. M., Mann, R., Nielsen, D. R., & Wanga, X., (2019). Bioprospecting of native efflux pumps to enhance furfural tolerance in ethanologenic *Escherichia coli*. *Applied and Environmental Microbiology, 85*(6), 1–11. https://doi.org/10.1128/AEM.02985-18.

182. Wang, W., Wu, B., Qin, H., Liu, P., Qin, Y., Duan, G., Hu, G., & He, M., (2019). Genome shuffling enhances stress tolerance of *Zymomonas mobilis* to two inhibitors. *Biotechnology for Biofuels, 12*(1), 1–12. https://doi.org/10.1186/s13068-019-1631-4.

183. Cao, H., Wei, D., Yang, Y., Shang, Y., Li, G., Zhou, Y., Ma, Q., & Xu, Y., (2017). Systems-level understanding of ethanol-induced stresses and adaptation in *E. coli*. *Scientific Reports*, pp. 1–15. https://doi.org/10.1038/srep44150.

184. Skupin, P., & Metzger, M., (2017). Stability analysis of the continuous ethanol fermentation process with a delayed product inhibition. *Applied Mathematical Modelling, 49*, 48–58. https://doi.org/10.1016/j.apm.2017.04.025.

185. Carreón-Rodríguez, O. E., Gutiérrez-Ríos, R. M., Acosta, J. L., Martinez, A., & Cevallos, M. A., (2019). Phenotypic and genomic analysis of *Zymomonas mobilis* ZM4 mutants with enhanced ethanol tolerance. *Biotechnology Reports, 23*. https://doi.org/10.1016/j.btre.2019.e00328.

186. Tan, F., Wu, B., Dai, L., Qin, H., Shui, Z., Wang, J., Zhu, Q., Hu, G., & He, M., (2016b). Using global transcription machinery engineering (gTME) to improve ethanol tolerance of *Zymomonas mobilis*. *Microbial Cell Factories, 15*(1), 1–9. https://doi.org/10.1186/s12934-015-0398-y.

187. Kim, D., (2018). Physico-chemical conversion of lignocellulose: Inhibitor effects and detoxification strategies: A mini review. *Molecules, 23*(2). https://doi.org/10.3390/molecules23020309.

188. Qiu, Z., Fang, C., Gao, Q., & Bao, J., (2020). A short-chain dehydrogenase plays a key role in cellulosic D-lactic acid fermentability of *Pediococcus acidilactici*. *Bioresource Technology, 297*, 122473. https://doi.org/10.1016/j.biortech.2019.122473.

189. Yi, X., Gao, Q., & Bao, J., (2019). Expressing an oxidative dehydrogenase gene in ethanologenic strain *Zymomonas mobilis* promotes the cellulosic ethanol fermentability. *Journal of Biotechnology, 303*, 1–7. https://doi.org/10.1016/j.jbiotec.2019.07.005.

190. Kopp, D., & Sunna, A. (2020). Alternative carbohydrate pathways–enzymes, functions and engineering. *Critical Reviews in Biotechnology, 40*(7), 895–912.

191. Mazaheri, D., & Pirouzi, A., (2020). Valorization of *Zymomonas mobilis* for bioethanol production from potato peel: Fermentation process optimization. *Biomass Conversion and Biorefinery.*
192. Palamae, S., Choorit, W., Chatsungnoen, T., & Chisti, Y., (2020). Simultaneous nitrogen fixation and ethanol production by *Zymomonas mobilis*. *Journal of Biotechnology, 314, 315,* 41–52.
193. Gao, X., Gao, Q., & Bao, J., (2018). Improving cellulosic ethanol fermentability of *Zymomonas mobilis* by overexpression of sodium ion tolerance gene ZMO0119. *Journal of Biotechnology, 282,* 32–37.
194. Wang, J. L., Wu, B., Qin, H., You, Y., Liu, S., Shui, Z. X., Tan, F. R., et al., (2016). Engineered *Zymomonas mobilis* for salt tolerance using EZ-Tn5-based transposon insertion mutagenesis system. *Microbial Cell Factories, 15*(1), 1–10. https://doi.org/10.1186/s12934-016-0503-x.
195. Duan, G., Wu, B., Qin, H., Wang, W., Tan, Q., Dai, Y., Qin, Y., Tan, F., Hu, G., & He, M., (2019). Replacing water and nutrients for ethanol production by ARTP derived biogas slurry tolerant *Zymomonas mobilis* strain. *Biotechnology for Biofuels, 12*(1), 1–12.
196. Robak, K., & Balcerek, M., (2018). Review of second-generation bioethanol production from residual biomass. *Food technology and Biotechnology, 56*(2), 174–187.
197. Valta, K., Papadaskalopoulou, C., Dimarogona, M., & Topakas, E., (2019). Bioethanol from waste-prospects and challenges of current and emerging technologies. *Byproducts from Agriculture and Fisheries: Adding Value for Food, Feed, Pharma, and Fuels,* 421–456.
198. Choi, K. R., Jiao, S., & Lee, S. Y., (2020). Metabolic engineering strategies toward production of biofuels. *Current Opinion in Chemical Biology, 59,* 1–14.
199. Sharma, B., Larroche, C., & Dussap, C. G., (2020b). Comprehensive assessment of 2G bioethanol production. *Bioresource Technology,* 123630.
200. Di Donato, P., Finore, I., Poli, A., Nicolaus, B., & Lama, L., (2019). The production of second-generation bioethanol: The biotechnology potential of thermophilic bacteria. *Journal of Cleaner Production, 233,* 1410–1417.
201. Lynd, L. R., Van, Z. W. H., McBride, J. E., & Laser, M., (2005). Consolidated bioprocessing of cellulosic biomass: An update. *Current Opinion in Biotechnology, 16*(5), 577–583.
202. Poudel, S., Giannone, R. J., Rodriguez, M., Raman, B., Martin, M. Z., Engle, N. L., & Ussery, D., (2017). Integrated omics analyses reveal the details of metabolic adaptation of *Clostridium thermocellum* to lignocellulose-derived growth inhibitors released during the deconstruction of switchgrass. *Biotechnology for Biofuels, 10*(1), 14.
203. Cui, J., Stevenson, D., Korosh, T., Amador-Noguez, D., Olson, D. G., & Lynd, L. R., (2020). Developing a cell-free extract reaction (CFER) system in *Clostridium thermocellum* to identify metabolic limitations to ethanol production. *Frontiers in Energy Research,* 8.
204. Holwerda, E. K., Olson, D. G., Ruppertsberger, N. M., Stevenson, D. M., Murphy, S. J. L., Maloney, M. I., Lanahan, A. A., et al., (2020). Metabolic and evolutionary responses of *Clostridium thermocellum* to genetic interventions aimed at improving ethanol production. *Biotechnology for Biofuels, 13*(1), 40. https://doi.org/10.1186/s13068-020-01680-5.
205. Garcia, S., Thompson, A. R., Giannone, R. J., Dash, S., Maranas, C., & Trinh, C. T., (2020). *Development of a Genome-Scale Metabolic Model of Clostridium thermocellum and its Applications for Integration of Multi-Omics Datasets and Strain Design.* BioRxiv.
206. Singh, N., Mathur, A. S., Gupta, R. P., Barrow, C. J., Tuli, D., & Puri, M., (2018). Enhanced cellulosic ethanol production via consolidated bioprocessing by *Clostridium thermocellum* ATCC 31924. *Bioresource Technology, 250,* 860–867.

207. Liu, K., Yuan, X., Liang, L., Fang, J., Chen, Y., He, W., & Xue, T., (2019b). Using CRISPR/ Cas9 for multiplex genome engineering to optimize the ethanol metabolic pathway in *Saccharomyces cerevisiae*. *Biochemical Engineering Journal, 145*, 120–126. https://doi. org/10.1016/j.bej.2019.02.017.

208. Kim, S. K., & Westpheling, J., (2018). Engineering a spermidine biosynthetic pathway in *Clostridium thermocellum* results in increased resistance to furans and increased ethanol production. *Metabolic Engineering, 49*, 267–274.

209. Cheng, C., Bao, T., & Yang, S. T., (2019). Engineering *Clostridium* for improved solvent production: Recent progress and perspective. *Applied Microbiology and Biotechnology, 103*(14), 5549–5566.

210. Lo, J., Olson, D. G., Murphy, S. J. L., Tian, L., Hon, S., Lanahan, A., & Lynd, L. R., (2017). Engineering electron metabolism to increase ethanol production in *Clostridium thermocellum*. *Metabolic Engineering, 39*, 71–79.

211. Rydzak, T., Garcia, D., Stevenson, D. M., Sladek, M., Klingeman, D. M., Holwerda, E. K., & Guss, A. M., (2017). Deletion of type I glutamine synthetase deregulates nitrogen metabolism and increases ethanol production in *Clostridium thermocellum*. *Metabolic Engineering, 41*, 182–191.

212. Riederer, A., Takasuka, T. E., Makino, S. I., Stevenson, D. M., Bukhman, Y. V., Elsen, N. L., & Fox, B. G., (2011). Global gene expression patterns in *Clostridium thermocellum* as determined by microarray analysis of chemostat cultures on cellulose or cellobiose. *Applied and Environmental Microbiology, 77*(4), 1243–1253.

213. Tian, L., Papanek, B., Olson, D. G., Rydzak, T., Holwerda, E. K., Zheng, T., & Hettich, R. L., (2016). Simultaneous achievement of high ethanol yield and titer in *Clostridium thermocellum*. *Biotechnology for Biofuels, 9*(1), 1–11.

214. Hon, S., Olson, D. G., Holwerda, E. K., Lanahan, A. A., Murphy, S. J., Maloney, M. I., & Lynd, L. R., (2017). The ethanol pathway from *Thermoanaerobacterium saccharolyticum* improves ethanol production in *Clostridium thermocellum*. *Metabolic Engineering, 42*, 175–184.

215. Deng, Y., Olson, D. G., Zhou, J., Herring, C. D., Shaw, A. J., & Lynd, L. R., (2013). Redirecting carbon flux through exogenous pyruvate kinase to achieve high ethanol yields in *Clostridium thermocellum*. *Metabolic Engineering, 15*, 151–158.

216. Argyros, D. A., Tripathi, S. A., Barrett, T. F., Rogers, S. R., Feinberg, L. F., Olson, D. G., & Caiazza, N. C., (2011). High ethanol titers from cellulose by using metabolically engineered thermophilic, anaerobic microbes. *Applied and Environmental Microbiology, 77*(23), 8288–8294.

217. Rydzak, T., Lynd, L. R., & Guss, A. M., (2015). Elimination of formate production in *Clostridium thermocellum*. *Journal of Industrial Microbiology & Biotechnology, 42*(9), 1263–1272.

218. Verbeke, T. J., Giannone, R. J., Klingeman, D. M., Engle, N. L., Rydzak, T., Guss, A. M., & Elkins, J. G., (2017). Pentose sugars inhibit metabolism and increase expression of an AgrD-type cyclic pentapeptide in *Clostridium thermocellum*. *Scientific Reports, 7*, 43355.

219. Xiong, W., Reyes, L. H., Michener, W. E., Maness, P. C., & Chou, K. J., (2018). Engi- neering cellulolytic bacterium *Clostridium thermocellum* to co-ferment cellulose-and hemicellulose-derived sugars simultaneously. *Biotechnology and Bioengineering, 115*(7), 1755–1763.

220. Van, D. V. D., Lo, J., Brown, S. D., Johnson, C. M., Tschaplinski, T. J., Martin, M., & Guss, A. M., (2013). Characterization of *Clostridium thermocellum* strains with

disrupted fermentation end-product pathways. *Journal of Industrial Microbiology & Biotechnology, 40*(7), 725–734.

221. Papanek, B., Biswas, R., Rydzak, T., & Guss, A. M., (2015). Elimination of metabolic pathways to all traditional fermentation products increases ethanol yields in *Clostridium thermocellum*. *Metabolic Engineering, 32*, 49–54.

222. Kannuchamy, S., Mukund, N., & Saleena, L. M., (2016). Genetic engineering of *Clostridium thermocellum* DSM1313 for enhanced ethanol production. *BMC Biotechnology, 16*(1), 1–6.

223. Selim, K. A., El-Ghwas, D. E., Easa, S. M., & Abdelwahab, H. M. I., (2018). Bioethanol a microbial biofuel metabolite; new insights of yeasts metabolic engineering. *Fermentation, 4*(1), 16. https://doi.org/10.3390/fermentation4010016.

224. Xue, T., Liu, K., Chen, D., Yuan, X., Fang, J., Yan, H., Huang, L., Chen, Y., & He, W., (2018). Improved bioethanol production using CRISPR/Cas9 to disrupt the ADH₂ gene in *Saccharomyces cerevisiae*. *World Journal of Microbiology and Biotechnology, 34*(10), 154. https://doi.org/10.1007/s11274-018-2518-4.

225. Fan, D., Li, C., Fan, C., Hu, J., Li, J., Yao, S., & Luo, K., (2020). MicroRNA6443-mediated regulation of ferulate 5-hydroxylase gene alters lignin composition and enhances saccharification in Populus tomentosa. *New Phytologist, 226*(2), 410–425.

226. Liu, S., Liu, Y. J., Feng, Y., Li, B., & Cui, Q., (2019a). Construction of consolidated bio-saccharification biocatalyst and process optimization for highly efficient lignocellulose solubilization. *Biotechnology for Biofuels, 12*(1), 35.

227. Toivari, M. H., Aristidou, A., Ruohonen, L., & Penttilä, M., (2001). Conversion of xylose to ethanol by recombinant *Saccharomyces cerevisiae*: Importance of xylulokinase (XKS1) and oxygen availability. *Metabolic Engineering, 3*(3), 236–249.

228. Fu, C., Mielenz, J. R., Xiao, X., Ge, Y., Hamilton, C. Y., Rodriguez, M., ... & Dixon, R. A. (2011). Genetic manipulation of lignin reduces recalcitrance and improves ethanol production from switchgrass. *Proceedings of the National Academy of Sciences, 108*(9), 3803–3808.

229. Bhatia, S. K., Kim, S. H., Yoon, J. J., & Yang, Y. H. (2017). Current status and strategies for second generation biofuel production using microbial systems. *Energy Conversion and Management, 148*, 1142–1156.

230. Javed, U., Ansari, A., Aman, A., & Qader, S. A. U. (2019). Fermentation and saccharification of agro-industrial wastes: A cost-effective approach for dual use of plant biomass wastes for xylose production. Biocatalysis and *Agricultural Biotechnology*, 21, 101341.

231. Ayodele, B. V., Alsaffar, M. A., & Mustapa, S. I. (2020). An overview of integration opportunities for sustainable bioethanol production from first-and second-generation sugar-based feedstocks. *Journal of Cleaner Production, 245*, 118857.

232. Ledesma-Amaro, R., Lazar, Z., Rakicka, M., Guo, Z., Fouchard, F., Crutz-Le Coq, A. M., & Nicaud, J. M. (2016b). Metabolic engineering of *Yarrowia lipolytica* to produce chemicals and fuels from xylose. *Metabolic Engineering, 38*, 115–124.

233. Hahn-Hägerdal, B., Galbe, M., Gorwa-Grauslund, M. F., Lidén, G., & Zacchi, G. (2006). Bio-ethanol–the fuel of tomorrow from the residues of today. *Trends in biotechnology, 24*(12), 549–556.

CHAPTER 6

Increasing Ethanol Tolerance in Industrially Important Ethanol Fermenting Organisms

KALYANASUNDARAM GEETHA THANUJA[1] and
SUBBURAMU KARTHIKEYAN[2]

[1]Department of Agricultural Microbiology,
Tamil Nadu Agricultural University, Coimbatore, Tamil Nadu, India

[2]Department of Renewable Energy Engineering, Tamil Nadu Agricultural University, Coimbatore, Tamil Nadu, India, E-mail: skarthy@tnau.ac.in

ABSTRACT

The growing population and increased urbanization have set great demand for energy. With the deprivation of fossil fuels, interest has focused on renewable sources. Among them, bioethanol is an attractive and green source being used as a promising alternative to reduce environmental pollution. In the process of ethanol production, varied ranges of feedstocks are being converted through microbial fermentation. Widespread industrial biocatalysts are exploited in bioethanol production like bacteria, yeast, fungi, and algae according to the type of raw material, environmental conditions, and resources available. However, the organisms leading the ethanol industries are *Saccharomyces cerevisiae, Pichia stipites, Zymomonas mobilis,* and *Clostridium thermocellum.* Their capacity to yield higher ethanol, ability to ferment sugars, growth in simple and inexpensive media, resistance to inhibitors, and contaminants makes them potential candidates in the fermentation process. While the maintenance of growth and metabolic efficiency of these microbes with resistance to various stresses is the desirable factor, several challenges predominate in the microbial fermentation inhibiting the overall ethanol production in which ethanol toxicity is a significant concern. The key advantages of microbial strain employed for industrial ethanol production are, ethanol tolerance limits and substrate concentration, ultimately decreasing productivity and increasing the costs of bioreactor. Ethanol toxicity is the complex and multi loci trait, affecting various cell functions and inhibits key glycolytic enzymes by denaturation mechanisms. The process renders the microbial cell to undergo reprogramming in cellular and metabolic activities with increased repair functions. However, it is crucial to understand the consequences of ethanol stress defense mechanism for ethanol tolerance improvement strategies. Various approaches have been attempted to circumvent the ethanol stress and enhance ethanol tolerance from optimizing its media to rewiring of genetic setup. Towards the establishment of ethanol tolerant phenotypes, integrative system, and the interplay of genes involving complex network is essential. With the assistance of molecular tools, the strain with improved ethanol tolerance can be achieved.

6.1 INTRODUCTION

The energy consumption across the globe is ever-widening for improved living standards. The demand for energy hunts mainly towards fossil fuels as the major resource which sets step for environmental pollution, global warming (GW), depleted natural resources, etc. With large depletion of the world's petroleum reserves and growing costs, interest vested on the production of ethanol. Ethanol has been widely acknowledged as an alternate fuel for transport which can be readily produced from various lignocellulosic feedstocks through microbial fermentation. Generation of high yield ethanol can be achieved through the implementation of consolidated bioprocessing (CBP). The feasibility of CBP gets hindered by various stress factors causing the industrial fermentation troublesome.

Various fermentative microbes, namely *S. cerevisiae, Z. mobilis,* and *C. thermocellum,* have been exploited successfully in the industrial production of alcohol-related products owing to their unique characteristics (Table 6.1). Although the organisms involved in the industrial processes have great sturdiness, they often lack tolerance to certain stress factors.

To ensure the viability of the industrial process, the end product has to accumulate with high yield and titer which becomes the near-universal stress factor. The end product of fermentation, ethanol interrupts with glycolytic enzymes harming the cell membrane of fermentative bacteria. To maintain the pace with escalating demand for ethanol, the pressing prerequisite is to overcome the barriers in the production and recovery process. This chapter provides an overview of different strategies to develop ethanol tolerant strains of *the above-mentioned organisms* and highlight specific strategies unique to individual organisms.

TABLE 6.1 Comparison of Physiological Characteristics of *S. cerevisiae, Z. mobilis, C. thermocellum*

Categories	S. cerevisiae	Z. mobilis	C. thermocellum
Growth condition	Facultative aerobe	Facultative anaerobe	Thermophilic anaerobe
Taxonomy	Eukaryotic	Gram-negative bacterium	Gram-positive bacterium
pH preference	2.0–6.5	3.5–7.5	6.0–8.5
Metabolic pathway	EMP	ED	ABE fermentation
Theoretical ethanol yield	>90%	>90%	2.5%
Ethanol tolerance (w/v)	15%	16%	1–2%
Safety	GRAS	GRAS	GRAS

The brewer's yeast, *Saccharomyces cerevisiae* is promising for industrial alcoholic beverage production and bioethanol fermentation. Ethanol begins to accumulate in the early stage of fermentation, and due to the depletion of essential nutrients, the quantity of ethanol outstretches to toxic levels towards the end of the process. Although, they have maximum tolerance between 115 and 200 gL^{-1}, their accumulation higher than 150 $g.L^{-1}$ in the culture broth limits yield and productivity. Ethanol is found to damage the mitochondrial DNA of yeast cells and inactivates key enzymes like dehydrogenase and hexokinase. Beyond the threshold limit, ethanol stress significantly influences lipid production, ion homeostasis, up-regulation of several genes, trehalose metabolism, and altered transcription behavior. Several strains of yeast have adapted mechanisms to tolerate a higher concentration of ethanol than most other organisms, yet their mechanism remains complex to elucidate. Ethanol tolerance in yeast is a complex phenotype influenced by various genetic, physiological, physical, and environmental factors.

Z. mobilis is an ethanologenic, gram-negative, and obligate fermentative bacterium recruiting unique metabolism to produce ethanol by the EntnerDoudoroff pathway. They are well recognized for ethanol and glucose tolerance and exhibit rapid fermentation than *S. cerevisiae*. Further, they own desirable industrial biocatalyst criteria such as increased specific-productivity, wide pH for production (pH 3.5–7.5), reduced cost for sophisticated aeration control, and safe status of production [1]. With more than 7% initial concentration of ethanol and at ≥35°C, the cell mass and ethanol production get disrupted.

The fermentation capacity of various *Z. mobilis* strains was assessed and optimized parameters for ethanol production [2] Compared with *S. cerevisiae*, the studies on the physiological and genetic basis of ethanol tolerance in *Z. mobilis* is relatively lower. Some ethanol tolerant strains have been developed by ethanol-adaption, mutagenesis, and genetic engineering [3–5]. Ethanol adaption (EA) in wild strains of *C. thermocellum* revealed a greater percentage of fatty acids and an increase in the percentage of plasmalogens significantly contribute to membrane rigidity counteracting the effects of ethanol [6]. The mechanism of ethanol tolerance was explored by systematic metabolomics to analyze the phenotypes of ethanol tolerant and wild type strains. The results revealed an accumulation of cellodextrin in the ethanol tolerant strain depicting its potential mechanism for stress resistance [7].

6.2   EFFECTS OF ETHANOL ON YEAST, *Zymomonas mobilis* AND *Clostridium thermocellum*

During the fermentation of ethanol, microbial cells experience various stresses like increasing ethanol concentration, high temperature, feedback inhibition, inappropriate substrate concentration, osmotic pressure, etc. The small size of ethanol and the functional group renders solubility to both aqueous and lipid environments. It easily passes through the plasma membrane and enhances the fluidity and permeability and causes a series of consequences at the cellular level [8]. Recent studies evidence that ethanol tolerance is a polygenic phenotype trait governed by multiple alleles with complex interaction and varies between the strains. As a response mechanism, saturated lipids, transmembrane lipids, sterols, and hopanoid gets enhanced in the membrane of *Clostridium, Zymomonas,* and *Saccharomyces* [9, 10]. In general, for the recovery of ethanol from classical downstream, *Saccharomyces* should possess the ability to grow and produce ethanol in 4% (v/v) ethanol [11]. It involves cellular dysfunctions, lowered respiratory rates and ATP, the formation of reactive oxygen species (ROS) which then induces DNA damage and oxidative stress. The flux of protons gets deregulated ultimately resulting in decreased proton motive force compromising the nutrient uptake. Among them, a high level of ethanol interrupts not only the physiology of the cell affecting growth, viability, and fermentation but also overall ethanol production. Table 6.2, briefs the effect of ethanol on microorganisms. Microarray and several expression tools revealed the genes and their complex network associated with ethanol tolerance disclosing its enigmatic nature. There are many challenges faced in developing strategies for enhancing ethanol tolerance. This necessitates the need for ethanol tolerant strains which could be achieved through several reprogramming strategies.

TABLE 6.2 Effect of Ethanol on the Physiology and Metabolism of Fermenting Microorganisms

	Effect	References
S. cerevisiae	At low concentration:	[17]
	Inhibits cell division, growth rate and decreases cell volume	
	At high concentration:	
	Decreases cell vitality and causes death	
	Function of ribosome is inhibited	[18]
	Governs vitality and viability of the cells, thereby terminating fermentation; Inhibits glucose and amino acids transport systems, denaturation key glycolytic enzymes	[19]
	Membrane permeabilization, cytosolic, and vacuolar acidification	[20]
	Dissipation of membrane integrity and increase in membrane permeability due to intercalation of lipid bilayer	[21]
Z. mobilis	Decreased cell mass and ethanol production, Inhibition of fermentation process	[22]
	Decreased rate of substrate conversion	[23]
	Inactivation of enzymes involved in ethanol production	[24]
C. thermocellum	Imbalance in NADPH	[3]
	Interruption in proton motive force and ATP production	[25]

Few studies revealed the genes involved in ethanol stress guiding the rational strategies to progress the process performance. The genome sequence of *Z. mobilis*ZM4 revealed the sigma factor (σE, ZMO4104) plays a crucial role in against ethanol stress [12]. Alcohol dehydrogenase II (ADH2) protein, encoded by the adhB gene was found to be the major protein involved in ethanol stress [13]. Further Pallach et al. [14] have proposed the exopolysaccharides (EPS) components of *Z. mobilis* contributes to its high ethanol tolerance mechanism.

Ethanol fermentation using thermophilic bacteria has been suggested as one of the novel systems. However, the membrane composition of *S. cerevisiae* and *Z. mobilis* provides higher tolerance to ethanol. *C. thermocellum* is a gram-positive, thermophilic anaerobe that serves as an efficient candidate in CBP due to its ability to rapidly fermenting cellulosic biomass. One of the challenges in the industrial application hindering their potential is low ethanol yield, titer, and tolerance producing acetate, lactate, H_2, formate as further fermentation products [15]. The complexity of ethanol tolerance in *C. thermocellum* was elucidated by eliminating the H_2 production by redirecting carbon flux towards only ethanol production, thereby rendering the electrons available for reduction of acetyl CoA to ethanol [16].

6.3 METHODS FOR IMPROVING ETHANOL TOLERANCE

Various tolerance mechanisms rendered by the organisms are depicted in Figure 6.1. Ethanol on passing into the plasma membrane increases membrane fluidity which in turn deteriorates membrane integrity through partitioning into a lipid bilayer. The aliphatic chains of ethanol get deeply inserted into the hydrophobic interior of a lipid bilayer and favor a high degree of permeability. As a tolerance mechanism, the cell wall lipid composition is increased with polyunsaturated fatty acids, ergosterol, and phosphatidylcholine. The cell wall architecture gets remodeled to attain robust nature through various signaling pathways [26]. The increased membrane permeability also results in cytosolic and vacuolar acidification for which vacuolar H^+-ATPase (V-ATPase) plays a crucial role in the maintenance of intracellular pH homeostasis. The intracellular transport of H^+ into vacuoles by H^+ V-ATPase assists in translocation across the vacuolar membrane through hydrolysis of ATP, thereby counteracts the ethanol stress. Further several genes would up regulate upon ethanol stress leading to an alteration in the transcription machinery which renders protection to the organism by various physiological changes. Therefore, the design of the ethanol tolerant organisms can be concentrated on the above-mentioned background with numerous techniques as follows. The creation of strains with augmented stress tolerance is however attainable through a range of approaches (Figure 6.2), combining with recent technology could limit time and tedious multistep. Traditional approaches for the development of ethanol tolerant phenotype include rational selection approaches for modification followed by probability-driven processes like adaptive evolution or random mutagenesis. The strain development for ethanol tolerance has certain gold standards like mutagenesis and selection traditionally in the industry. The advent of recombinant tools gives rise to more sophisticated alternate strategies to ameliorate alcohol toxicity. The advanced omics technologies such as genomics, proteomics, transcriptomics, and metabolomics have aided to understand the complex alcohol toxicity and tolerating mechanism. Exploring the molecular basis of ethanol tolerance en routes the development of rational approaches. The metabolomic analyzes of *E. coli* on the responses to ethanol and butanol tolerance elucidated several amino acids like valine, glycine, isoleucine, glutamic acid and osmoprotectants like trehalose. Most transcriptomic analyzes have shown the role of small molecular weight compounds on alcohol tolerance.

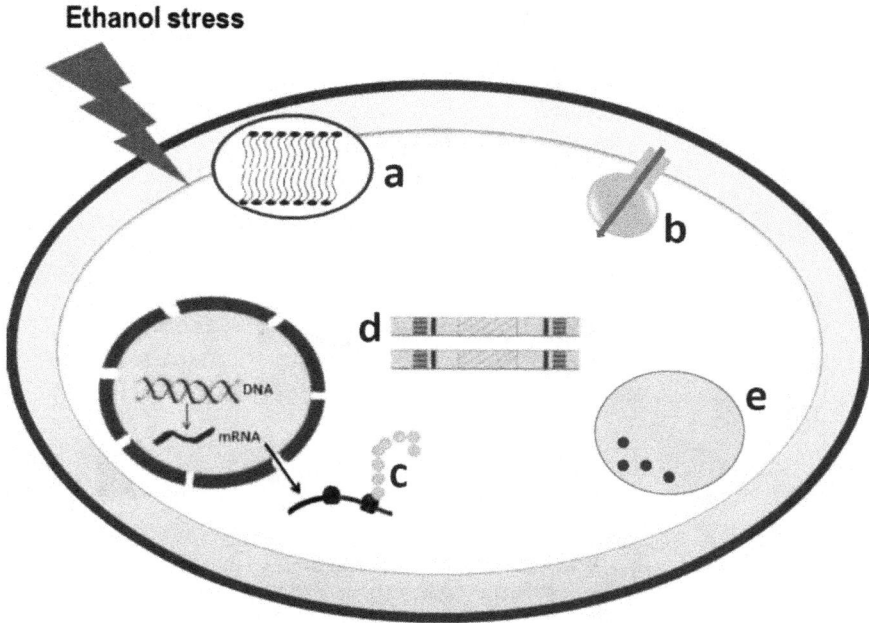

FIGURE 6.1 Ethanol tolerance mechanism of mentioned microbes. (a) change in membrane composition; (b) maintenance of intracellular pH by H⁺ ATPase pump; (c and d) production of stress-responsive proteins/genes; (e) vacuole acidification/vacuole mediated transport.

6.3.1   GENES INVOLVED

High ethanol concentration associates several 100 genes concerning cell wall and membrane organization, amino acid biosynthesis, lipids, and fatty acid metabolism, and few genes are presented in Table 6.3. The studies on comprehensive gene expression, regulatory networks, and pathway-based and integrated resistance have revealed numerous genes [27]. To explore the molecular mechanism beneath the tolerance of yeast cells, the cell wall integrity (CWI) pathway, and high-osmolarity glycerol (HOG) pathway was studied along with their genes involved. Upon ethanol exposure, CWI in collaboration with HOG triggers the transcription regulation of cell wall biosynthesis genes *FKS2*, *CRH1*, and *PIR3* thereby causes remodeling of the cell wall [28]. The mutants lacking CWI genes were found to be sensitive to ethanol.

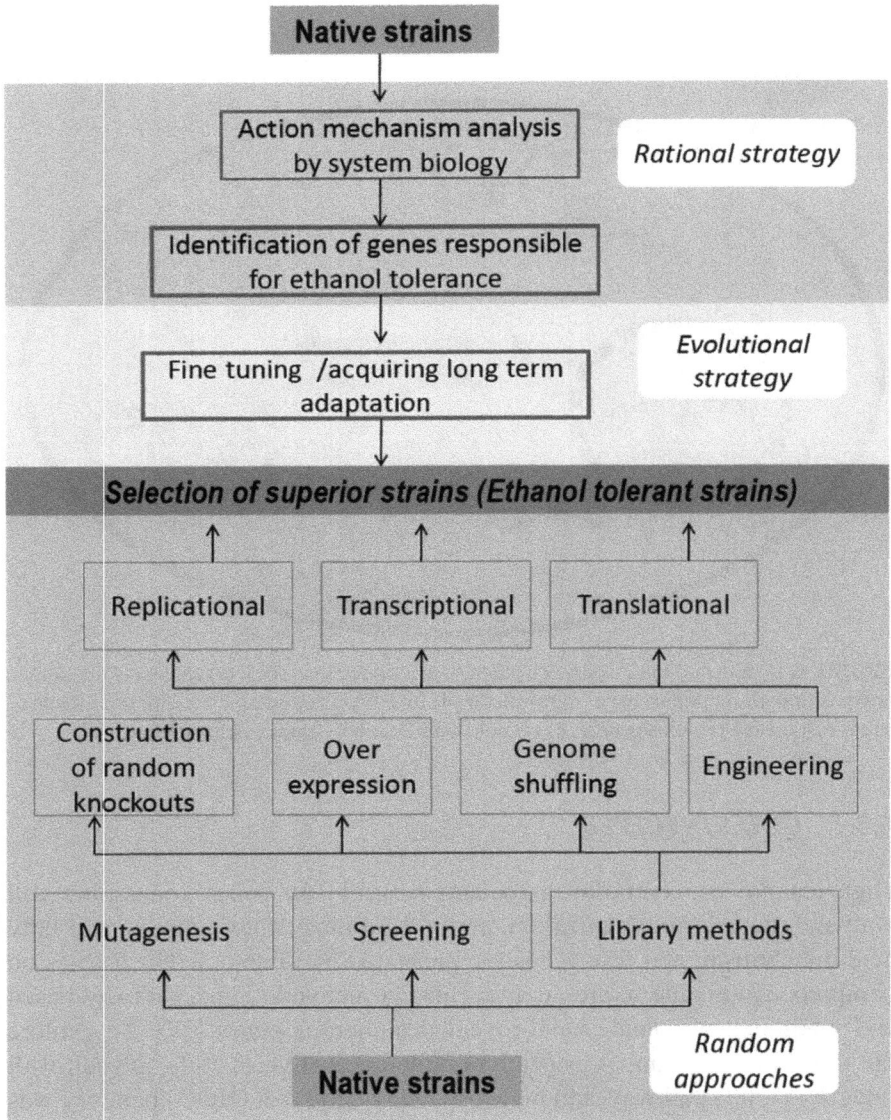

FIGURE 6.2   Various approaches are available to develop ethanol tolerant strains.

Msn2p is the transcription factor, whose activation is controlled by protein kinase A (PKA), which regulates the interconversion of glucose 1-phosphate and glucose-6-phosphate. Msn2p binds to the stress response elements, inducing up-regulation of more than 200 genes. The protein sequencing by *in silico* analysis of msn2 discovered, serine residue at position 625 as an effective target of PKA. Replacement of serine residue with alanine augmented the cell susceptibility to ethanol indicating its crucial role [29]. The transcriptomic and proteomic studies of *S. cerevisiae* 131 under ethanol stress revealed 937 differentially expressed genes and 457 differentially expressed proteins [18].

These expressions also revealed 10% (v/v) ethanol could induce sexual reproduction, signal transduction such as G-protein, silent information regulator (sir) proteins, and aromatic amino acids (AAA) helping in the survival of cells.

TABLE 6.3   Genes Involved in Ethanol Tolerance Mechanism

Genes	Function	Alteration in Gene	References
tps1	Catalyze trehalose synthesis	Increased accumulation of intracellular trehalose by over expression of *tps1*	[30]
CRH1	Cell wall biosynthesis	Transcriptional activation triggers remodeling of cell wall	[28]
SNF1	Develop cell resistance and consumption of glucose under stress conditions	Overexpression of *SNF1*	[31]
Rice metallothionein isoform OsMTI-1b	Protection against oxidative stress	Heterologous expression in yeast conferred tolerance	[32]
YAP1 and *SOD2*	Key stress responsible elements	Ensures signaling function desired for ethanol tolerance	[33]
ERG2, ERG3. ERG4	Ergosterol biosynthesis	Upregulation under ethanol stress	[34]
Hsp40, Hsp82, Hsp104, KAR2	Heat shock proteins		
TH14, TH15, and *TH13*	Biosynthesis of thiamine		
PRO1	Proline biosynthesis	Downregulation under ethanol stress	

6.3.2   *NUTRITIONAL STRATEGY*

Wide exploration of the mechanism is mandated to produce advanced yeast variants or mutants. The conventional methods of improving strains towards higher productivity have had partial accomplishment beyond the identification of medium optimization and various chemical protectants (Table 6.4).

TABLE 6.4   Nutritional Strategy for Enhancing Ethanol Tolerance

Optimized Conditions	Effects	References
Addition of 0.02 g/l zinc sulfate	Enhanced cell viability in ethanol shock treatment	[35]
Supplementation of zinc with other metal ions	Maximum viability, ethanol, and biomass production	[36]
Addition of KCl	Increased ergosterol and ATP content with improved biomass and cell viability	[37]
Supplementation of Mg^{2+} in the culture media	50% cells remained viable in exposure of 20% (v/v) ethanol for 9 h	[38]
Exponential feeding of vitamins	Increase in final ethanol titer and ethanol tolerance	[39]
Supplementation of $CaCl_2$ and $MgCl_2$	Protection against toxic effect of ethanol	[40]
Non-limiting oxygen condition	23% increase in viable cell mass	[41]
Supplementation of 10–20 mM magnesium	Reduction in cell mortality	[22]

6.3.3   EVOLUTIONARY ENGINEERING

It follows the engineering principle by variation and selection in genetic diversity. Random mutations in the genome induced by several stress factors are known to be adaptive evolution (domestication). They serve as a tool to develop strains through the continuous passage of strains and effectively change a few characteristics of physiological/phenotypic origin. They are also regarded as a capable approach for the non-recombinant alteration of industrial yeast strains. The existence of stress maintains the yeast cells under selective pressure and genotypes which survive are preserved. Stepwise increase of ethanol concentration in culture medium was reported to choose ethanol tolerant strains of the *S. cerevisiae* [42–44].

Change in the DNA sequence of six independent populations of genetically identical yeast upon exposure to gradually increasing concentration of ethanol was monitored over two years. The novel computational analyzes analyzed the mutational dynamics and molecular mechanisms of ethanol tolerance are achieved through complex and different mutational pathways [45]. Evolved populations exhibited intricate interaction of de novo single nucleotide mutations, copy number variation, ploidy level, and clonal interference significantly contributing to ethanol tolerance. Liquid nitrogen incubation of cell culture for 30 min followed by thawing at 30°C for 20 min (Freeze-thaw treatment) was found to enhance ethanol tolerance [46]. The mutant developed upon freeze-thaw treatment had a profound change in the dynamic structure of membrane lipids and capable of accumulating trehalose thereby contributing to ethanol tolerance. Further aneuploidy of chromosome III is attaining interest in industrial strains of *S. cerevisiae* as an adaptive trait [47]. Strains tolerating 12% (v/v) ethanol were developed by Turanli-Yildiz et al. [48] by *in vivo* evolutionary engineering and also hypothesized the link between ethanol tolerance and diploidization of yeast strain. Selective pressure of inhibitory concentration of hexanol evolved tolerant *S. cerevisiae* BY4741 [49]. The sequential transfer method with increasing ethanol concentration in *Z. mobilis* ZM4 mutant ER7ap exhibited better performance than wild type [50].

6.3.4 GENETIC APPROACHES

6.3.4.1 RANDOM MUTAGENESIS

The classical strain improvement of industrial microorganisms is using random mutagenesis and screening. Their well-established history of efficient application to make genetically diverse phenotype is an engrossing tool in the post- "omics" era. Although they have been used successfully in the strain improvement program, random methods and global techniques have to be combined to generate complex phenotypes. This method is often coupled with an adaptive laboratory evolution strategy since classical mutagenesis demands immense time for continuous screening and rarely yields elite strains. Furthermore, ethanol tolerance as a complex phenotype requires synergistic functions of a variety of genes thus making it complex to genetic engineering. For improving these complex phenotypes, random mutagenesis of TATA-binding protein is the strategy for transcription machinery engineering. Sufficient selection followed by screening is essential in industrial applications. Mathematical modeling serves as an adroit tool to achieve this selection. It was attempted with the help of transmission electron microscopy (TEM) in probiotic yeast *S. cerevisiae* var. *bouldardii* CNCM I-745 by Baranyi and Roberts model [51]. Hoponaid (hpn) transposon mutation, at elevated ethanol concentration the head group composition and abundance of hopanoids prevents membrane dissolution in *Z. mobilis* [52].

6.3.4.2   GENOME SHUFFLING

The combinatorial method which combines the primacy of multi-parental crossing enabled by DNA shuffling with the recombination of whole genomes generally connected with conventional breeding. This strain improvement technology however originated from protoplast fusion/electrofusion and greatly relies on them, it markedly differs in the recombination giving rise to stable and high-performance strain (Table 6.5). Genome shuffling is an evolutionary engineering technology that accelerates the strain development process by a sequence of three processes. First, the genetically divergent population is creation followed by inducing mutation. Second, the population is screened to identify the best performing isolate. Third, the genomes of supreme cells are shuffled either by sexual or asexual means following protoplast fusion or sporulation, respectively, in single strain recursive protoplast fusion between multi-parent strains.

TABLE 6.5   Different Genome Shuffling Techniques

Strain	Method	Profound Effects	References
S. cerevisiae SM-3	Recursive protoplast fusion	Tolerates 25% (v/v) ethanol	[53]
S. cerevisiae NCIM 3090	Recursive protoplast fusion followed by Random amplified polymorphic DNA	Resulted in hybrid strain (SP2-18) with increased cell viability than the parent strain	[54]
S. cerevisiae S310	EMS mutagenesis and recursive sporulation of diploid cells	Increased ethanol yield and tolerance after three rounds of genome shuffling	[55]
S. cerevisiae	Random mating and selection	Increased ethanol tolerance and fermentation capacity	[56]
S. cerevisiae Q and L	Inactive protoplast fusion	Higher ethanol tolerance in beer production-14%(v/v)	[57]

6.3.4.3   GENOME EDITING

In the past years, novel genome editing tools like transcription activator-like effector nuclease (TALEN's), zinc finger nuclease (ZFN) have been used to engineer strains with desirable traits in industrial settings. However, there are only a few studies applying the gene-editing tools for ethanol tolerance. Zhang et al. [58] have demonstrated TALEN assisted multiplex editing (TAME) for improving ethanol tolerance in yeast by multiple genetic perturbations in the genome. Mitsui, Yamada, and Ogino [59] have achieved higher cell viability in low pH and high ethanol concentration in *S. cerevisiae* using Clustered regularly interspaced short palindromic repeats (CRISPR)-CRISPR associated protein (Cas). The large-scale rearrangement of genomic DNA can be a promising strategy for the improvement of ethanol tolerant strains.

6.3.4.4 OVER-EXPRESSION

The iron-sulfur clusters (ISC) are the prosthetic groups responsible for catalytic and regulatory functions found in both prokaryotes and eukaryotes. The components of the mitochondrial ISC in *S. cerevisiae* were reported to be related in iron homeostasis due to ethanol tolerance and are encoded by the genes *NFS1, ISU1, ISU2, ISA1, ISA2, JAC1, SSQ1, YAH1, GRX5,* and *IBA5. The over-expression of Jac 1p and Isu 1p in S. cerevisiae UMArn3 was evaluated and revealed their counteracting ability of metabolic unbalance caused by an increase in ROS generation [60]. Further Snf complex of S. cerevisiae,* a member of AMP protein kinase family engages in a series of cellular functions and responds to environmental stresses. The effect of Snf over-expression revealed a 39% increase in cell survival rate over parental strain at 8% ethanol accompanied with altered expression of genes involved in glucose transport and accumulation of fatty acids [31].

6.3.4.5 GLOBAL TRANSCRIPTION MACHINERY (gTME)-MANIPULATION OF MULTIPLE GENES

In recent times, engineering the global transcription machinery has been investigated to reprogram the microbial transcription profile. It is regarded as an efficient tool in bringing a high degree of pleiotropy. The method of reprogramming transcription to elicit desired cellular phenotypes for industrial and technical applications has spurred interest in yeast engineering for stress tolerance. The transcriptional machinery of eukaryotes is quite complex with a substantial set of general and specific transcription factors (TFs). It modifies the key protein to synchronize the global transcription by error-prone polymerase chain reaction epPCR.

Engineering or mutagenesis in TF to modify DNA binding specificity and RNA polymerase II to induce perturbations have been implied successfully to develop ethanol tolerant strains of *S. cerevisiae*. In their transcription, two major complement complexes sharing TATA box-binding protein-associated factors (TBP), namely RNA polymerase II transcription factor D (TFIID) and SAGA (Spt-Ada-Gcn-5 acetyltransferase) are involved. Both SAGA and TFIID work together almost mRNA genes, however, all yeast mRNA strongly relies on TFIID [61]. Reprogramming global regulator RpoD via epPCR improved glucose consumption and ethanol yield under ethanol stress in *Z. mobilis* [54]. Further, incorporation of gene modules increased fatty acid production which served a crucial criterion for ethanol tolerance [62].

Beyond yeasts being physiologically ideal, the complex array of stress factors exhibited have to be encountered cooperatively. Conventional breeding techniques bestow bounded up-gradation of strain robustness. However, their painstaking standards assist "omics" technologies in recent research. Exploring genetic elements required to maintain ethanol tolerance would be a better approach rather than genetic modifications. On the other hand, alteration in the gene for ethanol tolerance could compromise several fermentation parameters. The ever-expanding knowledge on yeast genome and functions makes an obscure platform for commercial-scale ethanol tolerant strains. Recent precise technology for genome editing enabled quick engineering of yeast strains which would develop strains for numerous applications.

6.3.5   SMALL RNA (sRNA) ENGINEERING

sRNA is non-coding functional RNA with 50–300 nt capable of blocking translation or altering RNA's stability by various mechanisms thereby regulating protein expression. Hence, they represent strong tools in governing central pathways to engineer complex phenotypes [63]. The Discovery of novel sRNA candidates in *Z. mobilis* was attempted by Cho [64] using a union of computational approaches. The differential expression of selected sRNA under different ethanol concentrations was analyzed, followed by the effect of ethanol stress on the expression of sRNA. Their results revealed three sRNA namely Zms2, Zms6, and Zms18 were differentially expressed, suggesting their prime role in the regulatory mechanism of ethanol tolerance. Further, Zms4 and Zms6 were found to coordinate a large network of gene regulation including sRNA-sRNA interaction and the absence of identified sRNA made the cells ethanol sensitive [65]. Advancement in identifying novel sRNA using prevalent "omics" datasets with the bioinformatics pipeline, RNA seq examiner for phenotype-informed network engineering (REFINE) was proposed in *Z. mobilis*. REFINE approach combines existing computational tools with the new pipeline (sRNA phenoscore), to identify differentially expressed RNAs from different growth conditions REFINE sRNA prediction relies on scores assigned to differentially expressed intergenic regions suggesting their possible regulation relevant to the phenotype [66].

6.3.6   *AdhE MUTATION*

AdhE, a gene encodes for aldehyde dehydrogenase, the bifunctional and bidirectional enzyme. It is responsible for catalyzing two-terminal steps in the formation of ethanol: acetyl-CoA reduction to acetaldehyde by aldehyde dehydrogenase (ALDH) and reduction of acetaldehyde to ethanol by alcohol dehydrogenase (ADH) with two reduced electron donors, NADH, and NADPH. In *C. thermocellum,* the role of *adhE* is crucial and their deletion strain showed >90% activity loss of ALDH, ADH, and >95% reduction of ethanol production [67]. Point mutation resulted in the replacing certain amino acids at particular sites of ADH [3].

The genome of six isolates with n-butanol tolerance was re-sequenced and the mutations in the coding sequence of *Clo1313_0853* and *Clo1313_1798* were studied [5]. The gene *Clo1313_0853* catalyzes the hydrolysis of phosphatidylcholine and phospholipids to produce phosphatidic acid (PA). The mutation in the gene truncates the protein by frameshift and protects the membrane in the presence of n-butanol/ethanol. Mutation in the gene *Clo1313_1798*, encoding alcohol dehydrogenase (*adhE*) increases its ability to use NADPH as a co-factor yielding increased ethanol and titer. The increased NADPH activities are associated with *AdhE* mutant, where the reverse flux through ADH reaction affects NADH/NAD+ ratio and the NADPH/NADP+ ratio. Deletion of *adhE* strain (LL1111) increases tolerance and inhibits several primary alcohols by reverse flux through AdhE [68].

6.4   CONCLUSION

Ethanol tolerance in fermenting organisms is not a one-way approach and includes the interplay of events. It requires numerous pathways and networks to be recognized. Developing ethanol tolerant strains is often arduous due to a lack of in-depth perceptive of tolerance mechanism. Further ethanol tolerance research involves various strains of yeasts and other ethanol fermenting microorganisms under varied conditions that make difficulty in interpreting the results across the studies. In this regard, rational engineering and evolutionary engineering are crucial to identify genetic targets. Conventional methods of gene expression and evolutionary engineering can be used resourcefully with the aid of synthetic biology and omics techniques. The ethanol-related industries may take advantage of these findings to lower the cost of ethanol production.

KEYWORDS

- **bioethanol**
- *Clostridium thermocellum*
- **ethanol tolerance**
- **exopolysaccharides**
- **yeast**
- *Zymomonas mobilis*

REFERENCES

1. Yang, S., Fei, Q., Zhang, Y., Contreras, L. M., Utturkar, S. M., Brown, S. D., Himmel, M. E., & Zhang, M., (2016). *Zymomonas mobilis* as a model system for production of biofuels and biochemicals. *Microb. Biotechnol., 9*, 699–717.
2. Gunasekaran, P., Karunakaran, T., & Kasthuribai, M., (1986). Fermentation pattern of *Zymomonas mobilis* strains on different substrates—a comparative study. *J. Biosci., 10*, 181–186.
3. Brown, S. W., & Oliver, S. G., (1982). Isolation of ethanol-tolerant mutants of yeast by continuous selection. *Eur. J. Appl. Microbiol. Biotechnol., 16*, 119–122.
4. Rani, K. S., & Seenayya, G., (1999). High ethanol tolerance of new isolates of *Clostridium thermocellum* strains SS21 and SS22. *World J. Microbiol. Biotechnol., 15*, 173–178.
5. Shao, X., Raman, B., Zhu, M., Mielenz, J. R., Brown, S. D., Guss, A. M., & Lynd, L. R., (2011). Mutant selection and phenotypic and genetic characterization of ethanol-tolerant strains of *Clostridium thermocellum.Appl. Microbiol. Biotechnol., 92*, 641–652.
6. Timmons, M. D., Knutson, B. L., Nokes, S. E., Strobel, H. J., & Lynn, B. C., (2009). Analysis of composition and structure of *Clostridium thermocellum* membranes from wild-type and ethanol-adapted strains. *Appl. Microbiol. Biotechnol., 82*, 929–939.
7. Zhu, X., Cui, J., Feng, Y., Fa, Y., Zhang, J., & Cui, Q., (2013). Metabolic adaption of ethanol-tolerant *Clostridium thermocellum. PLoS One, 8*, e70631.
8. Albano, E., (2007). Alcohol, oxidative stress and free radical damage. *Proc. Nutr. Soc., 65*, 278–290.
9. Ding, J., Huang, X., Zhang, L., Zhao, N., Yang, D., & Zhang, K., (2009). Tolerance and stress response to ethanol in the yeast *Saccharomyces cerevisiae. Appl. Microbiol. Biotechnol., 85*, 253–263.

10. Rupcic, J., & Juresic, G. C., (2010). Influence of stressful fermentation conditions on neutral lipids of a *Saccharomyces cerevisiae* brewing strain. *World J. Microbiol. Biotechnol., 26*, 1331–1336.

11. Hahn-Hägerdal, B., Galbe, M., Gorwa-Grauslund, M. F., Lidén, G., & Zacchi, G., (2006). Bio-ethanol - the fuel of tomorrow from the residues of today. *Trends Biotechnol., 24*, 549–556.

12. Seo, J. S., Chong, H., Park, H. S., Yoon, K. O., Jung, C., Kim, J. J., Hong, J. H., et al., (2005). The genome sequence of the ethanologenic bacterium *Zymomonas mobilis* ZM4. *Nat. Biotechnol., 23*, 63–68.

13. An, H., Scopes, R. K., Rodriguez, M., Keshav, K. F., & Ingram, L. O., (1991). Gel electrophoretic analysis of *Zymomonas mobilis* glycolytic and fermentative enzymes: Identification of alcohol dehydrogenase II as a stress protein. *J. Bacteriol., 173*, 5975–5982.

14. Pallach, M., Marchetti, R., Di Lorenzo, F., Fabozzi, A., Giraud, E., Gully, D., Paduano, L., et al., (2018). *Zymomonas mobilis* exopolysaccharide structure and role in high ethanol tolerance. *Carbohydr. Polym., 201*, 293–299.

15. Ellis, L. D., Holwerda, E. K., Hogsett, D., Rogers, S., Shao, X., Tschaplinski, T., Thorne, P., & Lynd, L. R., (2012). Closing the carbon balance for fermentation by *Clostridium thermocellum* (ATCC 27405). *Bioresour Technol., 103*, 293–299.

16. Biswas, R., Zheng, T., Olson, D. G., Lynd, L. R., & Guss, A. M., (2015). Elimination of hydrogenase active site assembly blocks H_2 production and increases ethanol yield in *Clostridium thermocellum. Biotechnol. Biofuels, 8*, 1–8.

17. Birch, R. M., & Walker, G. M., (2000). Influence of magnesium ions on heat shock and ethanol stress responses of *Saccharomyces cerevisiae. Enzyme Microb. Technol., 26*, 678–687.

18. Li, R., Miao, Y., Yuan, S., Li, Y., Wu, Z., & Weng, P., (2019). Integrated transcriptomic and proteomic analysis of the ethanol stress response in *Saccharomyces cerevisiae* Sc131. *J. Proteomics, 203*, 103377.

19. Stanley, D., Bandara, A., Fraser, S., Chambers, P. J., & Stanley, G. A., (2010). The ethanol stress response and ethanol tolerance of *Saccharomyces cerevisiae. J. Appl. Microbiol., 109*, 13–24.

20. Charoenbhakdi, S., Dokpikul, T., Burphan, T., Techo, T., & Auesukaree, C., (2016). Vacuolar H^+ ATPase protects *Saccharomyces cerevisiae* cells against ethanol-induced oxidative and cell wall stresses. *Appl. Environ. Microbiol., 82*, 3121–3130.

21. Weber, F. J., & De Bont, J. A. M., (1996). Adaptation mechanisms of microorganisms to the toxic effects of organic solvents on membranes. *BBA. Rev. Biomembr., 1286*, 225–245.

22. Thanonkeo, P., Laopaiboon, P., Sootsuwan, K., & Yamada, M., (2007). Magnesium ions improve growth and ethanol production of *Zymomonas mobilis* under heat or ethanol stress. *Biotechnol., 6*, 112–119.

23. Laudrin, I., & Goma, G., (1982). Ethanol production by *Zymomonas mobilis*: Effect of temperature on cell growth, ethanol production and intracellular ethanol accumulation. *Biotechnol. Lett., 4*, 537–542.

24. Millar, D. G., Griffiths-Smith, K., Algar, E., & Scopes, R. K., (1982). Activity and stability of glycolytic enzymes in the presence of ethanol. *Biotech. Lett., 4*, 601–606.

25. Demain, A. L., Newcomb, M., & Wu, J. H. D., (2005). Cellulase, clostridia, and ethanol. *Microbiol. Mol. Biol. Rev., 69*, 124–154.

26. Levin, D. E., (2011). Regulation of cell wall biogenesis in *Saccharomyces cerevisiae*: The cell wall integrity signaling pathway. *Genetics, 189*(4), 1145–1175.

27. Ma, M., & Liu, Z. L., (2010). Mechanisms of ethanol tolerance in *Saccharomyces cerevisiae*. *Appl. Microbiol. Biotechnol., 87*, 829–845.

28. Udom, N., Chansongkrow, P., Charoensawan, V., & Auesukaree, C., (2019). Coordination of the cell wall integrity and high-osmolarity glycerol pathways in response to ethanol stress in *Saccharomyces cerevisiae*. *Appl. Environ. Microbiol., 85*, e00551–00519.

29. Vamvakas, S. S., Kapolosm, J., Farmakis, L., Koskorellou, G., & Genneos, F., (2019). Ser625 of msn2 transcription factor is indispensable for ethanol tolerance and alcoholic fermentation process. *Biotechnol. Prog., 35*, e2837.

30. Divate, N. R., Chen, G. H., Wang, P. M., Ou, B. R., & Chung, Y. C., (2016). Engineering *Saccharomyces cerevisiae* for improvement in ethanol tolerance by accumulation of trehalose. *Bioengineered, 7*, 445–458.

31. Meng, L., Liu, H. L., Lin, X., Hu, X. P., Teng, K. R., & Liu, S. X., (2020). Enhanced multi-stress tolerance and glucose utilization of *Saccharomyces cerevisiae* by overexpression of the SNF1 gene and varied beta isoform of Snf1 dominates in stresses. *Microbial Cell Factories, 19*, 134.

32. Ansarypour, Z., & Shahpiri, A., (2017). Heterologous expression of a rice metallothionein isoform (OsMTI-1b) in *Saccharomyces cerevisiae* enhances cadmium, hydrogen peroxide and ethanol tolerance. *Braz. J. Microbiol., 48*, 537–543.

33. Zyrina, A. N., Smirnova, E. A., Markova, O. V., Severin, F. F., & Knorre, D. A., (2017). Mitochondrial superoxide dismutase and Yap1p Act as a signaling module contributing to ethanol tolerance of the yeast *Saccharomyces cerevisiae*. *Appl. Environ. Microbiol., 83*, e02759–02716.

34. Li, R., Xiong, G., Yuan, S., Wu, Z., Miao, Y., & Weng, P., (2017). Investigating the underlying mechanism of *Saccharomyces cerevisiae* in response to ethanol stress employing RNA-seq analysis. *World J. Microbiol. Biotechnol., 33*, 206.

35. Xue, C., Zhao, X. Q., Yuan, W. J., & Bai, F. W., (2008). Improving ethanol tolerance of a self-flocculating yeast by optimization of medium composition. *World J. Microbiol. Biotechnol., 24*, 2257.

36. Ahmed, K., Zhang, M., Wang, M., & Wang, C., (2020). Effects of zinc combining with specific metal ions on ethanol tolerance of yeast *Saccharomyces cerevisiae*. *Int. J. Agric. Biol., 23*, 566–572.

37. Xu, Y., Yang, H., Brennan, C. S., Coldea, T. E., & Zhao, H., (2020). Cellular mechanism for the improvement of multiple stress tolerance in brewer's yeast by potassium ion supplementation. *Int. J. Food Sci Technol., 55*, 2419–2427.

38. Hu, C. K., Bai, F. W., & An, L. J., (2003). Enhancing ethanol tolerance of a self-flocculating fusant of *Schizosaccharomyces pombe* and *Saccharomyces cerevisiae* by Mg^{2+} via reduction in plasma membrane permeability. *Biotechnol. Lett., 25*:1191–1194.

39. Alfenore, S., Molina, J. C., Guillouet, S., Uribelarrea, J. L., Goma, G., & Benbadis, L., (2002). Improving ethanol production and viability of *Saccharomyces cerevisiae* by a vitamin feeding strategy during fed-batch process. *Appl. Microbiol. Biotechnol., 60*, 67–72.

40. Ciesarova, Z., Smogrovicova, D., & Domeny, Z., (1996). Enhancement of yeast ethanol tolerance by calcium and magnesium. *Folia Microbiol., 41*, 485–488.

41. Alfenore, S., Cameleyre, X., Benbadis, L., Bideaux, C., Uribelarrea, J. L., Goma, G., Molina, J. C., & Guillouet, S. E., (2004). Aeration strategy: A need for very high ethanol performance in *Saccharomyces cerevisiae* fed-batch process. *Appl. Microbiol. Biotechnol., 63*, 537–542.

42. Dinh, T. N., Nagahisa, K., Hirasawa, T., Furusawa, C., & Shimizu, H., (2008). Adaptation of *Saccharomyces cerevisiae* cells to high ethanol concentration and changes in fatty acid composition of membrane and cell size. *PLoS One, 3,* e2623.

43. Fiedurek, J., Skowronek, M., & Gromada, A., (2011). Selection and adaptation of *Saccharomyces cerevisiae* to increased ethanol tolerance and production. *Pol. J. Microbiol., 60,* 51–58.

44. Stanley, D., Fraser, S., Chambers, P. J., Rogers, P., & Stanley, G. A., (2010). Generation and characterization of stable ethanol-tolerant mutants of *Saccharomyces cerevisiae. J. Ind Microbiol. Biotechnol., 37,*139–149.

45. Voordeckers, K., Kominek, J., Das, A., Espinosa, C. A., De, M. D., Arslan, A., Van, P. M., et al., (2015). Adaptation to high ethanol reveals complex evolutionary pathways. *PLoS Genet., 11,* e1005635.

46. Wei, P., Li, Z., Lin, Y., He, P., & Jiang, N., (2007). Improvement of the multiple-stress tolerance of an ethanologenic *Saccharomyces cerevisiae* strain by freeze-thaw treatment. *Biotechnol. Lett., 29,* 1501–1508.

47. Morard, M., Macías, L. G., Adam, A. C., Lairón-Peris, M., Pérez-Torrado, R., Toft, C., & Barrio, E., (2019). Aneuploidy and ethanol tolerance in *Saccharomyces cerevisiae. Front. Genet., 10,* 82.

48. Turanli-Yildiz, B., Benbadis, L., Alkım, C., Sezgin, T., Akşit, A., Gökçe, A., Oztürk, Y., et al., (2017). In vivo evolutionary engineering for ethanol-tolerance of *Saccharomyces cerevisiae* haploid cells triggers diploidization. *J. Biosci. Bioeng., 124,* 309–318.

49. Davis, L. S. A., Griffith, D. A., Choi, B., Cate, J. H. D., & Tullman-Ercek, D., (2018). Evolutionary engineering improves tolerance for medium-chain alcohols in *Saccharomyces cerevisiae. Biotechnol. Biofuels, 11*(1), 90.

50. Carreón-Rodríguez, O. E., Gutiérrez-Ríos, R. M., Acosta, J. L., Martinez, A., & Cevallos, M. A., (2019). Phenotypic and genomic analysis of *Zymomonas mobilis* ZM4 mutants with enhanced ethanol tolerance. *Biotechnol. Rep., 23,* e00328.

51. Ramírez-Cota, G. Y., López-Villegas, E. O., Jiménez-Aparicio, A. R., & Hernández-Sánchez, H., (2020). Modeling the ethanol tolerance of the probiotic yeast *Saccharomyces cerevisiae* var. *boulardii* CNCM I-745 for its possible use in a functional beer. *Probiotics Antimicrob. Proteins.,* 1–8.

52. Brenac, L., Baidoo, E. E. K., Keasling, J. D., & Budin, I., (2019). Distinct functional roles for hopanoid composition in the chemical tolerance of *Zymomonas mobili. Mol. Microbiol., 112,* 1564–1575.

53. Shi, D. J., Wang, C. L., & Wang, K. M., (2009). Genome shuffling to improve thermo-tolerance, ethanol tolerance and ethanol productivity of *Saccharomyces cerevisiae. J. Ind. Microbiol. Biotechnol., 36,* 139–147.

54. Jetti, K. D., Gns, R. R., Garlapati, D., & Nammi, S. K., (2019). Improved ethanol productivity and ethanol tolerance through genome shuffling of *Saccharomyces cerevisiae* and *Pichia stipitis. Int Microbiol., 22,* 247–254.

55. Hou, L., (2010). Improved production of ethanol by novel genome shuffling in *Saccharomyces cerevisiae. Appl. Biochem. Biotechnol., 160,* 1084–1093.

56. Snoek, T., Picca, N. M., Van, D. B. S., Mertens, S., Saels, V., Verplaetse, A., Steensels, J., & Verstrepen, K. J., (2015). Large-scale robot-assisted genome shuffling yields industrial *Saccharomyces cerevisiae* yeasts with increased ethanol tolerance. *Biotechnol. Biofuels, 8,* 32.

57. Xin, Y., Yang, M., Yin, H., & Yang, J., (2020). Improvement of Ethanol Tolerance by Inactive Protoplast Fusion in *Saccharomyces cerevisiae*. *Biomed Res Int., 1979*318.

58. Zhang, G., Lin, Y., Qi, X., Li, L., Wang, Q., & Ma, Y., (2015). TALENs-assisted multiplex editing for accelerated genome evolution to improve yeast phenotypes. *ACS Synth. Biol., 4*, 1101–1111.

59. Mitsui, R., Yamada, R., & Ogino, H., (2019). Improved stress tolerance of *Saccharomyces cerevisiae* by CRISPR-Cas-mediated genome evolution. *Appl. Biochem. Biotechnol., 189*, 810–821.

60. Martínez-Alcántar, L., Madrigal, A., Sánchez-Briones, L., Díaz-Pérez, A. L., López-Bucio, J. S., & Campos-García, J., (2019). Over-expression of Isu1p and Jac1p increases the ethanol tolerance and yield by superoxide and iron homeostasis mechanism in an engineered *Saccharomyces cerevisiae* yeast. *J. Ind. Microbiol. Biotechnol., 46*, 925–936.

61. Warfield, L., Ramachandran, S., Baptista, T., Devys, D., Tora, L., & Hahn, S., (2017). Transcription of nearly all yeast RNA polymerase II-transcribed genes is dependent on transcription factor TFIID. *Mol. Cell, 68*, 118–129.e115.

62. Wang, H., Cao, S., Wang, W. T., Wang, K. T., & Jia, X., (2016). Very high gravity ethanol and fatty acid production of *Zymomonas mobilis* without amino acid and vitamin. *J. Ind. Microbiol. Biotechnol., 43*, 861–871.

63. Cho, S. H., Haning, K., & Contreras, L. M., (2015). Strain engineering via regulatory noncoding RNAs: Not a one-blueprint-fits-all. *Curr. Opin. Chemi. Eng., 10*, 25–34.

64. Cho, S. H., Lei, R., Henninger, T. D., & Contreras, L. M., (2014). Discovery of ethanol-responsive small RNAs in *Zymomonas mobilis*. *Appl. Environ. Microbiol., 80*,4189–4198.

65. Han, R., Haning, K., Gonzalez-Rivera, J. C., Yang, Y., Li, R., Cho, S. H., Huang, J., et al., (2020). Multiple small RNAs interact to co-regulate ethanol tolerance in *Zymomonas mobilis*. *Front. Bioeng. Biotechnol., 8*, 155.

66. Haning, K., Engels, S. M., Williams, P., Arnold, M., & Contreras, L. M., (2020). Applying a new REFINE approach in *Zymomonas mobilis* identifies novel sRNAs that confer improved stress tolerance phenotypes. *Front. Microbiol., 10*, 2987.

67. Lo, J., Zheng, T., Hon, S., Olson, D. G., & Lynd, L. R., (2015). The bifunctional alcohol and aldehyde dehydrogenase gene, *adhE*, is necessary for ethanol production in *Clostridium thermocellum* and *Thermoanaerobacterium saccharolyticum*. *J. Bacteriol., 197*, 1386–1393.

68. Tian, L., Cervenka, N. D., Low, A. M., Olson, D. G., & Lynd, L. R., (2019). A mutation in the AdhE alcohol dehydrogenase of *Clostridium thermocellum* increases tolerance to several primary alcohols, including isobutanol, n-butanol and ethanol. *Sci. Rep., 9*, 1736.

CHAPTER 7

Challenges in Developing Sustainable Fermentable Substrate for Bioethanol Production

CHARU SARAF and KAKOLI DUTT

Department of Bioscience and Biotechnology, Banasthali Vidyapith, Rajasthan–304022, India, Tel.: 9910331924,
E-mail: charusaraf17@gmail.com (C. Saraf), Tel.: 9887398384,
E-mail: kakoli_dutt@rediffmail.com (K. Dutt)

ABSTRACT

The ability to harness fire from burning wood was the first application of fuel-based energy. For a long time, wood and other derived products were the sources of energy to support various activities like cooking, metallurgy, pottery, etc. The discovery of fossil fuels and the development of the steam engine opened a new age of applications. But in this modern world, overuse of fossil fuels to sustain various technologies has started to take a heavy toll in the form of environmental pollution. Thus, the new search for alternative energy sources which can efficiently replace the fossil fuels and have minimal environmental issues has become mandatory, with several biofuels such as bioethanol and biodiesel gaining prominence. Ethanol production, a massive global commercial venture shows continuous evolution in the production process for yield enhancement. The developments observed in the characterization of different biomass as raw material for ethanol production has led to their categorization as first, second, third, and fourth-generation fuels. Numerous reviews and research articles account for the advantages and disadvantages of these substrates against each other. Additionally, the pretreatment and the fermentation processes play a pivotal role in the final yield obtained. Thus, in this chapter, an attempt has been made to present a comprehensive summary of ethanol production with emphasis on the different substrates, pretreatment, and fermentation processes. In the chapter, bottleneck of the bioethanol commercialization has also been briefly discussed.

7.1 INTRODUCTION

Over six decades ago, petroleum products were inarguably the source of energy that kept the world functional and expanding. Nevertheless, in the beginning of the 1970s, proclamation to control the production by the members of the Organization of Arab Petroleum Exporting Countries (OPEC) hiked the petroleum price, causing an energy crisis [1]. Additionally, climatic changes and greenhouse gas emissions (GHG) awakened the interest in alternative energy sources to achieve environmental sustainability [2, 3]. Biofuels (bioethanol and biodiesel) were welcomed as environmentally sustainable alternative energy sources [4]. Alcohol obtained from fermentation route quickly became a part of fuel industry as blended fuel 'gasohol' [5].

"Fuel for future" a term coined by Henry Ford for ethanol is a major contender among several key alternative energy sources. Obtained either from biomass through microbial fermentation or as a byproduct of several industrial processes [6], ethanol shows valid promise not only as feedstock for chemical industries but also as biofuel (Figure 7.1). The oxygenated nature of alcohol enhances its combustion efficiency which added to the properties of eco-friendly, less toxicity, higher heat of vaporization and flexibility of sugar substrate for alcohol fermentation increases its appeal [7–10]. Ethanol has higher octane number (106–110) than gasoline (91–96), resulting in enhancement of gasoline performance on blending [11–13].

Although, literature reported shows unlimited benefits of bioethanol with respect to environment and economy, its global development as a fuel and chemical feedstock is facing a serious bottle neck of usable and fermentable substrates [14, 15]. Significant researches are going on which are globally supported by private conglomerates, think tanks, NGOs, and Government funded projects. This chapter provides an overview of the evolution of ethanol as a fuel, fermentable substrates, technological approaches, and fermentation conditions for bioethanol formation with a brief outline on the hurdles in bioethanol commercialization.

7.2 BRIEF HISTORY OF ETHANOL

Uses of ethanol as fuel was first reported in 1826 in America and subsequently after 50 decades by Nikolaus Otto in 1876, who used ethanol to power an early internal combustion engine (Figure 7.2). Before the American Civil War, ethanol was used for lighting but this was severely curtailed due to liquor tax imposition. After the tax was repealed, ethanol resumed its function as fuel. In fact, Henry Ford's Model T in 1908 was designed to run on either gasoline or pure alcohol (adapted from: www.ott.doe.gov/biofuels/history.html).

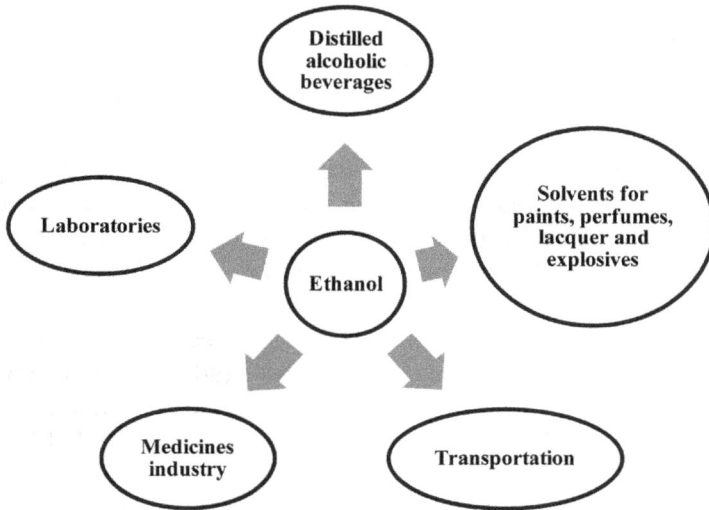

FIGURE 7.1   Various applications of ethanol.

FIGURE 7.2   The engine made by Nicolaus Otto in 1876.
Source: Public domain.

Prior to the Second World War, industrial use ethanol was produced by fermenting blackstrap molasses, a cheap, easy to handle, most economical and easily available substrate. After the war, the low-priced easy availability of petroleum-based fuel and natural gas curbed the interest in use of agricultural crops for fuel production which wiped out the economic incentives for production of liquid fuels from crops leading to the generation of a significant lacuna. Since, Government interest was lost, many wartime distillers were dismantled and others were converted to beverage alcohol plants [16].

The revival of alcohol-based fuel started in 1970s, when the supply of oil started being disrupted due to instability in the Middle East. The major oil companies of America began to market ethanol as a gasoline extender and octane booster (adapted from: www.ott.doe.gov/biofuels/history.html). Gasohol is a mixture of 9 volumes of gasoline and only 1 volume of ethanol. The total ethanol production in the USA has increased from 0.175 billion gallons in 1980 to 15.8 billion gallons in 2017 (adapted from: www.afdc. energy.gov/data/). In 2015–2016, Brazil had produced 8 billion gallons of ethanol and sold gasoline with a blend of 18 to 27.5% ethanol [17]. Among the global players, the United States, Brazil, and European Union (EU) are the major producers of ethanol followed by China among the developing countries, see Figure 7.3.

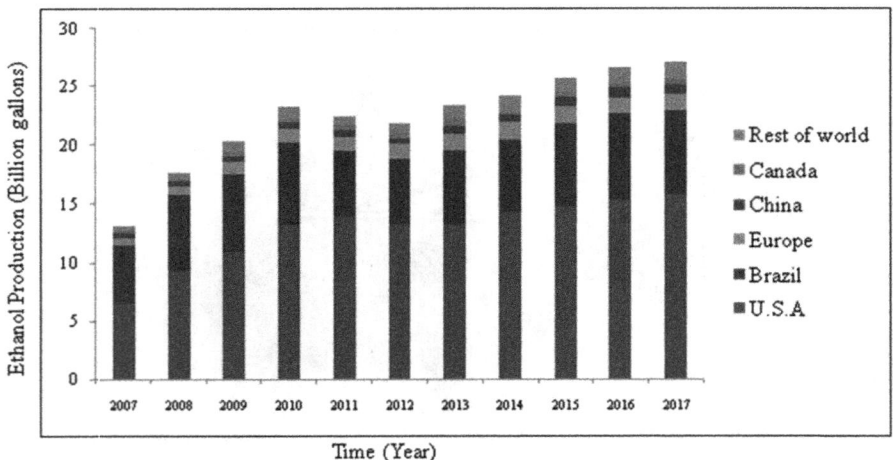

FIGURE 7.3   Global bioethanol production.
Source: Adapted from www.afdc.energy.gov/data/.

7.3 ORGANIC FEEDSTOCKS FOR ETHANOL PRODUCTION

Sugar rich substrates are required to produce alcohol and can be categorized into four generations [18–22]. Since alcohol is produced via the breakdown of hexoses and pentoses, the efficiency is related to the accessible amount of these monosaccharides. In the overall fermentative reaction using a mono-hexose (glucose) and a monopentose (xylose) represented, shows that for every 1 mole of glucose and 3 moles of xylose, 2 and 5 moles of ethanol may be obtained, respectively. The stoichiometry, neglecting NAD(P)H balance, is shown in Eqns. (1) and (2) [23].

$$1GLUCOSE + 2ATP \ 2 \ ETHANOL + 2CO_2 + 2ADP \tag{1}$$

$$3XYLOSE + 5ADP + 5Pi \ 5ETHANOL + 5CO_2 + 5ATP + 5H_2O \tag{2}$$

7.3.1 FIRST GENERATION BIOETHANOL

Cash crops like cereals and sugar crops were identified as readily fermentable substrates for first-generation ethanol (Table 7.1). It has carbon dioxide (CO_2) benefits and is commercially available even today. Sugar crops like sugarcane and sugar beet contribute 40% of the ethanol production in the tropical areas like India, Brazil, and Colombia, and the remaining 60% is contributed by the starch crops which is used by the United States, EU, and China [27, 50–52]. High corn and sugarcane harvests lead to an increase in global biofuel production by 9% in 2014, accounting for 74% (of the total) of ethanol followed by 23% biodiesel (adapted from: http://www.ren21.net/wpcontent/uploads/2015/07/REN12-GSR2015_Onli nebook_low1.pdf).

7.3.1.1 SUGAR BIOMASS

Before 1750, sugar production was limited to sugar cane (*Saccharum officinarum*) grown in the tropical and subtropical regions and shipped globally [53]. During 1880, a German chemist, Andreas Marggraf extracted sugar from beets (*Beta vulgaris*) and sugar beet soon replaced sugar cane as the main source of sugar in continental Europe as it was found that sugar beet contains enough amounts (16–20%) of sucrose than sugar cane [54, 55]. The other sources of sugar production are included in Table 7.2.

TABLE 7.1 Different Sources for Ethanol Production

SL. No.	Generation of Fuel	Source of Ethanol	Examples	References	Advantages	Disadvantages
1.	First-generation Biofuels	Sugar sources	Sugarcane (*Saccharum officinarum*)	[24, 25]	• High sugar yield	• Seasonal availability
			Sugar beet (*Beta vulgaris*)	[13, 24]	• Low conversion costs	• Threat to food security
			Sweet sorghum (*Sorghum bicolor*)	[13]	• Require less energy	
		Starch sources	Corn (*Zea mays*)	[24, 26, 27]	• Availability across the world	• Energy-intensive to produce
			Sorghum (grains) (*Sorghum bicolor*)	[28]	• Ease of conversion	• rise in price of the crops
			Wheat (*Triticum aestivum*)	[24, 29]	• Storage capability for a longer period	
			Cassava (*Monihot esculenta*)	[24, 30]	• High ethanol yield	
			Potatoes (*Solanum tuberosum*)	[31, 32]		
			Sweet potatoes (*Ipomoea batatas*)	[33, 34]		
2.	Second-generation Biofuels	Lignocellulosic sources	Perennial grasses (switchgrass, miscanthus, reed canarygrass, giant reed)	[35, 36]	• Abundant	• Costly
			Aquatic plants (*Eichhornia crassipes*)	[37, 38]	• Non-food biomass into fuel	• Difficult to make
			Forest material	[39–41]		• More energy-intensive
			Agricultural residues	[2, 26, 43, 44]		
			Organic portion of municipal wastes	[45–47]		
3.	Third-generation biofuels	Algae	Microalgae (*Pleurochrysis carterae*)	[24, 48]	• Renewable source for biofuels	• Costly
			Cyanobacteria (*Chlamydomonas* sp., *Cynaothece* sp., *Spirulina platensis*)	[49]	• High yield	• Chance of contamination
					• Easy to obtain large biomass	

TABLE 7.2 Other Sources of Sugar Substrates [56]

Sugar	Plants	Producing Countries
Date sugar	Date palm	Algeria, Iraq
Palm sugar	Palm species like Palmyra, saga or Toddy palm, coconut, and nipapalm	India, Sri Lanka, Malaysia, Philippines, etc.
Maple sugar	Maple tree (*Acer saccharum*)	North America (the USA and Canada), Japan

The sugar industry promotes solutions aiming at higher yield and no waste process for ethanol production from sugar beet as well as from intermediates and byproducts like molasses. Another byproduct of the industry, i.e., sugar beet pulp (SBP), is a potential feedstock for biofuels. It contains 20–25% cellulose, 25–36% hemicellulose, 20–25% pectin, 10–15% protein, and 1–2% lignin content on a dry weight basis [55, 57].

Sugar derived from plant biomass is a mixture of both hexoses and pentoses represented mainly by glucose and xylose, respectively. Since the wild-type strains of *Saccharomyces cerevisiae* do not metabolize xylose, researchers have developed two approaches to increase the fermentation yields of ethanol derived from sugar biomass. The first approach via genetic engineering is to add pentose metabolic pathways in yeast and other natural ethanologens. The second approach is to improve the ethanol production by modifying microorganisms having the ability to ferment both hexoses and pentoses [58, 59]. A general procedure for ethanol production from sugar biomass is illustrated in Figure 7.4. After fermentation, distillation is done which generate is vinasse as a byproduct, which can be used as an animal feed or as a fertilizer [56].

7.3.1.2 STARCH BIOMASS

Grain crops (corn, barley, wheat, or grain sorghum) and tuber crops (cassava, potato, sweet potato, Jerusalem artichoke, cactus, or arrowroot) contain large quantities of starch [60]. Isolated native starch from different sources can be used for further conversion into bio-based products or bioethanol production. The residue from starch isolation contains proteins and fiber, which has great potential for application in food and feed production (adapted from: http://www.bmbf.de/pub/Roadmap_Biorefineries_eng.pdf.). USA is the biggest corn starch producing countries with 80% of the worldwide market. There, 95% of ethanol is produced from corn, and the rest from barley, wheat, whey, and beverage residues [61]. Cassava in Thailand was also under investigation for bioethanol production. Cassava tubers contain by mass 80% starch and less than 1.5% proteins. Pretreatment of cassava tubers include cleaning, peeling, chipping, and drying. The dried cassava chips are used for bioethanol production [62]. There are two established procedures for processing starch biomass summarized in Table 7.3; Figures 7.5 and 7.6.

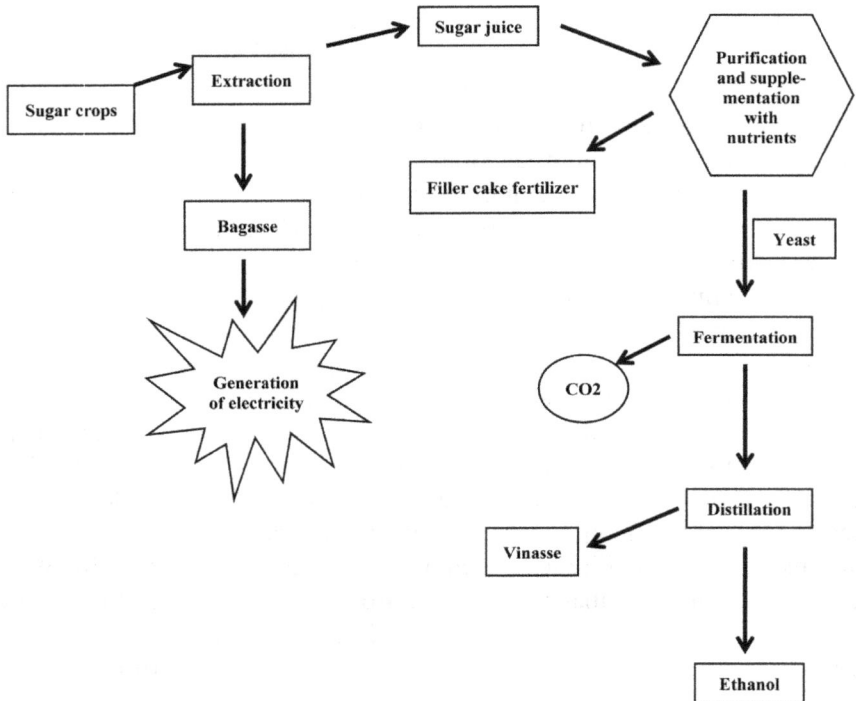

FIGURE 7.4　Diagram of ethanol production from sugar biomass [56].

TABLE 7.3　Dry-Grinded and Wet-Mill Method [63–65]

Dry-Grinded	Wet-Mill Method
Uses whole grains or tubers part of the starch crops	Involves separation of different components from the raw materials and only starch is used
It can be operated at a small-scale level with fewer requirements of equipment and capital investment	It requires extensive equipment and high capital investments for producing large amount of ethanol and a variety of co-products
It generates two major products: ethanol and distiller's dried grains with soluble (DDGS).	The various high value products like corn gluten meal (CGM) and corn gluten feed (CGF) are also produced, which can add to the commercial feasibility of the process.
The ethanol yield is 10.6 L/bushel	The ethanol yield is 9.5 L/bushel

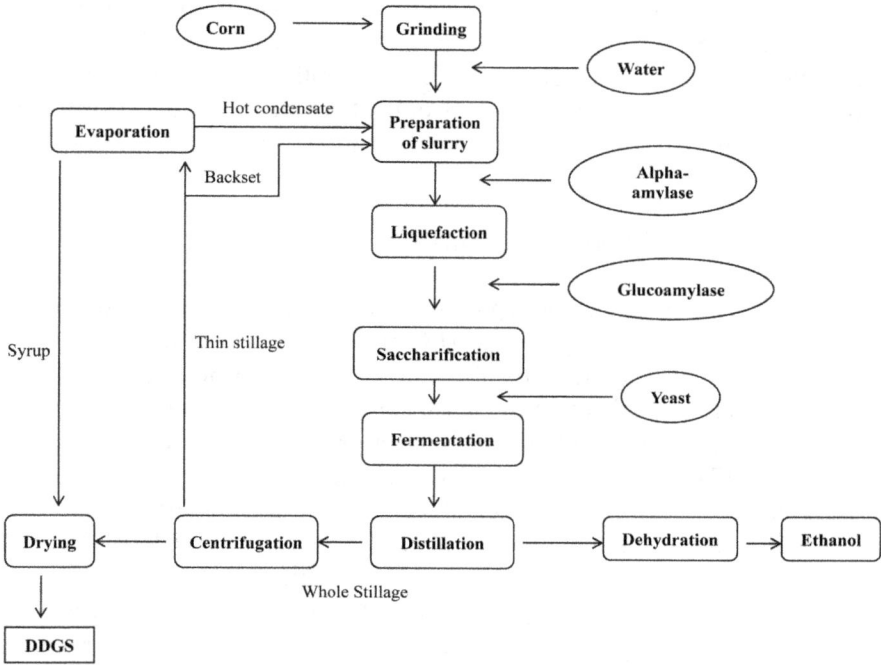

FIGURE 7.5   Diagram for dry-grind ethanol production from starch crops [65].

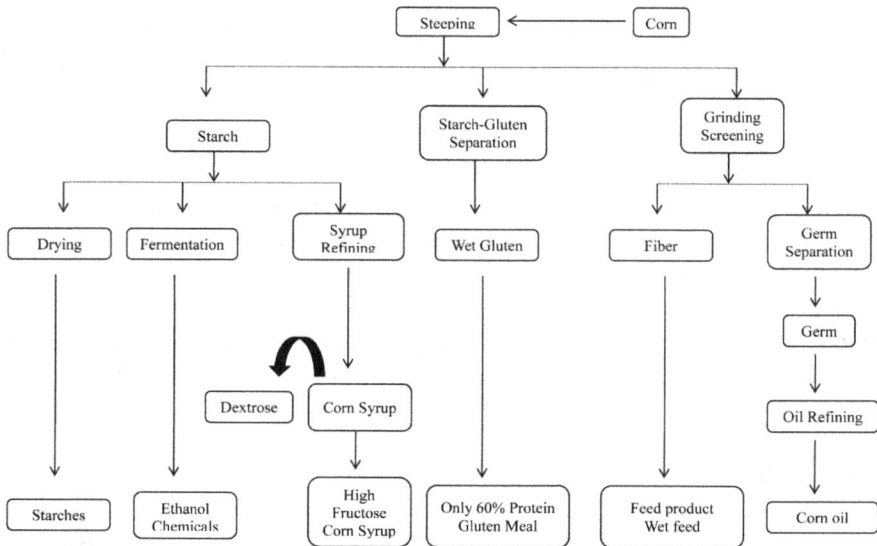

FIGURE 7.6   Diagram of wet-mill ethanol production from starch biomass [64].

The first-generation biofuels possessed two major issues, firstly, fuel vs. food trade-off, and secondly, biofuels potential of CO_2 reduction (biofuels production could release more carbon than CO_2 sequestration in the feedback growth process) [66]. This trade-off situation would affect the producers, distributors, related markets, and finally, the regional and national economies [67–69].

7.3.2 SECOND GENERATION ETHANOL

The use of food crops for ethanol production became debatable due to multiple reasons; the lignocellulosic biomass (LCB) (Figure 7.7 and Table 7.1) came into prominence as feedstock for the second-generation bioethanol [71–73]. Biomass of various agricultural wastes like rice husks, corn stalks, wheat, and barley straws, etc., and forest biomass such as grasses, wood, etc., are placed in this category [74, 75]. Some industrial wastes such as brewer's spent grains and spent grains from distillers, municipal solid wastes such as food waste, kraft paper and paper sludge containing cellulose were also considered [76–79]. They are inexpensive, abundant and non-competitive with the food crops [80]. However, an additional pretreatment step apart from the conventional three-step strategy (Figure 7.8), is required to break down the lignin present in the lignocellulosic material for subsequent hydrolysis and fermentation [81]. By using a suitable pretreatment method [82], the biomass is degraded into cellulose, hemicelluloses, and lignin polymer which can be further degraded into byproducts like furfurals, furans, acetic acid, etc. (Figure 7.9).

In general, recognition of new trending knowledge of energy-based biofuels (compared to fossil fuels) spurred the doubt of economic efficiency of biofuels [84]. As instance, cellulosic biofuel production is highly energy-intensive, which means the energy contained in this type of biofuels is lower than the energy required for its production [85]. Researchers in the past decades have attempted many experiments with limited success to lower the cellulosic ethanol production costs [66]. Study using microbial or fungal systems for more effective and faster cellulose breakdown and fermentation process has also been attempted [86, 87]. However, no wide-scale commercial solution has been found and the development in this field is still ongoing.

7.3.3 THIRD GENERATION ETHANOL

First and second-generation biofuels sources require large amount of arable land and are seasonally dependent, making their availability discontinuous for production. The three generations of biofuels have been compared with the petroleum products in Table 7.4. In a report by Kim and Day [89], the Louisiana sugar mills in the USA operate only quarterly as sugar cane availability is only from October to December, making capital investments for facility upgrade and maintenance very difficult [89]. Comparatively, microalgal cultivation will be a continuous process with less land requirement. Also, non-arable land can be used which provides a higher incentive [90–93]. Thus, a need for a fermentable substrate with all year availability emerges which if realized can change the entire economic of the industry (Table 7.5).

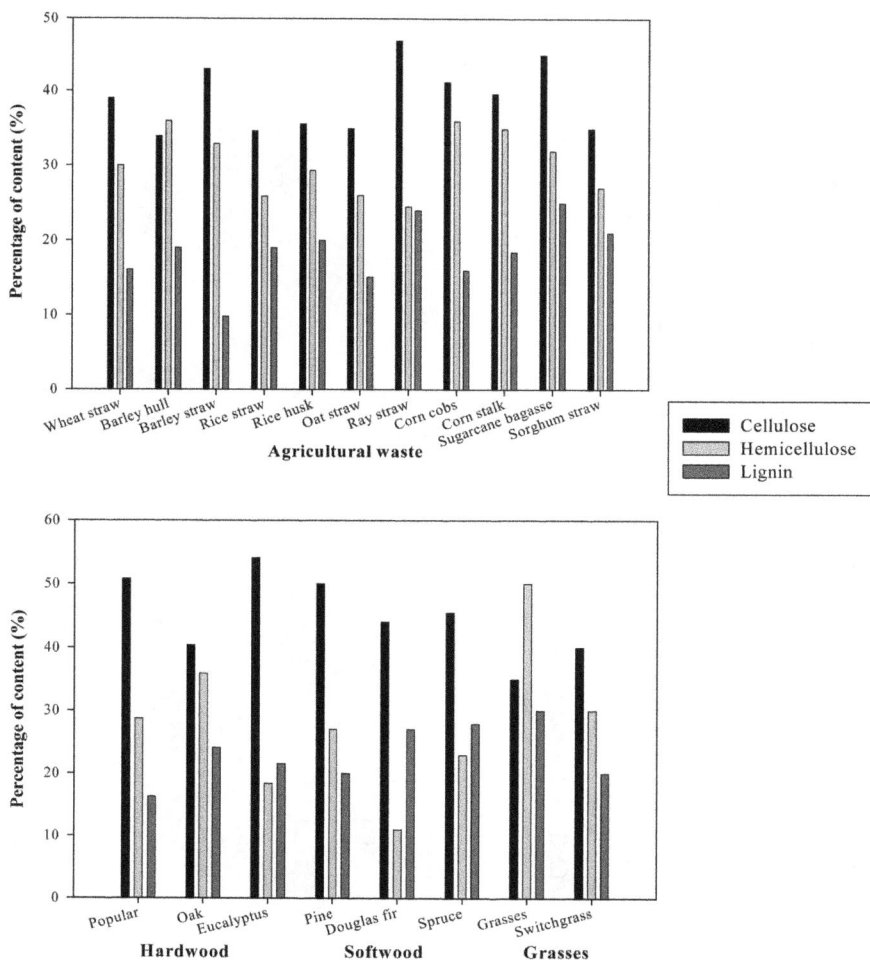

FIGURE 7.7 Types of lignocellulosic biomass and their composition [70].

TABLE 7.4 Comparison of Three Generations Biofuels with Petroleum Products [88]

	Petroleum Products	First Generation	Second Generation	Third Generation
Technology	Petroleum refinery	Microbial fermentation, chemical, and enzymatic transesterification	Pretreatment, hydrolysis, and fermentation, transesterification	Metabolic engineering for direct synthesis, fractionation of algal biomass
Feedstock	Crude petroleum	Vegetable oils and corn sugar feedstocks	Non-food, cheap, and abundant plant waste biomass (agricultural and forest residue, etc.)	Algae
Products	CNG, LPG, diesel, petrol, kerosene, and jet-fuel	Biodiesel, corn ethanol, sugar alcohol	Hydrotreating oil, bio-oil, FT oil, lignocellulose ethanol, butanol, mixed alcohols	Biodiesel, bioethanol, biohydrogen
Benefits	• High energy density: high compact portable source of energy used for most forms of mechanical transportation	• Environmental friendly • Economic and society security	• Environmental friendly • Not competing with food	• Environmental friendly • Not competing with food and agricultural land • Oil productivity is very high when compared with all other biomass • Algae is the most promising non-food source of biofuels • A rapid reproduction rate • Algae can grow in saltwater and harsh conditions • Algae thrive on CO_2 from gas and coal-fired power plants

TABLE 7.4   *(Continued)*

	Petroleum Products	First Generation	Second Generation	Third Generation
Problems	• Depletion • Declining of petroleum reserve • Environmental pollution • Economic and ecological problems	• Limited feedstock • Food *vs.* fuel competition • Blended partially with conventional fuel	• Agricultural land consumption • Complicated processes	• Algae biofuel contains no sulfur, is non-toxic and highly biodegradable. • Low product yield at large scale • Less biomass production

TABLE 7.5 Carbohydrate Content and Ethanol Yield in Different Feedstocks

SL. No.	Sources	Carbohydrate Content (%)	Ethanol Yield (gallons/acre)*	References
1.	Corn	70–72	370–430	[63]
2.	Sorghum	68–70.7	326–435	[28]
3.	Wheat	65.3–76	277	[29]
4.	Sugar beet	8–12	536–714	[95]
5.	Microalgae (*Chlorella vulgaris*)	37–55	5,000–15,000	[74]

*Ref. [93].

Microalgae are the oldest living organisms on Earth [96] which can grow 100-fold faster than terrestrial and can double their biomass in less than one day time period [97]. It is beneficial as microalgae can grow faster and fixes the CO_2 at a higher rate than terrestrial plants. Lignin is absent in microalgae making it easier to convert their starch and cellulose to monosaccharides [24, 98]. These microalgae can be used for ethanol production via various hydrolysis strategies and fermentation processes [24, 93, 99] (Tables 7.1 and 7.6; Figure 7.10). Nahak et al. [105] reported that seaweed and marine algae such as *Eneteromorpha* sp. contain 70% carbohydrate (dry weight basis), which can be explored for bioethanol production [105]. The reported microalgal species which are employed as a fermentable substrate for bioethanol production are listed in Table 7.6. The microalgal cells store carbohydrates in the outer layer of cell wall: pectin, agar, alginate; inner layer of the cell wall: cellulose, hemicelluloses; and inside the cell: starch [91]. These carbohydrates are hydrolyzed to fermentable sugars like glucose for the bioethanol production via microbial fermentation [48, 72]. The procedure includes cultivation of microalgae followed by the recovery of microalgal biomass from the medium, cell disruption for the release of biomolecules, saccharification (hydrolysis), fermentation, and finally separation by distillation [51, 91]. Chemical hydrolysis method facilitates the solvent extraction of lipids, thereby recovering both fermentable sugars and lipid from the microalgal biomass, where fermentable sugar can be converted to ethanol and lipid into biodiesel, increasing the commercial potential by two folds [106].

Algae-based fuel industries are still searching for innovative ways to reduce the production cost via reducing the costs of systems infrastructure and integration, algae biomass production process, harvesting, and dewatering techniques, extraction, and fractionation, and finally biofuels conversion process [107] (Figures 7.8–7.10).

TABLE 7.6 Carbohydrate Concentration and Microalgal Biomass Conversion to Ethanol

Microalgae	Carbohydrate Concentration (%w/w)			Ethanol Yield $(g_{ethanol}\ g^{-1}\ _{biomass})$	References
	Total	Starch	Glucose		
Chlamydomonas reinhardtii	59.7	43.6	44.7	0.235	[100]
Chlorococcum humicola	32.5	11.3	15.2	0.520	[101, 102]
Chlorococcum sp.	–	–	–	0.383	[101, 102]
Chlorococcum infusionum	32.5	11.3	15.2	0.261	[101, 102]
Chlorella vulgaris	50.9	–	48.0	0.209	[24]
Chlorella vulgaris	–	–	–	0.400	[103]
Dunaliella sp.	–	–	–	0.011*	[104]

*Theoretical conversion.

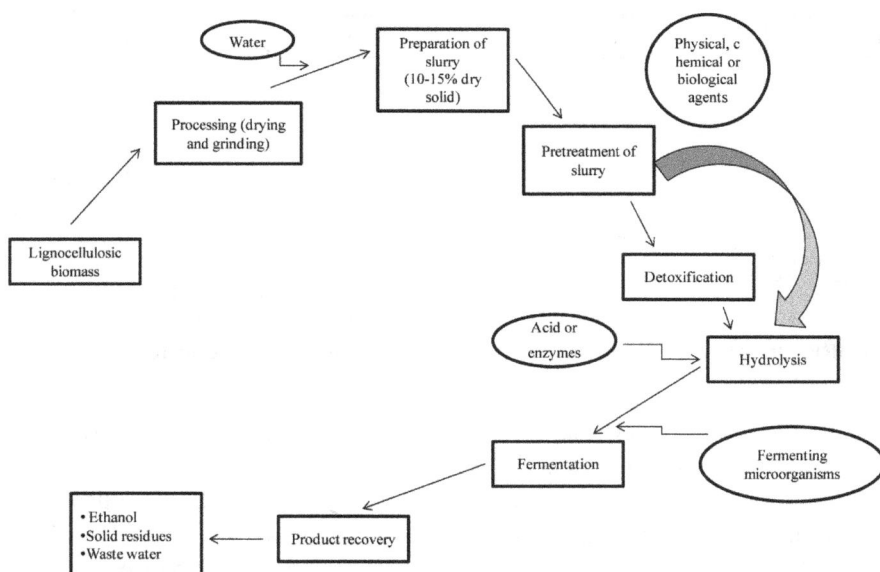

FIGURE 7.8 Ethanol production from lignocellulosic materials [27].

7.3.4 *FOURTH GENERATION BIOETHANOL*

This generation biofuel is under development and experimental stages. Thus, they combine a variety of technology, processing, and feedstock level [66].

The main feedstock for the fourth-generation biofuels production is genetically engineered, high biomass with low lignin and cellulose contents. This effectively eliminates the hurdles of the second-generation substrates. Another strong contender is metabolically engineered algae with high oil yield contents, increased carbon entrapment ability, and improved cultivation, harvesting, and fermentation processes, which is a vast improvement to the third-generation substrate [108].

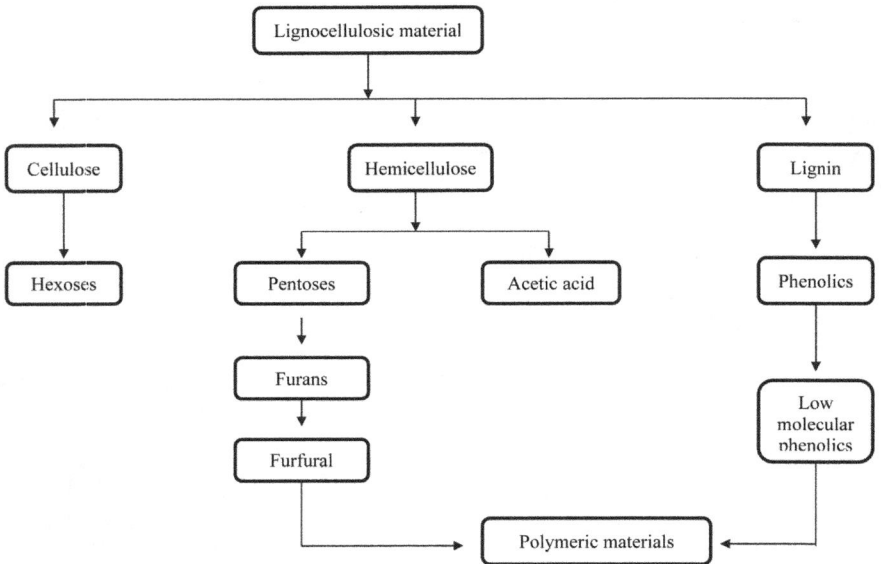

FIGURE 7.9 Degradation products occurring during hydrolysis of lignocellulosic material [83].

In order to enhance the ethanol production and tolerance towards it, certain microorganisms have been modified, for example, *E. coli*. This microorganism is extremely well studied, takes up genetic changes well, and additionally, it can grow three times faster than yeast and 100 times faster than most agricultural microbes [15]. Beside the strong efforts, the researchers have not reported a single commercially available CBP organism. However, Brethauer and Studer [109] have reported a microbial consortium which could be used instead of using a single microbe. The proposed model utilizes *Trichoderma reesei*, *S. cerevisiae* and *Scheffersomyces stipites*. The former organism secretes enzyme for hydrolysis; the middle and later microbes ferment hexoses and pentoses, respectively [109]. But the major obstacle in this approach is to control the consortium and it is also difficult to find microorganisms having identical fermentation conditions. Genetically modified microorganisms reported for ethanol production are listed in Table 7.7.

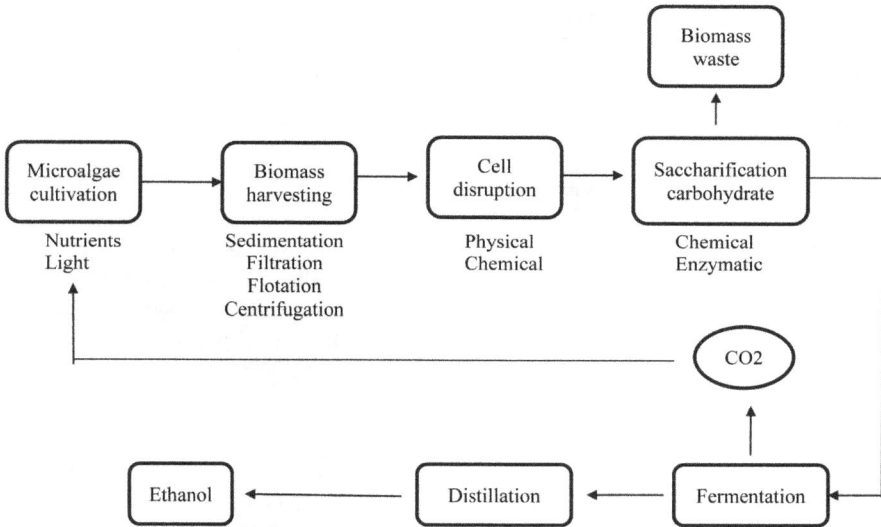

FIGURE 7.10   Ethanol production from microalgae [51].

7.4   MICROORGANISM USED FOR FERMENTATION

Fermentation is based on the discipline of chemistry, biochemistry, and microbiology, in which the fermentable sugars are converted to ethanol by microorganisms [90]. In this, *Saccharomyces cerevisiae* and *Zymomonas mobilis* are the most common and well-studied microorganisms with applications in the food and beverage industry, molecular genetic research, and biotechnological ethanol production [111, 112]. The theoretical values of ethanol production are 87–95% and 94–97%, respectively [113–117], with the difference in the metabolic pathways opted for the sugar catalysis as mentioned in Figures 7.11 and 7.12. Other microorganisms reported for ethanol production are *Aerobacter, Bacillus, Klebsiella, Thermoanerobacter, Aeromonas, Escherichia coli, S. bayanus, Candida* sp., *Pichia stipitis, P. angophorae, Pachysolen tannophilus, Fusarium, Mucor, Neurospora, Monilia,* and *Rhizopus* [90, 117–123].

TABLE 7.7   List of Genetically Modified Microorganisms Used in Ethanol Production [110]

Microorganisms	Strain	Features
Yeast	*Candida shehatae* NCL-3501	Co-ferment xylose and glucose
	Saccharomyces cerevisiae D5a	Improvement in yield of ethanol
	Saccharomyces cerevisiae 590E1	Ferment glucose and cellobiose
	Saccharomyces cerevisiae RWB217	Ferment glucose and xylose
	Saccharomyces cerevisiae RWB218	
Bacteria	*Zymomonas mobilis* ZM4	Ferment xylose and glucose
	Pichia stipitis BCC15191	
	Zymomonas mobilis AX101	Ferment arabinose, glucose, and xylose
	Thermoanaerobacterium saccharolyticum ALK2	Improved ethanol yield, have ability to ferment arabinose, glucose, xylose, and mannose
	Thermoanaerobacter mathranii BG1L1	Improved ethanol yield
	Clostridium thermocellum DSM1313 and YD01	
	Escherichia coli KO11	Ferment xylose and glucose
	Escherichia coli FBR5	Ferment xylose and arabinose
	Pichia stipites A	Adapted at hydrolysate increased concentration
	Pichia stipitis NRRL Y-7124	

7.5   PRETREATMENT

Pretreatment of biomass is needed to remove or modify the surrounding matrix of lignin and hemicellulose prior to hydrolysis of the polysaccharides like cellulose and hemicellulose [124]. The main aim of pretreatment is to increase the enzyme accessibility improving digestibility of cellulose. Different pretreatment methods have a different effect on the cellulose, hemicellulose, and lignin fraction, thus forming the basic criteria of method selection [80]. Here, the pretreatment methods are discussed according to the biomass used.

7.5.1   *LIGNOCELLULOSIC MATERIALS*

Pretreatment of lignocellulosic material is done to fractionate, solubilize, hydrolyze, and separate cellulose, lignin, and hemicellulose [70, 125]. It includes various categories: physical, chemical, physicochemical, and

Glucose

↓ (Glycolysis)

Dyhroxyacetone phosphate   Glyceraldehyde-3-phosphate

↓   ↓

Glycerol-3-phosphate   Pyruvate

NADH ⌐   ↓

⌐→NAD+

Glycerol   Acetaldehyde-TPP complex

NADH ⌐   NAD(P)H ⌐

⌐→NAD(P)+

Ethanol   NAD+   Acetate

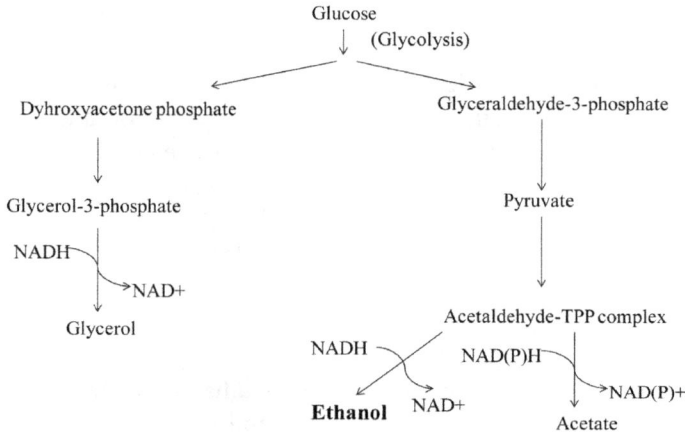

FIGURE 7.11   Ethanol biosynthesis pathway in *Saccharomyces cerevisiae* [111].

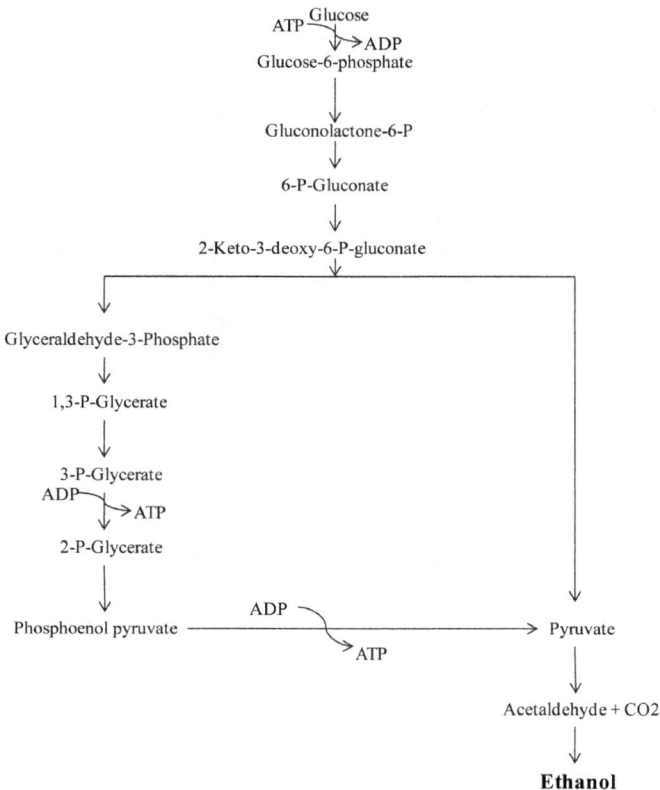

ATP ⌐ Glucose

⌐→ADP

Glucose-6-phosphate

↓

Gluconolactone-6-P

↓

6-P-Gluconate

↓

2-Keto-3-deoxy-6-P-gluconate

Glyceraldehyde-3-Phosphate

↓

1,3-P-Glycerate

↓

3-P-Glycerate

ADP ⌐

⌐→ATP

2-P-Glycerate

↓

Phosphoenol pyruvate   ADP ⌐   Pyruvate

⌐→ATP

↓

Acetaldehyde + CO2

↓

Ethanol

FIGURE 7.12   Entner-Doudoroff (ED) pathway in *Z. mobilis* [114].

biological [126]. These methods differ from each other in context to mode of action, reaction conditions and overall outcomes.

Physical method includes uncatalyzed steam explosion (SE), liquid hot water (LHW) pretreatment, mechanical comminution and high energy radiation [124, 127]. The methods increase the surface area and pore volume, decreases the degree of polymerization (DP) of cellulose and lignin [128]. By using uncatalyzed SE method, Grous et al. [129] achieved the enzymatic digestibility of untreated poplar chips from 15% to 90% after treatment [129]. Perez et al. [130] used LHW to pretreat wheat straw and obtained maximum hemicellulose-derived sugar recovery of 53% and enzymatic hydrolysis yield of 96% [130].

Although, the physical pretreatment method has some disadvantages like energy consuming, ecologically unhealthy, and non-viable for commercial process. Chemical pretreatments were initially exploited for the production of high-quality paper products. The improvement in biodegradability of cellulose by removing lignin and hemicellulose was the primary goal [124]. It includes techniques like alkali (e.g., NaOH, KOH, NH_4OH and $Ca(OH)_2$), acid (H_2SO_4, HCl, HNO_3), organic acids (fumaric and maleic acids) and several cellulose solvents such as alkaline H_2O_2, ozone, glycerol, dioxane, phenol, and ethylene glycol [124, 131, 132]. Millet et al. [133] reported an increase in digestibility of NaOH-treated hardwood from 14% to 55% with a decrease of lignin content from 24%–55% to 20% [133].

Physicochemical pretreatment is a combination of physical conditions and chemicals. It includes SE, ammonia fiber explosion (AFEX), soaking aqueous ammonia (SAA), ammonia recycling percolation (ARP), wet oxidation and CO_2 explosion. It increases the accessible surface area, decreases cellulose crystallinity, and removes hemicellulose and lignin from the lignocellulosic materials. The yields of enzymatic hydrolysis of AFEX-pretreated newspaper (18–30% lignin) and aspen chips (25% lignin) were reported by McMillan [23] as only 40% and below 50%, respectively [23].

Biological pretreatment is a friendly method used for lignin removal [22]. It is carried out using microorganisms, particularly fungi, which are white rot, brown rot, and soft rot fungi [134]. White and soft rot fungi attack on both cellulose and lignin, whereas, brown rot fungi attack only on cellulose. The biological pretreatment was studied by Hwang et al. [135] using four different white-rot fungi for 30 days [135]. The team found that glucose yield of pretreated wood by *Trametes versicolor* MrP 1 reached 45% by enzymatic hydrolysis while 35% solid was converted to glucose during fungi incubation. Beside this, the lower hydrolysis rate and longer incubation period as compared to the other pretreatment methods are some disadvantages of the biological pretreatment method. However, the benefits of using biological pretreatment method, such as are low energy requirement and mild environmental conditions, are unignorable [136].

7.5.2 MICROALGAL BIOMASS

To release carbohydrate trapped within microalgal cells, either chemical or enzymatic methods may be used to disrupt the microalgal cell wall. Microalgae has a simple cellular structure and lignin is also absent in them, thus mild reaction conditions are sufficient [24, 137]. This method is fast and inexpensive than the enzymatic method. However, the extreme conditions like high temperature and pressure can degrade carbohydrate into furfural, acetic acid, gypsum, vanillin, and aldehydes which are potential fermentation inhibitors leading to yield reduction [91, 132].

The enzymatic hydrolysis of the carbohydrate, particularly starch, by alpha-amylase (liquefaction) in a random manner generates oligosaccharides with three or more α-$(1\rightarrow4)$-linked D-glucose units as an intermediate product. Subsequently, starch saccharifying enzyme (amyloglucosidase) is introduced to the liquefied starch and simple reducing sugars are generated [83, 100]. This method has advantages over the chemical method such as higher conversion yield, minimal byproduct formation, mild operating condition, and low energy input [138].

7.6 FERMENTATION PROCESS

Both six-carbon (hexoses) and five-carbon (pentoses) sugars (from cellulose and hemicellulose hydrolysis) derived from the biomass can be used as a substrate for the fermentation process [82]. The fermentation parameters are: temperature range, pH range, alcohol tolerance, growth rate, productivity, osmotic tolerance, specificity, yield, genetic stability, and inhibitor tolerance. The selection of microorganisms is also dependent upon their compatibility to co-exist with products, processes, and equipment for bioethanol production. Though, *Saccharomyces cerevisiae* and *Zymomonas mobilis* are widely used for bioethanol fermentation, they show inability to efficiently convert xylose [23, 139, 140]. Thus, natural xylose-fermenting yeasts like *Pichia stipites*, *Candida shehatae* and *C. parapsilosis* are also used [23, 141]. Depending on the process design, the combination of the above-mentioned parameters may vary, being consecutive or simultaneous [142]. The most commonly used bioconversion processes are: separate hydrolysis and fermentation (SHF), simultaneous saccharification and fermentation (SSF), simultaneous saccharification and co-fermentation (SSCF) and consolidated bioconversion process (CBP) (Figure 7.13).

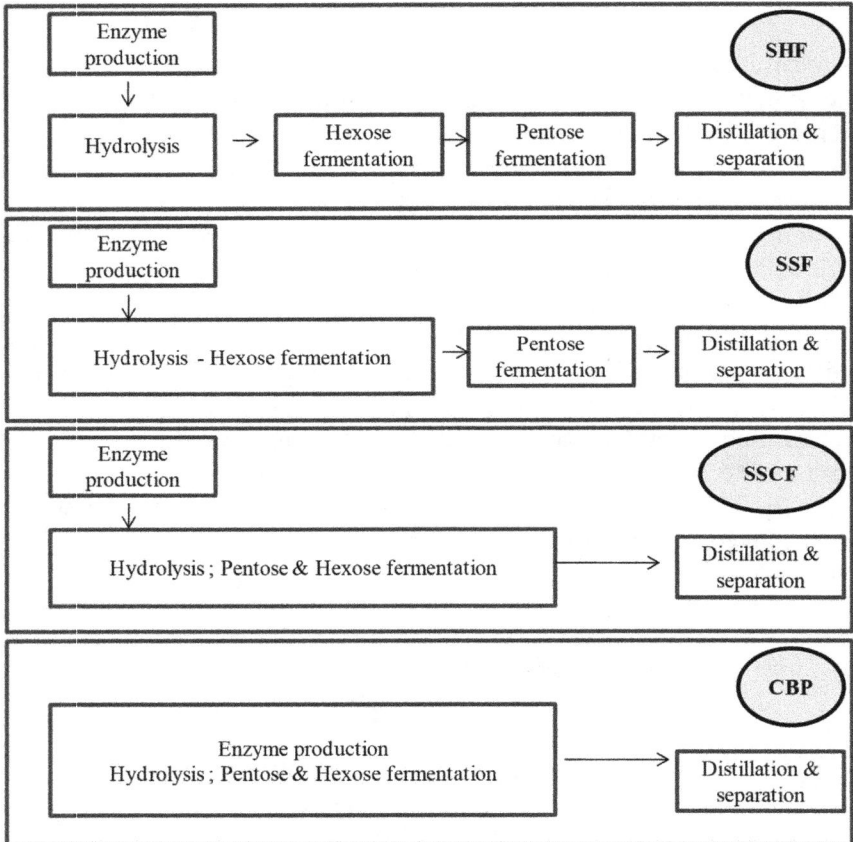

FIGURE 7.13   Types of fermentation processes [140].

7.6.1   *SEPARATE HYDROLYSIS AND FERMENTATION (SHF)*

Two different reactors used for the hydrolysis and fermentation process [142]. This has the advantage that both the process can be carried out under different conditions like temperature, pH, and time [101, 102, 142]. Thus, more substrate could be produced during the hydrolysis process by using the acid, alkali, or enzyme. However, neutralizing the acid or alkali will be an additional step. On the other hand, enzymatic hydrolysis leads to accumulation of glucose and cellobiose, which will inhibit the cellulases and thus reduce the hydrolysis rate and the overall yield of the process [142, 143].

7.6.2 SIMULTANEOUS SACCHARIFICATION AND FERMENTATION (SSF)

In a single reactor, both hydrolysis and fermentation process take place [144]. Since, yeast will directly convert the glucose and cellobiose into bioethanol, these compounds could not cause the inhibition [142]. SSF has advantages over the SHF that it requires lower enzyme load, higher bioethanol yield, shorter fermentation time, and reduced risk of contamination by external microflora [145]. However, compromises in the operating conditions are seen during the SSF process [146]. Additionally, it is difficult to recycle the enzymes and yeast, which makes the process challenging for the scaling up for commercialization purposes [147].

7.6.3 CONSOLIDATED BIOCONVERSION PROCESS

In the process, the same microorganisms carry out both saccharification and fermentation of biomass in a single stage [148]. CBP is also known as direct microbial conversion (DMC). CBP uses a single microorganism community for cellulase production, cellulose hydrolysis, and fermentation in a single step, has two major benefits: (i) avoid the cost of cellulase production, (ii) saves the energy [149, 150]. *Clostridium thermohydrosulfuricum, Thermoanaerobacter ethanolicus, Thermoanaerobacter mathranii, Thermoanaerobacter brockii* strain, etc., are some thermophilic cellulolytic anaerobic bacteria that have been studied for their potential as bioethanol producers [13].

7.6.4 SIMULTANEOUS SACCHARIFICATION AND CO-FERMENTATION (SSCF)

In this fermentation, microorganism can completely assimilate all the sugars released during the pretreatment and hydrolyze LCB. Sanchez and Cardona [151] reported that the mixed cultures of yeasts can assimilate both hexoses and pentoses [151]. However, hexose-utilizing microorganisms grow faster than pentoses-utilizing microorganisms and thus, hexoses conversion to ethanol gets elevated. In nature, there is an existence of microorganisms which can assimilate both hexoses and pentoses for ethanol production. This conversion efficiency can be increased by genetic modification of the microorganisms [81].

7.7   BOTTLENECKS IN COMMERCIALIZATION OF ETHANOL

The commercialization of the bioethanol is dependent on the economics of the process. Simultaneously, the ease of implementation is another important factor for the success of a new bioprocess technology. There are many other factors which also influence the commercialization of ethanol production including the availability of feedstock, cost of raw material and ethanol recovery cost. The prices of feedstock depend on the location, seasons, local state of the supply-demand conditions and transportation needed [152]. Studies are done which state that biomass feedstock made up 40% of the ethanol production cost [139, 152]. The main challenge is to economize the operating costs of biomass conversion processes that include pretreatment and enzymatic hydrolysis, toward a full-scale commercialization. Extensive studies are done with aim of developing a process which reduce bioconversion time, use of cellulose during fermentation and enhance the total ethanol yield [152, 153]. Another cost adding factor is the separation and recovery of ethanol from the fermented liquor of lignocelluloses hydrolysates [136, 151]. Although microalgae are rich in carbohydrates and proteins, limited reports are available on its use as a source for fermentation for ethanol production [42]. Thus, despite being a fully functional industry, ethanol production still has lacunae that need to be addressed to improve the efficiency of this commercial endeavor and its economics.

7.8   CONCLUSION

With the sharp increase in population and vehicles, only bioethanol will not be sufficient to match the energy demand. However, with each generation, the bioprocess technology is undergoing massive changes, through which we are slowly but contently increasing production potential. The changes initiate from the nature of substrate to the microorganisms and the process involved. Dealing with the current scenario of possible feedstock with their advantages and disadvantages, it becomes necessary to evolve better and sustainable methodologies for higher yield outputs. Feedstock generated from forest and agricultural wastes, algal biomass and engineered organisms could lead to the development of efficient technologies, thereby reducing the ethanol price and environmental pollution, and also preserving the natural resources. The fourth-generation biofuels are a new concept and could have a significant potential in the near future.

ACKNOWLEDGMENTS

We are thankful to Professor A. Shastri, Vice-Chancellor of Banasthali Vidyapith, Rajasthan, for kindly extending the facilities of "Banasthali Center of Education" for research in basic sciences sanctioned under CURIE program at the Department of Science and Technology.

CONFLICTS OF INTEREST

The authors declare that there is no conflict of interest to disclose.

KEYWORDS

- **bioethanol**
- **feedstock**
- **fermentation**
- **pretreatment**

REFERENCES

1. Zhang, Y., (2013). The links between the price of oil and the value of US dollar. *Int. J. Energy Econ. Policy, 3*, 341–351.
2. Lal, R., (2005). World crop residues production and implications of its use as a biofuel. *Environ. Int., 31*, 575–584. https://doi.org/10.1016/j.envint.2004.09.005.
3. Rosales-Calderon, O., & Arantes, V., (2019). A review on commercial-scale high-value products that can be produced alongside cellulosic ethanol. *Biotechnol. Biofuels, 12*, 240. https://doi.org/10.1186/s13068-019-1529-1.
4. Gaurav, N., Sivasankari, S., Kiran, G. S., Ninawe, A., & Selvin, J., (2017). Utilization of bioresources for sustainable biofuels: A review. *Renew. Sust. Energy Rev., 73*, 205–214. https://doi.org/ 10.1016/j.rser.2017.01.070.

5. Kremer, F. G., Jardim, J. L. F., & Maia, D. M., (1996). *Effects of Alcohol Composition on Gasohol Vehicle Emissions (No. 962094).* SAE Technical Paper.
6. Demirbas, A., (2005). Bioethanol from cellulosic materials: A renewable motor fuel from biomass. *Energ. Sources, 27*(4), 327–337. https://doi.org/10.1080/00908310390266643.
7. Green, K. R., & Lowenbach, W. A., (2001). MTBE contamination: Environmental, legal, and public policy challenges guest editorial. *Environ. Forensics, 2*(1), 3–6. https://doi.org/10.1006/enfo.2001.0030.
8. Kar, Y., & Deveci, H., (2006). Importance of P-series fuels for flexible-fuel vehicles (FFVs) and alternative fuels. *Energ. Sources Part A, 28*(10), 909–921. https://doi.org/10.1080/00908310600718841.
9. Pickett, J., Anderson, D., Bowles, D., Bridgwater, T., Jarvis, P., Mortimer, N., Poliakoff, M., & Woods, J., (2008). *Sustainable Biofuels: Prospects and Challenges.* The Royal Society. UK. London.
10. Yao, C., Yang, X., Roy, R. R., Cheng, C., Tian, Z., & Li, Y., (2009). The effects of MTBE/ethanol additives on toxic species concentration in gasoline flame. *Energ. Fuel, 23*(7), 3543–3548. https://doi.org/10.1021/ef900035q.
11. Minteer, S., (2006). Ethanol blends: E10 and E-diesel. *Alcoholic Fuels,* 125–136.
12. Nigam, P. S., & Singh, A., (2011). Production of liquid biofuels from renewable resources. *Prog. Energ. Combust., 37*(1), 52–68. https://doi.org/10.1016/j.pecs.2010.01.003.
13. Vohra, M., Manwar, J., Manmode, R., Padgilwar, S., & Patil, S., (2014). Bioethanol production: Feedstock and current technologies. *J. Environ. Chem. Eng., 2*, 573–584. https://doi.org/10.1016/j.jece.2013.10.013.
14. Banerjee, S., Mudliar, S., Sen, R., Giri, B., Satpute, D., Chakrabarti, T., & Pandey, R. A., (2010). Commercializing lignocellulosic bioethanol: Technology bottlenecks and possible remedies. *Biofuel. Bioprod. Bior., 4*(1), 77–93. https://doi.org/10.1002/bbb.188.
15. Niphadkar, S., Bagade, P., & Ahmed, S., (2017). Bioethanol production: Insight into past, present and future perspectives. *Biofuels.* http://dx.doi.org/10.1080/17597269.2017.1334338.
16. Hunt, D. V., (1981). *The Gasohol Handbook.* New York, NY: Industrial Press.
17. Barros, S. (2019). "*Brazil Biofuels Annual Report 2019.*" United States Department of Agriculture (USDA)–Foreign Agricultural Service (FAS). GAIN Report Number BR19029.
18. Ramos, J. L., & Duque, E., (2019). Twenty-first-century chemical odyssey: Fuels versus commodities and cell factories versus chemical plants. *Microb. Biotechnol., 12*(2), 200–209. https://doi.org/10.1111/1751-7915.13379.
19. Alonso-Gómez, L. A., & Bello-Pérez, L. A., (2018). Four generations of raw materials used for ethanol production: Challenges and opportunities. *Agrociencia (Montecillo), 52*(7), 967–990.
20. Dalena, F., Senatore, A., Iulianelli, A., Di Paola, L., Basile, M., & Basile, A., (2019). Ethanol from biomass: Future and perspectives. In: *Ethanol* (pp. 25–59). *Elsevier.* https://doi.org/10.1016/B978-0-12-811458-2.00002-X.
21. Alia, K. B., Rasul, I., Azeem, F., Hussain, S., Siddique, M. H., Muzammil, S., Riaz, M., et al., (2019). Microbial production of ethanol. In: *Microbial Fuel Cells: Materials and Applications* (pp. 307–334). Materials Research Forum LLC., Pennsylvania. https://doi.org/10.21741/9781644900116-12.
22. Ahorsu, R., Medina, F., & Constantí, M., (2018). Significance and challenges of biomass as a suitable feedstock for bioenergy and biochemical production: A review. *Energies, 11*(12), 3366. https://doi.org/10.3390/en11123366.

23. McMillan, J. D., (1993). *Xylose Fermentation to Ethanol: A Review* (No. NREL/ TP-421-4944). National Renewable Energy Lab., Golden, CO (United States).
24. Ho, D. P., Ngo, H. H., & Guo, W., (2014). A mini-review on renewable sources for biofuel. *Bioresour. Technol., 169*, 742–749. https://doi.org/10.1016/j.biortech.2014.07.022
25. Sindhu, R., Gnansounou, E., Binod, P., & Pandey, A., (2016). Bioconversion of sugarcane crop residue for value-added products: An overview. *Renew. Energ., 98*, 203–215. https:// doi.org/10.1016/j.renene.2016.02.057.
26. Kim, S., & Dale, B. E., (2004). Global potential bioethanol production from wasted crops and crop residues. *Biomass Bioenerg., 26*(4), 361–375. https://doi.org/10.1016/j. biombioe.2003.08.002.
27. Zabed, H., Sahu, J. N., Boyce, A. N., & Faruq, G., (2016). Fuel ethanol production from lignocellulosic biomass: An overview on feedstocks and technological approaches. *Renew. Sust. Energ. Rev., 66*, 751–774. https://doi.org/10.1016/j.rser.2016.08.038.
28. Ramírez, M. B., Ferrari, M. D., & Lareo, C., (2016). Fuel ethanol production from commercial grain sorghum cultivars with different tannin content. *J. Cereal Sci., 69*, 125–131. https://doi.org/10.1016/j.jcs.2016.02.019.
29. Murphy, J. D., & Power, N. M., (2008). How can we improve the energy balance of ethanol production from wheat?. *Fuel, 87*(10, 11), 1799–1806. https://doi.org/10.1016/j. fuel.2007.12.011.
30. Pervez, S., Aman, A., Iqbal, S., Siddiqui, N. N., & Qader, S. A. U., (2014). Saccharification and liquefaction of cassava starch: An alternative source for the production of bioethanol using amylolytic enzymes by double fermentation process. *BMC Biotechnol., 14*(1), 49. https://doi.org/10.1186/1472-6750-14-49.
31. Ziska, L. H., Runion, G. B., Tomecek, M., Prior, S. A., Torbet, H. A., & Sicher, R., (2009). An evaluation of cassava, sweet potato and field corn as potential carbohydrate sources for bioethanol production in Alabama and Maryland. *Biomass Bioenerg., 33*(11), 1503–1508. https://doi.org/10.1016/j.biombioe.2009.07.014.
32. Khan, R. A., Nawaz, A., Ahmed, M., Khan, M. R., Azam, F. D., Ullah, S., Sadullah, F., et al., (2012). Production of bioethanol through enzymatic hydrolysis of potato. *Afr. J. Biotechnol., 11*(25), 6739–6743. http://dx.doi.org/10.5897/AJB11.2791.
33. George, N. A., Pecota, K. V., Bowen, B. D., Schultheis, J. R., & Yencho, G. C., (2011). Root piece planting in sweet potato—A synthesis of previous research and directions for the future. *Horttechnology, 21*(6), 703–711. https://doi.org/10.21273/HORTTECH.21.6.703.
34. Duvernay, W. H., Chinn, M. S., & Yencho, G. C., (2013). Hydrolysis and fermentation of sweet potatoes for production of fermentable sugars and ethanol. *Ind. Crops Prod., 42*, 527–537. https://doi.org/10.1016/j.indcrop.2012.06.028.
35. Lewandowski, I., Scurlock, J. M., Lindvall, E., & Christou, M., (2003). The development and current status of perennial rhizomatous grasses as energy crops in the US and Europe. *Biomass Bioenerg., 25*(4), 335–361.https://doi.org/10.1016/S0961-9534(03)00030-8.
36. Rengsirikul, K., Ishii, Y., Kangvansaichol, K., Sripichitt, P., Punsuvon, V., Vaithanomsat, P., Nakamanee, G., & Tudsri, S., (2013). Biomass yield, chemical composition and potential ethanol yields of 8 cultivars of Napier grass (*Pennisetum purpureum* Schumach.) harvested 3-Monthly in central Thailand. *J. Sust. Bioenergy Syst., 3*(2), 107–112. https://doi.org/10.4236/jsbs.2013.32015.
37. Kumar, A., Singh, L. K., & Ghosh, S., (2009). Bioconversion of lignocellulosic fraction of water-hyacinth (*Eichhornia crassipes*) hemicellulose acid hydrolysate to ethanol by *Pichia stipitis*. *Bioresour. Technol., 100*(13), 3293–3297. https://doi.org/10.1016/j. biortech.2009.02.023.

38. Aswathy, U. S., Sukumaran, R. K., Devi, G. L., Rajasree, K. P., Singhania, R. R., & Pandey, A., (2010). Bio-ethanol from water hyacinth biomass: An evaluation of enzymatic saccharification strategy. *Bioresour. Technol., 101*(3), 925–930. https://doi.org/10.1016/j.biortech.2009.08.019.
39. Zhu, J. Y., & Pan, X. J., (2010). Woody biomass pretreatment for cellulosic ethanol production: Technology and energy consumption evaluation. *Bioresour. Technol., 101*(13), 4992–5002. https://doi.org/10.1016/j.biortech.2009.11.007.
40. Gonzalez, R. W., Treasure, T., Phillips, R. B., Jameel, H., & Saloni, D., (2011). Economics of cellulosic ethanol: Green liquor pretreatment for softwood and hardwood, greenfield and repurpose scenarios. *Bioresources, 6*(3), 2551–2567.
41. Zhu, J. Y., Luo, X., Tian, S., Gleisner, R., Negrone, J., & Horn, E., (2011). Efficient ethanol production from beetle-killed lodgepole pine using SPORL technology and *Saccharomyces cerevisiae* without detoxification. *Tappi J., 10*(5), 9–18.
42. Singh, J., & Gu, S., (2010). Commercialization potential of microalgae for biofuels production. *Renew. Sust. Energ. Rev., 14*(9), 2596–2610. https://doi.org/10.1016/j.rser.2010.06.014.
43. Limayem, A., & Ricke, S. C., (2012). Lignocellulosic biomass for bioethanol production: Current perspectives, potential issues and future prospects. *Prog. Energ. Comb., 38*(4), 449–467. https://doi.org/10.1016/j.pecs.2012.03.002.
44. Kreith, F., & Krumdieck, S., (2013). *Principles of Sustainable Energy Systems*. CRC Press.
45. Buah, W. K., Cunliffe, A. M., & Williams, P. T., (2007). Characterization of products from the pyrolysis of municipal solid waste. *Process Saf. Environ., 85*(5), 450–457. https://doi.org/10.1205/psep07024.
46. Prasad, S., Singh, A., & Joshi, H. C., (2007). Ethanol as an alternative fuel from agricultural, industrial and urban residues. *Resour. Conserv. Recy., 50*(1), 1–39. https://doi.org/10.1016/j.resconrec.2006.05.007.
47. Li, S., Zhang, X., & Andresen, J. M., (2012). Production of fermentable sugars from enzymatic hydrolysis of pretreated municipal solid waste after autoclave process. *Fuel, 92*(1), 84–88. https://doi.org/10.1016/j.fuel.2011.07.012.
48. Lam, M. K., & Lee, K. T., (2015). Bioethanol production from microalgae. In: *Handbook of Marine Microalgae* (pp. 197–208). https://doi.org/10.1016/B978-0-12-800776-1.00012-1.
49. Nguyen, M. A., & Hoang, A. L., (2016). *A Review on Microalgae and Cyanobacteria in Biofuel Production.* USTH: Hanoi, Vietnam.
50. Duden, A. S., Verweij, P. A., Faaij, A. P. C., Baisero, D., Rondinini, C., & Van, D. H. F., (2020). Biodiversity impacts of increased ethanol production in Brazil. *Land, 9*(1), 12. https://doi.org/10.3390/land9010012.
51. Mussatto, S. I., Dragone, G., Guimarães, P. M. R., Paulo, J., Silva, A., Carneiro, I. M., Roberto, I. C., et al., (2010). Technological trends, global market, and challenges of bio-ethanol production. *Biotechnol. Adv., 28*(6), 817–830. https://doi.org/10.1016/j.biotechadv.2010.07.001.
52. Cheng, J. J., & Timilsina, G. R., (2011). Status and barrier of advanced biofuel technologies: A review. *Renew. Energ., 36*, 3541–3549. https://doi.org/10.1016/j.renene.2011.04.031.
53. Erdal, G., Esengün, K., Erdal, H., & Gündüz, O., (2007). Energy use and economical analysis of sugar beet production in Tokat province of Turkey. *Energy, 32*(1), 35–41. https://doi.org/10.1016/j.energy.2006.01.007.
54. Cosyn, S., Woude, K. V. D., Sauvenier, X., & Evrard, J. N., (2011). Sugar beet: A complement to sugar cane for sugar and ethanol production in tropical and subtropical areas. *Int. Sugar J., 113*(1346), 120–123.

55. Duraisam, R., Salelgn, K., & Berekete, A. K., (2017). Production of beet sugar and bioethanol from sugar beet and it bagasse: A Review. *Int. J. Eng. Trends Technol., 43*, 222–233.

56. De La Piscina, P. R., & Homs, N., (2008). Use of biofuels to produce hydrogen (reformation processes). *Chem. Soc. Rev., 37*(11), 2459–2467. https://doi.org/10.1039/B712181B.

57. Grahovac, J. A., Dodić, J. M., Dodić, S. N., Popov, S. D., Vučurović, D. G., & Jokić, A. I., (2012). Future trends of bioethanol co-production in Serbian sugar plants. *Renew. Sust. Energ. Rev., 16*(5), 3270–3274. https://doi.org/10.1016/j.rser.2012.02.040.

58. Dien, B. S., Cotta, M. A., & Jeffries, T. W., (2003). Bacteria engineered for fuel ethanol production: Current status. *Appl. Microbiol. Biotechnol., 63*(3), 258–266. https://doi.org/10.1007/s00253-003-1444-y.

59. Jeffries, T. W., & Jin, Y. S., (2004). Metabolic engineering for improved fermentation of pentoses by yeasts. *Appl. Microbiol. Biotechnol., 63*(5), 495–509. https://doi.org/10.1007/s00253-003-1450-0.

60. Jobling, S., (2004). Improving starch for food and industrial applications. *Curr. Opin. Plant Biol., 7*(2), 210–218. https://doi.org/10.1016/j.pbi.2003.12.001.

61. Solomon, B. D., Barnes, J. R., & Halvorsen, K. E., (2007). Grain and cellulosic ethanol: History, economics, and energy policy. *Biomass Bioenerg., 31*(6), 416–425. https://doi.org/10.1016/j.biombioe.2007.01.023.

62. Khanal, S. K., (2008). anaerobic biotechnology for bioenergy production: Principles and applications. In: *Conference Proceedings* (pp. 43–63).

63. Bothast, R. J., & Schlicher, M. A., (2005). Biotechnological processes for conversion of corn into ethanol. *Appl. Microbiol. Biotechnol., 67*(1), 19–25. https://doi.org/10.1007/s00253-004-1819-8.

64. Erickson, G. E., Klopfenstein, T. J., Adams, D. C., & Rasby, R. J., (2005). General overview of feeding corn milling co-products to beef cattle. *Corn Processing Co-products Manual*, 3–12.

65. Kim, Y., Mosier, N., & Ladisch, M. R., (2008). Process simulation of modified dry grind ethanol plant with recycle of pretreated and enzymatically hydrolyzed distillers' grains. *Bioresour. Technol., 99*(12), 5177–5192. https://doi.org/10.1016/j.biortech.2007.09.035.

66. Ziolkowska, J. R., (2020). Biofuels technologies: An overview of feedstocks, processes, and technologies. In: *Biofuels for a More Sustainable Future* (pp. 1–19). Elsevier.

67. Baffes, J., (2013). A framework for analyzing the interplay among food, fuels, and biofuels. *Global Food Security, 2*(2), 110–116. https://doi.org/10.1016/j.gfs.2013.04.003.

68. Filip, O., Janda, K., Kristoufek, L., & Zilberman, D., (2017). Food versus fuel: An updated and expanded evidence. *Energy Econ.* https://doi.org/10.1016/j.eneco.2017.10.033.

69. Tomei, J., & Helliwell, R., (2016). Food versus fuel? Going beyond biofuels. *Land Use Policy, 56*, 320–326. https://doi.org/10.1016/j.landusepol.2015.11.015.

70. Isikgor, F. H., & Becer, C. R., (2015). Lignocellulosic biomass: A sustainable platform for the production of bio-based chemicals and polymers. *Polym. Chem., 6*(25), 4497–4559. https://doi.org/ 10.1039/C5PY00263J.

71. Balat, M., & Balat, H., (2009). Recent trends in global production and utilization of bioethanol fuel. *Appl. Energ., 86*, 2273–2282. https://doi.org/10.1016/j.apenergy. 2009.03.015.

72. Harun, R., Singh, M., Forde, G. M., & Danquah, M. K., (2010). Bioprocess engineering of microalgae to produce a variety of consumer products. *Renew. Sust. Energ. Rev., 14*, 1037–1047. https://doi.org/10.1016/j.rser.2009.11.004.

73. Naik, S. N., Goud, V. V., Rout, P. K., & Dalai, A. K., (2010). Production of first and second-generation biofuels: A comprehensive review. *Renew. Sust. Energ. Rev., 14*(2), 578–597. https://doi.org/10.1016/j.rser.2009.10.003.

74. Dragone, G., Fernandes, B. D., Vicente, A. A., & Teixeira, J. A., (2010). Third-generation biofuels from microalgae. *Curr. Res. Technol. Edu. Topics. Appl. Microbiol. Microbial. Biotechnol., 2*, 1355–1366.

75. Elshahed, M. S., (2010). Microbiological aspects of biofuel production: Current status and future directions. *J. Advanc. Res., 1*(2), 103–111. https://doi.org/10.1016/j.jare.2010.03.001.

76. Wilkinson, S., Smart, K. A., James, S., & Cook, D. J., (2017). Bioethanol production from brewers spent grains using a fungal consolidated bioprocessing (CBP) approach. *Bioenergy Res., 10*(1), 146–57. https://doi.org/10.1007/s12155-016-9782-7 31.

77. Kricka, W., James, T. C., Fitzpatrick, J., & Bond, U., (2015). Engineering *Saccharomyces pastorianus* for the co-utilization of xylose and cellulose from biomass. *Microb. Cell Fact, 14*, 61. https://doi.org/10.1186/s12934-015-0242-4 32.

78. Prasetyo, J., Naruse, K., Kato, T., Boonchird, C., Harashima, S., & Park, E. Y., (2011). Bioconversion of paper sludge to biofuel by simultaneous saccharification and fermentation using a cellulase of paper sludge origin and thermotolerant *Saccharomyces cerevisiae* TJ14. *Biotechnol. Biofuels, 4*, 35. https://doi.org/10.1186/1754-6834-4-35.

79. Robak, K., & Balcerek, M., (2018). Review of second-generation bioethanol production from residual biomass. *Food Technol. Biotechnol., 56*(2), 174–187. https://doi.org/10.17113/ftb.56.02.18.5428.

80. Alvira, P., Tomás-Pejó, E., Ballesteros, M., & Negro, M. J., (2010). Pretreatment technologies for an efficient bioethanol production process based on enzymatic hydrolysis: A review. *Bioresour. Technol., 101*(13), 4851–4861. https://doi.org/10.1016/j.biortech.2009.11.093.

81. Cardona, C. A., & Sánchez, Ó. J., (2007). Fuel ethanol production: Process design trends and integration opportunities. *Bioresour. Technol., 98*(12), 2415–2457. https://doi.org/10.1016/j.biortech.2007.01.002.

82. Tiwari, G., Sharma, S., & Prasad, R., (2015). Bioethanol production: Future prospects from non-traditional sources in India. *Int. J. Res. BioSci., 4*(4), 1–15.

83. Demirbas, A., (2008). Products from lignocellulosic materials via degradation processes. *Energ. Sources Part A, 30*(1), 27–37. https://doi.org/10.1080/00908310600626705.

84. Czekała, W., Bartnikowska, S., Dach, J., Janczak, D., & Mazurkiewicz, J., (2018). The energy value and economic efficiency of solid biofuels produced from digestate and sawdust. *Energy, 159*, 1118–1122. https://doi.org/10.1016/j.energy.2018.06.090.

85. Ge, Y., & Li, L., (2018). System-level energy consumption modeling and optimization for cellulosic biofuel production. *Appl. Energy, 226*(15), 935–946. https://doi.org/10.1016/j.apenergy.2018.06.020.

86. Ziolkowska, J. R., (2014). Prospective technologies, feedstocks and market innovations for ethanol and biodiesel production in the US. *Biotechnol. Rep., 4*, 94–98. https://doi.org/10.1016/j.btre.2014.09.001.

87. Bhatia, S. K., Kim, S. H., Yoon, J. J., & Yang, Y. H., (2017). Current status and strategies for second-generation biofuel production using microbial systems. *Energy Convers. Manag., 148*(15), 1142–1156. https://doi.org/10.1016/j.enconman.2017.06.073.

88. Suganya, T., Varman, M., Masjuki, H. H., & Renganathan, S., (2016). Macroalgae and microalgae as a potential source for commercial applications along with biofuels production: A biorefinery approach. *Renew. Sust. Energ. Rev., 55*, 909–941. http://dx.doi.org/10.1016/j.rser.2015.11.026.

89. Kim, M., & Day, D. F., (2011). Composition of sugar cane, energy cane, and sweet sorghum suitable for ethanol production at Louisiana sugar mills. *J. Ind. Microbiol. Biotechnol., 38*(7), 803–807. https://doi.org/10.1007/s10295-010-0812-8.

90. Özçimen, D., & İnan, B., (2015). An overview of bioethanol production from algae. *Biofuels-Status and Perspective*, 141–162.
91. Chen, C. Y., Zhao, X. Q., Yen, H. W., Ho, S. H., Cheng, C. L., Lee, D. J., Bai, F. W., & Chang, J. S., (2013). Microalgae-based carbohydrates for biofuel production. *Biochem. Eng. J., 78*, 1–10. https://doi.org/10.1016/j.bej.2013.03.006.
92. De Farias, S. C. E., & Bertucco, A., (2016). Bioethanol from microalgae and cyanobacteria: A review and technological outlook. *Process Biochem., 51*(11), 1833–1842. https://doi.org/10.1016/j.procbio.2016.02.016.
93. De Morais, E. G., Moraes, L., De Morais, M. G., & Costa, J. A. V., (2016). Biodiesel and bioethanol from microalgae. *Green Fuels Technol.*, 359–386. https://doi.org/10.1007/978-3-319-30205-8_14.
94. Chaudhary, L., Pradhan, P., Soni, N., Singh, P., & Tiwari, A., (2014). Algae as a feedstock for bioethanol production: New entrance in biofuel world. *Int. J. Chem. Technol. Res., 6*, 1381–1389.
95. Thatoi, H., Dash, P. K., Mohapatra, S., & Swain, M. R., (2016). Bioethanol production from tuber crops using fermentation technology: A review. *Int. J. Sust. Energ., 35*(5), 443–468. https://doi.org/10.1080/14786451.2014.918616.
96. Song, D., Fu, J., & Shi, D., (2008). Exploitation of oil-bearing microalgae for biodiesel. *Chin. J. Biotechnol., 24*(3), 341–348.https://doi.org/10.1016/S1872-2075(08)60016-3
97. Tredici, M. R., (2010). Photobiology of microalgae mass cultures: Understanding the tools for the next green revolution. *Biofuels, 1*, 143–162. https://doi.org/10.4155/bfs.09.10.
98. Harun, R., Yip, J. W., Thiruvenkadam, S., Ghani, W. A., Cherrington, T., & Danquah, M. K., (2014). Algal biomass conversion to bioethanol-a step-by-step assessment. *Biotechnol. J., 9*(1), 73–86. https://doi.org/10.1002/biot.201200353.
99. Varaprasad, D., Narasimham, D., Paramesh, K., Sudha, N. R., Himabindu, Y., Keerthi, K. M., Nazaneen, P. S., & Chandrasekhar, T., (2019). Improvement of ethanol production using green alga *Chlorococcum minutum*. *Env. Technol.*, 1–9. https://doi.org/10.1080/09593330.2019.1669719.
100. Choi, S. P., Nguyen, M. T., & Sim, S. J., (2010). Enzymatic pretreatment of *Chlamydomonas reinhardtii* biomass for ethanol production. *Bioresour. Technol., 101*(14), 5330–5336. https://doi.org/10.1016/j.biortech.2010.02.026.
101. Harun, R., & Danquah, M. K., (2011). Influence of acid pretreatment on microalgal biomass for bioethanol production. *Process Biochem., 46*(1), 304–309. https://doi.org/10.1016/j.procbio.2010.08.027.
102. Harun, R., Jason, W. S. Y., Cherrington, T., & Danquah, M. K., (2011). Exploring alkaline pretreatment of microalgal biomass for bioethanol production. *Appl. Energ., 88*(10), 3464–3467. https://doi.org/10.1016/j.apenergy.2010.10.048.
103. Lee, S., Oh, Y., Kim, D., Kwon, D., Lee, C., & Lee, J., (2011). Converting carbohydrates extracted from marine algae into ethanol using various ethanolic *Escherichia coli* strains. *Appl. Biochem. Biotechnol., 164*(6), 878–888. https://doi.org/10.1007/s12010-011-9181-7.
104. Shirai, F., Kunii, K., Sato, C., Teramoto, Y., Mizuki, E., Murao, S., & Nakayama, S., (1998). Cultivation of microalgae in the solution from the desalting process of soy sauce waste treatment and utilization of the algal biomass for ethanol fermentation. *World. J. Microbiol. Biotechnol., 14*(6), 839–842. https://doi.org/10.1023/A:1008860705434.
105. Nahak, S., Nahak, G., Pradhan, I., & Sahu, R. K., (2011). Bioethanol from marine algae: A solution to global warming problem. *J. Appl. Environ. Biol. Sci., 1*(4), 74–80.

106. Silva, C. E. D. F., & Bertucco, A., (2019). Bioethanol from microalgal biomass: A promising approach in biorefinery. *Braz. Arch. Biol. Technol., 62*. http://dx.doi.org/10.1590/1678-4324-2019160816.

107. US DOE, (2010). *National Algal Biofuels Technology Roadmap.* DOE, Washington, DC.

108. Dutta, K., Daverey, A., & Lin, J. G., (2014). Evolution retrospective for alternative fuels: First to fourth generation. *Renew. Energy, 69*, 114–122. https://doi.org/10.1016/j.renene.2014.02.044.

109. Brethauer, S., & Studer, M. H., (2014). Consolidated bioprocessing of lignocellulose by a microbial consortium. *Energy Environ. Sci., 7*(4), 1446–53. https://doi.org/10.1039/c3ee41753k.

110. Aditiya, H. B., Mahlia, T. M. I., Chong, W. T., Nur, H., & Sebayang, A. H., (2016). Second-generation bioethanol production: A critical review. *Renew. Sustain. Energy Rev., 66*, 631–653. https://doi.org/10.1016/j.rser.2016.07.015.

111. Macedo, N., & Brigham, C. J., (2014). From beverages to biofuels: The journeys of ethanol-producing microorganisms. *Int. J. Biotechnol. Wellness Ind., 3*(3), 79–87. http://dx.doi.org/10.6000/1927-3037.2014.03.03.1.

112. Xia, J., Yang, Y., Liu, C. G., Yang, S., & Bai, F. W., (2019). Engineering *Zymomonas mobilis* for robust cellulosic ethanol production. *Trends Biotechnol.,* (2019). https://doi.org/10.1016/j.tibtech.2019.02.002.

113. Karsch, T., Stahl, U., & Esser, K., (1983). Ethanol production by *Zymomonas* and *Saccharomyces*, advantages and disadvantages. *Eur. J. Appl. Microbiol. Biotechnol., 18*(6), 387–391. https://doi.org/10.1007/BF00504750.

114. Zhang, M., Eddy, C., Deanda, K., Finkelstein, M., & Picataggio, S., (1995). Metabolic engineering of a pentose metabolism pathway in ethanologenic *Zymomonas mobilis*. *Science, 267*(5195), 240–243. https://doi.org/10.1126/science.267.5195.240.

115. Quevedo-Hidalgo, B., Monsalve-Marín, F., Narváez-Rincón, P. C., Pedroza-Rodríguez, A. M., & Velásquez-Lozano, M. E., (2013). Ethanol production by *Saccharomyces cerevisiae* using lignocellulosic hydrolysate from *Chrysanthemum* waste degradation. *World. J. Microb. Biotechnol., 29*(3), 459–466. https://doi.org/10.1007/s11274-012-1199-7.

116. Diaz, C. E., Sierra, Y. K., & Hernández, J. A., (2018). Determination of the percentage of ethanol produced by *Saccharomyces cerevisiae* from semi-purified glycerin. In: *Journal of Physics: Conference Series* (Vol. 1126, No. 1, p. 012008). https://doi.org/10.1088/1742-6596/1126/1/012008.

117. Todhanakasem, T., Salangsing, O. L., Koomphongse, P., Kanokratana, P., & Champreda, V., (2019). *Zymomonas mobilis* biofilm reactor for ethanol production using rice straw hydrolysate under continuous and repeated batch processes. *Front. Microbiol., 10*, 1777. https://doi.org/10.3389/fmicb.2019.01777.

118. Horn, S. J., Aasen, I. M., & Østgaard, K., (2000). Ethanol production from seaweed extract. *J. Ind. Microbiol. Biotechnol., 25*(5), 249–254. https://doi.org/10.1038/sj.jim.7000065.

119. Harun, R., Danquah, M. K., & Forde, G. M., (2010). Microalgal biomass as a fermentation feedstock for bioethanol production. *J. Chem. Technol. Biotechnol., 85*(2), 199–203. https://doi.org/10.1002/jctb.2287.

120. Adams, J. M. M., Toop, T. A., Donnison, I. S., & Gallagher, J. A., (2011). Seasonal variation in *Laminaria digitata* and its impact on biochemical conversion routes to biofuels. *Bioresour. Technol., 102*(21), 9976–9984. https://doi.org/10.1016/j.biortech.2011.08.032.

121. Rastogi, M., & Shrivastava, S., (2017). Recent advances in second-generation bioethanol production: An insight to pretreatment, saccharification and fermentation processes. *Renew. Sust. Energ. Rev., 80*, 330–340. https://doi.org/10.1016/j.rser.2017.05.225.

122. Anisha, G. S., & John, R. P., (2015). Microalgae as an alternative feedstock for green biofuel technology. *Environ. Res. J., 9*(2), 223–240.

123. Simas-Rodrigues, C., Villela, H. D., Martins, A. P., Marques, L. G., Colepicolo, P., & Tonon, A. P., (2015). Microalgae for economic applications: Advantages and perspectives for bioethanol. *J. Exp. Bot., 66*(14), 4097–4108. https://doi.org/10.1093/jxb/erv130.

124. Zheng, Y., Pan, Z., & Zhang, R., (2009). Overview of biomass pretreatment for cellulosic ethanol production. *Int. J. Agr. Biol. Eng., 2*(3), 51–68.

125. Wyman, C. E., Dale, B. E., Elander, R. T., Holtzapple, M., Ladisch, M. R., & Lee, Y. Y., (2005). Coordinated development of leading biomass pretreatment technologies. *Bioresour. Technol., 96*(18), 1959–1966. https://doi.org/10.1016/j.biortech.2005.01.010.

126. Barakat, A., De Vries, H., & Rouau, X., (2013). Dry fractionation process as an important step in current and future lignocellulose biorefineries: A review. *Bioresour. Technol., 134*, 362–373. https://doi.org/10.1016/j.biortech.2013.01.169.

127. Taherzadeh, M. J., & Karimi, K., (2008). Pretreatment of lignocellulosic wastes to improve ethanol and biogas production: A review. *Int. J. Mol. Sci., 9*(9), 1621–1651. https://doi.org/10.3390/ijms9091621.

128. Szczodrak, J., & Fiedurek, J., (1996). Technology for conversion of lignocellulosic biomass to ethanol. *Biomass Bioenerg., 10*(5, 6), 367–375. https://doi.org/10.1016/0961-9534 (95)00114-X.

129. Grous, W. R., Converse, A. O., & Grethlein, H. E., (1986). Effect of steam explosion pretreatment on pore size and enzymatic hydrolysis of poplar. *Enzyme Microb. Technol., 8*(5), 274–280. https://doi.org/10.1016/0141-0229(86)90021-9.

130. Pérez, J. A., González, A., Oliva, J. M., Ballesteros, I., & Manzanares, P., (2007). Effect of process variables on liquid hot water pretreatment of wheat straw for bioconversion to fuel-ethanol in a batch reactor. *J. Chem. Technol. Biotechnol., 82*(10), 929–938. https://doi.org/10.1002/jctb.1765.

131. Sun, Y., & Cheng, J., (2002). Hydrolysis of lignocellulosic materials for ethanol production: A review. *Bioresour. Technol., 83*(1), 1–11. https://doi.org/10.1016/S0960-8524 (01)00212-7.

132. Bensah, E. C., & Mensah, M., (2013). Chemical pretreatment methods for the production of cellulosic ethanol: Technologies and innovations. *Int. J. Chem. Eng.* https://doi.org/ 10.1155/2013/719607.

133. Millett, M. A., Baker, A. J., & Satter, L. D., (1976). Physical and chemical pretreatments for enhancing cellulose saccharification. *Biotechnol. Bioeng. Symp., 6*, 125–153.

134. Sarkar, N., Ghosh, S. K., Bannerjee, S., & Aikat, K., (2012). Bioethanol production from agricultural wastes: An overview. *Renew. Energ., 37*(1), 19–27. https://doi.org/10.1016/j.renene.2011.06.045.

135. Hwang, S. S., Lee, S. J., Kim, H. K., Ka, J. O., Kim, K. J., & Song, H. G., (2008). Biodegradation and saccharification of wood chips of *Pinus strobus* and *Liriodendron tulipifera* by white-rot fungi. *J. Microbiol. Biotechnol., 18*, 1819–1825. https://doi.org/ 10.4014/jmb.0800.231.

136. Knauf, M., & Moniruzzaman, M., (2004). Lignocellulosic biomass processing: A perspective. *Int. Sugar J., 106*(1263), 147–150.

137. Miranda, J. R., Passarinho, P. C., & Gouveia, L., (2012). Pretreatment optimization of *Scenedesmus obliquus* microalga for bioethanol production. *Bioresour. Technol., 104*, 342–348. https://doi.org/10.1016/j.biortech.2011.10.059.

138. Taherzadeh, M. J., & Karimi, K., (2007). Enzyme-based hydrolysis processes for ethanol from lignocellulosic materials: A review. *BioResources, 2*(4), 707–738.

139. Hamelinck, C. N., Van, H. G., & Faaij, A. P., (2005). Ethanol from lignocellulosic biomass: Techno-economic performance in short-, middle-and long-term. *Biomass Bioenerg., 28*(4), 384–410. https://doi.org/10.1016/j.biombioe.2004.09.002.

140. Talebnia, F., Karakashev, D., & Angelidaki, I., (2010). Production of bioethanol from wheat straw: An overview on pretreatment, hydrolysis and fermentation. *Bioresour. Technol., 101*(13), 4744–4753. https://doi.org/10.1016/j.biortech.2009.11.080

141. Zaldivar, J., Nielsen, J., & Olsson, L., (2001). Fuel ethanol production from lignocellulose: A challenge for metabolic engineering and process integration. *Appl. Microbiol. Biotechnol., 56*(1, 2), 17–34. https://doi.org/10.1007/s002530100624

142. Xiros, C., Topakas, E., & Christakopoulos, P., (2013). Hydrolysis and fermentation for cellulosic ethanol production. *Energ. Environ., 2*(6), 633–654.

143. Tengborg, C., Galbe, M., & Zacchi, G., (2001). Reduced inhibition of enzymatic hydrolysis of steam-pretreated softwood. *Enzyme Microbiol. Technol., 28*, 835–844.

144. Hahn-Hägerdal, B., Galbe, M., Gorwa-Grauslund, M. F., Lidén, G., & Zacchi, G., (2006). Bio-ethanol-the fuel of tomorrow from the residues of today. *Trends Biotechnol., 24*(12), 549–556. https://doi.org/10.1016/j.tibtech.2006.10.004.

145. Lin, Y., & Tanaka, S., (2006). Ethanol fermentation from biomass resources: Current state and prospects. *Appl. Microbiol. Biotechnol., 69*(6), 627–642. https://doi.org/10.1007/s00253-005-0229-x.

146. Gupta, R. B., & Demirbas, A., (2010). *Gasoline, Diesel, and Ethanol Biofuels from Grasses and Plants*. Cambridge University Press.

147. Olofsson, K., Bertilsson, M., & Lidén, G., (2008). A short review on SSF–an interesting process option for ethanol production from lignocellulosic feedstocks. *Biotechnol. Biofuels, 1*(1), 7. https://doi.org/10.1186/1754-6834-1-7.

148. Carere, C., Sparling, R., Cicek, N., & Levin, D., (2008). Third-generation biofuels via direct cellulose fermentation. *Int. J. Mol. Sci., 9*(7), 1342–1360. https://doi.org/10.3390/ijms9071342.

149. Lynd, L. R., Weimer, P. J., Van, Z. W. H., & Pretorius, I. S., (2002). Microbial cellulose utilization: Fundamentals and biotechnology. *Microbiol. Mol. Biol. Rev., 66*(3), 506–577. https://doi.org/10.1128/MMBR.66.3.506-577.2002.

150. Lynd, L. R., Van, Z. W. H., McBride, J. E., & Laser, M., (2005). Consolidated bioprocessing of cellulosic biomass: An update. *Curr. Opin. Biotechnol., 16*(5), 577–583. https://doi.org/10.1016/j.copbio.2005.08.009.

151. Sanchez, O. J., & Cardona, C. A., (2008). Trends in biotechnological production of fuel ethanol from different feedstocks. *Bioresour. Technol., 99*(13), 5270–5295. https://doi.org/10.1016/j.biortech.2007.11.013.

152. Chandel, A. K., Singh, O. V., Chandrasekhar, G., Rao, L. V., & Narasu, M. L., (2010). Key drivers influencing the commercialization of ethanol-based biorefineries. *J. Commer. Biotechnol., 16*(3), 239–257. https://doi.org/10.1057/jcb.2010.5.

153. Gray, K. A., Zhao, L., & Emptage, M., (2006). Bioethanol. *Curr. Opin. Chem. Biol., 10*(2), 141–146.

CHAPTER 8

Emerging Strategies for Ethanol Purification

ALAN D. PEREZ,[1] JAVIER QUINTERO,[2] and JUAN A. LEÓN[3]

[1]*Sustainable Process Technology (SPT), University of Twente, Enschede, Overijssel, The Netherland, E-mail: a.d.perezavila@utwente.nl*

[2]*Department of Chemical Engineering, National University of Colombia, Manizales, Caldas, Colombia, E-mail: jaquinteroj@unal.edu.co*

[3]*Laboratory of Process Intensification and Hybrid System (LIPSH), Department of Chemical Engineering, National University of Colombia, Manizales, Caldas, Colombia, E-mail: jaleonma@unal.edu.co*

ABSTRACT

This chapter is focused on the separation technologies mainly applied for ethanol recovery from fermentation broths and aqueous solutions during the process of fuel alcohol production. Conventional technologies to separate ethanol such as distillation, azeotropic distillation, extractive distillation (ED), and adsorption using molecular sieves are described, in order to present the current separation equipment used in biorefineries and distilleries. On the other hand, novel and alternative technologies to recover, concentrate, and dehydrate ethanol, either from the treated wine or directly from the fermenter are presented. These alternative technologies are able to increase the performance of a fermentation, reduce the number of equipment, reduce the operation and capital costs, associating a higher energy efficiency, as compared to the conventional separation process to recover ethanol. Although there are several separation technologies proposed to recover ethanol, in the following chapter, only the most representative and researched technologies are described.

8.1 INTRODUCTION

The ethanol purification is an important stage in the production process, since the ethanol quality defines the use and price of this component (commodity) in the market. Although ethanol can be commercialized at its azeotropic composition, the ethanol is highly demanded in the energy market (fuel alcohol). As a result, ethanol must be dehydrated and its azeotrope must be broken, in order to achieve optimum condition to blend it to gasoline.

The processes commonly used for the ethanol purification are distillation, azeotropic distillation, and ED. These latter are mature, stablished, and applied technologies to recover ethanol in the current biorefineries. However, the thermodynamic limitations in distillation, the use of entrainer/solvents, and the need of several number of columns in the distillation trains make these technologies not very efficient to recover ethanol, especially to use it as fuel alcohol.

In addition, adsorption has been successfully applied in several biorefineries to recover ethanol from aqueous solutions. Extractive or azeotropic distillation columns can be replaced by adsorption to overcome the azeotrope (ethanol dehydration). For instance, the use of molecular sieves (from potassium aluminosilicates) has been applied in industrial cases, achieving to replace the azeotropic distillation [1]. Although adsorption with molecular sieves can associate high capital costs, it is a technology with a high separation performance with a lower operation cost as compared to traditional distillation schemes.

On the other hand, pressure-swing distillation has been extensively studied to separate ethanol by reducing the operation pressure inside a column in the separation train. At vacuum condition, the separation by distillation leads to the disappearance of the azeotrope. However, it is necessary the use of columns composed of a large number of plates, and high reflux ratios to achieve an ethanol with a high purity. Also, high capital and operation costs can be associated to maintain the vacuum condition for a large volume [2].

Moreover, several efforts have been made in order to propose separation technologies with a higher performance than those previously mentioned. For instance, the use of non-conventional and non-volatile solvents to increase the ethanol volatility, and consequently, to overcome the azeotropic condition. In this chapter, ED using dissolved salts, ionic liquid and hyperbranched polymers (HyPols) are presented.

Also, ethanol recovery through membrane technology and its integration to distillation is described. Pervaporation (PV) has been demonstrated to be able to separate ethanol without thermodynamic limitations, which is a significant advantage as compared to conventional distillation. However, the capital cost associated to PV are substantially high. The integration of PV to distillation has been approached in order to compensate the distillation economy with the PV performance, achieving successful industrial cases of a separation stage composed of hybrid PV-distillation systems.

On the other hand, ethanol is the primary metabolite produced in the fermentation process. However, it is the main inhibitor of the yeast cells as well, compromising the yeast cell activity at high ethanol concentration in the fermentation broths. Therefore, a simultaneous ethanol removal during its formation can improve the fermentation performance. In this chapter, it is described a pervaporative fermentation and a liquid-liquid extraction (LLX)-fermentation as alternative to increase the fermentation performance.

8.2 CONVENTIONAL SEPARATION TECHNOLOGY FOR ETHANOL RECOVERY

8.2.1 CONVENTIONAL DISTILLATION

In the first generation of alcoholic fermentation, ethanol is diluted in the fermentation broths (5–12 wt.%). The fermenter downstream is mainly composed of water [3, 4]. The free-biomass alcoholic solution (filtered fermentation broth), called "wine or beer" is processed to recover the ethanol produced in the fermentation stage.

Due to the high volatility of ethanol in aqueous solutions, distillation is a versatile separation technology to recover the ethanol from the wine [5]. Although ethanol separation by distillation is an energy-intensive process, it is an effective and profitable separation to recover ethanol from a bioprocess. At low ethanol composition in aqueous solutions, the constant in Henry's law presents values significantly higher than 1, allowing a high ethanol separation as a function of the liquid-vapor equilibrium [6]. A large capacity of processing is able in distillation columns since low reflux ratios are required.

On the other hand, ethanol promotes the formation of a minimum boiling azeotrope in an aqueous solution. As a result, the maximum ethanol composition attainable by conventional distillation (at 1.013 bar) is close to 95.6 wt.%, which is the azeotropic composition. Commonly, the ethanol-water mixture close to the azeotropic composition is called hydrated ethanol in distilleries. The distillation process to obtain hydrated ethanol is presented in Figure 8.1.

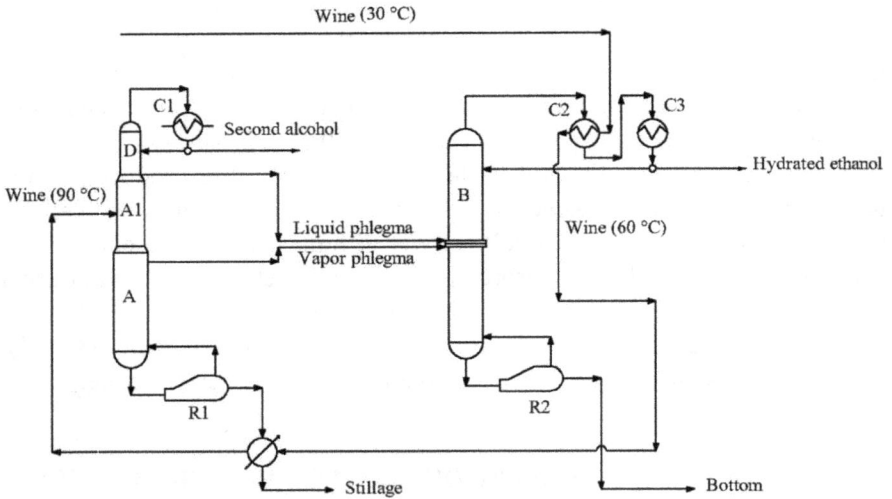

FIGURE 8.1 Distillation process to obtain hydrated ethanol from sugarcane molasses with thermal integration in the column condenser.
Source: Adapted from: Dias et al. [7].

In general, the ethanol composition in the hydrated ethanol in distilleries is in the range of 92.5–94.6 wt.%, and is produced by two-column arrangement, as Figure 8.1 shows [7]. The first column is used to recover the ethanol and a fraction of water from the wines (some organic components in very low compositions are separated as well). This column is divided in three sections. Section D is located at the top of the column and is commonly composed of 6 stages. Section A1, where the pre-heated wine is fed, is an intermediate section constituted by 8 stages. Section A at the bottom of the column, which can be composed of 16 to 24 stages. The distillation product of this column is a second alcohol stream (91 wt.%) produced at the top of Section A, and the stillage at the bottom of the column. In addition, from Section D and A are produced liquid and vapor Phlegma streams with an ethanol composition about to 30 wt.%, respectively.

Both phlegma streams are fed to a second distillation column B, which is composed of 40–50 stages (Figure 8.1). In this column, the mixture is rectified to completely recover the ethanol at the top of the column, achieving a distillate close to the azeotropic composition (hydrated ethanol). On the other hand, the excess of water in the phlegma streams is eliminated at the bottom of column B.

In almost all distilleries, the wine is pre-heated close to the bubble temperature of the mixture to reduce the vapor condensation, improving the performance and energy efficiency of the distillation column. An integrated heating system, involving the stillage stream and top product of column B is usually applied (Figure 8.1). The wine is used as a cooling service stream in the condenser C2 at the top of column B, producing a partial condensation of the hydrated ethanol, followed by a total condensation in condenser C3 with water. The wine leaves C2 at 60°C. The heated wine is used to reduce the stillage temperature as well. The wine achieves a temperature of 90°C, before being fed to the first distillation column.

The produced ethanol in biorefineries is mainly used as fuel alcohol. Therefore, hydrated ethanol is not still applicable for blending in gasoline in flex-fuel engines, which are the most common technology using fuel alcohol. Azeotropic ethanol must be dehydrated. At least an ethanol composition of 99.3 wt.% or higher must be produced to be added in gasoline [8]. Since it is not possible to overcome the azeotropic composition by conventional distillation, azeotropic distillation (AD) is used to produce anhydrous ethanol. AD is an economical technology widely applied in Brazilian distilleries [9].

8.2.2   AZEOTROPIC DISTILLATION (AD)

The separation of an ethanol-water mixture by azeotropic distillation (AD), involves adding a third volatile component denominated entrainer. This is made in order to form a new azeotrope with the components to be separated [10]. The relative volatilities are modified as compared to a binary distillation of ethanol-water mixture, allowing achieving an ethanol composition further than the azeotropic composition [11]. Due to an entrainer is involved in an AD, additional columns are necessary to recover individually each component, including the entrainer.

According to the thermodynamic and molecular structure of the entrainer, two types of AD can be considered. In a homogenous azeotropic distillation (HAD), a polar entrainer is used. Then, there is not a formation of a second liquid phase at the bubble temperature of the mixture. In a heterogeneous azeotropic distillation (HTAD), a second liquid phase is formed. This is due to a nonpolar entrainer is applied [11]. In addition, a decanter to split the organic and aqueous phases is required in a HTAD. Therefore, a liquid-liquid and liquid-liquid-vapor equilibria must be considered in a mixture separation by HTAD.

Figure 8.2 is presented the flow sheet of a HTAD for the ethanol dehydration. The hydrated ethanol and the entrainer are mixed and fed to the
azeotropic distillation column (1), where anhydrous alcohol is produced
at the bottom of the same column. The product at the top of the column
(1) is mostly composed of the entrainer, a fraction of water, and a little
of ethanol. Inside the decanter (2), the organic and aqueous phases are
split. The organic phase (reflux stream) is recycled to the column (1). The
aqueous phase (distillate stream) is fed to the entrainer recovery column
(3). Water, as the heavy component, accompanied by ethanol traces, is
separated at the bottom of the column. The entrainer and a small amount
of ethanol from the top of the column (3) are recycled to the column (1)
again.

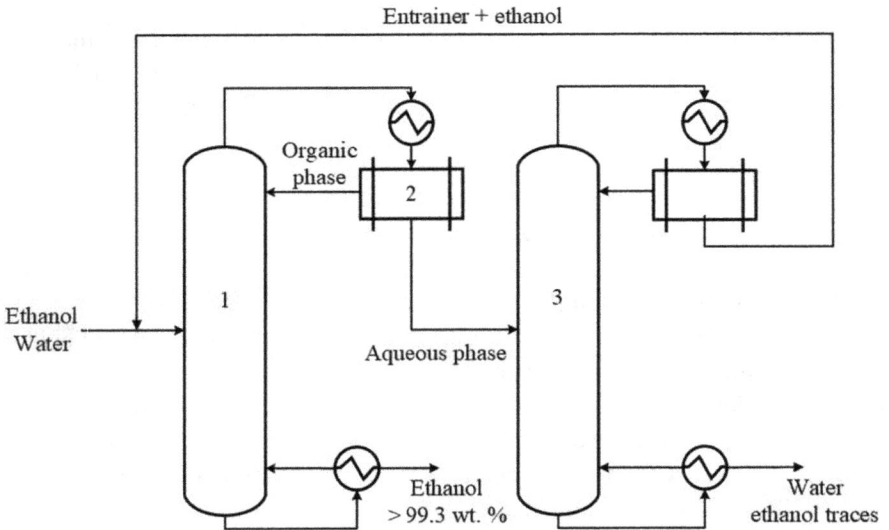

FIGURE 8.2   General scheme of azeotropic distillation for the separation of ethanol-water
mixtures: (1) azeotropic distillation column; (2) decanter; and (3) distillation column for the
entrainer recovery [12].

Benzene was widely used as an entrainer for the ethanol dehydration
by HTAD in the past [12]. Currently, the use of benzene has been replaced
for other entrainers, due to its carcinogenic effects. Cyclohexane is one of
the most used entrainer for the ethanol dehydration [13]. Although this is
widely applied in distilleries, this component is highly flammable. In addition, toluene can be used as entrainer as well, producing a similar separation

behavior of benzene [11]. The entrainer is not necessarily restricted to a pure component, a mixture of hydrocarbon (e.g., hexane-isooctane-cyclohexane, gasoline) are potential candidates as entrainer in an HTAD for the ethanol dehydration [14].

8.2.3 DEHYDRATION PROCESS BY MOLECULAR SIZE DIFFERENCE: ADSORPTION WITH MOLECULAR SIEVES

Molecular sieves are cylindrical or spherical granular substances called zeolites, which can be natural or made from potassium aluminosilicates. They are identified according to their nominal size of the internal pores whose diameter is generally measured in angstroms. Molecular sieves are materials that are characterized by their excellent capacity to retain on their surface defined types of chemical species. These species are usually solvents (water most of the cases), which are desired to be removed from a mixture to obtain a final product with its given specifications [15].

Adsorption has been applied as an alternative process for ethanol dehydration. Recently, new distilleries plants have opted to use adsorption rather than AD to produce anhydrous ethanol. Adsorption on molecular sieves is an established technology for the ethanol dehydration as an energy-efficient process. The anhydrous ethanol production by a pressure swing adsorption (PSA) process is shown in Figure 8.3. Commonly, a vapor-phase adsorption (vapor feed) is used to process directly the output stream from the conventional distillation stage (hydrated ethanol) [16]. Two zeolite adsorption units (A and B) in parallel arrangement are applied for a continuous ethanol dehydration process. While in one bed (e.g., A) is operating for dehydrating of superheated ethanol vapor at high pressure, the other bed (e.g., B) is being regenerated at vacuum conditions by recirculating (splitter valve C) a small portion of the anhydrous ethanol through the saturated sieves [17].

Adsorption using molecular sieve for ethanol dehydration can present a lower energy consumption as compared to azeotropic distillation, since the stream to be processed only must be vaporized one time. On the other hand, zeolite is a highly selective material. Water is strongly adsorbed, requiring high temperatures or a low operating pressure to regenerate the zeolite bed. In addition, the constant use of superheated vapor for the regeneration of the sieve bed, accelerates the sieve deterioration due to thermal shock [18].

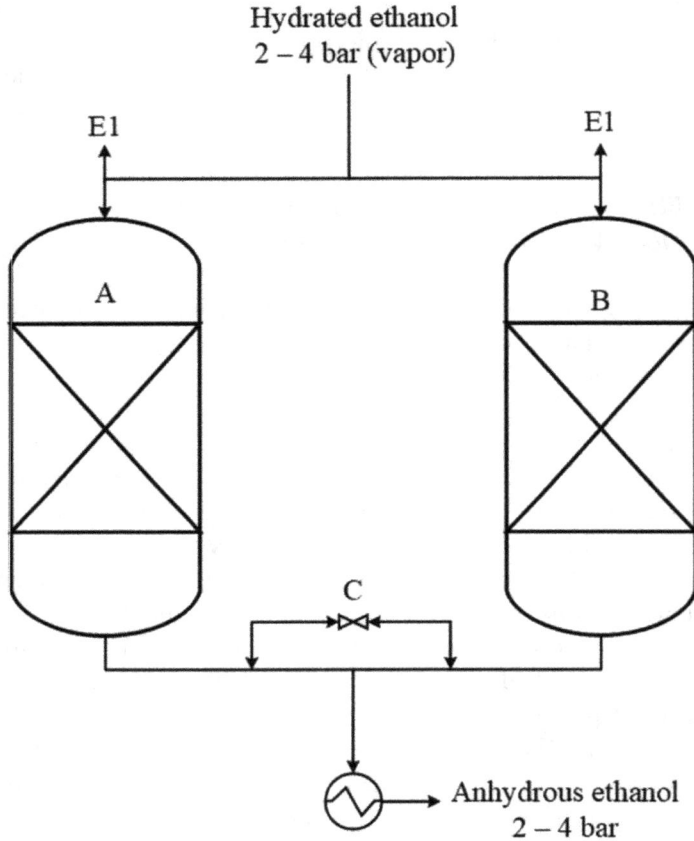

FIGURE 8.3   Schematic diagram of ethanol dehydration with molecular sieves.
Source: Adapted from: Ref. [15].

8.3   NONCONVENTIONAL TECHNOLOGIES TO RECOVER ETHANOL

8.3.1   *EXTRACTIVE DISTILLATION (ED) WITH NONCONVENTIONAL SOLVENTS*

Similar to AD, in an ED a third heavy component (solvent) is added to the ethanol-water mixture. Although the solvent can modify the relative volatility of the mixture, there is not an azeotrope formation as AD. Also, the solvent is commonly fed to the column in a different location than the hydrated ethanol feed. This is made in order to allow an extraction section in the extractive column. In addition, in ED some topological aspects differ from AD and can be found somewhere else [19].

8.3.1.1 ED WITH DISSOLVED SALTS

A dissolved salt can be used as solvent in an ED for the ethanol-water separation. The "salt effect" significantly alters the volatility of the binary mixture, overcoming the azeotropic composition inside the column. In the case of ethanol aqueous solutions, the liquid-vapor equilibrium is enhanced for ethanol. It means that the constant of Henry's law is increased, allowing a vapor with a higher ethanol composition as compared to a conventional distillation. In the case of water, its volatility is reduced by the salt effect. This effect is very effective for the separation of azeotropic mixtures, as in the case of ethanol-water mixture, promoting a vapor composition of ethanol higher than the azeotrope.

However, the use of salts as an extraction agent in ED has not reached a successful application due to the technical problems. The recrystallization of the salt (recovery step), its dissolution, and the special needs of the construction materials to avoid corrosion problems have limited the applicability of dissolved salts as a solvent in ED in the industry [20]. Although the limitation by technical issues in this technology promotes little interest in the industry, dissolved salts are a potential alternative to produce anhydrous ethanol at a low cost.

In Figure 8.4 is shown a simplified scheme of ED with dissolved salts for ethanol separation from aqueous solutions. The ethanol-water mixture is fed to the ED column (1), where the separation by distillation is governed by the modified liquid-vapor equilibrium due to the salt effect. The salt feed is located immediately after the splitter at the top of the column, where it is dissolved (2) in the reflux stream [22]. Since the salt is not a volatile component, it only remains inside the liquid phase and flows downward along the column. Then, the salt can be recovered at the bottom of the column and recycled by evaporation or drying, rather than by distillation.

The most common salts used in ED for ethanol dehydration are CH_3COONa, CH_3COOK, $CaCl_2$, $Ca(NO)_2$, KI, and NaI [22–24]. In addition, a mixture of these latter salts can be employed. For instance, a salt with a ratio 70/30 of CH_3COOK/CH_3COONa is able to produce ethanol with a composition of 99.8 wt.%. The energy consumption and capital costs are lower as compared to AD using benzene as entrainer and conventional ED with ethylene glycol as solvent [22]. On the other hand, a pilot plant for producing anhydrous ethanol using $CaCl_2$ was performed [21]. It was possible to reduce the energy consumption by more than 50% as compared to conventional ED and vacuum distillation. In other work, different salts (KCl, $CaCl_2$, NaCl, and KI) were theoretically evaluated to produce anhydrous ethanol [25]. In all cases, an ethanol composition higher than 99 wt.% was achieved. However, $CaCl_2$ is the most effective saline agent with the lowest energy consumption, even lower than a conventional ED with ethylene glycol.

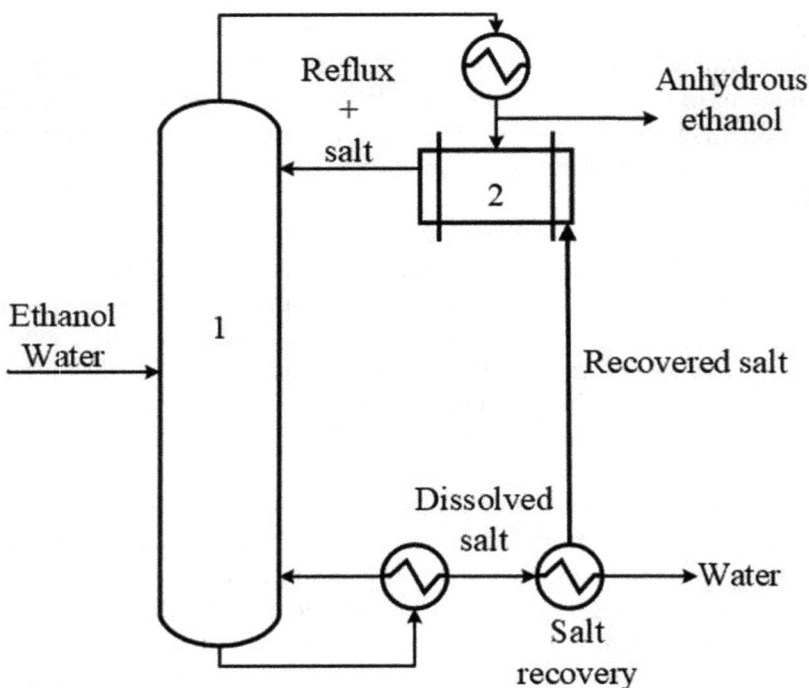

FIGURE 8.4 Typical separation of ethanol-water mixture by extractive distillation using dissolved salts. Extractive distillation column (1); and salt dissolver (2).

Source: Adapted from: Ref. [21].

8.3.1.2   ED USING IONIC LIQUIDS (ILS) AS SOLVENTS

An ED with ionic liquids (ILs) is similar to an ED with conventional solvents. However, the top products can be solvent-free streams due to the properties of ILs. Furthermore, this process presents several operative advantages as compared to ED with dissolved salts, since crystallization and an accelerated corrosion are avoided with ILs [26].

Typically, ILs are composed of an inorganic polyatomic anion and large organic cations. Further, at room temperature, the ILs are at liquid state. Therefore, as a separation agent, ILs presents the advantage of conventional solvents and the high separation ability of dissolved salts. Several combinations of anions and cations can be made, based on the separating agent structures and the separation performance. Also, several commercial ILs such as, 1-butyl-3-methylimidazolium tetrafluoroborate ([BMIM]$^+$[BF4]$^-$), 1-ethyl-3-methylimidazolium tetrafluoroborate ([EMIM]$^+$[BF4]$^-$) and 1-butyl-3-methylimidazolium chloride ([BMIM]$^+$[Cl]$^-$) can be found in the market [5].

In general, the ED with ILs presents a similar configuration of an ED using conventional solvents. According to Figure 8.5, hydrated ethanol (azeotropic) and the solvent (ILs) are fed to the ED column (1). The IL is fed at the top of the column to increase the separation performance. A solvent-free anhydrous ethanol is produced at the top of the ED column, while the bottom stream (solvent with water) is partially split by a flash evaporation (2). The solvent is recovered in the regeneration column (3), where water is produced at the top of the column, while the recovered solvent from the bottoms is recycled again to column (1).

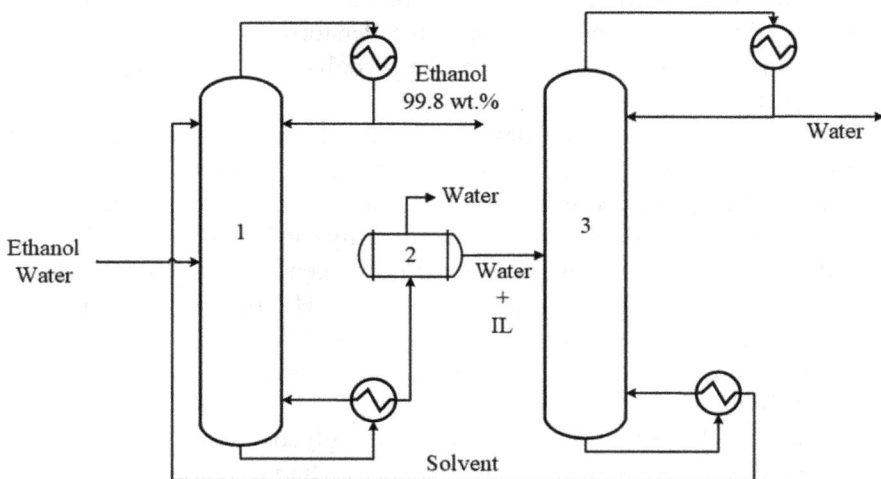

FIGURE 8.5 General scheme of an extractive distillation using ionic liquids as solvent. Extractive distillation column (1); flash drum (2); and solvent regeneration column (3).

Source: Adapted from: Seiler et al. [27].

Several studies have shown the potential of using IL for the ethanol dehydration. For instance, the liquid-vapor equilibrium for ethanol-water-solvent (IL) mixture was evaluated close to the azeotropic region [28]. The ILs used were 1-butyl-3-methylimidazolium bromide ([BMIM][Br]) and 1-ethyl-3 methylimidazolium bromide ([EMIM][Br]). In both cases, the ethanol composition in the vapor phase overcame the azeotrope, reaching a composition higher than 99 wt.%. Also, low energy demand for ethanol dehydration using 1-Ethyl-3-methylimidazolium tetrafluoroborate [EMIM] [BF$_4$] was achieved in other work [29]. The results showed a thermal energy reduction of the 14.6% and 28.6% as compared to glycerol and gasoline, respectively. Moreover, a cost reduction of 21%, 18%, and 59% in columns, reboilers, and cooler can be achieved using ILs as solvents in comparison to a conventional solvent ED [30].

8.3.1.3   *ED USING HYPERBRANCHED POLYMERS*

A hyperbranched polymer (HyPol) is a tridimensional macromolecule with a tree-based arrangement. The HyPols are characterized by presenting several random branched points with functional groups. These macromolecules can be manufactured with a low cost and can be designed with specific physical and chemical properties, which is favorable for industrial applications. Currently, different companies such as the Perstorp Group (Sweden), DSM Fine Chemicals (The Netherlands) and Hyperpolymers (Germany) are manufacturing HyPols on an industrial scale [31].

Since the HyPols can be designed with different polarity features by adjusting the functional group at the end of the polymer branch, selective solvents can be produced using HyPol technology. Also, properties as low viscosity, high thermal stability, and no volatility can be presented in HyPols. Based on these latter properties, HyPols have been attracting as solvents in ED. However, the works associated to the use of HyPols are mostly focused on the impact of the liquid-vapor equilibrium rather than the distillation itself, since there is not still much information about experimental data of the HyPol-mixture interaction.

Seiler et al. [32] tested poly(ethylene glycol) and hyperbranched polyester as solvents to identify the effect on the liquid-vapor equilibrium of the ethanol-water mixture. The relative volatility of ethanol can be increased from 1 to 1.1 and 1.25 at the azeotropic composition for hyperbranched polyester and poly(ethylene glycol), respectively. In other work, HyPol polyglycerol and ethanediol (conventional solvent) were compared in an

ED application [33]. The ethanol relative volatility is increased up to 1.8 at the azeotropic composition using polyglycerol, which is a similar value reached by ethanediol. However, the advantage of the non-volatile HyPol polyglycerol lies in the variety of separation and regeneration methods to recover the solvent. In addition, using HyPol polyglycerol as a solvent could produce a reduction of the energy requirement of 19% as compared to conventional ED.

8.3.2 ETHANOL SEPARATION USING MEMBRANE TECHNOLOGY

8.3.2.1 PERVAPORATION (PV)

A pervaporation (PV) process is an operation to separate liquid mixtures through a selective layer, called membrane. Unlike distillation, PV is not limited by the mixture thermodynamic. Therefore, PV is a potential technology to separate azeotropic mixtures as ethanol-water case. In addition, the thermal energy requirement in PV can be significantly lower as compared to distillation. The separation by PV depends on the intrinsic properties of the selective layer material. Considering that the components in the mixture present different affinities to the membrane material, the combination of their different permeation rate through the membrane produces the separation.

In Figure 8.6 is illustrated a PV process. A heated liquid mixture is fed to a PV unit, being contacted on one side of the membrane. The components of the mixture are transported through the membrane wall until the other side of the membrane is reached, the permeate side. The component transferred across the membrane wall is recovered as a vapor stream. The mass transfer through the membrane is driven by the difference of the vapor pressure between the feed and permeate. In order to maintain the driving force as high as possible, vacuum pressure is usually applied on the permeate side. Although the vacuum pressure is commonly generated by a pump in a lab-scale PV unit, in the industry, this is generated by the rapid condensation of the permeate itself [34].

The performance of PV membranes can be defined according to the selective component removed from the feed. The membrane selectivity is a measure of the permeability ratio for the components associated with the mixture, which can be calculated as the ratio of binary permeability. In the case of the ethanol-water mixture, the membrane selectivity can be determined as follows [35]:

$$\theta_{EtOH/w} = \frac{P_{EtOH}}{P_w} \qquad (1)$$

In Eqn. (1), P is the permeability coefficient of the components. P is a mass transfer parameter, in which is condensed the mass transfer mechanisms (sorption and diffusion) through a PV membrane based on the solution-diffusion theory [36]. The membrane selectivity is an independent parameter of the feed conditions, it depends exclusively on the membrane material [37].

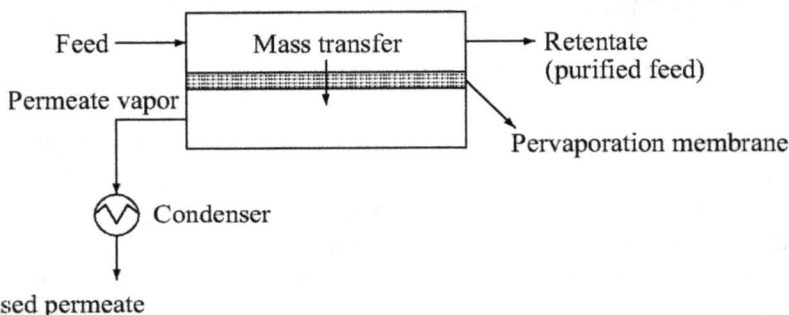

FIGURE 8.6   Sketch of a pervaporation process.

Source: Adapted from: Baker [34].

Two types of membranes can be identified for the ethanol-water mixture separation. Organophilic membranes (ethanol selective membranes), where $\theta_{EtOH/w}$ is higher than 1. It means that the ethanol affinity in the membrane material is higher than water affinity, producing a permeate enriched in ethanol ($P_{EtOH} > P_w$). Otherwise, water is selectively removed in hydrophilic PV membranes. The permeate is enriched with water, which means that $P_{EtOH} > P_w$.

8.3.2.1.1   *Organophilic Membranes for Ethanol Recovery*

According to the vapor-liquid equilibrium for the ethanol-water mixture, the highest constant in Henry's law is reached in the ethanol composition from 0 to 20 wt.%. The highest PV performance for ethanol recovery can be achieved in this composition range. Consequently, PV is highly applicable to recover ethanol from the fermentation downstream. Besides, the energy consumption in PV is lower as compared to distillation, due to only a fraction of the feed is evaporated on the permeate side [38].

Hereto, polydimethylsiloxane (PDMS) is the most researched membrane material for ethanol recovery from aqueous solutions. It is highly applicable to separate ethanol at low composition, due to the high hydrophobic nature of this material. Different companies manufacture commercial PDMS membranes, such as SolSep BV (Apeldoorn, Netherlands), Pervatech BV (Enter, Netherlands), Sulzer Chemtech (Neunkirchen, Germany), and Celanese Corp. (NC, United States) [39]. In general, PDMS membranes can present permeate fluxes in the range of $0.001–1$ kg m^{-2} h^{-1} and a selectivity from 1.8 to 4 [35, 40]. In many cases, the ethanol composition in the permeate is located under the liquid-vapor equilibrium curve for the ethanol-water mixture. Therefore, it is necessary a cascade arrangement of PV modules to achieve a similar separation duty of a conventional distillation. It involves a large membrane area, a meaningful increase in capital cost.

Also, poly(1-trimethylsily-1-propyne) (PTMSP) membranes have shown potential characteristics to recover ethanol from aqueous solution. In the case of a PV with PTMSP membranes, the permeate flux and the ethanol selectivity can be 3 and 2 times higher as compared to PDMS membranes, respectively [41]. Although PTMSP membranes initially showed a high performance, these are hardly applied for practical separation due to their instability. For instance, in a separation of an ethanol aqueous solution (50 wt.%) by PTMSP membranes, the permeate flux and ethanol selectivity can be reduced up to 60% and 22% after 200 h of operation [42]. PTMSP membranes have been submitted to different studies in order to improve the stability by modifying the polymer synthesis, as grafting copolymer to the selective layer. The synthesis of PTMSP using NbCl$_5$ and TaCl$_5$/Al(i-Bu)$_3$ as catalyst allowed a stable membrane, even with a low pH PV [43]. In a hybrid PTMSP/PDMS membranes, a higher permeate flux and ethanol selectivity were achieved as compared to a solely PTMSP membranes [44]. No membrane instability was reported. In addition, a permeate with an ethanol composition up to 70 wt.% was reached from a feed with 7 wt.% of ethanol.

Although PDMS has been the preferred choice to recover ethanol, inorganic membranes present a substantial improvement in the separation of ethanol from aqueous solution. Membranes based on hydrophobic zeolites have shown a higher performance for recovering ethanol in comparison to PDMS membranes [39]. A special characteristic of zeolite membranes is a lower swelling as compared to unmodified polymeric membranes, as a result, a high selectivity can be achieved in PV at high temperatures [45]. MFI membranes (silicate-1 and ZSM-5) are the representative zeolite-based membranes. It is possible to perform a PV at high temperature with MFI membranes, which cannot be made using polymeric membranes due to membrane swelling. The permeate flux and ethanol selectivity reached by a PV with an MFI membrane can be 2 and 5 times higher than a PDMS membrane [46]. However, inorganic membranes (especially zeolite-based membranes) can be up to 50 times more expensive than polymeric membranes [47]. In addition, it is difficult to manufacture this type of membranes commercially.

8.3.2.1.2   Hydrophilic Membranes for Ethanol Dehydration

In the field of ethanol dehydration, PV presents a significant advantage as compared to azeotropic distillation and other conventional distillation schemes. A small amount of water is needed to be separated. Consequently, the operation costs for the dehydration by PV can be significantly lower than an azeotropic distillation. An installed ethanol dehydration process by PV was compared to commercial dehydration with AD [48]. The ethanol losses in PV were neglected, and the alcohol efficiency in PV was about to 99.7%. On the other hand, the ethanol quality in the AD was lower as compared to PV. The permeate had not any impurity of the entrainer. Besides, the dehydration by PV presented a reduction of operation costs around 30%. This reduction was associated with low energy consumption (thermal utilities) in PV, and the entrainer make up in AD. This latter represented about 10% of the total operation cost.

For the application of PV and membrane selection in the ethanol dehydration, the liquid-vapor equilibrium and PV performance can be used. In Figure 8.7 is presented the comparison of the separations between distillation and PV with different membrane material. A proper dehydration membrane should be under the 45° line in the liquid-vapor equilibrium of the ethanol-water mixture. For this case, only the membrane selectivity is considered. In general, the three membrane types are able to remove water from the mixture selectively. For hydrated ethanol ($\approx$ 93 wt.%), PVA membranes are the most appropriate to produce anhydrous ethanol. In the range of 85–95 wt.%, the highest water selectivity is achieved by the PVA membrane. This reason makes that PVA membrane be widely used in the industry to dehydrate ethanol [48].

1. **Organic Membranes:** As it was mentioned above, PVA-based membranes are one of the most studied and applied PV membranes for alcohol dehydration. As a polymeric membrane, PVA is able to be swelled in ethanol solutions, decreasing its selectivity. However, due

to its highly hydrophilic character, superior abrasion resistance, and cost, PVA is an effective material to be applied in the water removal from hydrated ethanol. In general, the PVA membrane fluxes can be in the range of 0.06–6.33 kg m^{-2} h^{-1} and the water selectivity can vary from 43 to 890 for a maximum water content of 10 wt.% [50]. Besides, at low water content as the hydrated ethanol composition, PVA membrane can become glassy [51]. The water flux and selectivity can be significantly reduced. Therefore, recent works have been focused on improving the mechanical and physical properties of PVA membranes. This is made by crafting other copolymers and additives in the PVA matrix or preparing composite membranes [52–55].

Chitosan (CS) membranes are well-known hydrophilic membranes, which have been widely applied for ethanol dehydration. However, the performance of CS membranes can be decreased in aqueous solution due to its uncontrollable swelling [56]. Similar to PVA, CS membranes have been researched in order to improve their physical properties. For instance, a hybrid CS/PVA membrane was synthesized for an alcohol-water mixture (10 wt.% water) separation at low temperature [57]. A permeate flux of 0.113 kg m^{-2} h^{-1} was achieved and, the water selectivity was up to 17,000. In other work, a PV with hybrid CS/poly (sodium vinyl-sulfonate) was performed to dehydrate an ethanol solution at the azeotropic composition [58]. The experiments were carried out during 120 h, and the membrane presented a stable performance with a permeate flux of 1.98 kg m^{-2} h^{-1} and a water composition of 99.5 wt.% on the permeate side.

2. **Inorganic Membranes:** An advantage of inorganic membranes, for example, ceramic membranes, is the stability and mechanical properties in PV at high temperatures. Although the membrane selectivity is naturally given by the membrane material, the permeate flux and permeate composition are dependent on the vapor pressure of the component. In the case of a dehydration membrane, the water pressure and, consequently, the water flux can be enhanced by increasing the feed temperature. Since inorganic membranes do not present thermal instability, a wider operation range for PV can be presented as compared to organic membranes.

Microporous silica membrane is one of the most representative inorganic membranes for alcohol dehydration. A PV with silica membranes for an ethanol-water mixture (6 wt.% water) can achieve a permeate flux of 0.417 kg m^{-2} h^{-1} and a water selectivity of 207 [59]. Moreover, this type of membrane is widely manufactured and commercialized by several companies, such as, ECN, TNO, PervaTech, and SMS [60].

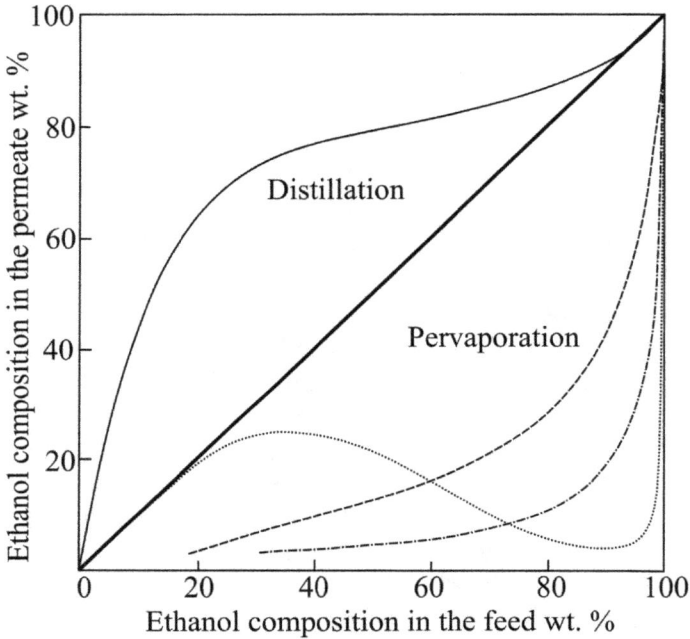

FIGURE 8.7   Comparison for the separation of ethanol-water mixtures by distillation based on the liquid-vapor equilibrium (—) and by three pervaporation membranes: (---) cellulose triacetate (CTA), an anionic polyelectrolyte membrane (– · –) and poly(vinyl alcohol) (PVA) (···).

Source: Adapted from: Ref. [49].

Another highly hydrophilic material is titanium dioxide (TiO$_2$). This material has been attracting the attention of researchers since it can be potentially applied in PV for alcohol dehydration. TiO$_2$ is commonly used as a filler in PV membranes, using nanoparticle of this, improving the mechanical and physical properties of organic membranes [61, 62]. A permeate flux of 0.340 kg m^{-2} h^{-1}, and a membrane selectivity of 196 can be reached for the separation of ethanol at 90 wt.% in an aqueous solution at 80°C. Similar to TiO$_2$, zeolites have been used as a filler in organic [63] and inorganic membranes [64].

Although PV presents significant advantages as compared to distillation, it is difficult to replace an entire distillery for a PV system completely.

Considering the high amount of material to process in a biorefinery, it would require a large membrane area to take the ethanol composition in the wine to the azeotropic composition. Currently, the membrane prices are considerably high, even for polymeric membranes. Although distillation involves high-energy consumptions, producing hydrated ethanol is more economically feasible by distillation than PV. Therefore, PV for ethanol recovery in biorefineries is mainly focused on using it as an assistant operation. It means that PV can be coupled to a main unit of the process to improve it. This is especially observed in hybrid processes as fermentation or distillation.

8.3.2.2 HYBRID DISTILLATION-PERVAPORATION (PV) COLUMNS

As it was previously mentioned, PV is a proper technology to produce anhydrous ethanol. In order to compensate the high capital costs of PV membranes, distillation, and PV can be coupled in a hybrid separation. Therefore, ethanol can be concentrated close to the azeotropic composition by distillation, while it is simultaneously dehydrated by PV. In this case, PV replaces the dehydration by AD or adsorption using molecular sieve driers.

In Figure 8.8 is shown the hybrid distillation-PV column for ethanol dehydration. The ethanol is concentrated in the first distillation column, similarly to the conventional distillation process (Figure 8.1). Liquid and vapor phlegma streams are produced with an ethanol composition about 30 wt.%. Unlike to the conventional distillation stage, a PV membrane is externally connected to the column B. The hydrated ethanol in the distillate stream produced by the column B is continuously fed to the PV unit, where the water removal takes place. In order to increase the driving force in the separation by PV, the feed temperature and pressure are increased to 130°C and 4 bar, respectively [34]. A retentate with an ethanol composition higher than 99.3 wt.% can be achieved. A permeate with a low amount of ethanol is recycled to the column B, specifically, to the stripping section where the water composition is high.

Although in Figure 8.8 is shown a single-stage PV, in the industry, three or four stages are required for ethanol dehydration. It is due to temperature drop during the PV. Therefore, inter-heating stages are necessary to compensate for the temperature drop inside each PV module. In some way, this allows maintaining relatively constant the PV temperature. In many cases, the thermal energy for PV is supplied by integrating the column condenser, stillage, and bottom stream with the PV feed stream.

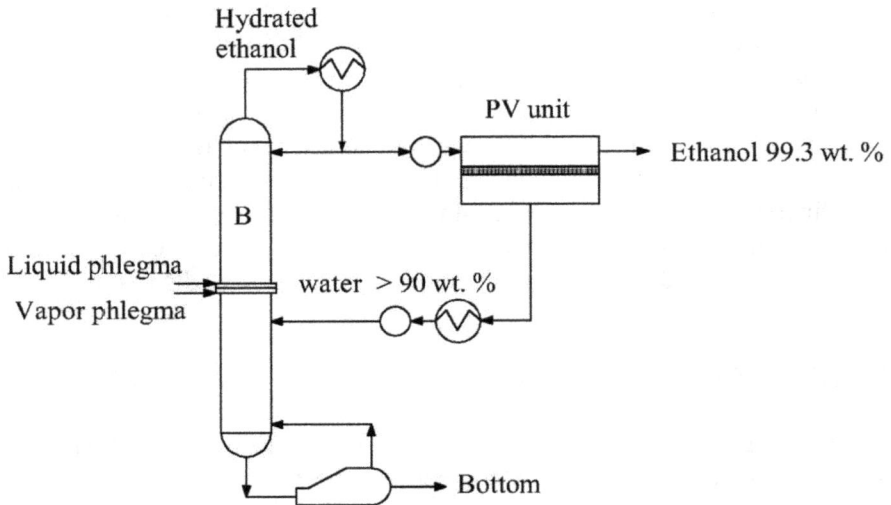

FIGURE 8.8 Hybrid distillation-pervaporation column to dehydrate ethanol from the phlegma streams generated in the first column in a conventional distillery.

In the case of large biorefinery, a different design of a hybrid distillation should be proposed. Vapor permeation is applied to improve the thermal integration in the process since the energy costs are significantly high. These have to be considered in the design. The main difference between a PV and vapor permeation is the physical state of the feed. In PV, a liquid feed is considered. While in vapor permeation, the feed is vaporized before entering to the membrane unit. In both cases, the driving force is the difference in the vapor pressure between the sides of the membrane. According to this, it is possible to reach higher permeate fluxes in a vapor permeation than PV. However, the vapor condensation inside the membrane module should be considered.

A hybrid distillation-vapor permeation process for ethanol separation and dehydration is shown in Figure 8.9. Initially, the ethanol concentration from the wine is concentrated in a vacuum stripper column (A) at 0.5 bar. The top and bottom products in column A are an overhead vapor (~65 wt.% of ethanol) and a liquid stream mostly composed of water (~0.1 wt.% of ethanol), respectively. The product of the top is compressed up to 3 bars (B), resulting in a temperature increase of the vapor. This vapor can supply a part of the energy required for the evaporation in the column A. The water content in the compressed vapor is reduced around 75% inside the vapor permeation unit 1 (C), producing a permeate with an ethanol composition of 7 wt.%. The permeate is recycled directly to the column A, taking advantage of the latent heat of this stream. Anhydrous ethanol is produced in the vapor permeation unit 2 (D). Since the ethanol composition is significantly high in the feed of D, the permeate generated in this unit (57 wt.% of ethanol) is condensed and mixed with the wine fed to the separation system. Finally, the anhydrous ethanol (vapor) stream is thermally integrated with the column A reboiler for its condensation (transfer of the latent heat). Approximately 30% of the thermal energy in the column reboiler is supplied by the retentate condensation leaving from D. Using the arrangement presented in Figure 8.9, it is able to reduce the energy consumption by up to 50% as compared to the hybrid distillation-PV system.

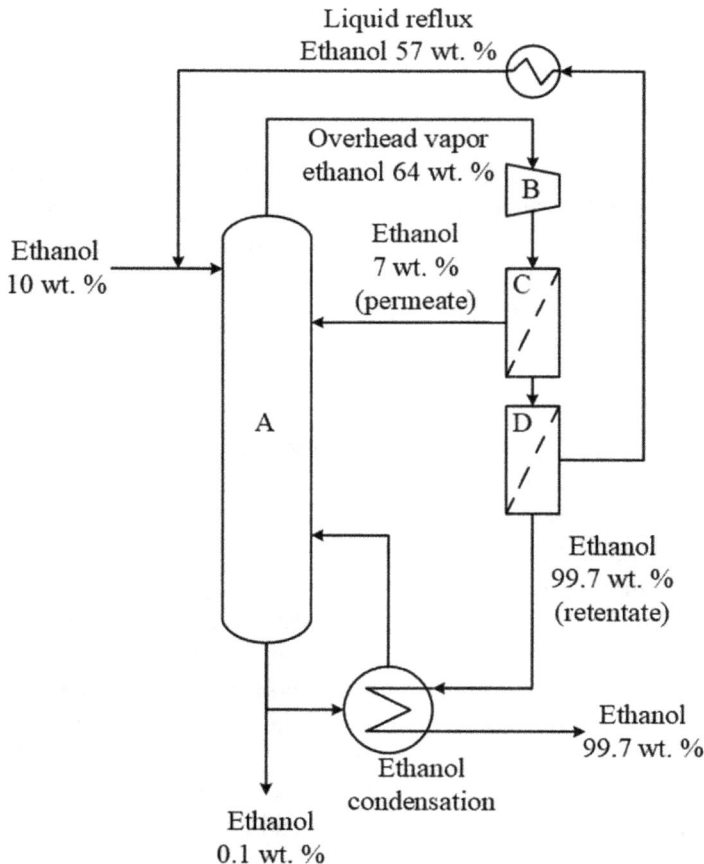

FIGURE 8.9 Hybrid distillation-vapor permeation process for ethanol-water separation in large-scale biorefineries [65]. Vacuum stripper column (A); compressor (B); pervaporation unit 1 (C); and pervaporation unit 2 (D).

Different works have presented the advantage of the hybrid distillation-PV system to recover ethanol. It was demonstrated that a hybrid system, using NaA zeolite membranes requires 52% less energy as compared to the azeotropic distillation, producing an anhydrous ethanol with a composition of 99.4 wt.% [66]. Besides, the ethanol dehydration by PV can be more energetic efficient than using molecular sieves in industrial cases. PV requires just 23% of the energy used in an adsorption process with molecular sieves [67]. In other work, a hybrid vapor stripping-vapor permeation process, termed membrane assisted vapor stripping (MAVS) was applied to separate alcohols in aqueous solutions [68]. Using MAVS technology allows a 65% more energy efficient separation of ethanol as compared to conventional distillation.

8.3.3 IN SITU ETHANOL RECOVERY: HYBRID FERMENTATION TECHNOLOGY

The conventional process for the production of ethanol by batch fermentation of sugars by yeast has dealt with several drawbacks such as the high freshwater consumption, the high energy consumption, and the inhibition by substrate or product [5, 69]. In order to avoid the product and substrate inhibition, low substrate concentration is used. Consequently, low reaction rates are achieved with slow cell growth, which hinders the downstream processing, increasing the energy consumption in the separation stage, and the total cost of the process [69–72].

Strains genetically modified with high tolerance to substrate and ethanol within the fermentation broth have been developed in order to overcome the inhibition drawback [69, 70, 73]. However, this approach to improve the conventional process for production of ethanol still is an uneconomic process as compared to fossil fuels.

In situ product recovery (ISPR) is a potential strategy to improve ethanol production [69]. ISPR involves ethanol removal while it is produced within the bioreactor, maintaining the ethanol concentration under the inhibition limit (90 kg m^{-3}). It provides several benefits to the process, such as, decreasing of the water usage, increase the reaction rate, and increase ethanol productivity, allowing a continuous fermentation process [69, 74].

Several separation methods for its integration to the ethanol fermentative process, such as PV, solvent extraction, and adsorption, have arisen as potential technologies for ISPR and overcome the abovementioned drawbacks of the conventional ethanol production.

8.3.3.1   INTEGRATED LIQUID-LIQUID EXTRACTION (LLX) AND ETHANOL FERMENTATION PROCESS

Liquid-liquid extraction (LLX), also called solvent extraction, was proposed as a separation method for in situ ethanol removal from fermentation broth in the 1980's, and remains as a potential separation process to be coupled to ethanol fermentation. The key considerations for the integration of the LLX to the ethanol fermentation involve the selection of the solvent and strategy of operation for the integrated LLX-fermentation system.

The main criteria for selection of the solvent to be used in a coupled LLX-fermentation system may be summarized in the distribution coefficient, selectivity for the solute, and biocompatibility with the yeast. It is desirable the solvent be low cost, low viscosity, completely immiscible with the aqueous phase, high selectivity and distribution coefficient for ethanol, and simple regeneration (for recovery of the product from the solvent phase, in this case, the ethanol).

Thermodynamically, the distribution coefficient at equilibrium condition in a liquid-liquid system (equality of chemical potential between both phases) is defined as the ratio between the equilibrium concentrations of solute in the organic phase and the aqueous phase [75]. A high distribution coefficient is convenient because it results in a decrease in the number of stages of the LLX process and reduces the amount of required solvent for removal of the desired solute [76, 77]. The polarity of the solvent influences on the solubility of ethanol. The distribution coefficient of polar solvents increases as the number of polar groups in the structure of the solvent increases. The distribution coefficient of ethanol ($K_{D,EtOH}$) in several solvents has been widely tested, and in Table 8.1 are shown the highest distribution coefficient values achieved.

The separation factor, defined as the ratio of the distribution coefficient of ethanol to water ($\alpha = K_{D,EtOH}/K_{D,water}$), is used to define the ability of the solvent to remove the ethanol from water. Values of the separation factor higher than the unity result in a selective extraction of ethanol from water. For the selection of the solvent, it is also important to know the selectivity on other components that are usually within the fermentation broths, specifically sugars and by-products that may achieve considerable concentrations during fermentation. Alcohols and esters have shown a high separation factor for ethanol [78]. Furthermore, the branched-chain solvents have higher values of the separation factor as compared to its counterpart, the linear-chain solvents [76]. In many cases, the solvent has shown high selectivity but not a high enough distribution coefficient. Looking for a compromise between selectivity and distribution coefficient, a mixture of solvents has been proposed [76]. It is usually used in the LLX of organic acids from fermentation broths [79, 80], where the mixture of solvents not only provides reasonable distribution coefficient and selectivity but also provide of a kind of tuning of the physicochemical properties of the solvent mixture, such as density, viscosity, and interfacial tension [79].

TABLE 8.1 Solvents with High Values of Distribution Coefficient (around 1 or higher than 1) for Ethanol Extraction

Solvent	K_D
1-Hexanol	1.0–1.2
3-Methylcyclo-hexanol	0.93
3-Methyl-3-pentanol	1.3
4-Methyl-2-pebtanol	1.1
2-Ethyl-butanol	0.69–1.03
3-Ethyl-3-pentanol	1.1
Phenol	2.15
o-Isopropylphenol	1.4
o-ter-butylphenol	1.4
Valeric acid	1.3
Hexanoic acid	0.944–1.1
Methyl acetate	0.91
Ethyl propionate	2.53
50% Hexan-1-ol + 50% 2-ethyl-1-ol (w/w)	1.03

Source: Adapted from: Ref. [76].

On the other hand, microorganisms are very sensitive to the conditions of their environment. These conditions refer to the pH, temperature, agitation, and composition of the fermentation broths [81]. During the fermentation, the optimal conditions for cell growth are set to avoid the stress on the cells. The presence of an organic solvent on the fermentation broths produces some stress on the cells, a toxic effect. The toxicity of organic solvents on the microorganism of the fermentation may arise in two ways. Direct contact between the microorganism with the solvent into the aqueous/organic interface (phase toxicity) and by the soluble portion of the organic solvent within the aqueous phase (molecular toxicity) [79, 82].

The stress of the organic solvents on yeast cells affects their morphology and physiological activity [81, 83]. The accumulation of organic solvents on the membrane cells modifies their fluidity and permeability [81], which involves hindrance of nutrients transport, leakage of metabolites, and in the most severe cases, lysis of the cell [76, 83, 84]. On the other hand, organic solvent within the cells may disturb the metabolic reactions decreasing the activity of enzymes because it blocks the enzymatic active sites [81, 83, 84]. As a response to the presence of organic solvents, both the membrane cell and into the cell, cells may change its lipid and protein composition in the membrane to modify the plasma membrane properties, such as rigidity, permeability, and fluidity [84]. However, any response of the cells to the organic solvent is energy demanding, which means that a portion of the consumed substrate must be used to supply this energetic demand [79, 83], decreasing both the cell growth and the metabolite production [76, 79].

Nonpolar solvents are nontoxic to cells [83]. However, nonpolar solvents provide of low distribution coefficient to ethanol. In general, solvents with a high distribution coefficient are toxic to cells [76, 79]. For instance, n-alcohols with high molecular weight have a low distribution coefficient for ethanol, while n-alcohols with low molecular weight have high toxicity and solubility in water [70]. According to several works, the efficiency on the in-situ ethanol removal and toxicity by using yeast cells follows [85, 86]: carboxylic acid > alcohols > esters > amines > ketones > ethers > hydrocarbons. However, the branch or length of the linear structure may affect [86].

Natural solvents, for instance, vegetable oils, which are very lipophilic, are nontoxic to the whole cells [83]. Solvents, such as n-amyl alcohol, 1-octanol, and 1-docecanol have been tested for in situ ethanol removal from the fermentation broth, resulting n-amyl alcohol the solvent with the highest recovery percentage, providing the highest ethanol productivity, despite its toxic effect [72]. Also, n-dodecanol has probed as an efficient solvent for removal of ethanol by LLX in a continuous fermentation [87, 88].

The toxicity of solvents on the yeast may be reduced and even avoided, by immobilizing the cells [71, 89]. Several compounds have been tested for immobilization of cells, such as alginates, collagen, carrageenan, agar, agarose, glutaraldehyde, and some materials as support such as CS, cellulosic material, natural zeolite, and γ-alumina [90, 91]. Calcium-alginate gel entrapping is the most used method due to it is low cost, provides high enzymatic activity and simple preparation [90]. Immobilized yeast has been tested in LLX-fermentation systems for ethanol production, achieving higher ethanol productivities and yields that its counterpart, the conventional ethanol fermentation process [72, 76, 88].

In addition, a proper LLX-fermentation configuration is able to reduce the toxicity of solvents, providing an efficient ethanol extraction with a minimum contact between yeast cells and solvent [71, 76]. Several configurations for continuous LLX-fermentation, considering direct and indirect contact of the yeast cell with the solvent are shown in Figure 8.10.

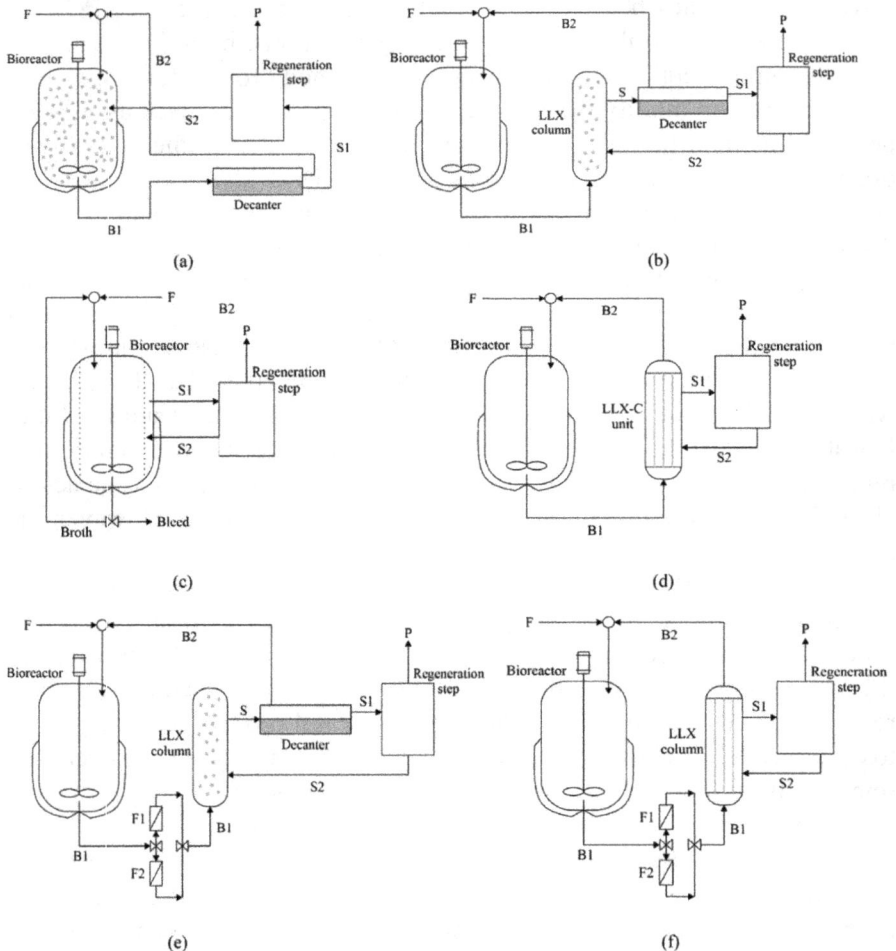

FIGURE 8.10   LLX-fermentation arrangements: (a) solvent directly mixed with the fermentation broths inside the fermenter; (b) solvent directly mixed with the fermentation broths in an external LLX unit; (c) solid porous support contactor inside the fermenter; (d) solid porous support contactor in an external LLX unit; (e) solvent directly mixed with the fermentation broths in an external LLX unit with yeast cells filtration; and (f) solid porous support contactor in an external LLX unit with yeast cells filtration.

In Figure 8.10(a), the ethanol fermentation and LLX are simultaneously carried out within the bioreactor. An output stream (B1) from the bioreactor is processed in a decanter, splitting the aqueous (B2) and organic (S1) phases. Therefore, B2 is the fermentation broth depleted in ethanol, and S1 is the solvent highly concentrated in ethanol. The B2 stream is recycled and mixed with the fermenter feed (F). The stream S1 is processed in a subsequent step for regeneration of solvent and recovery of the product. In the regeneration stage, the regenerated solvent (S2) is recycled to the bioreactor. P is an output stream from the regeneration stage, which is concentrated in ethanol. Worth noting that for this configuration, immobilized cells are recommended to reduce any toxic effect due to a direct contact yeast-solvent.

The ethanol fermentation also may carry out simultaneously with LLX, using separate units as Figures 8.10(b, d–f) show. In Figure 8.10(b), the fermentation broths (B1) from the bioreactor are fed to LLX column. Here, the solvent phase is in intimate contact with the fermentation broths (ethanol is transferred from fermentation broth to solvent phase), and both phases are split in a subsequent unit, the decanter. The aqueous phase from the decanter (B2) is returned to the bioreactor once is enriched of the substrate with the fermenter feed (F), while the solvent phase (S1) is fed in the unit for its regeneration. The clean solvent (S2) and the concentrated ethanol (P) are produced in the regeneration unit. Finally, the clean solvent (S2) is returned to the LLX column to perform a new extraction cycle. In this scheme, immobilization of cells is required in order to avoid the direct contact of the cells with the solvent phase.

An alternative scheme with a direct yeast-solvent contact is shown in Figure 8.10(c). However, the mechanism of contacting the fermentation broths and solvent is significantly different as compared to Figure 8.10(a). In this case, the yeast cells are directly contacted to the solvent by using a solid porous support submerged in the fermentation broths. The solid surface of the support spatially separates the fermentation broths from the solvent, while the porous, which are filled by the solvent, are exposed to the fermentation medium. Then, the solid porous support works as contact media (contactor) between fermentation broths and the solvent phase, providing a high interfacial area between the phases.

The configuration shown in Figure 8.10(d) is a modification of the scheme shown in Figure 8.10(c), since the LLX is externally performed using a contactor (LLX-C). An output stream from the bioreactor is fed to the LLX-C. In the LLX-C unit, the fermentation broths and solvent are partially separated by a contactor (solid porous support). Both phases are contacted through the filled porous of the contactor. Similar to the case of

Figure 8.10(c), the contractor provides a high interfacial area between the phases, but also, applying an external LLX unit.

The schemes presented in Figure 8.10(e) and f are variants to those previously showed in Figure 8.10(b) and (d), respectively. The difference lies in the use of a parallel arrangement of filters before the extraction unit, in order to avoid direct contact of the yeast cells with the solvent. It is an alternative for fermentation processes with non-immobilized yeast cells. Due to the filters suffer from fouling by the yeast cells and solids in the fermentation broths, two filters must be implemented (F1 and F2). While F1 is operating, F2 is in a cleaning stage or vice versa.

For all schemes shown in Figure 8.10, the molecular toxicity may be avoided if there is no recirculation of the fermentation broths (B2). However, recirculation of the fermentation broth reduces the use of fresh water in the process significantly, providing a positive impact on the cost of the process and in the environment. Immobilization of cells works to overcome this molecular toxicity drawback. Another option is finding a suitable solvent for ethanol removal, which be nontoxic on the yeast cells. Nowadays, the seeking of solvents for several processes still is an important field for the design of new reactive and separation processes. Ionic liquids (ILs), deep eutectic solvents (DESs) and aqueous two-phase systems (ATPS) are emerging as potential solvents to be applied in several processes. ILs and DESs have shown high efficiency for the extraction of several organic acids [76, 77, 92, 93]. These novel solvents have great potential for the removal of several metabolites of fermentations. However, it still requires research on the applicability of these solvents in the ISPR for ethanol fermentations.

Once the solvent is used to remove the ethanol from fermentation broth, it requires a regeneration step, as was shown in all configurations of the LLX-fermentation process. There are several choices for the regeneration of the solvent, such as vacuum flash vaporization, distillation, gas stripping, PV, and back extraction [76, 77]. Use one or another technique is highly related to the physicochemical properties of the solvent and ethanol, for example, the difference in density or differences in volatilities. However, the most used technique for the regeneration of the solvent rich in ethanol is vacuum flash.

8.3.3.2 *PERVAPORATIVE FERMENTATION (PV-FERMENTATION)*

PV is a promising technology to be applied in fermentation processes to recover ethanol from the fermentation broths. According to previous sections, organophilic membranes can be used to remove ethanol, while this is produced during the fermentation process. In this case, PV is not an invasive method. Unlike to other types of hybrid fermentation, where the fermentation medium can be contacted to solvent or, subjected to vacuum pressure, a solid membrane does not affect the fermentation medium. It maintains the ethanol under the inhibition concentration.

The general scheme of a PV-fermentation is shown in Figure 8.11. In a hybrid fermentation-PV system, the membrane (C) can be located inside or outside the fermenter (A). However, an internal membrane is more susceptible to be fouled due to the high solid concentration (mainly for yeast cells) in the fermentation broths. As a result, the operating time and membrane performance can be rapidly reduced. In order to minimize the membrane fouling due to the suspended solid, a fraction of the fermentation broths is withdrawn and passed through a rotatory filter (B). Around 99% of the yeast cell is retained in B, and recycled to the fermenter again [95]. In the case of a continuous fermentation, around 85% of retained yeast cells is feedback to the fermenter again [96]. In the membrane module, ethanol is continuously recovered by a PV at the fermentation conditions. The retentate is recycled to the fermenter.

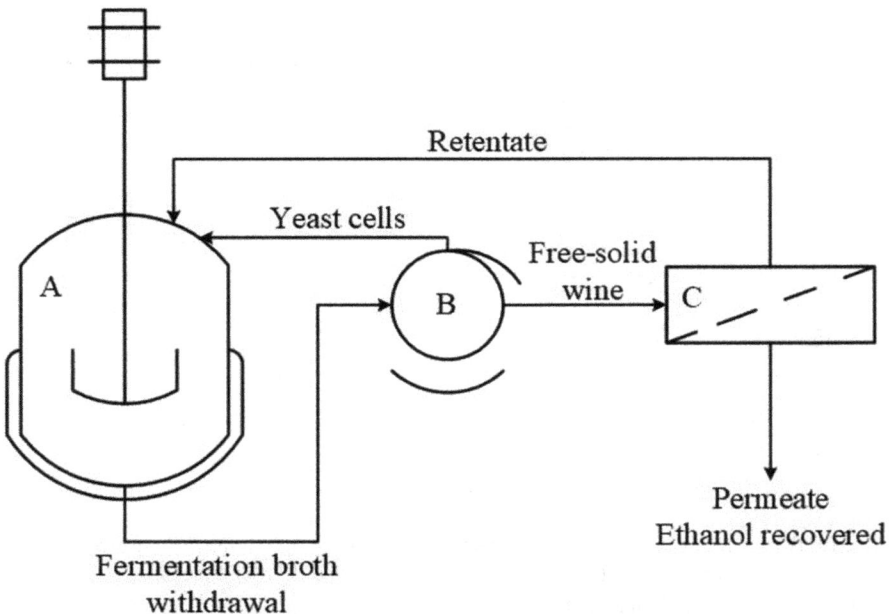

FIGURE 8.11 Pervaporative fermentation system with an ethanol selective membrane composed of (A) fermenter; (B) rotatory filter; and (C) external pervaporation membrane.

Source: Adapted from: Léon et al. [94].

Commonly, PDMS membranes are applied to recover ethanol from the fermentation broths due to their high stability [97]. For a fermentation process (29–35°C) is not significantly important the thermal stability of the membrane. However, PDMS membrane is swelled during the fermentation. Several studies have reported improvements in the fermentation process by coupling it to PDMS membranes. For instance, in a continuous membrane fermenter separator (CMFS) the ethanol production was increased up to 20% as compared to a conventional fermentation [98]. In other work, it was possible to reestablish high substrate consumption rates and ethanol productivity after the inhibition by ethanol was achieved in a continuous and closed-circulating fermentation system (CCCF) with PDMS membranes [99]. In Figure 8.12 is shown this latter impact of PV on the fermentation performance. Once PV is initiated, the cell activity arises, increasing the substrate consumption. Consequently, the ethanol productivity is increased, allowing an overall ethanol concentration up to 609.8 kg m^{-3}. This is almost 6-fold higher than a conventional fermentation. The reestablishment of the fermentation performance after the ethanol inhibition condition is reached, is a general effect of PV on the fermentation process [94, 100, 101].

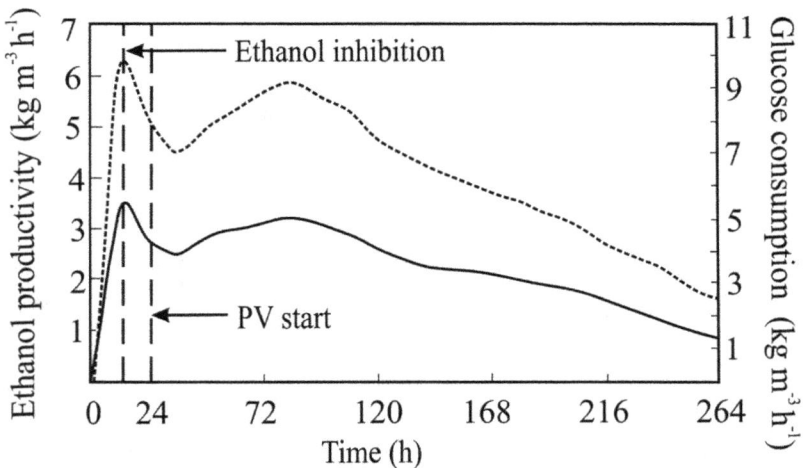

FIGURE 8.12   Ethanol productivity (—) and glucose consumption rate (···) in a continuous and closed-circulating fermentation system (CCCF) with PDMS membranes.
Source: Adapted from: Fan et al. [99].

On the other hand, the components in the fermentation broths influence the PV performance. As it was previously mentioned, yeast cells can cover the membrane surface, reducing the permeate flux by PV. In addition, a maximum glucose concentration of 100 kg m^{-3} increases the ethanol selectivity [102]. However, the permeate flux is reduced. In the case of a high concentration of dissolved salts (e.g., NaCl), the ethanol flux can be increased due to the increase of its vapor pressure [97]. Furthermore, the reduction of the membrane selectivity can occur by the presence of the main fermentation products at a high concentration [97]. Ethanol, acetone, n-butanol, and 2-propanol at 100 kg m^{-3}, 10 kg m^{-3}, 20 kg m^{-3} and 2–5 kg m^{-3}, respectively, reduce the PDMS membrane selectively. Also, the pH can modify the hydrophobicity of PDMS membranes [103]. At low pH, the water permeability is increased. The ethanol selectivity is reduced.

KEYWORDS

- chitosan
- extractive distillation
- hyperbranched polymer
- liquid-liquid extraction
- membrane assisted vapor stripping
- polydimethylsiloxane

REFERENCES

1. Zhang, K., Lively, R. P., Noel, J. D., Dose, M. E., McCool, B. A., Chance, R. R., & Koros, W. J., (2012). Adsorption of water and ethanol in MFI-type zeolites. *Langmuir, 28*, 8664–8673. doi: 10.1021/la301122h.
2. McAloon, A., Taylor, F., Yee, W., Ibsen, K., & Wooley, R., (2000). *Determining the Cost of Producing Ethanol from Corn Starch and Lignocellulosic Feedstocks*. Golden, CO (United States). doi: 10.2172/766198.
3. Phisalaphong, M., Srirattana, N., & Tanthapanichakoon, W., (2006). Mathematical modeling to investigate temperature effect on kinetic parameters of ethanol fermentation. *Biochem. Eng. J., 28*, 36–43. doi: 10.1016/j.bej.2005.08.039.
4. Lopes, M. L., De Paulillo, S. C. L., Godoy, A., Cherubin, R. A., Lorenzi, M. S., Giometti, F. H. C., Bernardino, C. D., et al., (2016). Ethanol production in Brazil: A bridge between science and industry. *Brazilian J. Microbiol., 47*, 64–76. doi: 10.1016/j.bjm.2016.10.003.
5. Huang, H. J., Ramaswamy, S., Tschirner, U. W., & Ramarao, B. V., (2008). A review of separation technologies in current and future biorefineries. *Sep. Purif. Technol. 62*, 1–21. doi: 10.1016/j.seppur.2007.12.011.

6. Banat, F. A., Al-Rub, F. A. A., & Simandl, J., (2000). Analysis of vapor-liquid equilibrium of ethanol-water system via headspace gas chromatography: Effect of molecular sieves. *Sep. Purif. Technol., 18,* 111–118. doi: 10.1016/S1383-5866(99)00057-X.

7. Dias, M. O. S., Modesto, M., Ensinas, A. V., Nebra, S. A., Filho, R. M., & Rossell, C. E. V., (2011). Improving bioethanol production from sugarcane: Evaluation of distillation, thermal integration and cogeneration systems. *Energy, 36,* 3691–3703. doi: 10.1016/j. energy.2010.09.024.

8. Costa, R. C., & Sodré, J. R., (2010). Hydrous ethanol vs. gasoline-ethanol blend: Engine performance and emissions. *Fuel, 89,* 287–293. doi: 10.1016/j.fuel.2009.06.017.

9. Dias, M. O. S., Cunha, M. P., Jesus, C. D. F., Scandiffio, M. I. G., Rossell, C. E. V., Filho, R. M., & Bonomi, A., (2010). Simulation of ethanol production from sugarcane in Brazil: Economic study of an autonomous distillery. *Comput. Aided Chem. Eng., 28,* 733–738. doi: 10.1016/S1570-7946(10)28123-3.

10. Widagdo, S., & Seider, W. D., (1996). Journal review: Azeotropic distillation. *AIChE J., 42,* 96–130. doi: 10.1002/aic.690420110.

11. Zhao, L., Lyu, X., Wang, W., Shan, J., & Qiu, T., (2017). Comparison of heterogeneous azeotropic distillation and extractive distillation methods for ternary azeotrope ethanol/toluene/water separation. *Comput. Chem. Eng., 100,* 27–37. doi: 10.1016/j. compchemeng.2017.02.007.

12. Chianese, A., & Zinnamosca, F., (1990). Ethanol dehydration by azeotropic distillation with a mixed-solvent entrainer. *Chem. Eng. J., 43,* 59–65. doi: 10.1016/0300-9467(90)80001-S.

13. Bastidas, P., Gil, I., & Rodríguez, G., (2010). Comparison of the main ethanol dehydration technologies through process simulation. In: *20th Eur. Symp. Comput. Aided Process Eng. ESCAPE 20.*

14. Gomis, V., Pedraza, R., Saquete, M. D., Font, A., & García-Cano, J., (2015). Ethanol dehydration via azeotropic distillation with gasoline fraction mixtures as entrainers: A pilot-scale study with industrially produced bioethanol and naphta. *Fuel Process. Technol., 140,* 198–204. doi: 10.1016/j.fuproc.2015.09.006.

15. Uyazán, A. M., Gil, I. D., Aguilar, J. L., Rodríguez, G., & Caicedo, L. A., (2004). Deshidratación del etanol. *Ing. e Investig.,* 49–59.

16. Karimi, S., Tavakkoli, Y. M., & Karri, R. R., (2019). A comprehensive review of the adsorption mechanisms and factors influencing the adsorption process from the perspective of bioethanol dehydration. *Renew. Sustain. Energy Rev., 107,* 535–553. doi: 10.1016/j.rser.2019.03.025.

17. Pedraza, B. R., (2012). *Deshidratación de Etanol Mediante Destilación Azeotrópica con Hidrocarburos Componentes de la Gasolina: Estudio de la Viabilidad del Proceso a Escala Semi-Planta Piloto.* Universidad de Alicante.

18. Jacques, K. A., Lyons, T. P., & Kelsall, D. R., (2003). *The Alcohol Textbook: A Reference for the Beverage, Fuel, and Industrial Alcohol Industries* (4th edn.).

19. Gerbaud, V., & Rodriguez-Donis, I., (2014). Extractive distillation. In: Andrzej, G., & Žarko, O., (eds.), *Distillation* (1st edn., pp. 201–245). Elsevier. doi: 10.1016/ B978-0-12-386878-7.00006-1.

20. Soares, R. B., Pessoa, F. L. P., & Mendes, M. F., (2015). Dehydration of ethanol with different salts in a packed distillation column. *Process Saf. Environ. Prot., 93,* 147–153. doi: 10.1016/j.psep.2014.02.012.

21. Barba, D., Brandani, V., & Di Giacomo, G., (1985). Hyperazeotropic ethanol salted-out by extractive distillation. Theoretical evaluation and experimental check. *Chem. Eng. Sci., 40,* 2287–2292. doi: 10.1016/0009-2509(85)85130-7.

22. Furter, W. F., (1992). Extractive distillation by salt effect. *Chem. Eng. Commun., 116,* 35–40. doi: 10.1080/00986449208936042.

23. Ligero, E., & Ravagnani, T. M., (2003). Dehydration of ethanol with salt extractive distillation—a comparative analysis between processes with salt recovery. *Chem. Eng. Process. Process Intensif., 42,* 543–552. doi: 10.1016/S0255-2701(02)00075-2.

24. Polka, H. M., & Gmehling, J., (1994). Effect of calcium nitrate on the vapor-liquid equilibria of ethanol + water and 2-propanol + water. *J. Chem. Eng. Data, 39,* 621–624. doi: 10.1021/je00015a053.

25. Pinto, R. T. P., Wolf-Maciel, M. R., & Lintomen, L., (2000). Saline extractive distillation process for ethanol purification. *Comput. Chem. Eng., 24,* 1689–1694. doi: 10.1016/S0098-1354(00)00455-5.

26. Lei, Z., & Chen, B., (2013). Distillation. In: *Sep. Purif. Technol. Biorefineries* (pp. 37–60). John Wiley & Sons, Ltd, Chichester, UK, (2013). doi: 10.1002/9781118493441.ch2.

27. Seiler, M., Jork, C., Kavarnou, A., Arlt, W., & Hirsch, R., (2004). Separation of azeotropic mixtures using hyperbranched polymers or ionic liquids. *AIChE J., 50,* 2439–2454. doi: 10.1002/aic.10249.

28. Tsanas, C., Tzani, A., Papadopoulos, A., Detsi, A., & Voutsas, E., (2014). Ionic liquids as entrainers for the separation of the ethanol/water system. *Fluid Phase Equilib., 379,* 148–156. doi: 10.1016/j.fluid.2014.07.022.

29. Zhu, Z., Ri, Y., Li, M., Jia, H., Wang, Y., & Wang, Y., (2016). Extractive distillation for ethanol dehydration using imidazolium-based ionic liquids as solvents. *Chem. Eng. Process. - Process Intensif., 109,* 190–198. doi: 10.1016/j.cep.2016.09.009.

30. Aniya, V., De, D., Singh, A., & Satyavathi, B., (2018). Design and operation of extractive distillation systems using different class of entrainers for the production of fuel-grade tert-butyl alcohol: A techno-economic assessment. *Energy, 144,* 1013–1025. doi: 10.1016/j.energy.2017.12.099.

31. Seiler, M., Köhler, D., & Arlt, W., (2002). Hyperbranched polymers: New selective solvents for extractive distillation and solvent extraction. *Sep. Purif. Technol., 29,* 245–263. doi: 10.1016/S1383-5866(02)00163-6.

32. Seiler, M., Buggert, M., Kavarnou, A., & Arlt, W., (2003). From alcohols to hyperbranched polymers: The influence of differently branched additives on the vapor–liquid equilibria of selected azeotropic systems. *J. Chem. Eng. Data, 48,* 933–937. doi: 10.1021/je025644w.

33. Seiler, M., Arlt, W., Kautz, H., & Frey, H., (2002). Experimental data and theoretical considerations on vapor-liquid and liquid-liquid equilibria of hyperbranched polyglycerol and PVA solutions. *Fluid Phase Equilib., 201,* 359–379. doi: 10.1016/S0378-3812(02)00082-1.

34. Baker, R. W., (2012). *Membrane Technology and Applications* (3rd edn.). Wiley & Sons, West Sussex.

35. León, J. A., & Fontalvo, J., (2019). PDMS modified membranes by 1-dodecanol and its effect on ethanol removal by pervaporation. *Sep. Purif. Technol., 210,* 364–370. doi: 10.1016/j.seppur.2018.08.019.

36. Wijmans, J. G. H., & Baker, R. W., (2006). The solution-diffusion model: A unified approach to membrane permeation. In: *Mater. Sci. Membr. Gas Vap. Sep.* (pp. 159–189). John Wiley & Sons, Ltd, Chichester, UK. doi: 10.1002/047002903X.ch5.

37. Baker, R. W., Wijmans, J. G., & Huang, Y., (2010). Permeability, permeance and selectivity: A preferred way of reporting pervaporation performance data. *J. Membr. Sci., 348,* 346–352.

38. Andre, A., Nagy, T., Toth, A. J., Haaz, E., Fozer, D., Tarjani, J. A., & Mizsey, P., (2018). Distillation contra pervaporation: Comprehensive investigation of isobutanol-water separation. *J. Clean. Prod., 187*, 804–818. doi: 10.1016/j.jclepro.2018.02.157.

39. Peng, P., Shi, B., & Lan, Y., (2010). Review of membrane materials for ethanol recovery by pervaporation. *Sep. Sci. Technol., 46*, 234–246.

40. O'Brien, D. J., & Craig, J. C., (1996). Ethanol production in a continuous fermentation/ membrane pervaporation system. *Appl. Microbiol. Biotechnol., 44*, 699–704. doi: 10.1007/BF00178605.

41. Schmidt, S. L., Myers, M. D., Kelley, S. S., McMillan, J. D., & Padukone, N., (1997). Evaluation of PTMSP membranes in achieving enhanced ethanol removal from fermentation by pervaporation. *Appl. Biochem. Biotechnol., 63–65*, 469–482. doi: 10.1007/BF02920447.

42. López-Dehesa, C., González-Marcos, J. A., & González-Velasco, J. R., (2007). Pervaporation of 50 wt % ethanol-water mixtures with poly(1-trimethylsilyl-1-propyne) membranes at high temperatures. *J. Appl. Polym. Sci., 103*, 2843–2848. doi: 10.1002/app.25375.

43. Volkov, V. V., Fadeev, A. G., Khotimsky, V. S., Litvinova, E. G., Selinskaya, Y. A., McMillan, J. D., & Kelley, S. S., (2004). Effects of synthesis conditions on the pervaporation properties of poly[1-(trimethylsilyl)-1-propyne] useful for membrane bioreactors. *J. Appl. Polym. Sci., 91*, 2271–2277. doi: 10.1002/app.13358.

44. Nagase, Y., Ishihara, K., & Matsui, K., (1990). Chemical modification of poly(substituted-acetylene): II. Pervaporation of ethanol / water mixture through poly(1-trimethylsilyl-1-propyne) / poly(dimethylsiloxane) graft copolymer membrane. *J. Polym. Sci. Part B Polym. Phys., 28*, 377–386. doi: 10.1002/polb.1990.090280309.

45. Korelskiy, D., Leppäjärvi, T., Zhou, H., Grahn, M., Tanskanen, J., & Hedlund, J., (2013). High flux MFI membranes for pervaporation. *J. Memb. Sci., 427*, 381–389. doi: 10.1016/j.memsci.2012.10.016.

46. Soydaş, B., Dede, Ö., Çulfaz, A., & Kalıpçılar, H., (2010). Separation of gas and organic/water mixtures by MFI type zeolite membranes synthesized in a flow system. Microporous *Mesoporous Mater., 127*, 96–103. doi: 10.1016/j.micromeso.2009.07.004.

47. Caro, J., Noack, M., & Kölsch, P., (2005). Zeolite membranes: From the laboratory scale to technical applications. *Adsorption, 11*, 215–227. doi: 10.1007/s10450-005-5394-9.

48. Sander, U., & Soukup, P., (1988). Design and operation of a pervaporation plant for ethanol dehydration. *J. Memb. Sci., 36*, 463–475. doi: 10.1016/0376-7388(88)80036-X.

49. Kaschemekat, J., Barbknecht, B., & Böddeker, K. W., (1986). Konzentrierung von ethanol Durch pervaporation. *Chemie Ing. Tech., 58*, 740–742. doi: 10.1002/cite.330580909.

50. Samei, M., Mohammadi, T., & Asadi, A. A., (2013). Tubular composite PVA ceramic supported membrane for bio-ethanol production. *Chem. Eng. Res. Des., 91*, 2703–2712. doi: 10.1016/j.cherd.2013.03.008.

51. Namboodiri, V. V., & Vane, L. M., (2007). High permeability membranes for the dehydration of low water content ethanol by pervaporation. *J. Memb. Sci., 306*, 209–215. doi: 10.1016/j.memsci.2007.08.050.

52. Penkova, A. V., Dmitrenko, M. E., Savon, N. A., Missyul, A. B., Mazur, A. S., Kuzminova, A. I., Zolotarev, A. A., et al., (2018). Novel mixed-matrix membranes based on polyvinyl alcohol modified by carboxy fullerene for pervaporation dehydration. *Sep. Purif. Technol., 204*, 1–12. doi: 10.1016/j.seppur.2018.04.052.

53. Castro-Muñoz, R., Buera-González, J., De La Iglesia, O., Galiano, F., Fíla, V., Malankowska, M., Rubio, C., et al., (2019). Towards the dehydration of ethanol using pervaporation cross-linked poly(vinyl alcohol)/graphene oxide membranes. *J. Memb. Sci., 582*, 423–434. doi: 10.1016/j.memsci.2019.03.076.

54. Dmitrenko, M. E., Penkova, A. V., Kuzminova, A. I., Morshed, M., Larionov, M. I., Alem, H., Zolotarev, A. A., et al., (2018). Investigation of new modification strategies for PVA membranes to improve their dehydration properties by pervaporation. *Appl. Surf. Sci., 450*, 527–537. doi: 10.1016/j.apsusc.2018.04.169.

55. Yuan, H., Wang, C., Liu, X., & Lu, J., (2020). Preparation of PVA-PFSA-Si pervaporative hybrid membrane and its dehydration performance. *Polym. Bull.* doi: 10.1007/s00289-020-03107-5.

56. Jyothi, M. S., Reddy, K. R., Soontarapa, K., Naveen, S., Raghu, A. V., Kulkarni, R. V., Suhas, D. P., et al., (2019). Membranes for dehydration of alcohols via pervaporation. *J. Environ. Manage, 242*, 415–429. doi: 10.1016/j.jenvman.2019.04.043.

57. Rao, K. S. V. K., Subha, M. C. S., Sairam, M., Mallikarjuna, N. N., & Aminabhavi, T. M., (2007). Blend membranes of chitosan and poly(vinyl alcohol) in pervaporation dehydration of isopropanol and tetrahydrofuran. *J. Appl. Polym. Sci., 103*, 1918–1926. doi: 10.1002/app.25078.

58. Zheng, P. Y., Ye, C. C., Wang, X. S., Chen, K. F., An, Q. F., Lee, K. R., & Gao, C. J., (2016). Poly(sodium vinyl sulfonate)/chitosan membranes with sulfonate ionic cross-linking and free sulfate groups: Preparation and application in alcohol dehydration. *J. Memb. Sci., 510*, 220–228. doi: 10.1016/j.memsci.2016.02.060.

59. Duque, S. A. C., Gómez, G. M. Á., Fontalvo, J., Jedrzejczyk, M., Rynkowski, J. M., & Dobrosz-Gómez, I., (2013). Ethanol dehydration by pervaporation using microporous silica membranes. *Desalin. Water Treat, 51*, 2368–2376. doi: 10.1080/19443994.2012.728053.

60. Peters, T. A., Fontalvo, J., Vorstman, M. A., Benes, N. E., Van, D. R. A., Vroom, Z. A. E. P., Soest-Vercammen, L. J. V., & Keurentjes, J. T. F., (2005). Hollow fibre microporous silica membranes for gas separation and pervaporation synthesis, performance and stability. *J. Membr. Sci., 248*, 73–80.

61. Sairam, M., Patil, M., Veerapur, R., Patil, S., & Aminabhavi, T., (2006). Novel dense poly(vinyl alcohol)–TiO$_2$ mixed matrix membranes for pervaporation separation of water-isopropanol mixtures at 30°C☆. *J. Memb. Sci., 281*, 95–102. doi: 10.1016/j.memsci.2006.03.022.

62. Yang, D., Li, J., Jiang, Z., Lu, L., & Chen, X., (2009). Chitosan/TiO$_2$ nanocomposite pervaporation membranes for ethanol dehydration. *Chem. Eng. Sci., 64*, 3130–3137. doi: 10.1016/j.ces.2009.03.042.

63. Sudhakar, H., Venkata, P. C., Sunitha, K., Chowdoji, R. K., Subha, M. C. S., & Sridhar, S., (2011). Pervaporation separation of IPA-water mixtures through 4A zeolite-filled sodium alginate membranes. *J. Appl. Polym. Sci., 121*, 2717–2725. doi: 10.1002/app.33695.

64. Ma, N., Wang, R., He, G., & Wang, Z., (2018). Preparation of high-performance zeolite NaA membranes in clear solution by adding SiO$_2$ into Al$_2$O$_3$ hollow-fiber precursor. *AIChE J., 64*, 2679–2688. doi: 10.1002/aic.16107.

65. Huang, Y., Baker, R. W., & Vane, L. M., (2010). Low-energy distillation-membrane separation process. *Ind. Eng. Chem. Res., 49*, 3760–3768. doi: 10.1021/ie901545r.

66. Kunnakorn, D., Rirksomboon, T., Siemanond, K., Aungkavattana, P., Kuanchertchoo, N., Chuntanalerg, P., Hemra, K., et al., (2013). Techno-economic comparison of energy

usage between azeotropic distillation and hybrid system for water-ethanol separation. *Renew. Energy, 51*, 310–316. doi: 10.1016/j.renene.2012.09.055.

67. Roza, M., & Maus, E., (2006). Industrial experience with hybrid distillation-pervaporation or vapor permeation applications. *Inst. Chem. Eng. Symp. Ser., 152*, 619–627.

68. Vane, L. M., & Alvarez, F. R., (2015). Effect of membrane and process characteristics on cost and energy usage for separating alcohol-water mixtures using a hybrid vapor stripping-vapor permeation process. *J. Chem. Technol. Biotechnol., 90*, 1380–1390. doi: 10.1002/jctb.4695.

69. Woodley, J. M., Bisschops, M., Straathof, A. J. J., & Ottens, M., (2008). Future directions for in-situ product removal (ISPR). *J. Chem. Technol. Biotechnol., 83*, 121–123. doi: 10.1002/jctb.1790.

70. Díaz, M., (1988). Three-phase extractive fermentation. *Trends Biotechnol., 6*, 126–130. doi: 10.1016/0167-7799(88)90102-3.

71. Zentou, H., Abidin, Z. Z., Yunus, R., Biak, D. R. A., & Korelskiy, D., (2019). Overview of alternative ethanol removal techniques for enhancing bioethanol recovery from fermentation broth. *Processes, 7*, 1–16. doi: 10.3390/pr7070458.

72. Widjaja, T., Altway, A., Permanasari, A. R., & Gunawan, S., (2014). Production of ethanol as a renewable energy by extractive fermentation. *Appl. Mech. Mater., 493*, 300–305. doi: 10.4028/www.scientific.net/AMM.493.300.

73. Honda, H., Taya, M., & Kobayashi, T., (1986). Ethanol fermentation associated with solvent extraction using immobilized growing cells of *saccharomyces cerevisiae* and its lactose-fermentable fusant. *J. Chem. Eng. Japan, 19*, 268–273. doi: 10.1252/jcej.19.268.

74. Kollerup, F., & Daugulis, A. J., (1986). Ethanol production by extractive fermentation-solvent identification and prototype development. *Can. J. Chem. Eng., 64*, 598–606. doi: 10.1002/cjce.5450640410.

75. Sandler, S. I., (2006). *Chemical, Biochemical, and Engineering Thermodynamics* (4th edn.). John Wiley & Sons Inc.

76. Kim, J. K., Iannotti, E. L., & Bajpai, R., (1999). Extractive recovery of products from fermentation broths. *Biotechnol. Bioprocess Eng., 4*, 1–11. doi: 10.1007/BF02931905.

77. Vane, L. M., (2008). Separation technologies for the recovery and dehydration of alcohols from fermentation broths. *Biofuels, Bioprod. Biorefining, 2*, 553–588. doi: 10.1002/bbb.108.

78. Matsumura, M., & Märkl, H., (1986). Elimination of ethanol inhibition by perstraction. *Biotechnol. Bioeng., 28*, 534–541. doi: 10.1002/bit.260280409.

79. Pérez, A. D., Rodríguez-Barona, S., & Fontalvo, J., (2018). Molecular toxicity of potential liquid membranes for lactic acid removal from fermentation broths using *Lactobacillus casei* ATCC 393. *Dyna, 85*, 360–366. doi: 10.15446/dyna.v85n207.72374.

80. Pérez, A. D., Gómez, V. M., Rodríguez-Barona, S., & Fontalvo, J., (2019). Liquid-liquid equilibrium and molecular toxicity of active and inert diluents of the organic mixture tri-iso-octylamine/dodecanol/dodecane as potential membrane phase for lactic acid removal. *J. Chem. Eng. Data.*

81. Xu, K., Lee, Y. S., Li, J., & Li, C., (2019). Resistance mechanisms and reprogramming of microorganisms for efficient biorefinery under multiple environmental stresses. *Synth. Syst. Biotechnol., 4*, 92–98. doi: 10.1016/j.synbio.2019.02.003.

82. Marták, J., Sabolová, E., Schlosser, Š., Rosenberg, M., & Kristofíková, L., (1997). Toxicity of organic solvents used in situ in fermentation of lactic acid by *Rhizopus arrhizus*. *Biotechnol. Tech., 11*, 71–75. doi: 10.1023/A:1018408220465.

83. Heipieper, H. J., Weber, F. J., Sikkema, J., Keweloh, H., & De Bont, J. A. M., (1994). Mechanisms of resistance of whole cells to toxic organic solvents. *Trends Biotechnol., 12*, 409–415. doi: 10.1016/0167-7799(94)90029-9.

84. Dyrda, G., Boniewska-Bernacka, E., Man, D., Barchiewicz, K., & Słota, R., (2019). The effect of organic solvents on selected microorganisms and model liposome membrane. *Mol. Biol. Rep., 46*, 3225–3232. doi: 10.1007/s11033-019-04782-y.

85. Offeman, R. D., Stephenson, S. K., Franqui, D., Cline, J. L., Robertson, G. H., & Orts, W. J., (2008). Extraction of ethanol with higher alcohol solvents and their toxicity to yeast. *Sep. Purif. Technol., 63*, 444–451. doi: 10.1016/j.seppur.2008.06.005.

86. Roy, S. R., Bhattacharya, P., & Sirkar, A., (2013). Studies on Toxicity Effect of Solvents on Growth of *Saccharomyces cerevisiae* (NCIM 3186). *Indian Chem. Eng., 55*, 247–257. doi: 10.1080/00194506.2013.851862.

87. Minier, M., & Coma, G., (1981). Production of ethanol by coupling fermentation and solvent extraction. *Biotechnol. Lett., 3*, 405–408. doi: 10.1007/BF01134098.

88. Gyamerah, M., & Glover, J., (1996). Production of ethanol by continuous fermentation and liquid-liquid extraction. *J. Chem. Technol. Biotechnol., 66*, 145–152. doi: 10.1002/(SICI)1097-4660(199606)66:2<145::AID-JCTB484>3.0.CO;2-2.

89. Kapucu, H., & Mehmetoğlu, Ü., (1998). Strategies for reducing solvent toxicity in extractive ethanol fermentation. *Appl. Biochem. Biotechnol. - Part A Enzym. Eng. Biotechnol., 75*, 205–214. doi: 10.1007/BF02787775.

90. Santos, E. L. I., Rostro-Alanís, M., Parra-Saldívar, R., & Alvarez, A. J., (2018). A novel method for bioethanol production using immobilized yeast cells in calcium-alginate films and hybrid composite pervaporation membrane. *Bioresour. Technol., 247*, 165–173. doi: 10.1016/j.biortech.2017.09.091.

91. Ivanova, V., Petrova, P., & Hristov, J., (2011). Application in the ethanol fermentation of immobilized yeast cells in matrix of alginate/magnetic nanoparticles, on chitosan-magnetite microparticles and cellulose-coated magnetic nanoparticles. *Int. Rev. Chem. Eng., 3*, 289–299.

92. Vanda, H., Dai, Y., Wilson, E. G., Verpoorte, R., & Choi, Y. H., (2018). Green solvents from ionic liquids and deep eutectic solvents to natural deep eutectic solvents. *Comptes. Rendus. Chim., 21*, 628–638. doi: 10.1016/j.crci.2018.04.002.

93. López-Porfiri, P., Gorgojo, P., & Gonzalez-Miquel, M., (2020). Green solvent selection guide for biobased organic acid recovery. *ACS Sustain. Chem. Eng., 8*, 8958–8969. doi: 10.1021/acssuschemeng.0c01456.

94. Leon, J., Palacios-Bereche, R., & Nebra, S. A., (2016). Batch pervaporative fermentation with coupled membrane and its influence on energy consumption in permeate recovery and distillation stage. *Energy, 109*, 77–91.

95. Bassetto, N. Z., (2006). *SEPPA-Specialist System for Ethanol Production Plant*. Universidade Estadual de Campinas.

96. Leon, J. A., (2015). Análise do desempenho de membranas de pervaporação no processo convencional de fermentação para produção de etanol. Universidade Federal do ABC.

97. Vane, L. M., (2005). A review of pervaporation for product recovery from biomass fermentation processes. *J. Chem. Technol. Biotechnol., 80*, 603–629. doi: 10.1002/jctb.1265.

98. Cho, C. W., & Hwang, S. T., (1991). Continuous membrane fermentor separator for ethanol fermentation. *J. Memb. Sci., 57*, 21–42. doi: 10.1016/S0376-7388(00)81160-6.

99. Fan, S., Xiao, Z., Zhang, Y., Tang, X., Chen, C., Li, W., Deng, Q., & Tao, P., (2014). Enhanced ethanol fermentation in a pervaporation membrane bioreactor with the convenient permeate vapor recovery. _Bioresour. Technol., 155_, 229–234.
100. Kargupta, K., Datta, S., & Sanyal, S. K., (1998). Analysis of the performance of a continuous membrane bioreactor with cell recycling during ethanol fermentation. _Biochem. Eng. J., 1_, 31–37. doi: 10.1016/S1369-703X(97)00006-5.
101. Nakao, S., Saitoh, F., Asakura, T., Toda, K., & Kimura, S., (1987). Continuous ethanol extraction by pervaporation from a membrane bioreactor. _J. Memb. Sci., 30_, 273–287. doi: 10.1016/S0376-7388(00)80123-4.
102. Groot, W. J., Van, D. Q. C. E., & Kossen, N. W. F., (1984). Pervaporation for simultaneous product recovery in the butanol/isopropanol batch fermentation. _Biotechnol. Lett., 6_, 709–714. doi: 10.1007/BF00133061.
103. Aguilar-Valencia, D. M., Gómez-García, M. A., & Fontalvo, J., (2012). Effect of pH, CO_2, and high glucose concentrations on polydimethylsiloxane pervaporation membranes for ethanol removal. _Ind. Eng. Chem. Res., 51_, 9328–9334.

CHAPTER 9

Design and Engineering Parameters of Bioreactors for Production of Bioethanol

DAVID FRANCISCO LAFUENTE-RINCÓN,[1]
INTY OMAR HERNÁNDEZ-DE LIRA,[2] HÉCTOR HERNÁNDEZ-ESCOTO,[3]
MARÍA ALEJANDRA SÁNCHEZ-MUÑOZ,[4]
HÉCTOR HUGO MOLINA CORREA,[4]
CRISTIAN EMANUEL GÁMEZ-ALVARADO,[4]
PERLA ARACELI MELÉNDEZ-HERNÁNDEZ,[3] and
JAVIER ULISES HERNÁNDEZ-BELTRÁN[4]

[1]*Bioprocesses and Bioprospecting Laboratory, Biological Sciences Faculty, Autonomous University of Coahuila, Torreón–27000, Coahuila, México*

[2]*Faculty of Mechanical and Electrical Engineering, Autonomous University of Coahuila, Torreón–27000, Coahuila, México*

[3]*Chemical Engineering Department, University of Guanajuato, Noria Alta s/n, Guanajuato, 36050, Mexico*

[4]*Bioremediation Laboratory, Biological Sciences Faculty, Autonomous University of Coahuila, Torreón–27000, Coahuila, México, Tel.: +52-871-7571-785, E-mail: ulises.hernandez@uadec.edu.mx (J. U. Hernández-Beltrán)*

ABSTRACT

As an effort to find renewable sources of energy, the ethanol production from lignocellulosic biomass (LCB) represents an environmentally friendly alternative to fossil fuels. However, the recalcitrant nature of lignin on holocellulose tackles the biomass use as feedstock and forces the use of pretreatment methods in a first step to switch the raw substrate into digestible cellulosic solids. In a second step, enzymatic hydrolysis process transforms the holocellulose into sugars, which in turn are converted to bioethanol by using yeast or bacteria in the final fermentation step. In these three steps, there are many challenges to face as the conversion of a high concentration of raw materials, saving chemical and biological reagents, and the reduction of the processing time. Therefore, it brings into play several engineering tasks such as the configuring of bioreactors in the aspects of geometry and equipment to achieve effective transfer of mass and heat, and the setting of process conditions. In this context, this chapter reviews the main engineering parameters and features to be considered in the design of bioreactors for the steps of pretreatment, enzymatic hydrolysis and fermentation.

9.1 INTRODUCTION

The fast growth of the human population has increased the demand for energy and consequently bringing into play concerns as global warming (GW) and reduction of natural sources [1]. One of the main aspects that the world faces is the high consumptions of fossil fuels which represent around 80% of the global energy usage [2]. Thus, the R&D in renewable sources of energy has become a priority. Biofuels are produced from biomass and have been evaluated as an alternative to reduce both the dependence of fossil fuels and environmental pollution [3]. In recent decades, they are emerged as an effective strategy to deal with the matter related to the decreasing of fossil fuels and the increasing of greenhouse gas (GHG) emissions [4]. Biofuels can be classified in terms of generation according to the raw material used. First generation biofuels are derived from food crops which unfortunately creates ethical concerns and the increase of food prices [5]. To solve these issues, the second-generation biofuels are produced from inedible LCB such as agro-residues, forestry wastes, herbaceous biomass, wood wastes or municipal wastes [6]. Among the most promising biofuels, the bioethanol of second generation is considered as a fuel substitute for road transportation due to it is an eco-friendly product with a higher octane number and emits fewer concentrations of GHGs than gasoline [7].

The LCB is converted into bioethanol through the steps of pretreatment, enzymatic hydrolysis, and fermentation that usually are carried out in a bioreactor which is a controlled reaction system where an optimum environment is provided to enable the highest possible yield [8]. Inherently, different configurations and operation modes of bioreactors have been explored, by studying the impact of different process conditions involved such as substrate concentration, catalyst concentration, temperature, pH, reaction volume, mixing, among others [9]. A bioreactor commonly used is the stirred tank reactor (STR) that can be operated under the different modes batch, fed-batch or continuous, and other innovative bioreactor designs have been proposed to process lignocellulosic materials [10]. It draws the convenience to consider the design and engineering parameters involved in the bioreactors for bioethanol production in the steps of pretreatment, enzymatic hydrolysis, and fermentation. Thus, in this book chapter, aspect of geometric configuration in the bioreactors of each step and operation conditions to perform the processes are discussed.

9.1.1 LIGNOCELLULOSIC BIOMASS (LCB)

Lignocellulosic biomass is an attractive renewable feedstock for bioethanol production due to its huge abundance, low cost and sustainable supply [11]. Lignocellulose is composed by three main biopolymers such as cellulose, hemicellulose, and lignin [12]. Table 9.1 shows the chemical composition for different LCB. In general, the fractions of cellulose, hemicellulose, and lignin corresponds to 25–55%, hemicellulose 5–35% and 5–30%, respectively.

TABLE 9.1 Chemical Composition of Different Lignocellulosic Biomass According to the Fraction of Cellulose, Hemicellulose, and Lignin

Type of Lignocellulosic Biomass	Chemical Composition (% w w^{-1})			References
	Cellulose	Hemicellulose	Lignin	
Bamboo residues	38.4 ± n.a.	21.3 ± n.a.	29.6 ± n.a.	[112]
Barley straw	45.7 ± 0.2	32.6 ± 0.5	5.2 ± 0.0	[11]
Oak sawdust	44.7 ± 0.3	14.8 ± 1.3	26.7 ± 0.8	[13]
Plantain pseudostem	34.7 ± 0.5	10.1 ± 0.1	30.8 ± 0.3	[14]
Rapeseed straw	30.0 ± 2.2	12.7 ± 0.7	21.1 ± 0.2	[113]
Rice straw	36.3 ± 1.2	27.9 ± 1.3	14.11 ± 0.5	[114]
Sorghum straw	26.9 ± 1.2	32.5 ± 1.9	10.1 ± 1.8	[15]
Spruce sawdust	55.4 ± 1.3	4.2 ± 0.4	28.7 ± 1.0	[13]
Sugarcane bagasse	54.2 ± 2.2	27.9 ± 3.6	10.3 ± 0.1	[16]
Wheat straw	39.25 ± 1.1	33.65 ± 1.1	7.35 ± 0.3	[115]

Cellulose is a linear polymer that is composed of D-Glucose subunits linked by β-1,4 glycosidic bonds, which form the cellobiose dimer. This long chain is linked together by hydrogen bonds and van der Waals forces [17]. Cellulose is generally present in crystalline form (80%) which is hardly to hydrolyze and the rest in amorphous form being more susceptible to enzymatic degradation [18]. On the other hand, hemicellulose is a polysaccharide with a lower molecular weight than cellulose and is composed of different sugars forming shorter and branched chains. These sugars can be divided into different groups such as pentoses (D-xylose, L-arabinose), hexoses (D-glucose, D-mannose, D-galactose), hexuronic acids (D-glucuronic acid, 4-O-methyl-glucuronic and D-galacturonic) and deoxyhexoses (rhamnose and fucose). The main chain of a hemicellulose consists of a single unit (homopolymer), such as xylan, or two or more units (heteropolymer), such as glucomannan. Sugars are linked by β-1,4 glycosidic bonds and sometimes by β-1,3 glycosidic bonds. Hemicellulose is amorphous, therefore, the differences in the composition of sugars, the presence of shorter chains and the ramifications of the main chains make the hemicellulose structure easier to hydrolyze than cellulose [19].

Lignin is the most abundant phenolic biopolymer in nature and is the third major component in lignocellulosic materials. Lignin is related to both hemicellulose and cellulose, forming a physical seal that is an impenetrable barrier in the plant cell wall. It is present in the cell wall to provide structural support, impermeability, and resistance against microbial attack and oxidative stress. Lignin is an optically inactive, non-water soluble, amorphous heteropolymer that is formed from phenylpropane bound through non-hydrolysable bonds. This polymer is generally synthesized by three different phenylpropanoid derivatives: coniferyl (guaiacyl propanol), coumaric (p-hydroxyphenyl propanol) and sinapyl (syringyl propanol) alcohols; synthesized from phenylalanine through various derivatives of cinnamic acid. Phenylpropanoid alcohols are bound in a polymer by the action of enzymes that generate intermediates in the form of free radicals. This heterogeneous structure is linked by C-C and aryl-ether bonds linked to β-aryl-aryl-glycerol ether, these being the predominant structures [20–22].

9.1.2 PROCESS OUTLINE OF BIOETHANOL PRODUCTION

Figure 9.1 shows the general process for bioethanol production from LCB, in which three main steps appear: (1) pretreatment methods have been identified as the first barrier in is used to break the lignin wall and disrupt the crystallinity of cellulose to allow the accessibility of the enzymes; (2) enzymatic hydrolysis process consists in the conversion of holocellulose (cellulose plus hemicellulose) to reducing sugars through the action of enzymes on pretreated biomass; (3) fermentation is the step in which the reducing sugars like glucose is converted to ethanol by using yeast or bacteria [23].

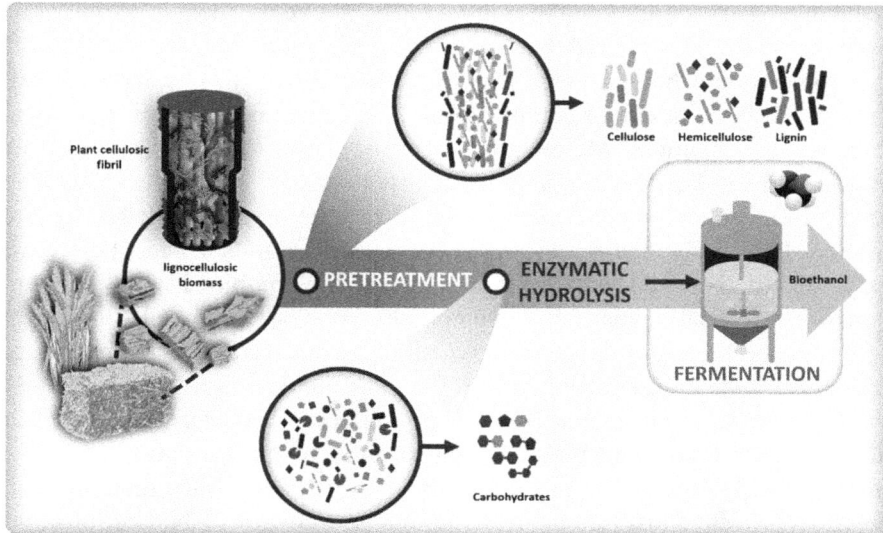

FIGURE 9.1 Outline of bioethanol production process.

9.1.2.1 PRETREATMENT

The conversion of LCB into added value bioproducts represent a prominent strategy to look forward in the circular bioeconomy. However, the pretreatment of LCB still be the most complex step in the conversion to biofuels from the economic point of view. LCB due to its chemical composition are consider as a recalcitrant biomass, resistant to the breakdown by biological, physical, and chemical process [24]. Various strategies have been explored to overcome these restrictions, including novel methods of pretreatments and the utilization of innovative bioreactors configuration. Several pretreatments' technics for biomass have been studied, which can be divided into, mechanical processes (particle size reduction), chemical processes (use of diluted acids or alkalis), physicochemical processes (steam explosion (SE) and hot water), biological processes (microbial consortium or enzymes) or the combination among those [25]. Albeit these techniques enhance significant cost to the process at commercial scale.

The most used physical method is a mechanical one which consists in the reduction of the particle size and it represents the first action in the processing of LCB because it has been established as an essential to increase the accessible surface area and the porosity of the particles, in addition to reducing the crystallinity of the cellulose [26]. However, there is no universal optimum particle size and before to start the processing of every LCB this issue should be addressed [6]. After that, chemical, physicochemical, biological pretreatment, or combination of them must be followed. Chemical and physicochemical pretreatment methods use the reagents at low concentrations, i.e., a common practice is the use of dilute acid like sulfuric acid on LCB and carry out at high temperature in an autoclave for several minutes [27]. As well, the combination of reagents results in a most effective way, i.e., an alkali reagent like sodium hydroxide combined with hydrogen peroxide as oxidative reagent which not only improves the depolymerization of lignin and the digestibility of lignocellulosic materials at a low operating cost but also maximizes the use of cellulose and hemicellulose [28]. Alkaline-oxidative is an environment-friendly pretreatment that can be carried out at mild temperature and pressure, which leads to the formation of minor inhibitors [16]. Regardless of the type of LCB, if ethanol is the final product, the pretreatment must guarantee a maximum recovery of holocellulose to achieve high concentration of reducing sugars. Furthermore, pretreatment has been identified as the crucial step, both technically and economically, in the conversion of global LCB due to its strong influences to restrict the other steps in the chain of bioethanol production.

9.1.2.2 ENZYMATIC HYDROLYSIS

Although lignocellulosic materials can be hydrolyzed using acids or enzymes, the process must consider the use of environmentally friendly and economically viable technologies. Enzyme-based hydrolysis is presented as the best option due to its high conversion efficiency and minimal production of by-products like inhibitors which negatively affect the performance of further fermentation process [29]. In addition, the operating conditions are moderate, i.e., temperatures used in enzymatic hydrolysis are between 45–60°C. It is not corrosive for the bioreactor used and the process requires a lower energy consumption. In contrast, acid hydrolysis requires a higher temperature like 120°C or more and produces furfurals and hydroxymethylfurfural as inhibitors for subsequent fermentation process [15].

Polysaccharide hydrolysis involves at least three different enzymes which are simultaneously working to degrade the polymer to its monomer. Enzymatic hydrolysis of cellulose is performed by cellulases or cellulolytic enzymes that catalyzes the cleavage of β-1,4 glycosidic bonds, present in the insoluble cellulose, to produce glucose. Cellulase enzymes comprise endo-1,4-β-D-glucanases, exo-1,4-β-D-glucanases (cellobiohydrolases) and β-glucosidases. The main reaction mechanism of cellulases is depicted in Figure 9.2. First, endoglucanases start breaking β-1,4 glycosidic bonds by the addition of a water molecule, releasing cellodextrin with a reducing and a non-reducing end from the amorphous regions of the cellulose backbone. Then, exoglucanases hydrolyzes cellodextrins, from its reducing and a non-reducing end, to produce cellobiose (disaccharide of two glucose units). Finally, β-glucosidases catalyzes the cleavage of cellobiose producing two soluble monomers of glucose [30–33]. Although enzymes are biological catalysts that perform chemical reactions efficiently, its activity is strongly regulated for different factors as temperature and pH, as well as, adsorption, mixing conditions, and solid-liquid ratio that will be further discussed.

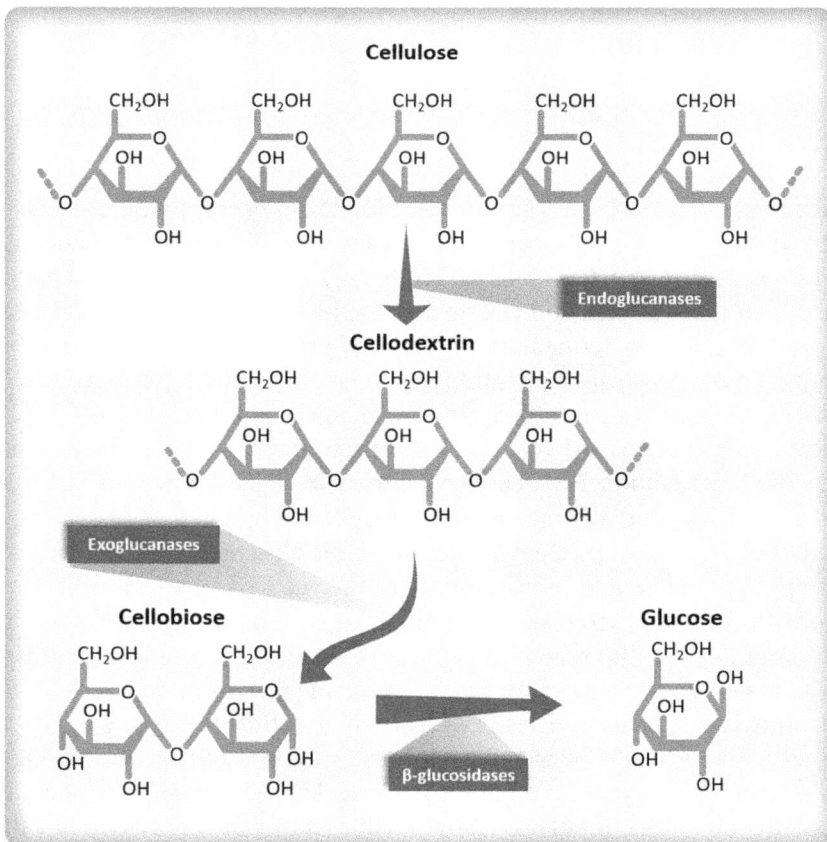

FIGURE 9.2 Action mechanism of cellulase enzymes on cellulose fraction.

On the other hand, enzymatic hydrolysis can be influenced by the concentrations of substrate and final product, the enzymatic activity, and the reaction conditions [33]. Recently, one of the main strategies to reach a high concentration of reducing sugars is by using a high concentration of substrate, however, this aspect affects directly to the mixing of the reaction and simultaneously to the process conditions such as pH and temperature cannot be constant along the time, even the adjustment of these parameters is difficult [34]. Thus, there is no possible to ensure the best performance considering that the mixing trouble begin from 7% w v^{-1} of substrate concentration [14]. Therefore, the research activities in enzymatic hydrolysis have been focused to reach high solids concentration of substrate between 20–40% w v^{-1} using engineering strategies as fed-batch or semi-continuous configurations for the reaction instead of a batch one and to deeply study the relationship of the several parameters which affect the enzymatic hydrolysis process to increase the profitability of fermentable sugars production [35].

9.1.2.3   FERMENTATION

In contrast to conventional ethanol production processes from starch and sugar raw materials, lignocellulosic sugars become a highly challenging environment due the fermentable broth may contain several compounds released in pretreatment and enzymatic hydrolysis steps which could affect the performance on fermentation process [36]. This fact requires robust fermenting microorganisms with high tolerance to inhibitors derived from biomass, ethanol concentration, and mechanical and osmotic stress [37]. *Saccharomyces cerevisiae* yeast is the most used microorganism for industrial alcoholic fermentation [38]. The most attractive characteristics of S. cerevisiae are: (i) it is generally recognized as safe (GRAS); (ii) it can consume all types of hexoses; (iii) it achieves ethanol yields close to theoretical; and (iv) it can produce ethanol at concentrations as high as 13% (v v^{-1}). Furthermore, the resistance of S. cerevisiae to lignocellulose-derived inhibitors is high, and therefore, it is the preferred microorganism for lignocellulosic ethanol production [39]. Despite showing all these interesting characteristics, the main drawback of S. cerevisiae is its inability to ferment xylose, which is the second most important sugar after glucose in lignocellulose. The challenge in the fermentation process to make large-scale production is the obtention of bioethanol broths of at least 4% w v^{-1} due from that concentration value, and the purification process starts to be feasible from an economical point of view. This means that fermentation process must be loaded with a broth of at least 80 g L^{-1} of six-carbon reducing sugar, considering the stoichiometric conversion of glucose to ethanol is 0.51 g g^{-1} [15].

9.2 BIOREACTORS FOR PRETREATMENT PROCESS

The reactor configuration, process conditions, and the operation mode during the pretreatment conversion process are essential to gap the high yields and productivity in the biomass conversion at industrial levels [40]. More recently, different studies have been conducted in several bioreactor configurations with the aim to obtain higher yields of glucose in the pretreatment stage [8].

The pretreatment step has many methods for its development and each one present their own phenomenology into the reaction. The mechanical method is the most used among the physical pretreatments [41] but does not require a bioreactor for reduction of particle size but rather a milling equipment, however, the reduction of particle size is a process condition that can limit the performance of the whole bioethanol production chain [42]. For this reason, a study of the optimum particle size for each LCB is a key issue [26]. Other common physical pretreatments are the microwave and hydrothermal methods. Microwave reactors (MWR) have noticed interest for lignocellulosic pretreatment. Microwaves are a kind of electromagnetic radiation that is shaped like energy propagating in a vacuum in the absence of any material in motion. Microwaves are originated by the separation of two charges positive and negative of equal magnitude which are separated by a fixed distance. MWR are mainly composed by a generator, a cavity section, a waveguide, and an advanced control monitoring system. The principle of these reactors is based on Maxwell's equation [43]. Hermiati et al. [44] studied microwave-assisted acid pretreatment of sugarcane trash, with a maximum work temperature of 180°C and biomass capacity of 100 mL. The sugar production from the pretreated LCB was 20.2 g L^{-1} of xylose, which offers a good yield scale under these conditions. In addition, Peng et al. [45] designed a pilot-scale continuous microwave irradiation reactor (MWIR). The temperature range of operation is between 0 to 300C, with a maximum power of 6 kW. MWIR ethanol yield indicated a value of 31.2 g 100 g^{-1} of pretreated corn straw. Table 9.2 describes the main characteristics of reactors system for conversion of LCB considering the reaction system, including the agitation method, reaction volume, reactor configuration, and the reducing sugars or ethanol yield. On the other hand, an emerging technology to convert wet biomass into valuable products due to the action of temperature (200–370°C) and high pressure (4–20 MPa) is hydrothermal liquefaction reactors

TABLE 9.2 Main Characteristics of Reactors System for Conversion of Lignocellulosic Biomass

Biomass	Reactor System	Agitation	Tank Volume	Configuration Process	Sugars[a] or Ethanol[b] Concentration Yield	References
Agave bagasse	Tubular reactor	Rushton-type stirred blades	1.0 L	Batch SSF	[b]37.6 g L^{-1}	[116]
Sugarcane bagasse	Hydrodynamic cavitation	Recirculation pump	2.0 L	Fed-batch SSF	[b]28.44 g L^{-1}	[48]
Dry corn	Hydrodynamic cavitation	Low-shear impeller	379 L	Batch SSF	[b]35.8% w w^{-1}	[49]
Reed grass	Hydrodynamic cavitation	Recirculation pump	150 ml	Batch SSF	[b]25.9 g L^{-1}	[50]
Apple pomace	Bubble column bioreactor	Air bubbles	–	Batch	[a]51.8 g L^{-1} fermentable sugar	[117]
Agave bagasse	High pressure stirred reactor	Anchor impeller	1.0 L	Batch	[a]3.7 g L^{-1} xylose	[8]
Sugarcane bagasse	Plasma in liquid reactor	N/A	1.0 L	Batch	[a]51.3% glucose	[118]
Sargassum horneri	Stirred tank bioreactor	Rushton impeller	2.0 L	Batch S-SSF	[b]2.89 g L^{-1}	[119]
Sugarcane bagasse	Supercritical CO$_2$ reactor	CO$_2$ bubbles	100 ml	Batch	[a]1.17 g L^{-1} fermentable sugars	[120]
Tobacco stalk	Rotating reactor	Rotation	2 L	Batch	[b]27.5 g Kg^{-1} TS	[121]
Sugarcane bagasse	Fluidized bed reactor	Air bubbles	2 L	Batch SSCF	[b]0.34 g g^{-1}	[122]
Birch biomass	Steam explosión react	N/A	4 L	Batch SSF	[b]80 g L^{-1}	[123]
Wheat Straw	High pressure reactor	Rotation	1 L	Batch	[a]72.7% glucose	[21]
Sugarcane trash	Microwave reactor	N/A	100 ml	Batch	[a]20.2 g L^{-1} xylose	[44]
Corn straw	Microwave irradiation reactor	N/A	4 m^3	Batch SSF	[b]31.2 g 100 g LCB^{-1}	[45]
Lignin-rich stream	Hydrothermal liquefaction reactor	N/A	43 ml	Batch	–	[47]

(HTLR) [46]. Several studies have revealed the potential of HTLR for bagasse pretreatment to produce a liquid product, known as bio-oil. Miliotti et al. [47] evaluated the valorization of lignin-rich stream from lignocellulosic ethanol production at industrial scale using hydrothermal liquefaction. The stainless-steel reactor was operated in batch mode with a volume capacity of 27 mL. It has been observed that reactors configuration and operation mode, in order to improve the LCB pretreatments, should consider the development of new technologies and novel methods to overcome the lab-scale into industrial level. Special attention might be focused on the energy reduction process and the improvement of sugars-bioethanol yields.

The design of bioreactors in chemical and physicochemical pretreatments differs in the temperature parameter. If the temperature is moderate, between 25–70°C, stirred tank bioreactors (STBR) are normally used. If the temperature to be used is higher, for example, more than 100°C, the bioreactors used also tend to withstand high pressures. The substrate concentration plays an important role to determine the type of stirring and can vary from a moderate concentration of 5% to a high concentration of >12% (w v^{-1}) [51]. This entails to the use of mechanical mixing and the increasing of speed of the stirring. The more concentration, the faster the stirring speed, which leads to higher energy requirement. For this reason, the engineering strategies used for processing high concentrations of substrate in chemical and physicochemical pretreatments are fed-batch or semi-batch and continuous modes of operation [10]. Besides, bioreactors must withstand to acid and alkaline values of, i.e., the value of pH using an acid reagent like sulfuric acid is <1 or by using an alkaline reagent like sodium hydroxide is >11 [27].

Another physical pretreatment but also can be performed as physio-chemical one is hydro-dynamic cavitation. It involves the generation of cavitation in freely flowing liquid which is constricted with a venture tube or orifice plate [52]. Cavitation phenomenon can be defined as the formation and subsequent growing and collapse of microbubbles due to the changes in pressure at constant temperature. Bernoulli's equation is the main principle in this technique, where the drop in pressure at the constriction below the vapor pressure of the flowing liquid generates bubbles or hydrodynamic cavities [53]. Hydrodynamic cavitation reactors (HCR) have been effectively applied for delignification of several LCB, such as reed grass [50], sugarcane bagasse [48], dry mill corn [49] among others. The great advantage of HCR is its much lower energy consumption compared with ultrasonic reactor, as well it has been observed that this pretreatment affects both delignification and crystallinity of cellulose to increase yields in bioethanol production. Terán et al. [48]

proposed hydrodynamic cavitation (HC)-assisted alkaline hydrogen peroxide pretreatment in a continuous process, the system contained two devices to generate cavitation. HCR consisted in a 2 L stainless steel reactor, provided by two centrifugal pumps of 1.3 hp.

Also, in the system was included two tanks for lignocellulosic biomass preparation and for the cavitation devices consisted in a plate of 8 to 24 orifices at different configuration. On the other hand, Ramirez-Cadavid et al. [49] built a pilot scale HCR. The system included a 379 L heat-jacketed vessel able to maintain temperatures from 82°C to 94°C. The agitation was provided by a low shear impeller mixer.

9.3   BIOREACTORS FOR ENZYMATIC HYDROLYSIS PROCESS

9.3.1   *OPERATION MODE AND GEOMETRIC CONFIGURATION*

In case of the saccharification bioreactor, operation mode and geometric configuration must be carefully selected in order to optimize the maximal lignocellulosic bioconversion into added-value products. Hence, the bioreactor must accomplish specific requirements to ensure the efficient use of the substrate and to fulfill the enzyme conditions which involves the prior study of the biological system. When the process works is optimized, it is expected to obtain high rates of glucose yield per volume of reactor by means of controlling certain parameters mainly temperature and pH [54]. However, enzyme-substrate ratio has to be guaranteed, as well as the agitation mode to ensure optimal interaction among cellulases and cellulose in order to increase the hydrolysis efficiency [55].

There are three operation modes according to the substrate feeding process: 1) discontinuous, semi-continuous, and continuous mode. Discontinuous or batch mode is a closed process where substrate is initially added without additional feed or removal of secondary products. This mode is commonly applied in many industries for bioethanol production [56]. Semi-continuous or fed-batch mode involves the intermittent addition of fresh nutrients at a certain period during the hydrolysis stage. This mode has several advantages as promoting a higher bioconversion process than batch mode, the constant mixing process prevents the rising of substrate viscosity, it decreases the enzymatic inhibition rates by reducing the unspecific enzyme binding and it provides a longer time to hydrolyze the biomass into fermentable sugars [57–59]. Continuous mode consists in the continuously feeding of the reactor with the concomitant removal of by-products from the reaction vessel, thus the liquid volume of the reactor keeps constant. However, this process can be affected by substrate inhibition [60].

Different types of configurations in the bioreactor fields had been employed including well-known configurations as STBR, which is the most commonly used for enzymatic processes, and the last advances on STRB which focus on the conversion of energy crops and residual materials into sugars and ethanol include separate hydrolysis and fermentation (SHF), simultaneous saccharification and fermentation (SSF), simultaneous saccharification and co-fermentation (SSCF) and consolidated bioprocessing (CBP) (Figure 9.3).

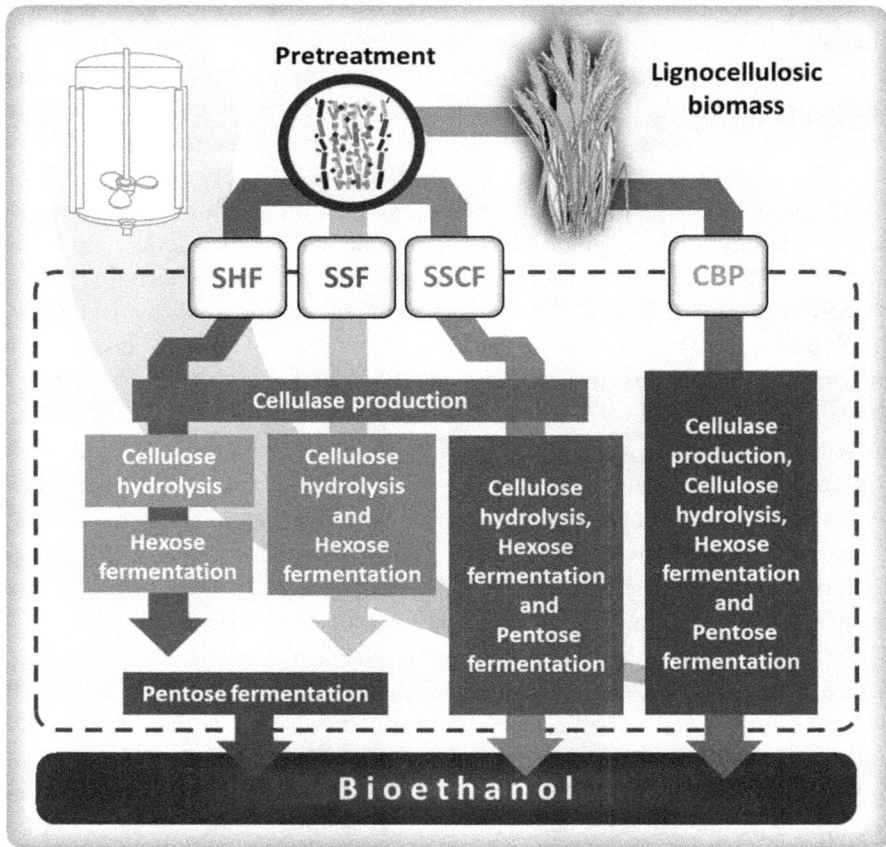

FIGURE 9.3 Different types of bioreactor configurations for sugars and bioethanol production. [*Abbreviations*: SHF: separate hydrolysis and fermentation; SSF: simultaneous saccharification and fermentation; SSCF: simultaneous saccharification and co-fermentation; CPB: consolidated bioprocessing].

STBR consist in cylindrical vessels with one or more impellers assembled with an external motor. This configuration is widely used in a batch mode enhancing the mixture homogeneity where biochemical reactions occur. These types of bioreactors are equipped with baffles to prevent vortexes, heating-cooling systems, or thermal jackets, and can be built, depending on the size, of glass (laboratory scale), stainless steel or carbonate [61].

SHF carries out the enzymatic hydrolysis and fermentation in two independent reactors for saccharification and fermentation processes. This process can be subject to different conditions since the reactors are separate which allows to enzymatic hydrolases work at optimal temperature, which is higher than fermentation, and to reduce the quantity of saccharifying enzymes than other simultaneous processes [62, 63].

SSF combines enzymatic hydrolysis and fermentation to obtain value added products in a single step, which can reduce significatively the time and cost of the process. Another advantage is the risk reduction of inhibitory compounds of saccharification. Hence, SSF has been widely used for production of biofuels from lignocellulosic materials. However, in comparison with SHF pH and temperature parameters cannot be individually set and an equilibrium point might be necessary for the process to work properly [64, 65].

SSCF is referred to the process where enzymatic hydrolysis can be conducted simultaneously with the co-fermentation of glucose and xylose in a single process. This process take advantage of the biochemical affinities due glucose and xylose compete for the same transport systems in the co-fermentation of xylose. Thus, SSCF is promising for ethanol production [66]. SSCF have some advantages over SHF which includes the removal of hydrolysis end-products, short processing time and low contamination risk. Besides, it has been proved its high efficiency and rates of ethanol yield [67].

CBP is a system where the enzyme production, substrate hydrolysis and fermentation are performed in one step by lignocellulolytic microorganisms. The production costs by using this process are lower for biofuel obtention, due implies lower energy inputs and higher conversion efficiencies in comparison to SHF. However, CBP has certain disadvantages, mainly involved with enzyme production because monocultures have shown some limitations. While co-cultures might have different enzyme production requirements due bacteria can ferment LCB at mesophilic or thermophilic temperatures and thus, different rates of cellulose hydrolysis and biofuel production can be observed [68, 69].

Jacket or coil for heating/cooling system is one of the most important instruments for laboratory-scale bioreactors, the vessel can be positioned into a thermal incubator or can be adapted to an electrical heater resistance, also the external wall of the vessel can be equipped with a thermal jacket in order to maintain the optimal temperature and equilibrium for the biological system. Otherwise, large-scale bioreactors possess an internal coil heating-cooling system.

On the other hand, as seen before, STRB is one of the most common systems for lignocellulosic bioconversion and in order to heat transfer and homogenization be adequate different types of agitation have been studied. It is worth to mention that agitation promotes the equal distribution and interaction among all the biological matter and the mixing efficiency will result in an optimal product yield. As seen in Figure 9.4, there are different types of impellers that can achieve different mixing patterns as axial or radial, depending on the movement direction of the components.

Three blade elephant ear impellers (Figure 9.4(A)) are commonly used in STRB. It has been showed that a configuration of two impellers of this type have two lowered energy consumption, diminished mixing time and enhanced homogenization of the medium. At the same time, this configuration has produced higher glucose conversion by diminishing the inhibition rate of soluble by-products [70]. Peg mixer (Figure 9.4(B)) has been recognized for being more convenient on enzymatic saccharification at high solid loadings where viscosity is challenging, due it promotes the enzymatic binding on the cellulose substrate and overcomes mass transfer limitations [71].

On the other hand, helical ribbon impellers (Figure 9.4(C)) are used for non-Newtonian fluids and substrates with high viscosity that present challenges for the mixing process. This impeller design allows to provide and efficient mixing not only for this type of fluids but an alternative for enzymatic hydrolysis at high solid loadings [72]. Besides, it has been demonstrated that this impeller enhances glucose yields and ethanol production by using less energy [73].

Finally, Rushton turbines are commonly present in most of the vessels which promotes a radial flow mixing pattern (Figure 9.4(D)). These impellers have been recently combined with other configurations as the elephant ear impeller due combines the axial and radial flow overcoming the disadvantage of Rushton turbines, which creates a staged mixing patters that produces an irregular enzyme-substrate interaction [70].

9.3.2 PROCESS CONDITIONS

The main parameters that must be under tight control and play a crucial role in the effectiveness of enzymatic catalysis are substrate concentration, enzyme concentration, temperature, pH, and mixing velocity.

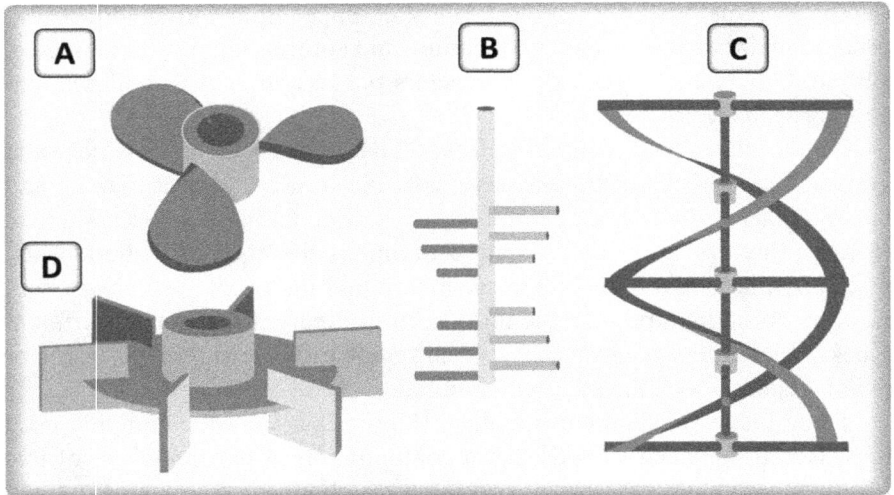

FIGURE 9.4 Different impeller designs for bioreactor systems. (A) Elephant ear impeller; (B) peg mixer; (C) helical ribbon impeller; and (D) Rushton turbine.

First, the enzymatic hydrolysis development depends on the adsorption of cellulases on the surface of the lignocellulosic material, most endoglucanases and exoglucanases have a carbohydrate-binding module that enhances the binding process. This process takes about 10 to 15 min, which is related to the reaction rate [74]. It is also known that cellulolytic enzymes activities are optimal at a range from 45 to 55°C and pH in a range of 4–5 [60, 75]. Besides, mixing conditions and solid-liquid ratio are essential in saccharification of polysaccharides due it is necessary to mass and heat transfer and to enhance the interaction among substrate-enzyme [76].

It is well-known that during the hydrolysis process where glucose is available to further fermentation, there is some risk of feedback inhibition by the end-product of fermentation. This problem has been overcome by SHF configuration reactors. However, enzyme and substrate concentrations also play an important role due to excessive substrate can cause an end-product inhibition of glucose that accumulates in the hydrolysis step [77]. Nonetheless, there are multiple studies that are based on the optimization process to reach the maximal enzymatic activities and obtain high yields, modifying several parameters during enzymatic lab-scale approaches, as well as large scale studies. Table 9.3 summarizes some studies related to some conditions for enzymatic hydrolysis of agricultural residues where high hydrolysis yields of product were reached.

TABLE 9.3 Conditions of Conversion of Agricultural Residues with High Hydrolysis Yield

Configuration mode/ Bioreactor System	Biomass Concentration	Culture/Enzyme Concentration	Hydrolysis Conditions	Hydrolysis Yield (%)	References
Corn Stover					
3L fed-batch steam-explosion reactor	Corn stover 30% (w v⁻¹)	*Trichoderma reesei*, 30 FPU g⁻¹ cellulose	50°C, pH 4.8, 150 rpm.	85.1%	[124]
Continuous pilot-scale reactor	Corn stover 12–15%	10.7 FPU g⁻¹ cellulose	45°C, pH 4.8, 400 rpm	80%	[59]
1L-batch reactor	Corn stover and dry distiller's grain and solubles 18%	15 FPU/g cellulose þ 64 pNPGU g⁻¹ cellulose	50°C, pH 4.8, 200 rpm.	>95%	[125]
Straw					
2L-batch reactor	Barley straw 15%	7.5 FPU g⁻¹ solids plus 13 CBU g⁻¹ solids	50°C, pH 5, 57 rpm.	81%	[126]
250 ml-SSF	Barley straw 15%	7 FPU g⁻¹ solids plus 8.4 IU β-glucosidase g⁻¹ solids plus 72 U xylanase g⁻¹ solids	50°C, pH 5.5, 150 rpm.	59%	[127]
250 ml batch reactor	Wheat straw 20%	7 FPU g⁻¹ DM plus β-glucosidase	50°C, pH 4.8, 6.6 rpm	60%	[128]
Batch reactor	Rye straw 17%	13 FPU g⁻¹ cellulose plus 35 CBU g⁻¹ cellulose	45°C, pH 5, 120 rpm	65%	[129]
Bagasse					
150 ml batch reactor	Sugarcane bagasse 9+8+7+6% (w v⁻¹)	*C. acetobutylicum* ATCC 824, 9.6 FPU g⁻¹ solids	50°C, pH 6.7, 120 rpm	55%	[130]
1L-fed batch hydrolytic reactor	Sweet sorghum bagasse 20%	30 FPU g⁻¹ cellulose	50°C, pH 4.8, 100 rpm	60%	[131]

9.4   BIOREACTORS FOR FERMENTATION PROCESS

The most common bioreactor used for fermentation experiments is a stirred tank operated in batch mode and the ideality is to achieve the continuous mode. Batch configuration is a discontinuous process where the bioreactor is initially loaded with a specific reaction volume that remains constant throughout the fermentation. The batch reactor is a simple but efficient from an operational point of view with low-cost, compared with other more sophisticated, usually the main components in batch reactors are the tank, the stirred and mixing system, and several controllers of variables as shown in Figure 9.5, i.e., in the industry prefer to the steel and concrete to avoid the corrosion issues that are materials expect a long lifetime [78]. Additionally, the batch system is the simplest operation mode which implies the least risk of contamination. Meanwhile, the continuous mode fresh culture medium is continuously added and extracted from the bioreactor [79]. The volume of the bioreactor is kept constant because both the inflow and outflow are equal. Normally for this kind of process is used a membrane for filtration of the ethanol [80]. In this way is easy to recover a higher amount of ethanol avoiding the inhibition of the ethanol that limits the growth of the microorganism [81].

9.4.1   GEOMETRIC CONFIGURATION AND INSTRUMENTATION

The configuration of the fermentation system is directly related to the sugar conversion process, where the separate processes of saccharification and fermentation and the simultaneous processes of saccharification and co-fermentation are the most common system [82]. Co-fermentation is based on the transformation of hexoses and pentoses contained in the fermentation broth. For which reason several strains of yeasts and bacteria have been studied with the aim of developing an efficient process of hydrolysis and co-fermentation of xylose and glucose. The advantage of developing a process to carry out SSF is the reduction in the number of vessels [83], in addition, a high ethanol yield can be obtained due to the relief of glucose inhibition in cellulase activities for a more complete cellulose hydrolysis [84], but the main drawback is finding a fermentative micro-organism that has operating conditions similar to those required by the enzymatic complex used during saccharification [83].

The design of the reactor structure and their components to reach the established process, in this case for second-generation bioethanol, some accessories in a STR operated under batch mode, are jackets or thermal blanket for temperature regulation, stirred system composed with propeller of type marine, rushton or helical, sensor for pH adjustment, and other components to maintaining a controlled environment (Figure 9.5). The heating-up control is an important operation in the industry; the jackets are considered the best way to control the temperature with practically no variation of temperature in comparison with other systems like thermal blankets. Jacketed reactors are extensively used for fermentation and other different industrial processes like crystallization. Nowadays, the use of new materials for the equipment in the industry becomes more versatile with the use of polymeric complementation with metallic parts in the structure of the reactor [85].

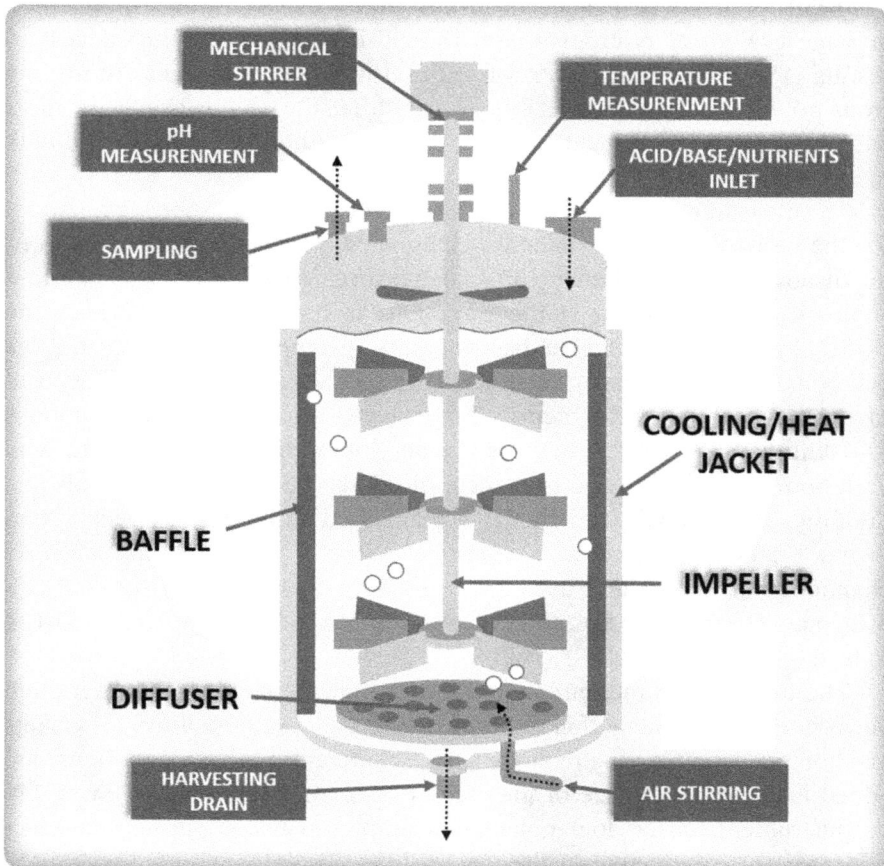

FIGURE 9.5 Design of stirred tank bioreactor detailing the process conditions involved in fermentation process.

By having a completely homogeneous stirred tank bioreactor, it can fulfill the characteristics of a chemostat in which the growth of the micro-organism is kept in a steady state by controlling the concentration of the substrate. This can be achieved by maintaining a continuous inflow and outflow where fresh medium is added and part of the inhibitor products is removed [86].

For the stirring system for the mixing of the substrate, we can identify several options of mixing from propeller to plates, and as well the baffles on the reactor walls. Other accessories on the bioreactor that support the stirring are the air diffusers, but also its combination with mechanical stirring is another strategy that provides a homogeneous in reaction when the broth presents resistance to mixing due to a saturation of substrate or huge growth of microorganism [87]. Stirred systems with successful results is using a blade stirrer for radial flow, it gives easiness of mixing from powders to fluids [88, 89]. Meanwhile, the air diffusers are most used in columns bioreactors in combination with continuous stirred tank reactors (CSTR).

The measurement of cells in fermentation is usually the cheaper analysis for the cell counting in Neubauer chamber via Microscopium. However, the disadvantage is still the no on-line measurement of cells [90, 91]. More sophisticated instrument is the use of fluorescent *in situ* hybridization (FISH) however is expensive. Isaka et al. [92] compared both analyzes, the cell counting by Neubauer chamber still more versatile for simplicity and low cost. In new development, several works apply different techniques to counting cells. One of them is the on-chip as high-speed camera with high optical resolution and flow technology record to analyze in real-time the flows [93]. Sobahi and Han [94] fabricated an on-chip system of two layers; a crystal substrate with electrode detection and the other microfluid channel layer on top and used *Saccharomyces cerevisiae* at 0.8×10^6 cell mL^{-1} and 1×10^6 cell mL^{-1} concentration diluted in peptone and dextrose to test.

The temperature measurement plays one of the key roles to keep high performance in fermentation, moreover, it needs to adjust flows on jackets and homogeneous stirring in the reactor. Temperature sensors commonly are added into a middle place of the reactor or close to the bottom lower, but the instrument sensor does not remain immersed inside the volume reac-tion, usually are external in the reactor [95]. For lab reactors also is used an electric heater, Roukas and Kotzekidou [96] used an electric heater in a batch reactor to adjust to 30°C a fermentation by *Saccharomyces cerevisiae*.

The monitoring of pH is crucial for optimum ethanol production because any variation in pH could make operational issues and inhibit the fermentation. Claros et al. [97] used a pH sensor in CSTR in real-time. Meanwhile, De Vleeschauwer et al. [98] reported the use of pH sensors JUMO® BlackLine to monitoring a sequencing batch reactor (SBR) wastewater could be applied in a fermentation reactor.

On the other hand, another design aspect to consider is the sterilization issue to keep an aseptic environment that through the sterilization of the bioreactor, medium, air, and auxiliary and accessories. The sterilization process can be carried out by thermal, physical, chemical, and radiation process being the first two the most used. The sterilization of industrial reactors is normally carried out by passing steam through pipes and a jacket for a certain time, for the sterilization of substrates the most common is to sterilize in the same way, passing steam through pipes and a jacket at a pressure of 0.1 mPa, or sterilize the substrate and solutions in industrial autoclaves and subsequently being injected through peristaltic pumps into the reactor, however, the sterilization potential can be reduced if the cleaning of the reactor is not efficient, if there are sediment residues from previous processes, the risk of contamination increases dramatically. However, some accessories cannot be sterilized many times or in the same way, i.e., if the method used is by moist heat or dry heat sterilization. Although the bioreactor design allows high temperatures for its sterilization, also with the use of several times, accessories like pH sensor can be suffer a sensitivity loss. The alternatives to not apply high temperatures are the use of gamma rays, ozone gas, or oxygen radicals [99, 100].

9.4.2 PROCESS CONDITIONS

The sugar contained in the hydrolysate are converted to ethanol by microbial fermentation [101]. For the industrial production of bioethanol, the microorganisms used must be resistant to environmental stress such as that generated by an acidic pH, high levels of sugars at the beginning of fermentation (which causes hyperosmotic stress) and can grow rapidly on various lignocellulosic substrates [83]. The most frequently microorganism used is the yeast *Saccharomyces cerevisiae* because it is a microorganism recognized as safe, capable of consuming all types of hexoses and reaching high theoretical ethanol yields, in addition to being able to produce ethanol concentrations of up to 18% (v v⁻¹) [101]. Other microorganisms also used are *Zymomonas mobilis, Pichia stipitis, Candida shehatae, Candida tropicalis, Pachysolan tannophilus* and *Kluyveromyces marixianus* (thermotolerant). Sometimes a co-culture of microorganisms is used to ferment the hexoses and pentoses contained in the fermentation broth. Several process conditions affect the performance of the microorganism into convert the reducing sugars into ethanol. In a first instance, the concentration of cells can be at moderate or high concentration, i.e., a concentration of 2×10^7 cell/mL is considered a moderate cell density, meanwhile 8×10^7 cell mL^{-1} is a high one. The fermentation process should be optimized to find the effective cell concentration taking into account the performance and reaction time. Phukoetphim, Salakkam, Laopaiboon, and Laopaiboon [132] tested the viability of microorganism at different ethanol concentration (v v^{-1}) produced in the fermentation process, by using a concentration of 5×10^7 cell mL^{-1} to ferment a broth of 100 g L^{-1} of reducing sugars and in a batch reactor system. In comparison with other bioreactor system such as, fed-batch reactor with the same concentration of cells obtained a high production of ethanol. Another alternative is the use of pellets with a certain volume of cells, Ju and Ho [102] reported a 0.66 g g^{-1} in cell pellet, with a cell density of 2.57×103.

The stoichiometric conversion of glucose to ethanol is 0.51 g g^{-1}. Achieving this conversion efficiency is difficult because microorganisms use a certain part of the carbon consumed into cellular metabolism and growth. The yeast S. cerevisiae is one of the microorganisms that manages to produce ethanol from high loads of substrate, being able to ferment in media containing up to 400 g L^{-1} of glucose. It is important to consider this parameter because microorganisms can be inhibited by high concentrations of substrate as well [103]. The concentration of reducing sugars depending on the resistance of S. cerevisiae, the concentration of ethanol could reach the 11–13% (w v^{-1}) [104, 105]. Pereira et al. [106] used cellulosic hydrolysates for alcoholic fermentation and it is carried out in 250 mL Erlenmeyer flasks. They use anhydrous glucose to reach the concentration of 100 g L^{-1}. These experiments were conducted for 8 h. Ariyajaroenwong et al. [107] used immobilized cells of S. cerevisiae in sweet sorghum stalk for the fermentation of sweet sorghum juice that had a concentration of between 120 and 280 g L^{-1} of total sugars. In this case, no agitation was used to prevent the detachment of the immobilized cells from the carriers.

Several substrates have different nutrient ratios, in the case of the C:N ratio, the relation of the carbon fermentable of lignocellulose substrates are interesting in the election for the facility to add less amount of nutrients to enhance the substrate. The supplementation of lignocellulose residues provides an acceptable range of carbon-nitrogen ratios, for the right development of yeast, the gain of energy, and the production of primary metabolites. Several biomasses increase their fermentable carbon after pretreatment, to become more available, as were mentioned in his chapter, different pretreatments are used to break the recalcitrant structure of lignin. Usually, the media for ethanol fermentation is supplemented with a limiting nutrient nitrogen nutrient as urea, extract yeast, peptone, and salts to enhance the fermentation potential of the lignocellulose residues.

On the other hand, the ethanol yield is linked to parameters like temperature and pH. An average operational temperature is 28–34° C using *Saccharomyces cerevisiae* [108, 109], although the temperature ratio can vary by the aim of growth of yeast, Lip et al. [110] reported the growth of yeast in several temperature conditions, where the maximum growth rate was between 30–34°C. Most of the fermenting broth used for bioethanol production has pH in the range of 4.5–5.5 tested in different sugars concentrations. The pH could improve de cell growth and ethanol production or can inhibit the microorganisms in the acid or alkali medium; however, several studies demonstrated an optimum range for pH conditions between 4–6 [111].

Table 9.4 shows yeast strains used in bioethanol production at different process conditions of agitation and temperature and as well the concentration of ethanol obtained from certain sugar concentration.

9.5 CONCLUSIONS

The literature reviewed in this chapter indicates that the field of novel reactors designs and operation process has been intensifying in pretreatment, enzymatic hydrolysis, and fermentation processes. This has made viable the use of a great range of LCB and cross the gap of lab-scale to biorefinery concept. Properly established bioreactor configuration and complex process conditions provides an important approach insight into the lignocellulosic bioethanol industry. The main enhancements to scale up biofuels production process, might be related to develop novel reactors and integrated process capable to reduce net energy consumption, as well as combined technologies for better yields in sugars and bioethanol concentration. Finally, tecno-economical scenarios of the design of new bioreactors configuration must be considered to evaluate the commercial value of bioethanol production through LCB since commercial point of view.

TABLE 9.4 Yeast Strains Used in Bioethanol Production at Different Process Conditions

Biomass	Inoculated Yeast Cells	Sugar Concentration (g L⁻¹)	Ethanol Yield (%)	T (°C)	Agitation (rpm)	References
Sugarcane bagasse	25 g L^{-1} (dry basis) S. cerevisiae	100 (glucose)	54.69	31	100	[106]
Sugarcane tops	25 g L^{-1} (dry basis) S. cerevisiae	100 (glucose)	65.52	31	100	
Sugarcane straw	25 g L^{-1} (dry basis) S. cerevisiae	100 (glucose)	74.70	31	100	[107]
Sweet sorghum	0.38 g L^{-1} S. cerevisiae NP 01	228	87.72	30	–	[16]
Sugarcane bagasse	7.8×10^{7} cell mL^{-1} S. cerevisiae AR5	78.6 (glucose)	95	33	120	[133]
Spent coffee grounds	1 g/L S. cerevisiae RL-11	52.5	26	30	200	[134]
Sorghum stover	S. cerevisiae MTCC 173	200	34	30	120	[135]
Giant reed	S. stipitis CBS 6054	33.4	33	30	150	[136]
Corn stover	S. cerevisiae ZU-10	66.9	41.6	30	120	[137]
Ipomea carnea	S. cerevisiae RPRT90	72.1	46.1	30	150	[138]
Wood	S. cerevisiae	37.47	49	30	150	[139]
Reed	S. cerevisiae ATCC	123	93	38	150	[140]
Water hyacinth	Kluyveromyces marxianus K213	23.3	16	42	–	[141]
Wheat straw	1 g/L Kluyveromyces marxianus CECT	22.8	45	42	150	[142]
Paper sludge	S. cerevisiae GIM-2	27.8	68.34	31	60	

KEYWORDS

- **bioethanol**
- **bioreactor design**
- **enzymatic hydrolysis**
- **fermentation**
- **pretreatment**
- **process parameters**

REFERENCES

1. Atelge, M. R., Atabani, A. E., Banu, J. R., Krisa, D., Kaya, M., Eskicioglu, C., & Duman, F., (2020). A critical review of pretreatment technologies to enhance anaerobic digestion and energy recovery. *Fuel, 270,* 117494. https://doi.org/10.1016/j.fuel.2020.117494 (accessed on 8 December 2021).
2. Alalwan, H. A., Alminshid, A. H., & Aljaafari, H. A. S., (2019). Promising evolution of biofuel generations. Subject review. *Renewable Energy Focus, 28,* 127–139. https://doi.org/10.1016/j.ref.2018.12.006 (accessed on 8 December 2021).
3. Li, J., Zhou, W., Fan, S., Xiao, Z., Liu, Y., Liu, J., & Wang, Y., (2018). Bioethanol production in vacuum membrane distillation bioreactor by permeate fractional condensation and mechanical vapor compression with polytetrafluoroethylene (PTFE) membrane. *Bioresource Technology, 268,* 708–714. https://doi.org/10.1016/j.biortech.2018.08.055 (accessed on 8 December 2021).
4. Toor, M., Kumar, S. S., Malyan, S. K., Bishnoi, N. R., Mathimani, T., Rajendran, K., & Pugazhendhi, A., (2020). An overview on bioethanol production from lignocellulosic feedstocks. *Chemosphere, 242,* 125080. https://doi.org/10.1016/j.chemosphere.2019.125080 (accessed on 8 December 2021).
5. Sharma, S., Kundu, A., Basu, S., Shetti, N. P., & Aminabhavi, T. M., (2020). Sustainable environmental management and related biofuel technologies. *Journal of Environmental Management, 273,* 111096. https://doi.org/10.1016/j.jenvman.2020.111096 (accessed on 8 December 2021).
6. Hernández-Beltrán, J. U., Hernández-De, L. I. O., Cruz-Santos, M. M., Saucedo-Luevanos, A., Hernández-Terán, F., & Balagurusamy, N., (2019). Insight into pretreatment methods of lignocellulosic biomass to increase biogas yield: Current state, challenges, and opportunities. *Applied Sciences, 9,* 3721. https://doi.org/10.3390/app9183721 (accessed on 8 December 2021).
7. Prasad, R. K., Chatterjee, S., Mazumder, P. B., Gupta, S. K., Sharma, S., Vairale, M. G., & Gupta, D. K., (2019). Bioethanol production from waste lignocelluloses: A review on microbial degradation potential. *Chemosphere, 231,* 588–606. https://doi.org/10.1016/j.chemosphere.2019.05.142 (accessed on 8 December 2021).

8. Pino, M. S., Rodríguez-Jasso, R. M., Michelin, M., Flores-Gallegos, A. C., Morales-Rodriguez, R., Teixeira, J. A., & Ruiz, H. A., (2018). Bioreactor design for enzymatic hydrolysis of biomass under the biorefinery concept. *Chemical Engineering Journal, 347*, 119–136. https://doi.org/10.1016/j.cej.2018.04.057 (accessed on 8 December 2021).

9. De La Fuente, S. N. M., & Alejo, A. A. M., (2019). Challenges of fermentation engineering. In: Chávez-González, M. L., Balagurusamy, N., & Aguilar, C. N., (eds.), *Advances in Food Bioproducts and Bioprocessing Technologies* (pp. 233–264). CRC Press. https://doi.org/10.1201/9780429331817-11 (accessed on 8 December 2021).

10. Saha, K., Maharana, A., Sikder, J., Chakraborty, S., Curcio, S., & Drioli, E., (2019). Continuous production of bioethanol from sugarcane bagasse and downstream purification using membrane integrated bioreactor. *Catalysis Today, 331*, 68–77. https://doi.org/10.1016/j.cattod.2017.11.031 (accessed on 8 December 2021).

11. Rocha-Meneses, L., Raud, M., Orupõld, K., & Kikas, T., (2019). Potential of bioethanol production waste for methane recovery. *Energy, 173*, 133–139. https://doi.org/10.1016/j.energy.2019.02.073 (accessed on 8 December 2021).

12. Ibarra-Gonzalez, P., & Rong, B. G., (2019). A review of the current state of biofuels production from lignocellulosic biomass using thermochemical conversion routes. *Chinese Journal of Chemical Engineering, 27*, 1523–1535. https://doi.org/10.1016/j.cjche.2018.09.018 (accessed on 8 December 2021).

13. Alayoubi, R., Mehmood, N., Husson, E., Kouzayha, A., Tabcheh, M., Chaveriat, L., & Gosselin, I., (2020). Low temperature ionic liquid pretreatment of lignocellulosic biomass to enhance bioethanol yield. *Renewable Energy, 145*, 1808–1816. https://doi.org/10.1016/j.renene.2019.07.091 (accessed on 8 December 2021).

14. Hernández-Beltrán, J. U., Fontalvo, J., & Hernández-Escoto, H., (2020). Fed-batch enzymatic hydrolysis of plantain pseudostem to fermentable sugars production and the impact of particle size at high solids loadings. *Biomass Conversion and Biorefinery*. https://doi.org/10.1007/s13399-020-00669-2 (accessed on 8 December 2021).

15. Hernández-Beltrán, J. U., & Hernández-Escoto, H., (2018). Enzymatic hydrolysis of biomass at high-solids loadings through fed-batch operation. *Biomass and Bioenergy, 119*. https://doi.org/10.1016/j.biombioe.2018.09.020 (accessed on 8 December 2021).

16. Meléndez-Hernández, P. A., Hernández-Beltrán, J. U., Hernández-Guzmán, A., Morales-Rodríguez, R., Torres-Guzmán, J. C., & Hernández-Escoto, H., (2019). Comparative of alkaline hydrogen peroxide pretreatment using NaOH and Ca(OH)$_2$ and their effects on enzymatic hydrolysis and fermentation steps. *Biomass Conversion and Biorefinery*. https://doi.org/10.1007/s13399-019-00574-3 (accessed on 8 December 2021).

17. Kucharska, K., Rybarczyk, P., Hołowacz, I., Łukajtis, R., Glinka, M., & Kamiński, M., (2018). Pretreatment of lignocellulosic materials as substrates for fermentation processes. *Molecules, 23*, 1–32. https://doi.org/10.3390/molecules23112937 (accessed on 8 December 2021).

18. Gusakov, A. V., Sinitsyn, A. P., & Klyosov, A. A., (1985). Kinetics of the enzymatic hydrolysis of cellulose: 1. A mathematical model for a batch reactor process. *Enzyme and Microbial Technology, 7*, 346–352. https://doi.org/10.1016/0141-0229(85)90114-0 (accessed on 8 December 2021).

19. Goodman, B. A., (2020). Utilization of waste straw and husks from rice production: A review. *Journal of Bioresources and Bioproducts, 5*, 143–162. https://doi.org/10.1016/j.jobab.2020.07.001 (accessed on 8 December 2021).

20. Ponnusamy, V. K., Nguyen, D. D., Dharmaraja, J., Shobana, S., Banu, J. R., Saratale, R. G., & Kumar, G., (2019). A review on lignin structure, pretreatments, fermentation reactions and biorefinery potential. *Bioresource Technology, 271*, 462–472. https://doi.org/10.1016/j.biortech.2018.09.070 (accessed on 8 December 2021).
21. Wang, H., Pu, Y., Ragauskas, A., & Yang, B., (2019). From lignin to valuable products-strategies, challenges, and prospects. *Bioresource Technology, 271*, 449–461. https://doi.org/10.1016/j.biortech.2018.09.072 (accessed on 8 December 2021).
22. Yoo, C. G., Meng, X., Pu, Y., & Ragauskas, A. J., (2020). The critical role of lignin in lignocellulosic biomass conversion and recent pretreatment strategies: A comprehensive review. *Bioresource Technology, 301*, 122784. https://doi.org/10.1016/j.biortech.2020.122784 (accessed on 8 December 2021).
23. Sharma, B., Larroche, C., & Dussap, C. G., (2020). Comprehensive assessment of 2G bioethanol production. *Bioresource Technology, 313*, 123630. https://doi.org/10.1016/j.biortech.2020.123630 (accessed on 8 December 2021).
24. Baruah, J., Nath, B. K., Sharma, R., Kumar, S., Deka, R. C., Baruah, D. C., & Kalita, E., (2018). Recent trends in the pretreatment of lignocellulosic biomass for value-added products. *Frontiers in Energy Research, 6*, 1–19. https://doi.org/10.3389/fenrg.2018.00141 (accessed on 8 December 2021).
25. Kumar, B., Bhardwaj, N., Agrawal, K., Chaturvedi, V., & Verma, P., (2020). Current perspective on pretreatment technologies using lignocellulosic biomass: An emerging biorefinery concept. *Fuel Processing Technology, 199*. https://doi.org/10.1016/j.fuproc.2019.106244 (accessed on 8 December 2021).
26. Kapoor, M., Semwal, S., Satlewal, A., Christopher, J., Gupta, R. P., Kumar, R., & Ramakumar, S. S. V., (2019). The impact of particle size of cellulosic residue and solid loadings on enzymatic hydrolysis with a mass balance. *Fuel, 245*, 514–520. https://doi.org/10.1016/j.fuel.2019.02.094 (accessed on 8 December 2021).
27. Shimizu, F. L., Monteiro, P. Q., Ghiraldi, P. H. C., Melati, R. B., Pagnocca, F. C., Souza De, W., & Brienzo, M., (2018). Acid, alkali and peroxide pretreatments increase the cellulose accessibility and glucose yield of banana pseudostem. *Industrial Crops and Products, 115*, 62–68. https://doi.org/10.1016/j.indcrop.2018.02.024 (accessed on 8 December 2021).
28. Niju, S., Nishanthini, T., & Balajii, M., (2020). Alkaline hydrogen peroxide-pretreated sugarcane tops for bioethanol production—A process optimization study. *Biomass Conversion and Biorefinery, 10*, 149–165. https://doi.org/10.1007/s13399–019–00524-z (accessed on 8 December 2021).
29. Wang, C., Zhang, X., Liu, Q., Zhang, Q., Chen, L., & Ma, L., (2020). A review of conversion of lignocellulose biomass to liquid transport fuels by integrated refining strategies. *Fuel Processing Technology, 208*, 106485. https://doi.org/10.1016/j.fuproc.2020.106485 (accessed on 8 December 2021).
30. De Oliveira, A. P. V., Sepulchro, A. G. V., & Polikarpov, I., (2020). Enzymes for lignocellulosic biomass polysaccharides valorization and production of nanomaterials. *Current Opinion in Green and Sustainable Chemistry*, 100397. https://doi.org/10.1016/j.cogsc.2020.100397 (accessed on 8 December 2021).
31. Maitan-Alfenas, G. P., Visser, E. M., & Guimarães, V. M. (2015). Enzymatic hydrolysis of lignocellulosic biomass: Converting food waste in valuable products. *Current Opinion in Food Science, 1*, 44–49. https://doi.org/10.1016/j.cofs.2014.10.001 (accessed on 8 December 2021).

32. Ruiz, H. A., Rodríguez-Jasso, R. M., Fernandes, B. D., Vicente, A. A., & Teixeira, J. A., (2013). Hydrothermal processing, as an alternative for upgrading agriculture residues and marine biomass according to the biorefinery concept: A review. *Renewable and Sustainable Energy Reviews, 21,* 35–51. https://doi.org/10.1016/j.rser.2012.11.069 (accessed on 8 December 2021).

33. Paz, A., Outeiriño, D., Pérez, G. N., & Domínguez, J. M., (2019). Enzymatic hydrolysis of brewer's spent grain to obtain fermentable sugars. *Bioresource Technology, 275,* 402–409. https://doi.org/10.1016/j.biortech.2018.12.082 (accessed on 8 December 2021).

34. Liu, Y., Yu, Q., Xu, J., & Yuan, Z., (2020). Evaluation of structural factors affecting high solids enzymatic saccharification of alkali-pretreated sugarcane bagasse. *Cellulose, 27,* 1441–1450. https://doi.org/10.1007/s10570-019-02890-3 (accessed on 8 December 2021).

35. Xu, C., Zhang, J., Zhang, Y., Guo, Y., Xu, H., Xu, J., & Wang, Z., (2019). Enhancement of high-solids enzymatic hydrolysis efficiency of alkali pretreated sugarcane bagasse at low cellulase dosage by fed-batch strategy based on optimized accessory enzymes and additives. *Bioresource Technology, 292,* 121993. https://doi.org/10.1016/j.biortech.2019.121993 (accessed on 8 December 2021).

36. Lorenci, W. A., Dalmas, N. C. J., Porto De, S., Vandenberghe, L., De Carvalho, N. D. P., Novak, S. A. C., Letti, L. A. J., & Soccol, C. R., (2020). Lignocellulosic biomass: Acid and alkaline pretreatments and their effects on biomass recalcitrance - conventional processing and recent advances. *Bioresource Technology, 304,* 122848. https://doi.org/10.1016/j.biortech.2020.122848 (accessed on 8 December 2021).

37. Kumar, D., Juneja, A., & Singh, V., (2018). Fermentation technology to improve productivity in dry-grind corn process for bioethanol production. *Fuel Processing Technology, 173,* 66–74. https://doi.org/10.1016/j.fuproc.2018.01.014 (accessed on 8 December 2021).

38. Germec, M., & Turhan, I., (2018). Ethanol production from acid-pretreated and detoxified rice straw as sole renewable resource. *Biomass Conversion and Biorefinery, 8,* 607–619. https://doi.org/10.1007/s13399-018-0310-1 (accessed on 8 December 2021).

39. Amadi, P. U., & Ifeanacho, M. O., (2016). Impact of changes in fermentation time, volume of yeast, and mass of plantain pseudo-stem substrate on the simultaneous saccharification and fermentation potentials of African land snail digestive juice and yeast. *Journal of Genetic Engineering and Biotechnology, 14,* 289–297. https://doi.org/10.1016/j.jgeb.2016.09.002 (accessed on 8 December 2021).

40. Rezania, S., Oryani, B., Cho, J., Talaiekhozani, A., Sabbagh, F., Hashemi, B., & Mohammadi, A. A., (2020). Different pretreatment technologies of lignocellulosic biomass for bioethanol production: An overview. *Energy, 199,* 117457. https://doi.org/10.1016/j.energy.2020.117457 (accessed on 8 December 2021).

41. Tsapekos, P., Kougias, P. G., Egelund, H., Larsen, U., Pedersen, J., Trénel, P., & Angelidaki, I., (2017). Mechanical pretreatment at harvesting increases the bioenergy output from marginal land grasses. *Renewable Energy, 111,* 914–921. https://doi.org/10.1016/j.renene.2017.04.061 (accessed on 8 December 2021).

42. Gu, H., An, R., & Bao, J., (2018). Pretreatment refining leads to constant particle size distribution of lignocellulose biomass in enzymatic hydrolysis. *Chemical Engineering Journal, 352,* 198–205. https://doi.org/10.1016/j.cej.2018.06.145 (accessed on 8 December 2021).

43. Aguilar-Reynosa, A., Romaní, A., Ma Rodríguez-Jasso, R., Aguilar, C. N., Garrote, G., & Ruiz, H. A., (2017). Microwave heating processing as alternative of pretreatment

in second-generation biorefinery: An overview. *Energy Conversion and Management, 136*, 50–65. https://doi.org/10.1016/j.enconman.2017.01.004 (accessed on 8 December 2021).

44. Hermiati, E., Laksana, R. P. B., Fatriasari, W., Kholida, L. N., Thontowi, A., Yopi, & Watanabe, T., (2020). Microwave-assisted acid pretreatment for enhancing enzymatic saccharification of sugarcane trash. *Biomass Conversion and Biorefinery.* https://doi.org/10.1007/s13399-020-00971-z (accessed on 8 December 2021).
45. Peng, H., Luo, H., Jin, S., Li, H., & Xu, J., (2014). Improved bioethanol production from corn stover by alkali pretreatment with a novel pilot-scale continuous microwave irradiation reactor. *Biotechnology and Bioprocess Engineering, 19*, 493–502. https://doi.org/10.1007/s12257-014-0014-8 (accessed on 8 December 2021).
46. Ariyawansha, T., Abeyrathna, D., Ahamed, T., & Noguchi, R., (2020). Integrated bagasse utilization system based on hydrothermal liquefaction in sugarcane mills: Theoretical approach compared with present practices. *Biomass Conversion and Biorefinery, 1.* https://doi.org/10.1007/s13399-020-00958-w (accessed on 8 December 2021).
47. Miliotti, E., Dell'Orco, S., Lotti, G., Rizzo, A. M., Rosi, L., & Chiaramonti, D., (2019). Lignocellulosic ethanol biorefinery: Valorization of lignin-rich stream through hydrothermal liquefaction. *Energies, 12*, 723. https://doi.org/10.3390/en12040723 (accessed on 8 December 2021).
48. Terán, H. R., Dionízio, R. M., Sánchez, M. S., Prado, C. A., De Sousa, J. R., Da Silva, S. S., & Santos, J. C., (2020). Hydrodynamic cavitation-assisted continuous pre-treatment of sugarcane bagasse for ethanol production: Effects of geometric parameters of the cavitation device. *Ultrasonics Sonochemistry, 63*, 104931. https://doi.org/10.1016/j.ultsonch.2019.104931 (accessed on 8 December 2021).
49. Ramirez-Cadavid, D. A., Kozyuk, O., Lyle, P., & Michel, F. C., (2016). Effects of hydrodynamic cavitation on dry mill corn ethanol production. *Process Biochemistry, 51*, 500–508. https://doi.org/10.1016/j.procbio.2016.01.001 (accessed on 8 December 2021).
50. Kim, I., Lee, I., Jeon, S. H., Hwang, T., & Han, J. I., (2015). Hydrodynamic cavitation as a novel pretreatment approach for bioethanol production from reed. *Bioresource Technology, 192*, 335–339. https://doi.org/10.1016/j.biortech.2015.05.038 (accessed on 8 December 2021).
51. Hernández-Guzmán, A., Navarro-Gutiérrez, I. M., Meléndez-Hernández, P. A., Hernández-Beltrán, J. U., & Hernández-Escoto, H., (2020). Enhancement of alkaline-oxidative delignification of wheat straw by semi-batch operation in a stirred tank reactor. *Bioresource Technology, 312*, 123589. https://doi.org/10.1016/j.biortech.2020.123589 (accessed on 8 December 2021).
52. Nakashima, K., Ebi, Y., Shibasaki-Kitakawa, N., Soyama, H., & Yonemoto, T., (2016). Hydrodynamic cavitation reactor for efficient pretreatment of lignocellulosic biomass. *Industrial and Engineering Chemistry Research, 55*, 1866–1871. https://doi.org/10.1021/acs.iecr.5b04375 (accessed on 8 December 2021).
53. Terán, H. R., Ramos, L., Da Silva, S. S., Dragone, G., Mussatto, S. I., & Santos, J. C. D., (2018). Hydrodynamic cavitation as a strategy to enhance the efficiency of lignocellulosic biomass pretreatment. *Critical Reviews in Biotechnology, 38*, 483–493. https://doi.org/10.1080/07388551.2017.1369932 (accessed on 8 December 2021).
54. Andrić, P., Meyer, A. S., Jensen, P. A., & Dam-Johansen, K., (2010). Reactor design for minimizing product inhibition during enzymatic lignocellulose hydrolysis. II.

Quantification of inhibition and suitability of membrane reactors. *Biotechnology Advances, 28*, 407–425. https://doi.org/10.1016/j.biotechadv.2010.02.005 (accessed on 8 December 2021).

55. Agrawal, R., Satlewal, A., Kapoor, M., Mondal, S., & Basu, B., (2017). Investigating the enzyme-lignin binding with surfactants for improved saccharification of pilot-scale pretreated wheat straw. *Bioresource Technology, 224*, 411–418. https://doi.org/10.1016/j.biortech.2016.11.026 (accessed on 8 December 2021).

56. Chang, Y. H., Chang, K. S., Huang, C. W., Hsu, C. L., & Jang, H. D., (2012). Comparison of batch and fed-batch fermentations using corncob hydrolysate for bioethanol production. *Fuel, 97*, 166–173. https://doi.org/10.1016/j.fuel.2012.02.006 (accessed on 8 December 2021).

57. Chen, H. Z., & Liu, Z. H., (2017). Enzymatic hydrolysis of lignocellulosic biomass from low to high solids loading. *Engineering in Life Sciences, 17*, 489–499. https://doi.org/10.1002/elsc.201600102 (accessed on 8 December 2021).

58. Du, J., Zhang, F., Li, Y., Zhang, H., Liang, J., Zheng, H., & Huang, H., (2014). Enzymatic liquefaction and saccharification of pretreated corn stover at high-solids concentrations in a horizontal rotating bioreactor. *Bioprocess and Biosystems Engineering, 37*, 173–181. https://doi.org/10.1007/s00449-013-0983-6 (accessed on 8 December 2021).

59. Hodge, D. B., Karim, M. N., Schell, D. J., & McMillan, J. D., (2009). Model-based fed-batch for high-solids enzymatic cellulose hydrolysis. *Applied Biochemistry and Biotechnology, 152*, 88–107. https://doi.org/10.1007/s12010-008-8217-0 (accessed on 8 December 2021).

60. Al-Zuhair, S., Al-Hosany, M., Zooba, Y., Al-Hammadi, A., & Al-Kaabi, S., (2013). Development of a membrane bioreactor for enzymatic hydrolysis of cellulose. *Renewable Energy, 56*, 85–89. https://doi.org/10.1016/j.renene.2012.09.044 (accessed on 8 December 2021).

61. Garcia-Ochoa, F., Santos, V. E., & Gomez, E., (2019). Stirred tank bioreactors. *Comprehensive Biotechnology* (3rd edn., Vol. 2). Elsevier. https://doi.org/10.1016/B978-0-444-64046-8.00078-1 (accessed on 8 December 2021).

62. Burman, N. W., Sheridan, C. M., & Harding, K. G., (2019). Lignocellulosic bioethanol production from grasses pre-treated with acid mine drainage: Modeling and comparison of SHF and SSF. *Bioresource Technology Reports, 7*, 100299. https://doi.org/10.1016/j.biteb.2019.100299 (accessed on 8 December 2021).

63. Maslova, O., Stepanov, N., Senko, O., & Efremenko, E., (2019). Production of various organic acids from different renewable sources by immobilized cells in the regimes of separate hydrolysis and fermentation (SHF) and simultaneous saccharification and fermentation (SFF). *Bioresource Technology, 272*, 1–9. https://doi.org/10.1016/j.biortech.2018.09.143 (accessed on 8 December 2021).

64. Loaces, I., Schein, S., & Noya, F., (2017). Ethanol production by *Escherichia coli* from Arundo donax biomass under SSF, SHF or CBP process configurations and in situ production of a multifunctional glucanase and xylanase. *Bioresource Technology, 224*, 307–313. https://doi.org/10.1016/j.biortech.2016.10.075 (accessed on 8 December 2021).

65. Mithra, M. G., Jeeva, M. L., Sajeev, M. S., & Padmaja, G., (2018). Comparison of ethanol yield from pretreated lignocellulose-starch biomass under fed-batch SHF or SSF modes. *Heliyon, 4*, e00885. https://doi.org/10.1016/j.heliyon.2018.e00885 (accessed on 8 December 2021).

66. Jin, M., Lau, M. W., Balan, V., & Dale, B. E., (2010). Two-step SSCF to convert AFEX-treated switchgrass to ethanol using commercial enzymes and *Saccharomyces cerevisiae* 424A(LNH-ST). *Bioresource Technology, 101,* 8171–8178. https://doi.org/10.1016/j.biortech.2010.06.026 (accessed on 8 December 2021).
67. Liu, Z. H., & Chen, H. Z., (2016). Simultaneous saccharification and co-fermentation for improving the xylose utilization of steam-exploded corn stover at high solid loading. *Bioresource Technology, 201,* 15–26. https://doi.org/10.1016/j.biortech.2015.11.023 (accessed on 8 December 2021).
68. M'Barek, H. N., Arif, S., Taidi, B., & Hajjaj, H., (2020). Consolidated bioethanol production from olive mill waste: Wood-decay fungi from central Morocco as promising decomposition and fermentation biocatalysts. *Biotechnology Reports.* https://doi.org/10.1016/j.btre.2020.e00541 (accessed on 8 December 2021).
69. Rastogi, M., & Shrivastava, S., (2017). Recent advances in second-generation bioethanol production: An insight to pretreatment, saccharification and fermentation processes. *Renewable and Sustainable Energy Reviews, 80,* 330–340. https://doi.org/10.1016/j.rser.2017.05.225 (accessed on 8 December 2021).
70. Corrêa, L. J., Badino, A. C., & Cruz, A. J. G., (2016). Mixing design for enzymatic hydrolysis of sugarcane bagasse: Methodology for selection of impeller configuration. *Bioprocess and Biosystems Engineering, 39,* 285–294. https://doi.org/10.1007/s00449-015-1512-6 (accessed on 8 December 2021).
71. Caspeta, L., Caro-Bermúdez, M. A., Ponce-Noyola, T., & Martinez, A., (2014). Enzymatic hydrolysis at high-solids loadings for the conversion of agave bagasse to fuel ethanol. *Applied Energy, 113,* 277–286. https://doi.org/10.1016/j.apenergy.2013.07.036 (accessed on 8 December 2021).
72. Modenbach, A. A., & Nokes, S. E., (2013). Enzymatic hydrolysis of biomass at high-solids loadings - A review. *Biomass and Bioenergy, 56,* 526–544. https://doi.org/10.1016/j.biombioe.2013.05.031 (accessed on 8 December 2021).
73. Zhang, J., Chu, D., Huang, J., Yu, Z., Dai, G., & Bao, J., (2010). Simultaneous saccharification and ethanol fermentation at high corn stover solids loading in a helical stirring bioreactor. *Biotechnology and Bioengineering, 105,* 718–728. https://doi.org/10.1002/bit.22593 (accessed on 8 December 2021).
74. Várnai, A., Siika-Aho, M., & Viikari, L., (2013). Carbohydrate-binding modules (CBMs) revisited: Reduced amount of water counterbalances the need for CBMs. *Biotechnology for Biofuels, 6,* 30. https://doi.org/10.1186/1754-6834-6-30 (accessed on 8 December 2021).
75. Michelin, M., Ruiz, H. A., Silva, D. P., Ruzene, D. S., Teixeira, J. A., & M, M. L. T., (2014). Cellulose from lignocellulosic waste. In: Ramawat, K. G., (ed.), *Polysaccharides.* Springer International Publishing Switzerland. https://doi.org/10.1007/978-3-319-03751-6 (accessed on 8 December 2021).
76. Saini, J. K., Patel, A. K., Adsul, M., & Singhania, R. R., (2016). Cellulase adsorption on lignin: A roadblock for economic hydrolysis of biomass. *Renewable Energy, 98,* 29–42. https://doi.org/10.1016/j.renene.2016.03.089 (accessed on 8 December 2021).
77. Wood, B. E., Aldrich, H. C., & Ingram, L. O., (1997). Ultrasound stimulates ethanol production during the simultaneous saccharification and fermentation of mixed waste office paper. *Biotechnology Progress, 13,* 232–237. https://doi.org/10.1021/bp970027v (accessed on 8 December 2021).
78. Ketchum, L. H., (1997). Design and physical features of sequencing batch reactors. *Water Science and Technology, 35*(1), 11–18. https://doi.org/10.1016/S0273-1223(96)00873-6 (accessed on 8 December 2021).

79. Kropp, C., Massai, D., & Zweigerdt, R., (2017). Progress and challenges in large-scale expansion of human pluripotent stem cells. *Process Biochemistry, 59*, 244–254. https://doi.org/10.1016/j.procbio.2016.09.032 (accessed on 8 December 2021).

80. Ylitervo, P., Franzén, C. J., & Taherzadeh, M. J., (2014). Continuous ethanol production with a membrane bioreactor at high acetic acid concentrations. *Membranes, 4*, 372–387. https://doi.org/10.3390/membranes4030372 (accessed on 8 December 2021).

81. Fu, C., Cai, D., Hu, S., Miao, Q., Wang, Y., Qin, P., & Tan, T., (2016). Ethanol fermentation integrated with PDMS composite membrane: An effective process. *Bioresource Technology, 200*, 648–657. https://doi.org/10.1016/j.biortech.2015.09.117 (accessed on 8 December 2021).

82. Zafar, S., (2019). *Ethanol from Lignocellulosic Biomass*. Retrieved, from https://www.bioenergyconsult.com/production-cellulosic-ethanol/ (accessed on 28 October 2021).

83. Zhang, C., (2020). Lignocellulosic ethanol: Technology and economics. In: Yun, Y., (ed.), *Alcohol Fuels: Current Technologies and Future Prospect*. Intechopen. https://doi.org/10.5772/intechopen.86701 (accessed on 8 December 2021).

84. Liu, C. G., Xiao, Y., Xia, X. X., Zhao, X. Q., Peng, L., Srinophakun, P., & Bai, F. W., (2019). Cellulosic ethanol production: Progress, challenges and strategies for solutions. *Biotechnology Advances, 37*, 491–504. https://doi.org/10.1016/j.biotechadv.2019.03.002 (accessed on 8 December 2021).

85. Leuteritz, A., Döring, K. D., Lampke, T., & Kuehnert, I., (2016). Accelerated ageing of plastic jacket pipes for district heating. *Polymer Testing, 51*, 142–147. https://doi.org/10.1016/j.polymertesting.2016.03.012 (accessed on 8 December 2021).

86. Pumphrey, B., & Julien, C., (1996). An introduction to fermentation: Fermentation basics. *Journal of the American Society of Brewing Chemists, 34*, 103–104. https://doi.org/10.1080/03610470.1976.12006198 (accessed on 8 December 2021).

87. Jin, B., & Lant, P., (2004). Flow regime, hydrodynamics, floc size distribution and sludge properties in activated sludge bubble column, air-lift and aerated stirred reactors. *Chemical Engineering Science, 59*(12), 2379–2388. https://doi.org/10.1016/j.ces.2004.01.061 (accessed on 8 December 2021).

88. Ameur, H., (2016). Mixing of complex fluids with flat and pitched bladed impellers: Effect of blade attack angle and shear-thinning behavior. *Food and Bioproducts Processing, 99*, 71–77. https://doi.org/10.1016/j.fbp.2016.04.004 (accessed on 8 December 2021).

89. Stewart, R. L., Bridgwater, J., & Parker, D. J., (2001). Granular flow over a flat-bladed stirrer. *Chemical Engineering Science, 56*(14), 4257–4271. https://doi.org/10.1016/S0009-2509(01)00104-X (accessed on 8 December 2021).

90. Caligiore-Gei, P. F., & Valdez, J. G. (2015). Adjustment of a rapid method for quantification of *Fusarium spp.* spore suspensions in plant pathology. *Revista Argentina de Microbiologia, 47*(2), 152–154. https://doi.org/10.1016/j.ram.2015.03.002 (accessed on 8 December 2021).

91. De Souza, L. A. T., Da Silva, F. E. A., De Morais, J. O. F., Simões, D. A., & De Morais, M. A., (2005). Contaminant yeast detection in industrial ethanol fermentation must by rDNA-PCR. *Letters in Applied Microbiology, 40*, 19–23. https://doi.org/10.1111/j.1472-765X.2004.01618.x (accessed on 8 December 2021).

92. Isaka, K., Date, Y., Sumino, T., Yoshie, S., & Tsuneda, S., (2006). Growth characteristic of anaerobic ammonium-oxidizing bacteria in an anaerobic biological filtrated reactor. *Applied Microbiology and Biotechnology, 70*, 47–52. https://doi.org/10.1007/s00253-005-0046-2 (accessed on 8 December 2021).

93. Goda, K., Ayazi, A., Gossett, D. R., Sadasivam, J., Lonappan, C. K., Sollier, E., & Jalali, B., (2012). High-throughput single-microparticle imaging flow analyzer. *Proceedings of the National Academy of Sciences of the United States of America, 109*, 11630–11635. https://doi.org/10.1073/pnas.1204718109 (accessed on 8 December 2021).

94. Sobahi, N., & Han, A., (2020). High-throughput and label-free multi-outlet cell counting using a single pair of impedance electrodes. *Biosensors and Bioelectronics, 166*, 112458. https://doi.org/10.1016/j.bios.2020.112458 (accessed on 8 December 2021).

95. Darnoko, D., & Cheryan, M., (2000). Kinetics of palm oil transesterification in a batch reactor. *JAOCS, Journal of the American Oil Chemists' Society, 77*, 1263–1267. https://doi.org/10.1007/s11746-000-0198-y (accessed on 8 December 2021).

96. Roukas, T., & Kotzekidou, P., (2020). Rotary biofilm reactor: A new tool for long-term bioethanol production from non-sterilized beet molasses by *Saccharomyces cerevisiae* in repeated-batch fermentation. *Journal of Cleaner Production, 257*, 120519. https://doi.org/10.1016/j.jclepro.2020.120519 (accessed on 8 December 2021).

97. Claros, J., Serralta, J., Seco, A., Ferrer, J., & Aguado, D., (2012). Real-time control strategy for nitrogen removal via nitrite in a SHARON reactor using pH and ORP sensors. *Process Biochemistry, 47*, 1510–1515. https://doi.org/10.1016/j.procbio.2012.05.020 (accessed on 8 December 2021).

98. De Vleeschauwer, F., Caluwé, M., Dobbeleers, T., Stes, H., Dockx, L., Kiekens, F., & Dries, J., (2020). A dynamic control system for aerobic granular sludge reactors treating high COD/P wastewater, using pH and DO sensors. *Journal of Water Process Engineering, 33*, 101065. https://doi.org/10.1016/j.jwpe.2019.101065 (accessed on 8 December 2021).

99. Nagatsu, M., Terashita, F., Nonaka, H., Xu, L., Nagata, T., & Koide, Y., (2005). Effects of oxygen radicals in low-pressure surface-wave plasma on sterilization. *Applied Physics Letters, 86*(21), 1–3. https://doi.org/10.1063/1.1931050 (accessed on 8 December 2021).

100. Takatsuji, Y., Ishikawa, S., & Haruyama, T., (2017). Efficient sterilization using reactive oxygen species generated by a radical vapor reactor. *Process Biochemistry, 54*, 140–143. https://doi.org/10.1016/j.procbio.2017.01.002 (accessed on 8 December 2021).

101. Godbey, W. T., (2014). *Fermentation, Beer, and Biofuels. An Introduction to Biotechnology*. Elsevier Ltd. https://doi.org/10.1016/b978-1-907568-28-2.00016-2 (accessed on 8 December 2021).

102. Ju, L. K., & Ho, C. S., (1988). Correlation of cell volume fractions with cell concentrations in fermentation media. *Biotechnology and Bioengineering, 32*, 95–99. https://doi.org/10.1002/bit.260320113 (accessed on 8 December 2021).

103. Moreno, A. D., Alvira, P., Ibarra, D., & Tomás-Pejó, E. (2017). Production of ethanol from lignocellulosic biomass. In Z. Fang, R. L. Smith, & J. X. Qi (Eds.), Production of Platform Chemicals from Sustainable Resources (pp. 375–410). Springer Nature Singapore Pte Ltd. https://doi.org/10.1007/978-981-10-4172-3_12 (accessed on 8 December 2021).

104. Ghareib, M., Youssef, K. A., & Khalil, A. A., (1988). Ethanol tolerance of *Saccharomyces cerevisiae* and its relationship to lipid content and composition. *Folia Microbiologica, 33*(6), 447–452. https://doi.org/10.1007/BF02925769 (accessed on 8 December 2021).

105. Stanley, D., Bandara, A., Fraser, S., Chambers, P. J., & Stanley, G. A., (2010). The ethanol stress response and ethanol tolerance of *Saccharomyces cerevisiae*. *Journal of Applied Microbiology, 109*, 13–24. https://doi.org/10.1111/j.1365-2672.2009.04657.x (accessed on 8 December 2021).

106. Pereira, S. C., Maehara, L., Machado, C. M. M., & Farinas, C. S., (2015). 2G ethanol from the whole sugarcane lignocellulosic biomass. *Biotechnology for Biofuels, 8,* 1–16. https://doi.org/10.1186/s13068-015-0224-0 (accessed on 8 December 2021).

107. Ariyajaroenwong, P., Laopaiboon, P., Salakkam, A., Srinophakun, P., & Laopaiboon, L., (2016). Kinetic models for batch and continuous ethanol fermentation from sweet sorghum juice by yeast immobilized on sweet sorghum stalks. *Journal of the Taiwan Institute of Chemical Engineers, 66,* 210–216. https://doi.org/10.1016/j.jtice.2016.06.023 (accessed on 8 December 2021).

108. De Souza, D. M. O., Maciel, F. R., Mantelatto, P. E., Cavalett, O., Rossell, C. E. V., Bonomi, A., & Leal, M. R. L. V., (2015). Sugarcane processing for ethanol and sugar in Brazil. *Environmental Development, 15,* 35–51. https://doi.org/10.1016/j. envdev.2015.03.004 (accessed on 8 December 2021).

109. Veloso, I. I. K., Rodrigues, K. C. S., Sonego, J. L. S., Cruz, A. J. G., & Badino, A. C., (2019). Fed-batch ethanol fermentation at low temperature as a way to obtain highly concentrated alcoholic wines: Modeling and optimization. *Biochemical Engineering Journal, 141,* 60–70. https://doi.org/10.1016/j.bej.2018.10.005 (accessed on 8 December 2021).

110. Lip, K. Y. F., García-Ríos, E., Costa, C. E., Guillamón, J. M., Domingues, L., Teixeira, J., & Van, G. W. M., (2020). Selection and subsequent physiological characterization of industrial *Saccharomyces cerevisiae* strains during continuous growth at sub- and- supra optimal temperatures. *Biotechnology Reports, 26.* https://doi.org/10.1016/j.btre.2020. e00462 (accessed on 8 December 2021).

111. Izmirlioglu, G., & Demirci, A., (2010). Ethanol production from waste potato mash by using *Saccharomyces cerevisiae. American Society of Agricultural and Biological Engineers Annual International Meeting 2010, 2,* 1571–1581. https://doi.org/10.3390/ app2040738 (accessed on 8 December 2021).

112. Lin, W., Xing, S., Jin, Y., Lu, X., Huang, C., & Yong, Q. (2020). Insight into understanding the performance of deep eutectic solvent pretreatment on improving enzymatic digestibility of bamboo residues. *Bioresource Technology, 306,* 123163. https://doi. org/10.1016/j.biortech.2020.123163 (accessed on 8 December 2021).

113. Tan, L., Zhong, J., Jin, Y. L., Sun, Z. Y., Tang, Y. Q., & Kida, K. (2020). Production of bioethanol from unwashed-pretreated rapeseed straw at high solid loading. *Bioresource Technology, 303,* 122949. https://doi.org/10.1016/j.biortech.2020.122949 (accessed on 8 December 2021).

114. Kainthola, J., Shariq, M., Kalamdhad, A. S., & Goud, V. V. (2019). Enhanced methane potential of rice straw with microwave assisted pretreatment and its kinetic analysis. *Journal of Environmental Management, 232,* 188–196. https://doi.org/10.1016/j.jenvman. 2018.11.052 (accessed on 8 December 2021).

115. Shen, J., Zheng, Q., Zhang, R., Chen, C., & Liu, G. (2019). Co-pretreatment of wheat straw by potassium hydroxide and calcium hydroxide: Methane production, economics, and energy potential analysis. *Journal of Environmental Management, 236,* 720–726. https://doi.org/10.1016/j.jenvman.2019.01.046 (accessed on 8 December 2021).

116. Pérez-Pimienta, J. A., Papa, G., Gladden, J. M., Simmons, B. A., & Sanchez, A. (2020). The effect of continuous tubular reactor technologies on the pretreatment of lignocellulosic biomass at pilot-scale for bioethanol production. *RSC Advances, 10,* 18147–18159. https://doi.org/10.1039/d0ra04031b (accessed on 8 December 2021).

117. Niglio, S., Procentese, A., Russo, M. E., Piscitelli, A., & Marzocchella, A. (2019). Integrated enzymatic pretreatment and hydrolysis of apple pomace in a bubble column bioreactor. *Biochemical Engineering Journal, 150,* 107306. https://doi.org/10.1016/j. bej.2019.107306 (accessed on 8 December 2021).

118. Miranda, F. S., Rabelo, S. C., Pradella, J. G. C., Carli, C. Di, Petraconi, G., Maciel, H. S., … Vieira, L. (2020). Plasma in-Liquid Using Non-contact Electrodes: A Method of Pretreatment to Enhance the Enzymatic Hydrolysis of Biomass. *Waste and Biomass Valorization, 11,* 4921–4931. https://doi.org/10.1007/s12649-019-00824-5 (accessed on 8 December 2021).

119. Zeng, G., You, H., Wang, K., Jiang, Y., Bao, H., Du, M., … Gu, Z. (2020). Semi-simultaneous Saccharification and Fermentation of Ethanol Production from Sargassum horneri and Biosorbent Production from Fermentation Residues. *Waste and Biomass Valorization, 11,* 4743–4755. https://doi.org/10.1007/s12649-019-00748-0 (accessed on 8 December 2021).

120. de Carvalho Silvello, M. A., Martínez, J., & Goldbeck, R. (2020). Application of Supercritical CO_2 Treatment Enhances Enzymatic Hydrolysis of Sugarcane Bagasse. *Bioenergy Research, 13,* 786–796. https://doi.org/10.1007/s12155-020-10130-x (accessed on 8 December 2021).

121. Yuan, Z., Wei, W., Wen, Y., & Wang, R. (2019). Comparison of alkaline and acid-catalyzed steam pretreatments for ethanol production from tobacco stalk. *Industrial Crops and Products, 142,* 111864. https://doi.org/10.1016/j.indcrop.2019.111864 (accessed on 8 December 2021).

122. Antunes, F. A. F., Chandel, A. K., Brumano, L. P., Terán Hilares, R., Peres, G. F. D., Ayabe, L. E. S., … Da Silva, S. S. (2018). A novel process intensification strategy for second-generation ethanol production from sugarcane bagasse in fluidized bed reactor. *Renewable Energy, 124,* 189–196. https://doi.org/10.1016/j.renene.2017.06.004 (accessed on 8 December 2021).

123. Matsakas, L., Nitsos, C., Raghavendran, V., Yakimenko, O., Persson, G., Olsson, E., … Christakopoulos, P. (2018). A novel hybrid organosolv: Steam explosion method for the efficient fractionation and pretreatment of birch biomass. *Biotechnology for Biofuels, 11,* 160. https://doi.org/10.1186/s13068-018-1163-3 (accessed on 8 December 2021).

124. Yang, J., Zhang, X., Yong, Q., & Yu, S. (2011). Three-stage enzymatic hydrolysis of steam-exploded corn stover at high substrate concentration. *Bioresource Technology, 102,* 4905–4908. https://doi.org/10.1016/j.biortech.2010.12.047 (accessed on 8 December 2021).

125. Lau, M. W., Dale, B. E., & Balan, V. (2008). Ethanolic fermentation of hydrolysates from ammonia fiber expansion (AFEX) treated corn stover and distillers grain without detoxification and external nutrient supplementation. *Biotechnology and Bioengineering, 99,* 529–539. https://doi.org/10.1002/bit.21609 (accessed on 8 December 2021).

126. Rosgaard, L., Andric, P., Dam-Johansen, K., Pedersen, S., & Meyer, A. S. (2007). Effects of substrate loading on enzymatic hydrolysis and viscosity of pretreated barley straw. *Applied Biochemistry and Biotechnology, 143,* 27–40. https://doi.org/10.1007/ s12010-007-0028-1 (accessed on 8 December 2021).

127. García-Aparicio, M. P., Oliva, J. M., Manzanares, P., Ballesteros, M., Ballesteros, I., González, A., & Negro, M. J. (2011). Second-generation ethanol production from steam exploded barley straw by *Kluyveromyces marxianus* CECT 10875. *Fuel, 90,* 1624–1630.

128. Jørgensen, H., Vibe-Pedersen, J., Larsen, J., & Felby, C. (2006). Liquefaction of ligno-cellulose at high-solids concentrations. *Biotechnology and Bioengineering, 96*, 862–870. https://doi.org/10.1002/bit.21115 (accessed on 8 December 2021).
129. Ingram, T., Wörmeyer, K., Lima, J. C. I., Bockemühl, V., Antranikian, G., Brunner, G., & Smirnova, I. (2011). Comparison of different pretreatment methods for lignocellulosic materials. Part I: Conversion of rye straw to valuable products. *Bioresource Technology, 102*, 5221–5228. https://doi.org/10.1016/j.biortech.2011.02.005 (accessed on 8 December 2021).
130. Zhang, Y., Han, B., & Ezeji, T. C. (2012). Biotransformation of furfural and 5-hydroxy-methyl furfural (HMF) by *Clostridium acetobutylicum* ATCC 824 during butanol fermentation. *New Biotechnology, 29*, 345–351. https://doi.org/10.1016/j.nbt.2011.09.001 (accessed on 8 December 2021).
131. Wang, W., Zhuang, X., Yuan, Z., Yu, Q., Qi, W., Wang, Q., & Tan, X. (2012). High consistency enzymatic saccharification of sweet sorghum bagasse pretreated with liquid hot water. *Bioresource Technology, 108*, 252–257. https://doi.org/10.1016/j.biortech.2011.12.092 (accessed on 8 December 2021).
132. Phukoetphim, N., Salakkam, A., Laopaiboon, P., & Laopaiboon, L. (2017). Improvement of ethanol production from sweet sorghum juice under batch and fed-batch fermentations: Effects of sugar levels, nitrogen supplementation, and feeding regimes. *Electronic Journal of Biotechnology, 26*, 84–92. https://doi.org/10.1016/j.ejbt.2017.01.005 (accessed on 8 December 2021).
133. Mussatto, S. I., Machado, E. M. S., Carneiro, L. M., & Teixeira, J. A. (2012). Sugars metabolism and ethanol production by different yeast strains from coffee industry wastes hydrolysates. *Applied Energy, 92*, 763–768. https://doi.org/10.1016/j.apenergy.2011.08.020 (accessed on 8 December 2021).
134. Sathesh-Prabu, C., & Murugesan, A. G. (2011). Potential utilization of sorghum field waste for fuel ethanol production employing *Pachysolen tannophilus* and *Saccharomyces cerevisiae*. *Bioresource Technology, 102*(3), 2788–2792. https://doi.org/10.1016/j.biortech.2010.11.097 (accessed on 8 December 2021).
135. Scordia, D., Cosentino, S. L., Lee, J. W., & Jeffries, T. W. (2012). Bioconversion of giant reed (Arundo donax L.) hemicellulose hydrolysate to ethanol by *Scheffersomyces stipitis* CBS6054. *Biomass and Bioenergy, 39*, 296–305. https://doi.org/10.1016/j.biombioe.2012.01.023 (accessed on 8 December 2021).
136. Zhao, J., & Xia, L. (2010). Bioconversion of corn stover hydrolysate to ethanol by a recombinant yeast strain. *Fuel Processing Technology, 91*(12), 1807–1811. https://doi.org/10.1016/j.fuproc.2010.08.002 (accessed on 8 December 2021).
137. Kumari, R., & Pramanik, K. (2013). Bioethanol production from *Ipomoea Carnea* biomass using a potential hybrid yeast strain. *Applied Biochemistry and Biotechnology, 171*(3), 771–785. https://doi.org/10.1007/s12010-013-0398-5 (accessed on 8 December 2021).
138. Gupta, R., Sharma, K. K., & Kuhad, R. C. (2009). Separate hydrolysis and fermentation (SHF) of Prosopis juliflora, a woody substrate, for the production of cellulosic ethanol by Saccharomyces cerevisiae and Pichia stipitis-NCIM 3498. *Bioresource Technology, 100*(3), 1214–1220. https://doi.org/10.1016/j.biortech.2008.08.033 (accessed on 8 December 2021).
139. Li, H., Kim, N. J., Jiang, M., Kang, J. W., & Chang, H. N. (2009). Simultaneous saccharification and fermentation of lignocellulosic residues pretreated with phosphoric acid-acetone for bioethanol production. *Bioresource Technology, 100*(13), 3245–3251. https://doi.org/10.1016/j.biortech.2009.01.021 (accessed on 8 December 2021).

140. Yan, J., Wei, Z., Wang, Q., He, M., Li, S., & Irbis, C. (2015). Bioethanol production from sodium hydroxide/hydrogen peroxide-pretreated water hyacinth via simultaneous saccharification and fermentation with a newly isolated thermotolerant *Kluyveromyces marxianu* strain. *Bioresource Technology, 193*, 103–109. https://doi.org/10.1016/j.biortech.2015.06.069 (accessed on 8 December 2021).

141. Tomás-Pejó, E., Oliva, J. M., González, A., Ballesteros, I., & Ballesteros, M. (2009). Bioethanol production from wheat straw by the thermotolerant yeast *Kluyveromyces marxianus* CECT 10875 in a simultaneous saccharification and fermentation fed-batch process. *Fuel, 88*(11), 2142–2147. https://doi.org/10.1016/j.fuel.2009.01.014 (accessed on 8 December 2021).

142. Peng, L., & Chen, Y. (2011). Conversion of paper sludge to ethanol by separate hydrolysis and fermentation (SHF) using *Saccharomyces cerevisiae*. *Biomass and Bioenergy, 35*(4), 1600–1606. https://doi.org/10.1016/j.biombioe.2011.01.059 (accessed on 8 December 2021).

CHAPTER 10

Integrated Production of Ethanol from Starch and Sucrose

C. A. PRADO,[1] S. SÁNCHEZ-MUÑOZ,[2] R. T. TERÁN-HILARES,[3] L. T. CARVALHO,[2] L. G. DE ARRUDA,[4] M. L. SILVA DA CUNHA,[2] P. ABDESHAHIAN,[1] S. S. DA SILVA,[1] N. BALAGURUSAMY,[5] and J. C. SANTOS[2]

[1]*Biopolymers, Bioprocesses, Process Simulation Laboratory, Department of Biotechnology, Engineering School of Lorena, University of São Paulo (EEL-USP), Lorena–12.602.810, SP, Brazil*

[2]*Bioprocesses and Sustainable Products Laboratory, Department of Biotechnology, Engineering School of Lorena, University of São Paulo (EEL-USP), Lorena–12.602.810, SP, Brazil, E-mail: jsant200@usp.br (J. C. Santos)*

[3]*Material Laboratory, Catolic University of Santa Maria (UCSM), Yanahuara–04013, AR, Perú*

[4]*Biopolymers, Bioprocesses, Process Simulation Laboratory, Department of Biotechnology, Engineering School of Lorena, University of São Paulo (EEL-USP), Lorena–12.602.810, SP, Brazil*

[5]*Bioremediation Laboratory, Biological Science Faculty, Autonomous University of Coahuila (UAdeC), Torreón Campus, Coahuila–27276, México*

ABSTRACT

In recent years, several scientific and technological advances in bio-refinery have greatly contributed to the evolution of the ethanol industry around the world. These contributions have increased our view and knowledge about technologies developed for first-generation (1G) and, more recently, second-generation (2G) ethanol. Nowadays, advanced technologies are employed to generate this alcohol in top ethanol-producing countries. In Brazil, for instance, a new innovation is the flex biorefinery in the first-generation model, which has been considered as a primary concept for the use of corn in the off-season of sugarcane, thus increasing the annual production of bioethanol. In this concept, there is also the possibility of flex biorefinery with 1G and 2G ethanol production. In fact, the conversion of sugarcane biomass into the fermentable sugars for the production of second-generation ethanol is a promising alternative to meet the future demands for this biofuel and could be successfully integrated with sugarcane and corn-based 1G facilities. However, in order to attain such success and to take advantage of those flexible industries, the construction of a bridge between science and industry is essential. In this view, investments in research and development with the transfer of new technologies from the academy to the industry are required. Furthermore, the training of skilled labor to deal with new technological challenges is essential.

10.1   INTRODUCTION

Petroleum-based energy sources are being exhausted, because of increasing world energy demand and the non-renewable global crude oil reserves [1–3]. As a consequence of the depletion of this conventional energy source, oil prices are unstable and directly bring about price fluctuation in the transportation fuel [4]. For these reasons, the change of the current model to the use of biofuels as a sustainable option is highly necessary, as a matter of energy security [5, 172] and it causes to mitigate the GHG emissions resulting from the burning of fossil fuels [2, 173].

Considering the aforementioned issues, many attempts are being made to use research and new technologies for the developments of traditional industries of biofuels, particularly ethanol.

Indeed, although a number of cars have been designed to run on ethanol, gasoline is still the primary fuel option for cars in many countries. In this regard, because of the 1970's oil crisis, the Brazilian government launched the "Proalcool" program in 1975 with the aim of reducing the country's dependence on oil imports. Nowadays, many cars in Brazil are flex with the possibility of using ethanol or gasoline. In this context, there are different policies in many countries for increasing ethanol blended with commercial gasoline [174, 175].

Currently, there are many projects for the production of bioethanol across the world, which is known greatly in line with the corn biorefinery in USA and sugarcane biorefinery in Brazil [176], while there are other countries investing in the utilization of this biofuel. For example, in America, Mexico has invested in studies focused on the first-generation biomass such as sugarcane, agave, and sorghum [177–180]. Moreover, the expansion of the corn crop in this country for bioethanol production was evaluated [181]; however, the current legislation prohibits the use of corn as an energy raw material, due to its importance as the main source of food in the country.

The share of ethanol in the consumption of liquid fuels in bioethanol production improved from 55% in 2012 to 75% in 2017 in the world [182]. The global ethanol production in 2019 was 37.8 billion gallons in which USA and Brazil had 83% and 26% of the total, respectively [183]. In the United States, corn is the main raw material to produce ethanol so that in 2017 it represented 10% of the demand for vehicular fuel. In Brazil, on the other hand, sugarcane (juice or molasses) is the main raw material for ethanol production [176].

Raw material choice is the fundamental step for the ethanol industry, which can represent up to 42% of the production costs [184, 185]. In addition, the commercial adoption of biofuels depends on the evaluation of the broad levels of efficiency [185, 186] considering several economic, social, environmental, and strategic criteria, such as national energy security. Hence, in order to use a source of bioenergy, the chosen biomass needs to meet these requirements. The ideal characteristics for using a biomass feedstock as a source of energy include high agricultural production, favorable natural cycles, low energy consumption in its cultivation, a low production cost, the low levels of contaminants, and a low demand for nutrients [6]. Furthermore, it is important that the selection of biomass can favor the carbon balance when considering life cycle of the biofuel, thus it can take into account the entire production and use in the production chain [175, 176, 184, 185].

Currently, ethanol can be classified as the first generation (1G) when it is produced from food crops such as sucrose and starch. Ethanol is known as the second generation (2G) when it is produced from biomass residues and byproducts such as sugarcane bagasse and corn cob, while third-generation (3G) of ethanol is referred to the ethanol produced from microbial biomass, namely microalgae. In this context, some authors indicate ethanol as the fourth generation (4G) of ethanol when it is produced from genetically modified organism, e.g., cyanobacteria through a process named 'photo-fermentation' (direct conversion of light and carbon dioxide (CO_2) into ethanol) [7]. Moreover, ethanol has been called as 1.5G when it is produced in the 1G industry, but additional sugars from cellulosic fibers is utilized, for example, from corn grain [8].

These concepts are important to a systematic study of new options for the integration of biorefineries and raw materials. In Brazil, the main raw material for ethanol production is sugarcane juice that is produced only for the alcohol, or as more commonly observed, sugarcane molasses used in integrated sugar and ethanol industries [9]. However, Brazil is a great producer of corn, mainly in some regions such as central west of the country. Thus, sugarcane has an interseason period from December to April, which is an interesting opportunity for 1G biorefineries using sugarcane with corn as raw materials in flex biorefineries [184].

Indeed, corn and sugarcane biorefineries are established processes used in the world to produce ethanol [3, 4, 10] such as Brazil so that such countries could integrate those raw materials and take advantage of their characteristics. The aim of the present chapter is to discuss the possibility of the integration of the use of both raw materials (corn and sugarcane) and to give future perspectives of the integration in flex biorefineries. In this way, a brief general overview is presented in the first sections.

10.2 MAIN CARBON-RICH RAW MATERIALS FOR ETHANOL PRODUCTION IN DIFFERENT COUNTRIES

Several countries have evaluated the production of ethanol from biomass feedstocks available in their region. There are a variety of renewable carbon sources with the appropriate potential for bioethanol production in integrated biorefineries [187]. Corn, for example, is produced in large quantities for ethanol generation in the world compared to any other crop. Around 850 million tons of maize kernels are produced worldwide, indicating an average productivity of 5.2 t/ha [11].

Vegetable biomass is classified as different carbon sources according to the type of carbohydrate. They can be divided into starchy compounds such as corn and cassava, sucrose-based substances (sugarcane juice) and lignocellulosic materials such as sugarcane bagasse, corn stover and corn straw [12].

Table 10.1 shows a number of potential biomass for ethanol production in the world. It is noteworthy that lignocellulosic biomass (LCB) represents appropriate raw materials for the production of bioethanol. Agro-industrial residues and woody by-products include residues derived from forest, the paper and cellulose industries (sepilho, shavings, and declassified chips of eucalyptus and pine), sawmills (sawdust), and agricultural residues produced from crops such as cereal straw, corn, wheat, corn cob, rice husks and oats [188]. In this context, it has been estimated that more than 75.73 million tons of dry biomass are generated worldwide from crops and agro-industrial residues.

TABLE 10.1 Main World Carbon Sources (Starchy and Sucrose-based Ones) for Ethanol Production

Countries	Biomass Production (million tons)	Ethanol Production (million gallons)	Biomass Residual in Biorefinery (million tons)	References
United States (data of 2018)	342 (corn) 149 (biomass from corn for ethanol production)	75 (corn ethanol production)	68.25 (corn straw)	[13, 14]
Brazil (data of 2019)	642.7 (sugarcane) 95.60 (corn)	26 (sugarcane ethanol production) 2.5 (corn ethanol production	96.3 (sugarcane bagasse lignocellulosic)	[11, 15, 189]
Europe Union (data of 2019)	61.60 (corn) 168.28 (wheat) 35.82 (beet)	13 (corn) 21 (beet)	64.28 (wheat straw)	[7, 14]
China (data of 2015)	215.00 (corn)	8.45 (corn)	–	[16]
Canada (data of 2019)	203 (corn)	4.36 (corn)	–	[17]
India (data of 2019)	115.6 (rice) 85.72 (corn) 21.16 (wheat)	2.23 (rice)	22.57 (rice straw)	[18, 190]

Considering the large quantity of raw materials, ethanol production industries could be established as flexible, allowing the utilization of different biomass feedstocks according to the seasonal availability. Indeed, the seasonality of the crops is an important issue to be considered for the production of biofuels. Sugarcane production, for example, has a well-defined off-season, from December to April. In addition, as a new solution, integrated biorefineries have been installed in the Central-West region of Brazil, considering the production of alcohol from corn in those months [187]. Therefore, ethanol production will cease, if no raw material is supplied during December to April. This fact occurs due to the perishability of sugarcane, which cannot be stored higher than 48 hours [191]. This problem is not common for corn, and thus, it is a suitable alternative to sugarcane in the interseason period of sugarcane cultivation in Brazil [19, 192]. Thus, the utilization of sugarcane and corn reveals an interesting example of a highly flexible and productive biorefinery.

10.2.1 CORN

The raw materials used for the first-generation fuel ethanol are mainly crops rich in starch or sucrose such as corn, cassava, and sugarcane [2, 20]. Corn (*Zea mays*) is a cereal grain, which is widely grown all over the world. It is widely used as a staple food for humans or as animal feed, due to its numerous nutritional properties. All studies suggest that its origin is Mexican since its domestication started 7,500 to 12,000 years ago in the center of Mexico [21]. Maize contains facilities for its cultivation and therefore has a great potential for production and adaptation to technology. It has a mechanized cultivation that benefits greatly from modern planting and harvesting techniques [20]. The world production of corn was 850 million tons in 2019, which was more than rice (678 million tons) and wheat (682 million tons) [184]. Corn is grown in different regions of the world. The largest producers are the USA and China, which have produced 37 and 21% of the total world production, respectively [11, 22].

10.2.2 SUGARCANE

The sugarcane cycle plant has an average of six years, in which four or five harvests occur. Sugarcane belongs to a group of tall perennial grass species of the genus *Saccharum*. This plant is native in the tropical regions such as the South of Asia and North of Brazil, and can be used for the production of ethanol and sugar [193]. Its stems are robust, articulated, and fibrous. Sugarcane height can be two to six meters. Researchers have been mixed sugarcane species to develop complex hybrid for enhancing its performance in different areas. Sugarcane belongs to economically important plants such as corn, wheat, rice, and sorghum [194, 195]. Sucrose is the principal sugar obtained and can be extracted and purified in biorefineries for further use as a raw material in the food industry, or it would be fermented to produce ethanol [196]. In 2019, sugarcane grew on about 26.0 million hectares of agricultural lands in more than 90 countries, with a worldwide harvest of 174 million tons [184]. Brazil is the largest producer of sugarcane in the world. Other largest producers are Thailand, India, China, Pakistan, and Mexico [11].

10.3 PROCESS IN BIOREFINERIES FOR ETHANOL PRODUCTION FROM CORN AND SUGARCANE

The development of biorefineries leads to the efficient utilization of the LCB, which can be used directly in the generation of bioenergy or serve as raw materials for chemical, enzymatic or fermentative processes [23, 197]. Those integrated facilities can work with different raw materials and technological routes, aiming to simultaneous production of different valuable products.

The research on new methods for biorefinery integration in power plants for production of biofuels from corn and sugarcane can reduce costs and increase profits. For this purpose, the understanding of how each biorefinery works and what are their bottlenecks of production is crucial.

10.3.1 CORN TO ETHANOL BIOREFINERY

The production of ethanol from corn is an established process used in American biorefineries [198]. Commercial ethanol fuel is obtained through the following main steps including the conversion of starch into simple sugars (fermentable sugars), the run of fermentation process and distillation process [14, 24–26]. The process of breaking the starch polymer into fermentable sugars consists of three sequential unit operations, namely milling, liquefaction, and enzyme-based saccharification [14, 21, 27].

Milling is a physical process that aims to reduce particle size, in which corn is processed through a hammer mill to produce corn flour. There are two well-established methods to produce bioethanol from corn and the major differences between them are related to this step. Wet-milling ethanol production is a large-scale capital-intensive process that consists of steeping corn grains in water to facilitate the separation of the various fractions of the corn kernel (starch, fiber, germ), allowing them to be processed to obtain various by-products. Dry milling is the most common process, corresponding to 90% of the US operational plants [14, 20, 28, 199]. In this case, the compounds of the grain are not separated and they are processed with the starchy fraction. Proteins, oils, and other composts remain unchanged in fermentation and will be separated further [21]. Since the latter is more common, it will be more discussed in detail.

In the conventional dry-grind process, after grinding grains in hammer mills, the cornflour is mixed with water to form corn slurry with 27–37% of solid content [29]. Then, the pH value of the mixture is adjusted and α-amylase, a thermostable enzyme, is added [24, 200]. The next step is called liquefaction or cooking, in which the slurry is heated to high temperatures to soften the tightly bonded grain of starch that becomes more digestible. The jet-cookers inject steam at high temperatures and high pressures. In this step, the process time depends on how the high temperature is utilized. The slurry can be cooked at 165°C for 3–5 minutes or at 90–105°C for 1–3 hours [14, 27], resulting in a mixture with gelatinized starch that is converted to dextrins and oligosaccharides by -amylase [20, 30].

The output from liquefaction is called corn mash which is rich in short-chain saccharides, showing partial solubility with no fermentative characteristic [30]. The corn mash is cooled approximately to 30°C [14, 21], and it proceeds to the saccharification process, in which the conversion of dextrins into fermentable sugars occurs. This hydrolysis happens in the presence of glucoamylase (GA, also called amyloglucosidase or AMG) for converting dextrins into glucose [10, 201]. In the past, the saccharification was carried by acid hydrolysis in a hazardous operation that occurred in extreme conditions of temperature and pH with sugar production yields of 85% [20]. The utilization of enzymes led to reduce the disadvantages and raise the yield up to 95–97% [30].

The corn hydrolysate is subjected to ethanol fermentation in which yeast *Saccharomyces cerevisiae* consumes glucose and produces ethanol and CO_2 under anaerobic conditions [1, 31, 200]. The yeast grows in seed tanks and then is transferred to the sugary syrup in the fermentation tanks [21]. Fermentation is the main operation in a distillery plant. It runs for 42–55 h using multiple fermenters as batch operation of the plant [32].

There are many different possible configurations for this process. The most common modes are SHF (separate hydrolysis and fermentation) and SSF (simultaneous saccharification and fermentation) [202]. The main benefit of the SSF process is to avoid the osmotic shock to yeast due to the high glucose concentrations. In this process, glucose is slowly released through the saccharification process and immediately consumed by the yeast to produce ethanol, preventing a glucose concentration that can cause inhibition [33, 203, 204]. The SSF technique can provide up to 8% more ethanol than the SHF for the same amount of grain [24, 32].

The fermentation product is composed by a liquid portion containing 14–20% of ethanol [29, 199]. Thus, the distillation goal is to obtain a product with higher ethanol purity (92–98%) [21, 34]. Ethanol is separated from water and non-fermentable residues which form the stillage [201]. This process can be separated in two stages. In the first column, unconverted solids and heavy compounds are removed at the bottom, while the top outlet feeds the second column with a liquid portion containing 30–40% ethanol. In the second column, ethanol is purified in which hydrous ethanol is obtained [32]. Pure ethanol can be recovered by combining distillation and molecular sieves which capture residual water to achieve anhydrous ethanol [35].

The stillage includes fiber, oil, and protein components of the grain, as well as non-fermented starch. The stillage can be processed to obtain byproducts [24]. It can be centrifuged to separate the solid and liquid fractions and produce a variety of coproducts known as distiller's grains [36]. The most popular is distillers dried grains (DDGs) which is used as animal feed [14]. In addition to animal feed, another coproduct obtained from the dry grind operational plants is distillers corn oil. It is recovered from the concentrated stillage liquid portion obtained from the centrifugation of whole stillage [10, 198].

The requirement of high temperatures is an unfavorable factor in bioethanol industries, since it increases the energy demand, which adversely affects the cost-effectiveness of the product, even though the elevated temperatures help contamination control [37]. In the context of 1G corn ethanol, some of the research's address reducing operational costs such as the energy demand [38]. Other issues that also draw attention are crop improvement [5], new enzymes [3], genetic engineering in yeast strains [39, 40] and coproduct recovery (Figure 10.1) [41].

10.3.2 SUGARCANE TO ETHANOL BIOREFINERY

In Brazil, this biomass is largely used for 1st generation ethanol [6, 199].

The 1st generation ethanol is produced by processing sugarcane via fermentation of sucrose-rich juice or molasses [42, 43]. In integrated sugar and alcohol industries (the most common), there is a strict relation between ethanol production and other main products (sugar and energy generated by burning bagasse) which is influenced by market demand, within the biorefinery concept [44, 45]. Figure 10.2 exemplifies a common industrial processes of a sugarcane biorefinery (1st generation) with joint production of ethanol, sugar, and energy.

At the reception, the harvested cane is aimed at cleaning tables and the feeding system. Furthermore, in this stage the impurities, such as vegetable and mineral residues are removed. If the harvested cane is chopped, a dry wash is adopted to minimize the loss of sugars during the reception process. Then, the clean cane is sent for preparation [46, 195].

In the preparation stage, the sugarcane is crushed by a series of crushing knives, allowing a uniform plant material, facilitating the next stage of juice extraction. Before extracting the juice, the chopped cane is submitted to a magnet treatment to remove metallic residues [189]. In the juice extraction stage in a usual process, the chopped cane is crushed via mills (a set of three to five cylindrical rolls), where the juice is separated from the vegetable fibrous fraction (bagasse). They are usually sequential mill systems. The last set of milled sugarcane is soaked into warm water, a way to increase the extraction of residual sugars [200]. Despite the separation that has been already done, a fibrous fraction still remains in the juice. In order to remove it, sieves are used, and the solid material is sent for recirculation in the mill system. The juice from the first group of mills is often utilized for sugar production, due to its great degree of purity and concentration of sugars. On the other hand, the juice of the other groups of mills which are combined with molasses is used for ethanol production [202–206].

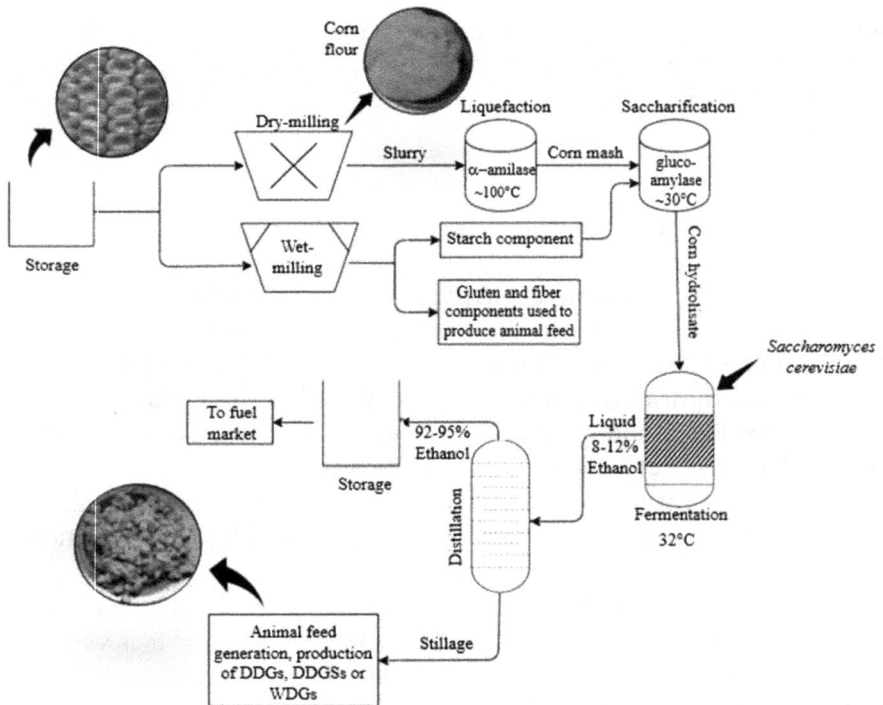

FIGURE 10.1   General scheme for the production of 1G corn ethanol in biorefinery.

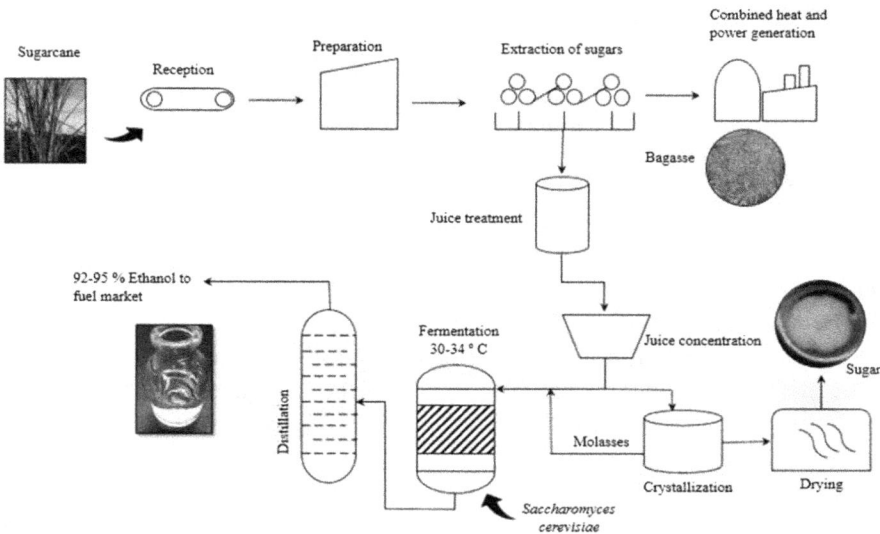

FIGURE 10.2 General scheme for the production of 1G ethanol, sugar, and energy from sugarcane in biorefinery.

The juice treatment is aimed at removing residual impurities in the juice. It consists of a set of physical-chemical treatments, ranging from the separation of fibers, sand, and particulate material with the use of sieves, heating of the juice (30 to 70°C), the addition of lime and phosphoric acid, followed by a second heating (105°C), and an addition of flocculating agent until clarification of the juice [47] [184]. Depending on the manufacturing, this process may be simpler once it is adapted according to the type of mill used and the purity of the juice after extraction. In some cases, the impurities removed in the juice treatment are still used for the recovery of sugars by filtration [6, 48].

The concentration of the treated juice is adjusted (juice concentration stage) to an ideal value for fermentation [7]. Typically, a molasses mixture obtained from the production of sugar is firstly combined with the treated broth, and when necessary, it is secondly used by evaporators for concentration adjustments [49]. Therefore, the juice is prepared for fermentation, which consists of a biological conversion of sugars into ethanol by the yeast *Saccharomyces cerevisiae* with the generation of CO_2 and other by-products (organic acids, higher alcohols, etc.), [48, 50]. The fermentation is conducted in the batch mode and is fed by cell recycling from the previous fermentation. The process is carried out in closed reactors continuously or discontinuously at a temperature of 30 to 34°C for a few hours [6, 51].

As a result, a liquid with the low ethanol content (maximum 10° GL) is obtained, and CO_2 which is still washed in adsorption columns is used to recover transported ethanol. After fermentation, the juice extracted is submitted to centrifugation, and the cells are removed in this process, followed by treating with water and sulfuric acid solution as a method to reduce future contamination [207, 208]. It is then stored and sent to the following fermentation batches [13, 52].

The liquid proceeds to the distillation/rectification stage, where ethanol is purified in distillation and rectification columns, resulting in hydrated ethanol. This ethanol can also increase the degree of purity and it is dehydrated in alternative purification systems such as azeotropic distillation with cyclohexane, extraction using monoethylene glycol, or adsorption in sieves [53, 209].

The first generation biorefineries are considered a consolidated technology with a low-cost production line. However, they are still dependent on agricultural resources, arable land, climatic conditions, domestic supply market and investment capital which should be implemented in new regions or countries [184].

The industrial facilities for the production of 1G ethanol are also designed for joint production of sugar and energy cogeneration. This has been resulted from the burning of existing bagasse and straw, as well as the generation of other products such as biogas obtained from fermented broth and proteins in the form of yeast cells [13].

The production of sugar within the biorefinery follows common lines for the production of ethanol. In this approach, the concentration of the treated juice is adjusted so that heated processes of evaporation of the broth, crystallization, and drying of the sugars occur (humidity between 0.5 and 2%) [52, 205]. Molasses obtained after crystallization can be used to produce ethanol. In the energy cogeneration system, bagasse (for mechanized harvests) is burned in a boiler and produced steam is driven to turbines, which in turn are connected to electric generators [49, 54]. Thus, a part of the generated electricity can be sold to distributing companies, and the exhausted steam can also be used in various operations at process units that require thermal energy [23, 55, 56, 208].

10.4　2G AND 1.5G ETHANOL FROM BY-PRODUCTS OF CORN AND SUGARCANE

The first-generation bioethanol is already a well know and established technology [57]. However, the increasing ethanol demand results in the feedstock production, affecting the food and land usage concerns [19, 57]. In order to better apply the feedstock instead of the focus on the increase of its production, the second-generation biofuel has been a prominent technology since it relies on lignocellulosic materials [58]. Lignocellulosic substances are variable and available materials, which can be divided into agricultural residues, forest residues, municipal solid waste, and energy crops (herbaceous or woody plants) [2]. The main components of this biomass are cellulose, hemicellulose, and lignin with small quantities of extractives and ashes [59]. Some examples of biomass applied for the second-generation bioethanol are sugarcane bagasse [60], eucalyptus [61], hazelnut shell [62], corncob [63], banana crop [64], rice straw [65], coconut husk [58], and sugar beet [210]. Each biomass presents specific quantities of cellulose, hemicellulose, and lignin so that sugarcane bagasse, for example, is composed of 20–25% lignin, 40–50% cellulose and 30–35% hemicellulose, making it a potential feedstock for 2G biofuel [57, 66]. Another example of potential biomass is sugar beet which is also used for bioethanol production and presents a high quantity of hemicelluloses (24–32%), and cellulose (22–30%) with a very low lignin amount [210].

Even though the lignocellulosic residues are abundant and do not interfere in food production, it is necessary to optimize the biomass bioconversion in order to produce an economically viable second-generation bioethanol [66]. Bioconversion of lignocellulosic materials consists of four main steps including pretreatment, enzymatic hydrolysis, fermentation, and product recovery [67]. In general, pretreatment contributes to the second-generation bioethanol production by facilitating the subsequent hydrolysis through the modification of the amorphous region and porosity of matrix with separating cellulose from hemicellulose and the lignin [68]. Many technologies have already been developed for lignocellulose pretreatment, and they can be divided into physical, chemical, physical-chemical, and biological [211]. Each type of the pretreatment has demonstrated advantages and disadvantages, acting in different methods to facilitate cellulose hydrolysis.

A large number of technologies have been focusing on the development of an efficient and economic lignocellulosic pretreatment. Pretreatment is one of the main macro-steps of the process in biorefineries and is closely related to the subsequent stages of hydrolysis and fermentation. Therefore, it is necessary to use a method that presents, in addition to high efficiency, low energy consumption and reduced chemical catalysts to result in a high recovery of carbohydrate fractions, low or no formation of fermentation inhibitors and high simplicity for use in a larger scale.

Physical methods include milling [69], extrusion [70], and irradiation [71]. Different methods of irradiation such as gamma rays and microwave have demonstrated to make significant changes in the lignocellulosic material including the increase of surface area, expansion of pore volume and the decrease of degree of polymerization (DP) [211].

Chemical pretreatment is also commonly applied to enhance delignification and reduce cellulose crystallinity. Some examples of chemical pretreatment are acid [72], alkaline [212], organosolv [73] and ionic liquids (ILs) [74]. A combination of two different chemical pretreatment, such as acid and alkali pretreatment can be applied as demonstrated by Gomes et al. [213]. The authors developed a the two-stages sugarcane bagasse pretreatment with sulfuric acid followed by sodium hydroxide which was combined with high-solid enzymatic hydrolysis, resulting in more than 150 g/L of glucose [213].

Other examples of combinations are steam explosion (SE) [75], ammonia recycle percolation [214], microwave-chemical [76], and hydrodynamic cavitation (HC) [77]. For example, in an optimized condition of NaOH pretreatment, a maximum glucose production equivalent to 85% of enzymatic digestibility is observed. For the pretreatment of CH, it is possible that this number reaches 96% digestibility. This is a new solution for the bottleneck regarding 2 G produce ethanol [78]. Pretreatment based on HC has advantages compared to other methods, such as the requirement of milder pretreatment conditions and shorter process times [77]. Microorganisms such as fungi and bacteria are also used in lignocellulosic pretreatment. They are capable of modifying and degrading the complex structure into simpler substrates by enzyme digestion. Some examples of microorganism are white-rot, brown-rot, and soft-rot fungi. White-rot fungi stand out for providing better sugar yields [215, 216]. However, microbiological pretreatment has considerable disadvantages, especially the low hydrolysis rate caused by the presence of inhibitors and the broth conditions [79].

After the pretreatment, it is necessary to convert the polymeric carbohydrate obtained from lignocellulosic material into fermentable sugars (Figure 10.3), due to the yeast inability to process carbohydrate polymers [217]. Therefore, two hydrolysis methods can be applied in order to provide second-generation bioethanol production, including acid and enzymatic hydrolysis, which are chosen according to the biomass composition [67]. Acid hydrolysis can be performed by dilute acid or concentrated acid with demanding different temperature and pressure conditions [17]. Sulfuric acid (H_2SO_4) [80] is mostly applied for acid hydrolysis, however, researchers have also developed different methods, including hydrochloric acid (HCl), nitric acid (HNO_3), and phosphoric acid (H_3PO_4) [81]. Enzymes can also be used to hydrolyze both cellulose and hemicellulose components, in order to produce fermentable sugars. Enzymatic hydrolysis (enzymatic saccharification) has

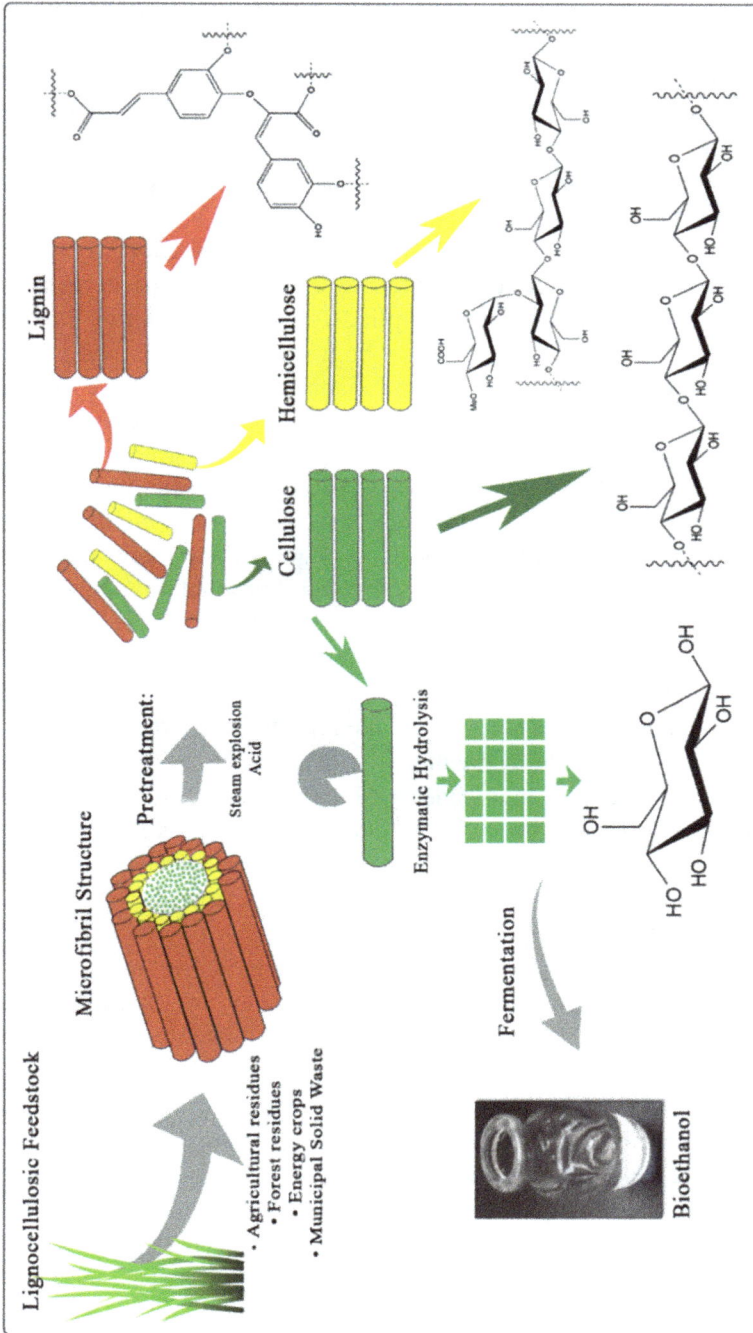

FIGURE 10.3 Schematic process of ethanol conversion for 2nd generation of biomass (stages for 2G pretreatment, enzymatic hydrolysis, and fermentation).

high advantages such as the requirement of mild process conditions (low temperature and atmospheric pressure) with avoiding acid corrosion [17, 82]. This type of hydrolysis enhances the soluble sugar production by cellulase, xylanase or amylase [82]. In order to reduce the cost of 2G bioethanol and make it large scale commercialization, it is necessary to improve the process economics, for example, by decrease in the cost of cellulase production and enzymatic hydrolysis [83].

Fermentation is the main process for bioethanol production. It converts soluble sugars into the alcohol by the metabolic process of microorganisms. A range of microorganisms can be applied for the fermentation process, however, not all of them can process both pentoses (e.g., xylose and arabinose) and hexoses (e.g., glucose and mannose) [67, 218]. A co-culture system has been reported in the literature, favoring both pentose and hexoses fermentation. Farias and Filho [84] evaluated a co-culture fermentation using hexoses-fermenting yeasts (such as *Saccharomyces cerevisiae*) and xylose-fermenting yeasts (*Scheffersomyces stipitis* and *Spathaspora passalidarum*), resulting in a maximum ethanol titer of $49.2\,\mathrm{gL^{-1}}$. Four main alternative processes are commonly developed for bioethanol production, namely SSF, SHF, SSCF, and consolidated bioprocessing (CBP) [7]. Each alternative interferes in the process with advantages and disadvantages, giving the lignocellulosic bioethanol production options according to a variety of the processes, especially by consideration of the feedstock, microorganism, and final product [41, 208]. Due to the many advantages of SSCF process, this process was investigated for sugarcane bagasse pretreated with alkali assisted HC. With the developed methodology, 62.33% of the total hydrolyzed carbohydrate and 17.26 g/L of ethanol production were obtained, indicating 0.48 g of ethanol/g of glucose and xylose consumed by *Scheffersomyces stipitis* NRRL-Y7124 [219].

Overall, the use of LCB for the second-generation bioethanol has presented many advantages. The sustainable pathway, fossil fuel substitution, and the use of a variety of feedstock are some benefits of the second-generation ethanol production [19]. Despite technological barriers, plants of 2G bioethanol are already operating in the world [6]. Some examples of the second-generation commercial ethanol plants are GranBio, Raízen, Poet-DSM, BetaRenewables, Abengoa (plant bought by Synata Bio), and DuPont [85]. Among them, two Brazilian companies, namely GranBio and Raizen have reached commercial scale by predominantly using sugarcane bagasse as the LCB [86]. In Nevada, Iowa, USA, the cellulosic ethanol plant of DuPont has developed a projected production of 30 million US gallons of bioethanol from corn stover as a lignocellulosic material [87].

Considering a hybridization mode of the first and second-generation technologies, the 1.5 generation technology uses the sugars remained in the corn grain fiber in corn-based ethanol plants [88, 220]. The 1.5G is a new ethanol production technology, which uses the same industrial facilities for the production of the first-generation ethanol, in particular corn ethanol facilities (the platform where this technology was first implemented) [89].

This technology was incorporated at the end of 2016, in corn ethanol refineries in the United States (USA), due to the ethanol and corn oil production increase and the environmental interest for using a cellulosic origin [14]. The 1.5 G technology was proven and tried in the pilot fermenters (15,000 gals) for production (585,000 gals) up to 1,000 hours [221]. The first 1.5 G biorefinery was built in March 2017. In this first plant, the ethanol produced up to 5 million gallons of cellulosic ethanol per year was attained using this technology [185]. In the corn ethanol refineries, this form of ethanol production is combined with the 1G. Moreover, the 1.5G technology is still attractive for the usage of cellulosic raw material, and it allows a reduction in GHG emissions and indulges companies with producer titles. Therefore, renewable products can receive technological and government incentives [14, 90].

The corn grain fiber, which is used in biorefineries for the production of 1.5 G ethanol, has some particular characteristics. Firstly, it includes its pure and homogeneous composition formed by cellulose, hemicellulose, residual starch, and little lignin, being a minimally recalcitrant source. Furthermore, this material requires fewer pretreatment steps which support the diminishing of the coast of ethanol production [220, 222].

Production of 1.5 G integrates a process that transforms corn fiber to cellulosic ethanol with existing ethanol plants. In this way, the pathway to cellulosic ethanol is accomplished by combining mechanical, chemical, and biological processes. The fiber stream is submitted to an acid pretreatment that deconstructs it; thus, it can access the cellulose. Additionally, the cellulose stream is broken down into sugars with the enzyme cocktail (Novozymes), and then the C5 and C6 sugars are transformed into ethanol with other process detailed in the following parts [8].

The 1.5 G process was developed through the collaborations with the two world-leading biotechnology companies. Nowadays, two different ways are adopted for the bioconversion of the corn grain fiber into ethanol. One way is to adopt the process separately, where the hydrolysis of starch and grain fiber occurs separately. On the other hand, a second way is to adopt combined systems of pretreatment of the grain (starch and fiber) before fermentation [14]. The companies that start this process have with ICM's patented technologies, selective milling technology (SMT), and fiber separation technology (FST). SMT selectively grinds corn slurry to make the starch and oil more accessible in the entire process. Novozymes provided the enzyme cocktail which converts the cellulose stream into accessible sugars. DSM has developed yeast that ferment both the C5 and C6 sugars.

Fiber enzymatic hydrolyzes and fermentation processes can be carried out separately or in the SSF process [223]. In both processes, enzymatic cocktails are used in the low doses composed of cellulase for converting cellulose, hemicellulose, and residual starch into sugars (C5 and C6), which are used for fermentation [224]. Subsequently, the liquid obtained is sent to distillation and dehydration, following the same processes as the production of 1G corn ethanol. Protein and non-fermentable residues are removed and presented in the aqueous fraction after distillation/dehydration, which are still utilized for animal feed [222].

The 1.5 G technology demonstrates a great potential for ethanol generation, and its production process has a great potential to be adapted in a smoothie way for the facilities already existing in a corn starch ethanol plant [14]. Some future considerations can be suggested for the improvement of lignocellulosic bioethanol, such as process optimization to minimize water and energy usage, the development of engineered microorganisms and the combination of different microorganisms, which increases the fermentation capacity for bioethanol production [91]. In addition to these strategies, a significant reduction in the enzymes and pretreatment costs is necessary to improve the yield and productivity of the bioethanol for increasing its commercial scales around the world [92, 93]. Therefore, integration, and consolidation to improve the whole second-generation bioethanol production will make this a competitive market with petroleum fuel in the biorefinery process.

10.5　INTEGRATED CORN AND SUGARCANE TO ETHANOL BIOREFINERY (1ST GENERATION FLEX INDUSTRIES)

Even with the high performance of the ethanol production process, there is still a way for top countries such as Brazil to increase its productivity even more. This could occur due to the approach of several crops with higher productivity in terms of ethanol per unit of area [94]. For example, in Brazil, sugarcane can be produced between 6,000–7,500 L/ha, compared to sugar beet (EU) 5,500 L/ha and corn (USA) 3,800 L/ha [95–97]. This higher productivity of sugarcane implies the lower ethanol production costs. However, even with a harvest cycle higher than that of corn or sugar beet, there is an off-season, in which sugarcane cannot be processed, since it is not possible to store it.

Nevertheless, this is different from corn, which can be stored all year round [96]. Both biomass has differences in relation to their derived products. For example, in addition to ethanol, sugarcane gives a rise in electricity and sugar. This underproduction has a strategic value for the sugarcane mills, as it allows the capture of value in different markets [48]. In this context, corn can raise ethanol and other food products, such as oil and protein for animal feed (BBGDS). According to RFA [225], for each corn unit converted to ethanol, a third part returns to the animal nutrition. However, corn bioprocess plant is not self-sufficient in energy terms as the Brazilian one.

Therefore, con is the biggest gain for the Brazilian bioenergy process to invest in new biorefinery models, such as the "flex" ones, which are capable of integrating two different feedstocks, namely corn and sugarcane, or the use of their residues as an alternative feedstock [9]. In the last case, one strategy for biorefinery model occurs when it integrates residues of second-generation biomass in first-generation process, such as sugarcane bagasse and straw, which could help the sugar-energy sector to overcome low levels of profitability of ethanol [98].

Fuel ethanol production generally consists of five-stage process: (i) raw material collection; (ii) pretreatment; (iii) hydrolysis; (iv) fermentation; (v) separation and (vi) dehydration (downstream). However, there are some differences when it is produced from different feedstocks [99]. In the case of flex biorefineries, some steps are adapted and currently these are key points of challenges to increase conversion yields. Some of these integrations and flex modes are presented in Figure 10.4.

Moreover, the application of alternative biorefinery processes help to reduce the momentum for investments in the construction of new plants in the region. At the same time, this situation needs intense search for innovations with the potential to increase the profitability levels of these young companies. This incentive could increase the production of corn ethanol in flex mills to 1.3 billion liters [226]. Other phenomenon that this process integrates is fermentation of sugarcane juice (or molasses) and starchy feedstocks, providing a number of advantages over conventional distilleries [94]. This process has a faster fermentation (34–36 h) in comparison with the traditional process (45–60 h) adopted by distilleries in the USA. In this approach, less sugar is deviated for yeast multiplication and production of cellular biomass [227]. Thus, this mode of "flex" biorefinery uses the same distillation system applied for sugarcane and extends the period of ethanol production to 345 days a year, reducing initial investments and fixed costs. For this new process each corn ton allows to produce 415 L of ethanol and 250 kg of DDGS (dry distillation grain) [94, 96, 228].

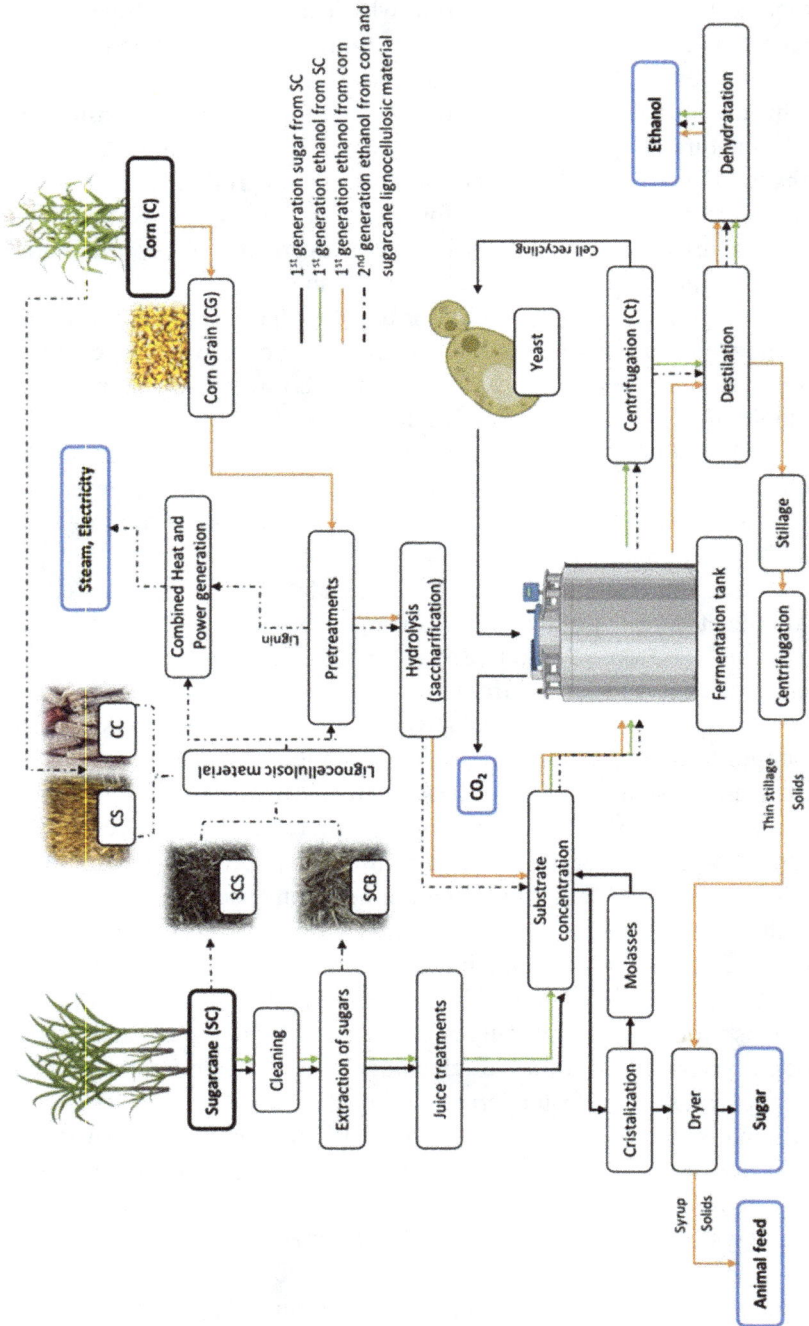

FIGURE 10.4 Biorefinery integration of 1st and 2nd generation of sugarcane and corn as feedstock. [*Abbreviations:* CS: corn stover; CC: corn cob; SCS: sugarcane straw; SCB: sugarcane bagasse).

Therefore, the integration of ethanol production with sugarcane and corn has the potential to offer the most significant economic advantages in comparison to standalone units, since important operations such as inoculum, feedstock, laboratories, and technical personnel may be shared between them [9]. Flex-fuel plants of corn and sugarcane have an energy balance and reduction of GHG emissions that do not affect the fermentation performance. This process is economically viable in regions with corn supply at low prices and high demand of DDGS for animal feed. Thus, it is an opportunity for Brazilian biorefineries in corn producing regions [96]. In addition, ethanol production from starchy gives two feedstock options to reduce the end problem of biofuels, and to enable the use of LCB-derived sugars for the other best production [23, 100, 227].

In the case of lignocellulosic materials, the cellulose and hemicellulose content account for more than 60% of dry weight (ex. sugarcane bagasse) and can be converted into fermentable sugars either by acid or enzymatic hydrolysis [7, 100]. These sugars could provide substrate to produce not only ethanol, but also other biofuels such as butanol and 1-methlypropanol. All those strategies could make the possible development of a new economically and energetically panorama for flex biorefineries [227].

For example, in Brazil, there are two industrial plants integrated into operation for the production of second-generation ethanol that use a semi-industrial process. These processes were originated from investment programs of the Brazilian government in the second-generation ethanol launched in 2011 [96, 101]. Nevertheless, the production costs of the second-generation ethanol are still high, mainly concerning equipment handling at bagasse pretreatment and the use of enzymes [96, 102].

Comparing characteristics between the different types of biomass, researchers have found that sugarcane is currently the most promising source for the production of biofuels [229]. However, the flex biorefinery shows that it is possible to have one process more promising that sugarcane biorefinery [94, 96]. Table 10.2 shows different situations corresponding to the first- and second-generation plant scenarios dedicated exclusively to 1G processing (situation 1), and corn processing (situation 2). This case shows consolidated ethanol production in the world with different biomass. Situation 3 shows the lowest amounts of GHG compared to all other processes studied in Table 10.2. Moreover, this situation could impact in a positive way, when corn is integrated into a sugarcane flex plant. The main factor that contributes to decrease GHG emissions is the inclusion of corn in integrated biorefineries (situation 3 and 4) which in this process exists the higher ethanol

TABLE 10.2 Examples of Different Processes of Ethanol Production in Biorefineries Flex with Two Different Biomass Feedstocks (Sugarcane and Corn/1G and 2G)

Biorefinery Process	Characteristics, Advantages, and Limitations	References
1. Sugarcane ethanol 1G	• Conventional soil preparation • Mechanized planting and harvesting • Without pre-harvest burning • Requirement of a shorter planting area • High yield of sugarcane ethanol production (77 ton/ha) • Industry does not work in sugarcane crop off-season	[98, 106, 107, 231]
2. Corn ethanol 1G	• O straw collection • High mechanized farming and livestock • Chemical pest control. • Corn is a popular and highly successful agricultural crop available worldwide. • It requires a larger planting area.	[7, 14, 108]
3. Flex corn and sugarcane ethanol (Flex-Example 1)	• Flex plant with the processing of corn only in the off-season of sugarcane. • Cell recycling. • The inclusion of corn in the system increases the production of bioethanol and decreases the pollution emission in the air. • There is a loss of sugars in solids by removal operation before the fermentation step.	[109, 110]
4. Integrated biorefinery of corn and sugarcane ethanol in Midwest Brazil (Flex-Example 2)	• This biorefinery work with the corn and sugarcane in all 12 months of the year • In this process, there is a six-year cycle (planting + five cuts). • Possibility for co-fermentation of the broth of the two raw materials during the sugarcane harvest. It can solve probable operational problems (for example, stoppages). • This biorefinery has reduced the impetus for investments in the construction of new plants in the Midwest Brazilian region.	[7, 98, 104]

TABLE 10.2 *(Continued)*

Biorefinery Process	Characteristics, Advantages, and Limitations	References
5. Corn and sugarcane ethanol integrated production off-season (Flex-Example 3)	• In this biorefinery, corn is stored and used together with sugarcane to produce ethanol for 8 months per year. • This type of biorefinery, with the integration of the sugarcane and corn culture, can increase the production of Brazilian ethanol.	[48, 96, 108]
6. Flex sugarcane 1 G, 2 G	• Cellulosic materials such as bagasse which are already present in the 1G ethanol production chain could be used as raw material for 2G production by conventional infrastructure. • High pretreatment price • High enzyme cost • Low pentose fermentation yield	[7, 48, 103, 105]

production and it is possible to achieve this yield because of the amount of corn processed [9, 230].

Another possibility is the co-fermentation of sugarcane and corn broth produced during harvesting, which makes it possible to solve operational problems (for example, stoppages at the sugarcane milling), incorporating gains in the production system. In this context, the integration of the sugarcane culture with the harvest of up to eight months, into other energy crops such as the corn could increase the sustainability of ethanol as a bioenergy [9]. Other flex plants are capable of processing sugarcane and bagasse from sugarcane (situation 5), and they could help the sugar-energy sector to overcome the current adverse environmental situation and decrease the levels of ethanol profitability [94, 227].

Furthermore, the situation 6 could reduce the investments in the construction of new plants in the world, because of the use of 2 G biomass [103]. At the same time, this situation sparks an intense research for innovations with a potential to increase the profitability levels of these young sectors, and for the investment in more flex bioprocess [104]. One disadvantage is that this process needs an overwork on the logistics side, because freight and storage is not a simple issue for industry [13]. In addition, the 1G + 2G configuration has the greatest efficiency (44.4%) compared to 1G biorefinery configuration [105]. Another problem for this situation is that investors demand a return of 5-fold investment. However, the most important factor in situation 6 is the low investment (50 million to adapt an existing biorefinery) of these biorefineries, which opens many possibilities for developing countries [14, 103].

10.6 TECHNOLOGICAL CHALLENGES AND PERSPECTIVES FOR SUCROSE AND STARCH INTEGRATION INTO BIOREFINERY (2nd GENERATION AND FLEX)

Currently, the demand for biofuels has increased due to environmental, social, and technological care [13]. In this view, there is a great worldwide interest in the use of biomass residues, such as cellulosic residues (for example, sugarcane bagasse and corn straw) for the production of biofuels 2 G [7, 103].

In the context of scaling up production, there is the integrated biorefinery configuration, called flex biorefinery, in which there is a possibility of integrating different types of biomass for the production of 1G ethanol. For example, in Brazil, there is a high rate of productivity in this biorefinery configuration during a four-month period in which there is no ethanol production, due to the lack of raw material and thus the plant goes through a long period of maintenance [104, 111]. This fact occurs because of the perishability of sugarcane [232]. As seen before, this problem is not common for corn, and the best option is the use of both feedstocks for ethanol production [23, 105].

All biorefineries have the potential to meet the demand for the production of bioethanol, but there are still many challenges in the integration of 1G-1G (different biomasses), and 1G and 2G biorefineries. For example, economic problems, substrate approach, costs in main steps (pretreatment and saccharification), low total sugar (C5 and C6 sugars) conversion yield in wild type of microorganisms are prominent complications [13, 233].

The second generation has many challenges that hinder the integrated process of 1G and 2G. In the case of 2G, it works with lignocellulosic materials for biofuel production [234]. These feedstocks could contain little protein or other nutrients (corn) that fermentation microbes require, however, not all feedstocks have these advantages, and it is important to add some supplementation [98, 103].

Other challenges that need to be considered are harvest, transportation costs and storage. For example, sugar crops (sugarcane) are especially difficult to store without losses, and this might also be a problem with lignocellulosic feedstocks [104]. Another challenge in the process is the stage of saccharification, which is important because current enzymes for hydrolysis are more expensive and less effective than corresponding enzymes used with starch [111, 208]. Another alternative for saccharification process is to conduct SSF of lignocellulose due to the feedback inhibition of the enzymes [23, 197].

Pretreatment is the most significant bottleneck in biorefinery process; however, pretreatment is not needed for all sugar crops. It is used for lignocellulosic materials where it is an overriding feedstock. However, the lignocellulose pretreatments are often sufficiently intensive and generate inhibitory compounds, and this is prejudicial to the total process. In addition, it is difficult for the industrial process, since pretreatments have high costs and do not have short process time options to be applied for the continuous industrial process [98, 112, 113].

Another challenge is that microbial growth is limited due to the presence of inhibitors in the fermentation process generated in previous stages (pretreatment). In addition, different sugars (C5 and C6 sugars) in substrates obtained from LCB are difficult to be metabolized by many wild type microorganisms because it comes from pretreated biomass. In fermentation, there are difficulties with the concentration of sugars and solids content in the medium which are also unsatisfactory factors for the system [111, 236].

The cost of integrated 1G + 2G biorefineries requires two to four times more investment than 2G ethanol prices to ensure competitiveness, which is one of the main problems in flex production. However, this can be fixed when pretreatment and hydrolysis technologies are chosen [8, 23]. As a result, it is necessary to investigate more developed schemes, particularly with staged hydrolysis that can provide more competitiveness for the 1G alternatives [114, 237–240]. On the other hand, the operational unit is important due to its cost is the heat exchange process which can be reduced by adopting developed schemes for evaporation and backward integrated evaporation that is a new perspective for flex biorefinery [98, 241–244].

1G and 2G biorefineries have higher investments in costs than 1G+ 2G + electricity integration, due to the hydrolysis step with cogeneration which presents a lower yield. The 1G + electricity schemes have provided better results and it is superior to 1G+2G + electricity equivalents, mainly because of its highest investment costs [234, 241]. Other problems of sugarcane 1 G and 2 G plants are related to enzyme parameter costs and reduced electricity production. However, the advantage of this process lies in ethanol prices with lower capital costs, presenting the greatest potential to increase the economic utilization of 1G + 2G (flex) schemes. In this regard, the cost is lower than 2G biorefinery process and it has greater productivity than 1G [197].

Other option in the biorefinery integration is the development of plants with corn and sugarcane as feedstocks that results in 1G and 2G processes. This biorefinery could have high productivity which could help solving the global demand of ethanol [112, 245]. Thus, sugarcane biomass is more versatile to be used in this kind of biorefinery, because it could be integrated with different biomasses (sugarcane and corn) in 1G or 2G integrated [98, 241, 246].

Energy integration in a sugarcane biorefinery can provide economical advantage, environmental benefits, and increased ethanol production. The last factor is related to the lower steam consumption in the plant due to energy integration and consequently, and less bagasse needs to be burnt in the electricity generation. Hence, its surplus can be made available for the production of second-generation ethanol [51, 115, 234].

There are other integration processes with different advantages, such as, 1G, 2G and cogeneration system (1G+2G+COGEN), 1G and 2G from corn, 1G and 2G for sugarcane, 1G (biomass A) and 2G (biomass B) that could be the future of fuels from biomass [111, 113]. These integrations could give us positive results with 2G, which involves the production of 1G and 2G ethanol in a combined distillery, enzymatic hydrolysis, and cogeneration plant [112]. These kind of biorefineries offer many techno-economic and environmental alternatives.

Total production cost calculations for these kinds of biorefineries can be resulted in 74 settings, covering 5 fuel output types, 8 feedstock types, 12 countries, and 8 combinations of agricultural management systems between 2010 and 2030 worldwide. This shows that many countries have interested in this 1G and 2G flex technology [112]. However, this integration is not able to meet the international demand and it is important to modify the production of bioethanol according to the international market demand for these agricultural products [13, 104]. Biorefinery integration modes could meet the demand for biofuels and its use which will be a part of a big energy program for each country by considering its necessities and approach to its resources.

10.7 CONCLUSION

First and second-generation bioethanol production from sugarcane and corn are successful processes established in several countries, and it is currently analyzed to find out the possibility for the integration of other biorefinery modes to enhance some particular bottlenecks. Those bottlenecks are the seasonal lack of the feedstocks and approach of second-generation sugars. Several strategies, such as the alternation of feedstocks in off-seasons and the integration of second-generation into the first-generation processes are deeply evaluated. However, the integration of these modes in biorefinery is a hard-analytical work to understand the role that plays each step-in bioethanol production from two feedstocks with differences in structure, sugar content, and processing. Even though the correct application of those strategies could result in a successful integration from both biomass, the main steps in bioprocesses such as pretreatments, saccharification, and fermentation need more work to obtain a cost-effective competitive conversion.

KEYWORDS

- **consolidated bioprocessing**
- **distillers dried grains**
- **hydrochloric acid**
- **saccharification and fermentation**
- **separate hydrolysis and fermentation**
- **simultaneous saccharification and co-fermentation**

REFERENCES

1. Bai, F. W., Anderson, W. A., & Moo-Young, M., (2008). Ethanol fermentation technologies from sugar and starch feedstocks. *Biotechnology Advances, 26*(1), 89–105.
2. Zabed, H., Sahu, J. N., Suely, A., Boyce, A. N., & Faruq, G., (2017). Bioethanol production from renewable sources: Current perspectives and technological progress. *Renewable and Sustainable Energy Reviews, 71*, 475–501.
3. Akram, F., Ul Haq, I., Imran, W., & Mukhtar, H., (2018). Insight perspectives of thermostable endoglucanases for bioethanol production: A review. *Renewable Energy, 122,* 225–238.
4. Mussatto, S. I., Machado, E. M., Martins, S., & Teixeira, J. A., (2011). Production, composition, and application of coffee and its industrial residues. *Food and Bioprocess Technology, 4*(5), 661.
5. Gumienna, M., Szwengiel, A., Lasik, M., Szambelan, K., Majchrzycki, D., Adamczyk, J., & Czarnecki, Z., (2016). Effect of corn grain variety on the bioethanol production efficiency. *Fuel, 164*, 386–392.
6. Vallejos, M. E., Kruyeniski, J., & Area, M. C., (2017). Second-generation bioethanol from industrial wood waste of South American species. *Biofuel Research Journal, 4*(3), 654–667.
7. Milanez, A. Y., Nyko, D., Valente, M. S., Xavier, C. E. O., Kulay, L. A., Donke, A. C. G., & Capitani, D. H. D., (2014). Ethanol production by integrating off-season corn into mills Final report 2019/20 crop, south-central region of Brazil. Sugarcane: Environmental, Economic Assessment and Policy Suggestions. https://web.bndes.gov.br/bib/jspui/handle/1408/1921 (accessed on 28 October 2021).
8. Sydney, E. B., Letti, L. A. J., Karp, S. G., Sydney, A. C. N., De Souza, V. L. P., De Carvalho, J. C., & Soccol, C. R., (2019). Current analysis and future perspective of reduction in worldwide greenhouse gases emissions by using first and second-generation bioethanol in the transportation sector. *Bioresource Technology Reports, 7,* 100234.
9. Cabral, F. F. R., (2019). Agronomic performance and nutritional balance in the corn plant fertigated with concentrated vinasse and potassium chloride.
10. Kumar, D., & Singh, V., (2019). Bioethanol production from corn. In: *Corn* (pp. 615–631). AACC International Press.
11. CONAB, (2019). Retrieved from https://www.conab.gov.br// (accessed on 28 October 2021).
12. Vanier, N. L., Vamadevan, V., Bruni, G. P., Ferreira, C. D., Pinto, V. Z., Seetharaman, K., & Berrios, J. D. J., (2016). Extrusion of rice, bean and corn starches: Extrudate structure and molecular changes in amylose and amylopectin. *Journal of Food Science, 81*(12), E2932–E2938.
13. Guo, M., Song, W., & Buhain, J., (2015). Bioenergy and biofuels: History, status, and perspective. *Renewable and Sustainable Energy Reviews, 42*, 712–725.
14. Mohanty, S. K., & Swain, M. R., (2019). Bioethanol production from corn and wheat: Food, fuel, and future. In: *Bioethanol Production from Food Crops* (pp. 45–59). Academic Press.
15. EMBRAPA, (2019). Retrieved from: https://www.embrapa.br/agua-na-agricultura/links (accessed on 8 December 2021).
16. Zhao, L., Ou, X., & Chang, S., (2016). Life-cycle greenhouse gas emission and energy use of bioethanol produced from corn stover in China: Current perspectives and future prospectives. *Energy, 115,* 303–313.

17. Torabi, S., Satari, B., & Hassan-Beygi, S. R., (2020). Process optimization for dilute acid and enzymatic hydrolysis of waste wheat bread and its effect on aflatoxin fate and ethanol production. *Biomass Conversion and Biorefinery, 1–9.*

18. Agarwal, M., Rampure, M., Todkar, A., & Sharma, P., (2019). Ethanol from maize: An entrepreneurial opportunity in agrobusiness. *Biofuels, 10*(3), 385–391.

19. Ayodele, B. V., Alsaffar, M. A., & Mustapa, S. I., (2020). An overview of integration opportunities for sustainable bioethanol production from first- and second-generation sugar-based feedstocks. *Journal of Cleaner Production, 245*(1).

20. Chen, S., Xu, Z., Li, X., Yu, J., Cai, M., & Jin, M., (2018). Integrated bioethanol production from mixtures of corn and corn stover. *Bioresource Technology, 258*, 18–25.

21. Mosier, N. S., & Ileleji, K. E., (2020). How fuel ethanol is made from corn. In: *Bioenergy* (pp. 539–544). Academic Press.

22. Carvalho, J. L. N., Nogueirol, R. C., Menandro, L. M. S., Bordonal, R. D. O., Borges, C. D., Cantarella, H., & Franco, H. C. J., (2017). Agronomic and environmental implications of sugarcane straw removal: A major review. *GCB Bioenergy, 9*(7), 1181–1195.

23. Sharma, B., Larroche, C., & Dussap, C. G., (2020). Comprehensive assessment of 2G bioethanol production. *Bioresource Technology, 123630.*

24. Bothast, R. J., & Schlicher, M. A., (2005). Biotechnological processes for conversion of corn into ethanol. *Applied Microbiology and Biotechnology, 67*(1), 19–25.

25. Eskicioglu, C., Kennedy, K. J., Marin, J., & Strehler, B., (2011). Anaerobic digestion of whole stillage from dry-grind corn ethanol plant under mesophilic and thermophilic conditions. *Bioresource Technology, 102*(2), 1079–1086.

26. Yu, J., Xu, Z., Liu, L., Chen, S., Wang, S., & Jin, M., (2019). Process integration for ethanol production from corn and corn stover as mixed substrates. *Bioresource Technology, 279*, 10–16.

27. Lamsal, B. P., Wang, H., & Johnson, L. A., (2011). Effect of corn preparation methods on dry-grind ethanol production by granular starch hydrolysis and partitioning of spent beer solids. *Bioresource Technology, 102*(12), 6680–6686.

28. Wang, M., Wu, M., & Huo, H., (2007). Life-cycle energy and greenhouse gas emission impacts of different corn ethanol plant types. *Environmental Research Letters, 2*(2), 024001.

29. Zhao, Y., Damgaard, A., & Christensen, T. H., (2018). Bioethanol from corn stover-a review and technical assessment of alternative biotechnologies. *Progress in Energy and Combustion Science, 67*, 275–291.

30. Power, R. F., (2003). Enzymatic conversion of starch to fermentable sugars. *The Alcohol Textbook, 23–32.*

31. Bonassa, G., Schneider, L. T., Cremonez, P. A., De Oliveira, C. D. J., Teleken, J. G., & Frigo, E. P., (2015). < b> Optimization of first-generation alcoholic fermentation process with< i> *Saccharomyces cerevisiae. Acta Scientiarum. Technology, 37*(3), 313–320.

32. Saville, B. A., Griffin, W. M., & MacLean, H. L., (2016). Ethanol Production technologies in the Us: Status and future developments. In: *Global Bioethanol* (pp. 163–180). Academic Press.

33. Chen, H., & Li, G., (2013). An industrial level system with nonisothermal simultaneous solid state saccharification, fermentation and separation for ethanol production. *Biochemical Engineering Journal, 74*, 121–126.

34. Madson, P. W., (2003). Ethanol distillation: The fundamentals. *The Alcohol Textbook, 4.*

35. Chum, H. L., Warner, E., Seabra, J. E., & Macedo, I. C., (2014). A comparison of commercial ethanol production systems from Brazilian sugarcane and US corn. *Biofuels, Bioproducts and Biorefining, 8*(2), 205–223.

36. Sharma, A., & Bhargava, R., (2016). Production of Biofuel (Ethanol) from Corn and co-product evolution: A review. *IRJET, 3*(12), 745–749.

37. Cripwell, R. A., Favaro, L., Viljoen-Bloom, M., & Van, Z. W. H., (2020). Consolidated bioprocessing of raw starch to ethanol by *Saccharomyces cerevisiae*: Achievements and challenges. *Biotechnology Advances,* 107579.

38. Brown, A., Waldheim, L., Landälv, I., Saddler, J., Ebadian, M., McMillan, J. D., & Bonomi, A., (2020). Advanced biofuels-potential for cost reduction. *IEA Bioenergy,* 88.

39. Gombert, A. K., & Van, M. A. J., (2015). Improving conversion yield of fermentable sugars into fuel ethanol in 1st generation yeast-based production processes. *Current Opinion in Biotechnology, 33,* 81–86.

40. Duan, S. F., Shi, J. Y., Yin, Q., Zhang, R. P., Han, P. J., Wang, Q. M., & Bai, F. Y., (2019). Reverse evolution of a classic gene network in yeast offers a competitive advantage. *Current Biology, 29*(7), 1126–1136.

41. Sharma, V., Sharma, S., & Kuila, A., (2016). A review on current technological advancement of lignocellulosic bioethanol production. *Journal of Applied Biotechnology Bioengineering,* 1(2), 61–66.

42. Klein, B. C., Chagas, M. F., Watanabe, M. D. B., Bonomi, A., & Maciel, F. R., (2019). Low carbon biofuels and the new Brazilian national biofuel policy (RenovaBio): A case study for sugarcane mills and integrated sugarcane-microalgae biorefineries. *Renewable and Sustainable Energy Reviews, 115,* 109365.

43. Emori, E. Y., Ferreira, J., Secchi, A. R., Ravagnani, M. A., & Costa, C. B., (2020). Dynamic study of the evaporation stage of an integrated first and second-generation ethanol sugarcane biorefinery using EMSO software. *Chemical Engineering Research and Design, 153,* 613–625.

44. Moraes, M. A. F. D., Oliveira, F. C. R., & Diaz-Chavez, R. A., (2015). Socio-economic impacts of Brazilian sugarcane industry. *Environmental Development, 16,* 31–43.

45. Júnior, J. C. F., Palacio, J. C. E., Leme, R. C., Lora, E. E. S., Da Costa, J. E. L., Reyes, A. M. M., & Del, O. O. A., (2020). Biorefineries productive alternatives optimization in the Brazilian sugar and alcohol industry. *Applied Energy, 259,* 113092.

46. Junqueira, T. L., Dias, M. O. S., Jesus, C. D. F., Mantelatto, P. E., Cunha, M. P., Cavalett, F. O., & Bonomi, A., (2011). Simulation and evaluation of autonomous and annexed sugarcane distilleries. *Chemical Engineering Transactions, 25,* 941–946.

47. Dias, F. M. D. S., RM, M. P., Cavalett, O., Rossell, C. E. V., Bonomi, A., & Leal, M. R. L. V., (2015). Sugarcane processing for ethanol and sugar in Brazil. *Environmental Development, 15*(2015), 35–51.

48. Cavalett, O., Junqueira, T. L., Dias, M. O., Jesus, C. D., Mantelatto, P. E., Cunha, M. P., & Bonomi, A., (2012). Environmental and economic assessment of sugarcane first generation biorefineries in Brazil. *Clean Technologies and Environmental Policy, 14*(3), 399–410.

49. Khatiwada, D., Leduc, S., Silveira, S., & McCallum, I., (2016). Optimizing ethanol and bioelectricity production in sugarcane biorefineries in Brazil. *Renewable Energy, 85,* 371–386.

50. Andrade, L. P., Crespim, E., De Oliveira, N., De Campos, R. C., Teodoro, J. C., Galvão, C. M. A., & Maciel, F. R., (2017). Influence of sugarcane bagasse variability on sugar recovery for cellulosic ethanol production. *Bioresource Technology, 241,* 75–81.

51. Macrelli, S., Mogensen, J., & Zacchi, G., (2012). Techno-economic evaluation of 2nd generation bioethanol production from sugarcane bagasse and leaves integrated with the sugar-based ethanol process. *Biotechnology for Biofuels, 5*(1), 22.

52. Chandel, A. K., Junqueira, T. L., Morais, E. R., Gouveia, V. L. R., Cavalett, O., Rivera, E. C., & Da Silva, S. S., (2014). Techno-economic analysis of second-generation ethanol in Brazil: Competitive, complementary aspects with first-generation ethanol. In: *Biofuels in Brazil* (pp. 1–29).
53. Junqueira, T. L., Dias, M. O., Maciel, F. R., Maciel, M. R., & Rossell, C. E., (2009). Simulation of the azeotropic distillation for anhydrous bioethanol production: Study on the formation of a second liquid phase. In: *Computer-Aided Chemical Engineering* (Vol. 27, pp. 1143–1148). Elsevier.
54. Khatiwada, D., Seabra, J., Silveira, S., & Walter, A., (2012). Power generation from sugarcane biomass—A complementary option to hydroelectricity in Nepal and Brazil. *Energy, 48*(1), 241–254.
55. Serra, L. M., Lozano, M. A., Ramos, J., Ensinas, A. V., & Nebra, S. A., (2009). Polygeneration and efficient use of natural resources. *Energy, 34*(5), 575–586.
56. Leal, M. R. L., Walter, A. S., & Seabra, J. E., (2013). Sugarcane as an energy source. *Biomass Conversion and Biorefinery, 3*(1), 17–26.
57. Bezerra, T. L., & Ragauskas, A. J., (2016). A review of sugarcane bagasse for second-generation bioethanol and biopower production. *Bioproducts and Biorefining, 10*(5), 634–647.
58. Bolivar-Telleria, M., Turbay, C., Favarato, L., Carneiro, T., De Biasi, R. S., Fernandes, A. A. R., & Fernandes, P., (2018). Second-generation bioethanol from coconut husk. *BioMed Research International*.
59. Siqueira, J. G. W., Rodrigues, C., De Souza, V. L. P., Woiciechowski, A. L., & Soccol, C. R., (2020). Current advances in on-site cellulase production and application on lignocellulosic biomass conversion to biofuels: A review. *Biomass and Bioenergy, 132*, 105419.
60. González-Bautista, E., Alarcón-Gutiérrez, E., Dupuy, N., Gaime-Perraud, I., Ziarelli, F., Foli, L., & Farnet-Da-Silva, A. M., (2020). Preparation of a sugarcane bagasse-based substrate for second-generation ethanol: Effect of pasteurization conditions on dephenolization. *Renewable Energy*.
61. Schneider, W. D. H., Fontana, R. C., Baudel, H. M., De Siqueira, F. G., Rencoret, J., Gutiérrez, A., & Dillon, A. J. P., (2020). Lignin degradation and detoxification of eucalyptus wastes by on-site manufacturing fungal enzymes to enhance second-generation ethanol yield. *Applied Energy, 262*, 114493.
62. Uyan, M., Alptekin, F. M., Cebi, D., & Celiktas, M. S., (2020). Bioconversion of hazelnut shell using near critical water pretreatment for second-generation biofuel production. *Fuel, 273*, 117641.
63. Kleingesinds, E. K., José, Á. H., Brumano, L. P., Silva-Fernandes, T., Rodrigues, Jr. D., & Rodrigues, R. C., (2018). Intensification of bioethanol production by using Tween 80 to enhance dilute acid pretreatment and enzymatic saccharification of corncob. *Industrial Crops and Products, 124*, 166–176.
64. Guerrero, A. B., Ballesteros, I., & Ballesteros, M., (2018). The potential of agricultural banana waste for bioethanol production. *Fuel, 213*(1), 176–185.
65. Fonseca, B. G., Mateo, S., Moya, A. J., & Roberto, I. C., (2018). Biotreatment optimization of rice straw hydrolyzates for ethanolic fermentation with *Scheffersomyces stipitis*. *Biomass and Bioenergy, 112*, 19–28.
66. Anwar, Z., Gulfraz, M., & Irshad, M., (2014). Agro-industrial lignocellulosic biomass a key to unlock the future bio-energy: A brief review. *Journal of Radiation Research and Applied Sciences, 7*(2), 163–173.

67. Fennouche, I., Khellaf, N., Djelal, H., & Amrane, A., (2019). An effective acid pretreatment of agricultural biomass residues for the production of second-generation bioethanol. *SN Applied Sciences, 1*(11), 1460.
68. Aditiya, H. B., Mahlia, T. M. I., Chong, W. T., Nur, H., & Sebayang, A. H., (2016). Second-generation bioethanol production: A critical review. *Renewable and Sustainable Energy Reviews, 66*, 631–653.
69. Robak, K., & Balcerek, M., (2018). Review of second-generation bioethanol production from residual biomass. *Food Technol Biotechnol, 56*(2),174–187.
70. Liu, L., An, X., Zhang, H., Lu, Z., Nie, S., Cao, H., & Liu, H., (2020). Ball milling pretreatment facilitating α-amylase hydrolysis for production of starch-based bio-latex with high performance. *Carbohydrate Polymers,* 116384.
71. Eckard, A. D., Muthukumarappan, K., & Gibbons, W., (2012). Modeling of pretreatment condition of extrusion-pretreated prairie cordgrass and corn stover with poly (oxyethylene) 20 sorbitan monolaurate. *Applied Biochemistry and Biotechnology, 167*, 377–393.
72. Kristiani, A., Effendi, N., Styarini, D., Aulia, F., & Sudiyani, Y., (2016). The effect of pretreatment by using electron beam irradiation on oil palm empty fruit bunch. *Atom Indonesia, 42*(1), 9–12.
73. Marinho, C. C. P., Pereira, L. V. B., & Vasconcelos, S. M., (2020). Acid pretreatment of sugarcane straw aiming at the second-generation ethanol production. *Revista Tecnologia e Sociedade. 16*(41), 1–14.
74. Chopda, R., Ferreira, J. A., & Taherzadeh, M. J., (2020). Biorefining oat husks into high-quality lignin and enzymatically digestible cellulose with acid-catalyzed ethanol organosolv pretreatment. *Processes, 8*(4), 435.
75. Equihua-Sanchez, M., & Barahona-Perez, L. F., (2019). Physical and chemical characterization of *Agave tequilana* bagasse pretreated with the ionic liquid 1-ethyl-3-methylimidazolium acetate. *Waste and Biomass Valorization.* 10, 1285–1294.
76. Damay, J., Duret, X., Ghislain, T., Lalonde, O., & Lavoie, J. M., (2018). Steam explosion of sweet sorghum stems: Optimization of the production of sugars by response surface methodology combined with the severity factor. *Industrial Crops and Products, 111*, 482–493.
77. Mikulski, D., & Kłosowski, G., (2020). Microwave-assisted dilute acid pretreatment in bioethanol production from wheat and rye stillages. *Biomass and Bioenergy, 136*, 105528.
78. Terán-Hilares, R., Dionízio, R. M., Prado, C. A., Ahmed, M. A., Da Silva, S. S., & Santos, J. C., (2019). Pretreatment of sugarcane bagasse using hydrodynamic cavitation technology: Semi-continuous and continuous process. *Bioresource Technology, 290*, 121777.
79. Terán-Hilares, R., Dionízio, R. M., Muñoz, S. S., Prado, C. A., De Sousa, J. R., Da Silva, S. S., & Santos, J. C., (2020). Hydrodynamic cavitation-assisted continuous pre-treatment of sugarcane bagasse for ethanol production: Effects of geometric parameters of the cavitation device. *Ultrasonics Sonochemistry, 63*, 104931.
80. Kucharska, K., Rybarczyk, P., Hołowacz, I., Łukajtis, R., Glinka, M., & Kamiński, M., (2018). Pretreatment of lignocellulosic materials as substrates for fermentation processes. *Molecules, 23*(11), 2937.
81. Gebrehiwot, G., Birhane, K., & Gebrekidan, T., (2020). Optimization of sulphuric acid hydrolysis process for fermentable sugars from lignocellulosic content of wood sawdust for production of cellulosic ethanol. *Emerging Trends in Chemical Engineering, 7*(1).
82. Horst, D. J., Behainne, J. J. R., & Petter, R. R., (2011). Analysis of hydrolysis yields by using different acids for bioethanol production from Brazilian woods. *XVII International Conference on Industrial Engineering and Operations Management.* 1–14.

83. Anindyawati, T., Triwahyuni, E., Maryana, R., & Sudiyani, Y., (2020). The enzymatic process of lignocellulosic biomass for second generation bioethanol production, the benefits and challenges: a review. *International Journal of Agricultural Technology, 16*(3), 529–544.
84. Kim, J. K., Yang, J., Park, S. Y., Yu, J. H., & Kim, K. H., (2019). Cellulase recycling in high-solids enzymatic hydrolysis of pretreated empty fruit bunches. *Biotechnology for Biofuels, 12*(1), 1–9.
85. Farias, D., & Filho, F. M., (2019). Co-culture strategy for improved 2G bioethanol production using a mixture of sugarcane molasses and bagasse hydrolysate as substrate. *Biochemical Engineering Journal, 147*, 29–38.
86. Neto, A. C., Guimarães, M. J. O. C., & Freire, E., (2018). Business models for commercial-scale second-generation bioethanol production. *Journal of Cleaner Production, 184*, 168–178.
87. Grassi, M. C. B., & Pereira, G. A. G., (2019). Energy-cane and RenovaBio: Brazilian vectors to boost the development of biofuels. *Industrial Crops and Products, 129*, 201–205.
88. Lynd, L. R., Liang, X., Biddy, M. J., Allee, A., Cai, H., Foust, T., & Wyman, C. E., (2017). Cellulosic ethanol: status and innovation. *Current Opinion in Biotechnology, 45*, 202–211.
89. Morán, K., Moriarty, K. L., Milbrandt, A. R., Warner, E., Lewis, J. E., & Schwab, A. A., (2017). *2016 Bioenergy Industry Status Report (No. NREL/TP-5400-70397).* National Renewable Energy Lab. (NREL), Golden, CO (United States).
90. Stolark, J., (2017). *"1.5 Gen Technologies Could Boost Cellulosic Ethanol Production by Nearly 2 Billion Gallons".* Ideas, Insights, Sustainable, Solutions. Environmental and Energy Study Institute (EESI).
91. Mueller, K., & Krissek, G., (2014). *Future Opportunities and Challenges for Ethanol Production and Technology.*
92. Liu, C. G., Xiao, Y., Xia, X. X., Zhao, X. Q., Peng, L., Srinophakun, P., & Bai, F. W., (2019). Cellulosic ethanol production: Progress, challenges and strategies for solutions. *Biotechnology Advances, 37*(3), 491–504.
93. Ramos, L. P., Suota, M. J., Fockink, D. H., Pavaneli, G., Da Silva, T. A., & Łukasik, R. M., (2020). Enzymes and biomass pretreatment. In: *Recent Advances in Bioconversion of Lignocellulose to Biofuels and Value-Added Chemicals within the Biorefinery Concept* (pp. 61–100). Elsevier.
94. Carrillo-Nieves, D., Saldarriaga-Hernandez, S., Gutiérrez-Soto, G., Rostro-Alanis, M., Hernández-Luna, C., Alvarez, A. J., & Parra-Saldívar, R., (2020). Biotransformation of agro-industrial waste to produce lignocellulolytic enzymes and bioethanol with a zero waste. *Biomass Conversion and Biorefinery.*
95. De Lucca, M. R. Z., (2020). Empirical Analysis of Production Decision Determinants of Sugar and Ethanol in the Sugarcane Agroindustry Paulista (Doctoral dissertation).
96. Soccol, C. R., De Souza, V. L. P., Medeiros, A. B. P., Karp, S. G., Buckeridge, M., Ramos, L. P., & Da Silva, B. E. P., (2010). Bioethanol from lignocelluloses: Status and perspectives in Brazil. *Bioresource Technology, 101*(13), 4820–4825.
97. Lopes, M. L., De Lima, P. S. C., Godoy, A., Cherubin, R. A., Lorenzi, M. S., Giometti, F. H. C., & De Amorim, H. V., (2016). Ethanol production in Brazil: A bridge between science and industry. *Brazilian Journal of Microbiology, 47*, 64–76.
98. UNICA, (2020). Final report 2019/20 crop year in the center-south region of Brazil. http://unicadata.com.br/listagem.php?idMn=118 (accessed on 28 October 2021).
99. Dias, M. O., Junqueira, T. L., Cavalett, O., Cunha, M. P., Jesus, C. D., Rossell, C. E., & Bonomi, A., (2012). Integrated versus stand-alone second-generation ethanol production from sugarcane bagasse and trash. *Bioresource Technology, 103*(1), 152–161.

100. Quintero, J. A., Montoya, M. I., Sánchez, O. J., Giraldo, O. H., & Cardona, C. A., (2008). Fuel ethanol production from sugarcane and corn: Comparative analysis for a Colombian case. *Energy, 33*(3), 385–399.
101. Borges, J. R., (2018). Obtenção de bioetanol por fermentação a partir do milho.
102. Paula, C. E. D. E. T., & Biaggi, D. E., (2017). Technological innovations and trends in the production of second-generation ethanol from sugarcane through the enzymatic hydrolytic route: A technological prospection study.
103. Pereira, Jr. N., Couto, M. A. P. G., & Santa, A. L. M., (2008). Biomass of lignocellulosic composition for fuel ethanol production and the context of biorefinery. *Series on Biotechnology, 2*, 2–45.
104. Bechara, R., Gomez, A., Saint-Antonin, V., Schweitzer, J. M., & Maréchal, F., (2016). Methodology for the design and comparison of optimal production configurations of first and first and second generation ethanol with power. *Applied Energy, 184*, 247–265.
105. Manochio, C., (2014). Bioethanol production from sugarcane, corn and sugar beet: A comparison of technological, environmental and economic indicators. Course completion work (Chemical Engineering)–Federal University of Alfenas. Wells of Caldas, MG.
106. Miguel, J. V. P., (2013). *Integrated production of first and second generation sugarcane bioethanol: energy, environmental and economic analysis* (Doctoral dissertation, Universidade De São Paulo).
107. Lopes, E. M., (2006). Energy analysis and technical feasibility of producing biodiesel from beef tallow. Itajubá: UNIFEI.
108. PECEGE- Continuing education program in economics and business management, (2012). Sugarcane, sugar and ethanol production costs in Brazil: 2011/2012 harvest closing. Piracicaba: University of São Paulo, Superior School of Agriculture "Luiz de Queiroz," Continuing Education Program in Business Economics and Management/ Department of Economics, Administration and Sociology. P. 50. Report presented to the Confederation of Agriculture and Livestock of Brazil (CNA).
109. Donke, A. C. G., et al., (2013). Energo and exergo-environmental analysis of a multipurpose process for ethanol production from sugarcane and corn. In: *International Exergy, Life Cycle Assessment, and Sustainability Workshop & Symposium; Nisyros. Anais. Nisyros: Ecost, 1.305–1.312*, 437–448.
110. Silveira, R., (2017). *Brazil Opens First Ethanol Plant Fueled 100% with Corn.* Canal Rural.
111. Montipó, et al., (2018). Integrated production of second generation ethanol and lactic acid from stem exploded elephant grass. *Bioresource Technology, 249*, 1017–1024.
112. Dias, M. O., Da Cunha, M. P., Maciel, F. R., Bonomi, A., Jesus, C. D., & Rossell, C. E., (2011). Simulation of integrated first and second generation bioethanol production from sugarcane: Comparison between different biomass pretreatment methods. *Journal of Industrial Microbiology & Biotechnology, 38*(8), 955–966.
113. Oliveira, C. M., Cruz, A. J., & Costa, C. B., (2014). Comparison among proposals for energy integration of processes for 1G/2G ethanol and bioelectricity production. In: *Computer Aided Chemical Engineering* (Vol. 33, pp. 1585–1590). Elsevier.
114. Bergmann, J. C., Trichez, D., Sallet, L. P., E Silva, F. C. D. P., & Almeida, J. R., (2018). Technological advancements in 1G ethanol production and recovery of by-products based on the biorefinery concept. In: *Advances in Sugarcane Biorefinery* (pp. 73–95). Elsevier.
115. Sun, X. Z., Fujimoto, S., & Minowa, T., (2013). A comparison of power generation and ethanol production using sugarcane bagasse from the perspective of mitigating GHG emissions. *Energy Policy, 57*, 624–629.

116. Sharma, A., Singh, Y., Ansari, N. A., Pal, A., & Lalhriatpuia, S., (2020b). Experimental investigation of the behavior of a DI diesel engine fueled with biodiesel/diesel blends having effect of raw biogas at different operating responses. *Fuel, 279*, 118460.

117. Araújo, R. M. D., (2019). Food Acquisition Program (2013–2017): Evaluation of the implementation by Conab in Rio Grande do Norte.

118. Asghar, U., Irfan, M., Nadeem, M., Nelofer, R., & Syed, Q., (2017). Effect of KOH pretreatment on lignocellulosic waste to be used as substrate for ethanol production. *Iranian Journal of Science and Technology, Transactions A: Science, 41*(3), 659–663.

119. Astolfi, A. L., Rempel, A., Cavanhi, V. A. F., Alves, M., Deamici, K. M., Colla, L. M., & Costa, J. A. V., (2020). Simultaneous saccharification and fermentation of *Spirulina* sp. and corn starch for the production of bioethanol and obtaining biopeptides with high antioxidant activity. *Bioresource Technology, 301*, 122698.

120. Bacchi, M. R. P., & Caldarelli, C. E., (2015). Socioeconomic impacts of the expansion of the sugar-energy sector in the State of São Paulo, between 2005 and 2009. *Nova Economia, 25*(1), 209–224.

121. Bechara, R., Gomez, A., Saint-Antonin, V., Albarelli, J., Ensinas, A., Schweitzer, J. M., & Maréchal, F., (2014). Methodology for minimizing the utility consumption of a 2G ethanol process. *Chemical Engineering, 39*.

122. Blümmel, M., et. al., (2018). Ammonia Fiber Expansion (AFEX) as spin off technology from 2nd generation biofuel for upgrading cereal straws and stover's for livestock feed. *Animal Feed Science and Technology, 236*, 178–186.

123. BNDES, (2019). (Vol. 41, pp. 148–208). http://www.bndes.gov.br/SiteBNDES/export/sites/default/bndespt/Galerias/Arquivos/conhecimento/revista/rev4174 (accessed on 28 October 2021).

124. Borsatto, J. L. Jr., Souza, R. F., Weiss, L., & Dal, V. D. G., (2015). Theoretical essay on the feasibility of producing ethanol from corn: A possibility for sugar and ethanol mills and grain producers in the state of Mato Grosso. In: *Annals of IV International Symposium on Project Management, Innovation and Sustainability*. Available from: http://www.singep.org.br/4singep/resultado/319.pdf.102.CorriganME (accessed on 28 October 2021).

125. Chen, E., Song, H., Li, Y., Chen, H., Wang, B., Che, X., & Zhao, S., (2020). Analysis of aroma components from sugarcane to non-centrifugal cane sugar using GC-O-MS. *RSC Advances, 10*(54), 32276–32289.

126. Da Conceição, G. A., Moysés, D. N., Santa, A. L. M. M., & De Castro, A. M., (2018). Fed-batch strategies for saccharification of pilot-scale mild-acid and alkali pretreated sugarcane bagasse: Effects of solid loading and surfactant addition. *Industrial Crops and Products, 119*, 283–289.

127. De Farias, S. C. E., & Bertucco, A., (2016). Bioethanol from microalgae and cyanobacteria: A review and technological outlook. *Process Biochemistry, 51*(11), 1833–1842.

128. Dias, M. O., Junqueira, T. L., Cavalett, O., Pavanello, L. G., Cunha, M. P., Jesus, C. D., & Bonomi, A., (2013). Biorefineries for the production of first and second generation ethanol and electricity from sugarcane. *Applied Energy, 109*, 72–78.

129. Dias, M. O., Junqueira, T. L., Jesus, C. D., Rossell, C. E., Maciel, F. R., & Bonomi, A., (2012b). Improving second generation ethanol production through optimization of first generation production process from sugarcane. *Energy, 43*(1), 246–252.

130. Dias, M. O., Junqueira, T. L., Rossell, C. E. V., Maciel, F. R., & Bonomi, A., (2013b). Evaluation of process configurations for second generation integrated with first generation bioethanol production from sugarcane. *Fuel Processing Technology, 109*, 84–89.

131. Dutta, S., & Pal, S., (2014). Promises in direct conversion of cellulose and lignocellulosic biomass to chemicals and fuels: Combined solvent-nanocatalysis approach for biorefinary. *Biomass and Bioenergy, 62*, 182–197.
132. Ernsting, Y. R. M., Stevens, L., Wardle, A. R., & Hall, J. C., (2016). *The Renewable Fuel Standard's Effect on Corn Belt Income and Unemployment* (No. 16–30).
133. Esmaeili, S. A. H., Szmerekovsky, J., Sobhani, A., Dybing, A., & Peterson, T. O., (2020). Sustainable biomass supply chain network design with biomass switching incentives for first-generation bioethanol producers. *Energy Policy,* 111222.
134. Faria, D., Machado, G., Eichler, P., Boneberg, B., Fernanda, R. A. Y. E., Vilares, M., & Santos, F., (2016). Scenarios and perspectives of the main cultures of Rio Grande do Sul in biorefinery processes. *UERGS Scientific Electronic Journal, 2*(3), 291–306.
135. Furlan, F. F., Tonon, F. R., Pinto, F. H., Costa, C. B., Cruz, A. J., Giordano, R. L., & Giordano, R. C., (2013). Bioelectricity versus bioethanol from sugarcane bagasse: Is it worth being flexible. *Biotechnology for Biofuels, 6*(1), 142.
136. Ge, S., Wu, Y., Peng, W., Xia, C., Mei, C., Cai, L., & Tsang, Y. F., (2020). High-pressure CO_2 hydrothermal pretreatment of peanut shells for enzymatic hydrolysis conversion into glucose. *Chemical Engineering Journal, 385*, 123949.
137. Gibbons, W. R., & Hughes, S. R., (2009). Integrated biorefineries with engineered microbes and high-value co-products for profitable biofuels production. *In Vitro Cellular & Developmental Biology-Plant, 45*(3), 218–228.
138. Kang, A., & Lee, T. S., (2015). Converting sugars to biofuels: Ethanol and beyond. *Bioengineering, 2*(4), 184–203.
139. Kumar, A. K., & Sharma, S., (2017). Recent updates on different methods of pretreatment of lignocellulosic feedstocks: A review. *Bioresources and Bioprocessing, 4*(7).
140. Lennartsson, P. R., Erlandsson, P., & Taherzadeh, M. J., (2014). Integration of the first and second generation bioethanol processes and the importance of by-products. *Bioresource Technology, 165*, 3–8.
141. Lin, M., & Gi-Hyung, R., (2014). Effects of thermomechanical extrusion and particle size reduction on bioconversion rate of corn fiber for ethanol production. *Cereal Chemistry, 91*(4), 366–373.
142. Liu, W., Wu, R., Wang, B., Hu, Y., Hou, Q., Zhang, P., & Wu, R., (2020). Comparative study on different pretreatment on enzymatic hydrolysis of corncob residues. *Bioresource Technology, 295*, 122244.
143. Luiza, A. A., Rempel, A., Cavanhi, V. A. F., Alves, M., Deamici, K. M., Colla, L. M., & Costa, J. A. V., (2020). Simultaneous saccharification and fermentation of *Spirulina* sp. and corn starch for the production of bioethanol and obtaining biopeptides with high antioxidant activity.
144. Mass, R. A. (2018). Utilization of distiller's dried grains with solubles (DGGS) by cattle. In: Ingledew, W. M., Austin, G. D., Kelsall, D. R., & Kluhspies, C., (eds.), *The Alcohol Textbook* (5th edn.). A reference for the beverage, fuel, and industrial alcohol industries, Nottingham.
145. Maurya, D. P., Singla, A., & Negi, S., (2015). An overview of key pretreatment processes for biological conversion of lignocellulosic biomass to bioethanol. *3 Biotech, 5*, 597–609.
146. Mendes, F. M., Dias, M. O. S., Ferraz, A., Milagres, A. M. F., Santos, J. C., & Bonomi, A., (2017). Techno-economic impacts of varied compositional profiles of sugarcane experimental hybrids on a biorefinery producing sugar, ethanol and electricity. *Chemical Engineering Research and Design, 125*, 72–78.

147. Modesto, M., Zemp, R. J., & Nebra, S. A., (2009). Ethanol production from sugarcane: Assessing the possibilities of improving energy efficiency through exergetic cost analysis. *Heat Transfer Engineering, 30*(4), 272–281.

148. Mojović, L., Nikolić, S., Rakin, M., & Vukasinović, M., (2006). Production of bioethanol from corn meal hydrolyzates. *Fuel, 85*(12, 13), 1750–1755.

149. Nielsen, F., Galbe, M., Zacchi, G., & Wallberg, O., (2019). The effect of mixed agricultural feedstocks on steam pretreatment, enzymatic hydrolysis, and cofermentation in the lignocellulose-to-ethanol process. *Biomass Conversion and Biorefinery*, 1–14.

150. Nikolić, S., Mojović, L., Rakin, M., Pejin, D., & Pejin, J., (2011). Utilization of microwave and ultrasound pretreatments in the production of bioethanol from corn. *Clean Technologies and Environmental Policy, 13*(4), 587–594.

151. Nogueira, K. M. V., Mendes, V., Carraro, C. B., Taveira, I. C., Oshiquiri, L. H., Gupta, V. K., & Silva, R. N., (2020). Sugar transporters from industrial fungi: Key to improving second-generation ethanol production. *Renewable and Sustainable Energy Reviews, 131*, 109991.

152. Ferment Technologies in Sugar and Alcohol Ltd. (2009). Process of reusing yeast biomass in integration Alcoholic fermentation of sugarcane and starchy substrates (Vol. 103, pp. 323–335). 2015. BR10 2015 021056 6.104. Nottingham University Press.

153. Oliveira, C. M., Cruz, A. J. G., & Costa, C. B. B., (2016). Improving second generation bioethanol production in sugarcane biorefineries through energy integration. *Applied Thermal Engineering, 109*, 819–827.

154. Palacios-Bereche, R., Mosqueira-Salazar, K. J., Modesto, M., Ensinas, A. V., Nebra, S. A., Serra, L. M., & Lozano, M. A., (2013). Exergetic analysis of the integrated first-and second-generation ethanol production from sugarcane. *Energy, 62*, 46–61.

155. Petersen, A. M., Aneke, M. C., & Görgens, J. F., (2014). Techno-economic comparison of ethanol and electricity coproduction schemes from sugarcane residues at existing sugar mills in Southern Africa. *Biotechnology for Biofuels, 7*(1), 105.

156. Rajak, R. C., & Banerjee, R., (2020). An innovative approach of mixed enzymatic venture for 2G ethanol production from lignocellulosic feedstock. *Energy Conversion and Management, 207*(1).

157. Ramos, L. P., Da Silva, L., Ballem, A. C., Pitarelo, A. P., Chiarello, L. M., & Silveira, M. H. L., (2015). Enzymatic hydrolysis of steam-exploded sugarcane bagasse using high total solids and low enzyme loadings. *Bioresource Technology, 175*, 195–202.

158. Rezania, S., Oryani, B., Cho, J., Talaiekhozani, A., Sabbagh, F., Hashemi, B., & Mohammadi, A. A., (2020). Different pretreatment technologies of lignocellulosic biomass for bioethanol production: An overview. *Energy*, 117457.

159. Rezic, T., Oros, D., Markovic, I., Kracher, D., Ludwig, R., & Santek, B., (2013). Integrated hydrolyzation and fermentation of sugar beet pulp to bioethanol. *J Microbiol Biotechnol, 23*(9), 1244–1252.

160. Saini, J. K., Saini, R., & Tewari, L., (2015). Lignocellulosic agriculture wastes as biomass feedstocks for second-generation bioethanol production: Concepts and recent developments. *3 Biotech, 5*(4), 337–353.

161. Santos, F., Eichler, P., De Queiroz, J. H., & Gomes, F., (2020). Production of second-generation ethanol from sugarcane. In: *Sugarcane Biorefinery, Technology and Perspectives* (pp. 195–228). Academic Press.

162. Seabra, J. E., & Macedo, I. C., (2011). Comparative analysis for power generation and ethanol production from sugarcane residual biomass in Brazil. *Energy Policy, 39*(1), 421–428.

163. Sewsynker-Sukai, Y., & Kana, E. G., (2018). Simultaneous saccharification and bioethanol production from corn cobs: Process optimization and kinetic studies. *Bioresource Technology, 262*, 32–41.

164. Shen, J., & Agblevor, F. A., (2010). Modeling semi-simultaneous saccharification and fermentation of ethanol production from cellulose. *Biomass and Bioenergy, 34*(8), 1098–1107.

165. Terán, H. R., Ramos, L., Da Silva, S. S., Dragone, G., Mussatto, S. I., & Santos, J. C. D., (2018). Hydrodynamic cavitation as a strategy to enhance the efficiency of lignocellulosic biomass pretreatment. *Critical Reviews in Biotechnology, 38*(4), 483–493.

166. Téran-Hilares, R. T., Ienny, J. V., Marcelino, P. F., Ahmed, M. A., Antunes, F. A., Da Silva, S. S., & Dos, S. J. C., (2017). Ethanol production in a simultaneous saccharification and fermentation process with interconnected reactors employing hydrodynamic cavitation-pretreated sugarcane bagasse as raw material. *Bioresource Technology, 243*, 652–659.

167. Testoni, F. M., & Expressão, P. E. C. D., (2017). National Energy Research Center in Materials – CNPEM National Bioethanol Science and Technology Laboratory-CTBE.

168. Van, Z. W. H., Bloom, M., & Viktor, M. J., (2012). Engineering yeasts for raw starch conversion. *Applied Microbiology and Biotechnology, 95*(6), 1377–1388.

169. Vargas, F., Domínguez, E., Vila, C., Rodríguez, A., & Garrote, G., (2015). Agricultural residue valorization using a hydrothermal process for second generation bioethanol and oligosaccharides production. *Bioresource Technology, 191*, 263–270.

170. Zhang, H., Zhang, J., Xie, J., & Qin, Y., (2020). Effects of NaOH-catalyzed organosolv pretreatment and surfactant on the sugar production from sugarcane bagasse. *Bioresource Technology*, 123601.

171. Zhao, J., Xu, Y., Wang, W., Griffin, J., Roozeboom, K., & Wang, D., (2020). Bioconversion of industrial hemp biomass for bioethanol production: A review. *Fuel, 281*, 118725 (biomass).

172. Zhao, Y., Damgaard, A., Liu, S., Chang, H., & Christensen, T. H., (2020). Bioethanol from corn stover-Integrated environmental impacts of alternative biotechnologies. *Resources, Conservation and Recycling, 155*, 104652.

173. Lin, C. C., Kang, J. R., Huang, G. L., & Liu, W. Y. (2020). Forest biomass-to-biofuel factory location problem with multiple objectives considering environmental uncertainties and social enterprises. *Journal of Cleaner Production, 262*, 121327.

174. Macrelli, S., Galbe, M., & Wallberg, O. (2014). Effects of production and market factors on ethanol profitability for an integrated first and second generation ethanol plant using the whole sugarcane as feedstock. *Biotechnology for Biofuels, 7*(1), 1–16.

175. Dantas, G. A., Legey, L. F., & Mazzone, A. (2013). Energy from sugarcane bagasse in Brazil: An assessment of the productivity and cost of different technological routes. *Renewable and Sustainable Energy Reviews, 21*, 356–364.

176. Sartana & Afriyeni, N. (2017). Perundungan maya (*Cyber Bullying*) pada remaja awal. *Jurnal Psikologi Insight, 1*(1), 25–39.

177. Markel, E., Sims, C., & English, B. C. (2018). Policy uncertainty and the optimal investment decisions of second-generation biofuel producers. *Energy Economics, 76*, 89–100.

178. Gómez, Á & Aristizábal, V., (2010). Biorefineries based on coffee cut-stems and sugarcane bagasse: Furan-based compounds and alkanes as interesting products. *Bioresource technology, 196*, 480–489.

179. García, C. A., Fuentes, A., Hennecke, A., Riegelhaupt, E., Manzini, F., & Masera, O. (2011). Life-cycle greenhouse gas emissions and energy balances of sugarcane ethanol production in Mexico. *Applied Energy, 88*(6), 2088–2097.

180. García, C. A., & Manzini, F. (2012). Environmental and economic feasibility of sugarcane ethanol for the Mexican transport sector. *Solar Energy*, *86*(4), 1063–1069.
181. Garcia, M. L. et al. (2015). Bioenergy from stillage anaerobic digestion to enhance the energy balance ratio of ethanol production. *Journal of Environmental Management*, *162*, 102–114.
182. Hennecke, A. M., Mueller-Lindenlauf, M., García, C. A., Fuentes, A., Riegelhaupt, E., & Hellweg, S. (2016). Optimizing the water, carbon, and land-use footprint of bioenergy production in Mexico-Six case studies and the nationwide implications. *Biofuels, Bioproducts and Biorefining*, *10*(3), 222–239.
183. Shuba, E. S., & Kifle, D. (2018). Microalgae to biofuels:'Promising' alternative and renewable energy, review. *Renewable and Sustainable Energy Reviews*, *81*, 743–755.
184. Renewable Fuels Association (RFA). (2019). World Fuel Ethanol Production. URL https://ethanolrfa.org/resources/industry/statistics/.
185. IEA, (2018). IEA Agricultural Outlook 2018–2020.
186. BNDES–National Bank for Economic and Social Development. "BNDES the economic effects of the ethanol first-generation." BNDES Electronic Portal (2018).
187. CGGE-Bioetanol Combustível: Uma Oportunidade para o Brasil (2009), UNICAMP. Brasília (2009).
188. García-Olivares, A., Solé, J., & Osychenko, O. (2018). Transportation in a 100% renewable energy system. *Energy Conversion and Management*, *158*, 266–285.
189. Ramos, J. M., Simó-Alfonso, E. F., Ramis-Ramos, G., Gelfi, C., & Righetti, P. G. (2000). Determination of cow's milk and ripening time in nonbovine cheese by capillary electrophoresis of the ethanol-water protein fraction. *ELECTROPHORESIS: An International Journal*, *21*(3), 633–640.
190. Eijck, J., Romijn, H., Balkema, A., & Faaij, A. (2014). Global experience with jatropha cultivation for bioenergy: an assessment of socio-economic and environmental aspects. *Renewable and Sustainable Energy Reviews*, *32*, 869–889.
191. Milanez, D. H., Amaral, R. M. D., Faria, L. I. L. D., & Gregolin, J. A. R. (2013). Assessing nanocellulose developments using science and technology indicators. *Materials Research*, *16*, 635–641.
192. Schütt, K. T., Arbabzadah, F., Chmiela, S., Müller, K. R., & Tkatchenko, A. (2017). Quantum-chemical insights from deep tensor neural networks. *Nature communications*, *8*(1), 1–8.
193. Kumar, D., Jansen, M., Basu, R., & Singh, V. (2020). Enhancing ethanol yields in corn dry grind process by reducing glycerol production. *Cereal Chemistry*, *97*(5), 1026–1036.
194. Pereira, L. G., Cavalett, O., Bonomi, A., Zhang, Y., Warner, E., & Chum, H. L. (2019). Comparison of biofuel life-cycle GHG emissions assessment tools: The case studies of ethanol produced from sugarcane, corn, and wheat. *Renewable and Sustainable Energy Reviews*, *110*, 1–12.
195. Manochio, C., Andrade, B. R., Rodriguez, R. P., & Moraes, B. S. (2017). Ethanol from biomass: A comparative overview. *Renewable and Sustainable Energy Reviews*, *80*, 743–755.
196. Pachón, E. R., Mandade, P., & Gnansounou, E. (2020). Conversion of vine shoots into bioethanol and chemicals: Prospective LCA of biorefinery concept. *Bioresource technology*, *303*, 122946.
197. FAO, (2018). FAO Agricultural Outlook 2018–2020. https://www.fao.org/3/BT092e/BT092e.pdf.

198. Morandi, F., Perrin, A., & Østergård, H. (2016). Miscanthus as energy crop: Environmental assessment of a miscanthus biomass production case study in France. *Journal of Cleaner Production, 137*, 313–321.
199. Zhan, C., Feng, Z., Zhang, M., Tang, C., & Huang, Z. (2016). Experimental investigation on effect of ethanol and di-ethyl ether addition on the spray characteristics of diesel/biodiesel blends under high injection pressure. *Fuel, 218*, 1–11.
200. Xuan, Z., Naimi, T. S., Kaplan, M. S., Bagge, C. L., Few, L. R., Maisto, S., ... & Freeman, R. (2010). Alcohol policies and suicide: a review of the literature. *Alcoholism: clinical and experimental research, 40*(10), 2043–2055.
201. Mojović, L., Nikolić, S., Rakin, M., & Vukasinović, M. (2006). Production of bioethanol from corn meal hydrolyzates. *Fuel, 85*(12–13), 1750–1755.
202. Agarwal, A., Rana, M., Qiu, E., AlBunni, H., Bui, A. D., & Henkel, R. (2018). Role of oxidative stress, infection and inflammation in male infertility. *Andrologia, 50*(11), e13126.
203. Cripwell, R. A., Favaro, L., Viljoen-Bloom, M., & van Zyl, W. H. (2020). Consolidated bioprocessing of raw starch to ethanol by Saccharomyces cerevisiae: Achievements and challenges. *Biotechnology Advances, 42*, 107579.
204. Shen, J., & Agblevor, F. A. (2010). Modeling semi-simultaneous saccharification and fermentation of ethanol production from cellulose. *Biomass and Bioenergy, 34*(8), 1098–1107.
205. Kurambhatti, C., Kumar, D., & Singh, V. (2019). Impact of fractionation process on the technical and economic viability of corn dry grind ethanol process. *Processes, 7*(9), 578.
206. Sadhukhan, J., Martinez-Hernandez, E., Amezcua-Allieri, M. A., & Aburto, J. (2019). Economic and environmental impact evaluation of various biomass feedstock for bioethanol production and correlations to lignocellulosic composition. *Bioresource Technology Reports, 7*, 100230.
207. Pachón, E. R., Mandade, P., & Gnansounou, E. (2020). Conversion of vine shoots into bioethanol and chemicals: Prospective LCA of biorefinery concept. *Bioresource Technology, 303*, 122946.
208. Peterson, V. L., McCool, B. A., & Hamilton, D. A. (2015). Effects of ethanol exposure and withdrawal on dendritic morphology and spine density in the nucleus accumbens core and shell. *Brain research, 1594*, 125–135.
209. Ramos, W. B., Figueiredo, M. F., Brito, K. D., Ciannella, S., Vasconcelos, L. G., & Brito, R. P. (2016). Effect of solvent content and heat integration on the controllability of extractive distillation process for anhydrous ethanol production. *Industrial & Engineering Chemistry Research, 55*(43), 11315–11328.
210. Mandira, M., Rampure, M., Todkar, A., & Sharma, P. (2019). Ethanol from maize: an entrepreneurial opportunity in agrobusiness. *Biofuels, 10*(3), 385–391.
211. Rezic, T., Oros, D., Markovic, I., Kracher, D., Ludwig, R., & Santek, B. (2013). Integrated hydrolyzation and fermentation of sugar beet pulp to bioethanol. *Journal of Microbiology and Biotechnology, 23*(9), 1244–1252.
212. Cheah, W. Y., Sankaran, R., Show, P. L., Ibrahim, T. N. B. T., Chew, K. W., Culaba, A., & Jo-Shu, C. (2020). Pretreatment methods for lignocellulosic biofuels production: current advances, challenges and future prospects. *Biofuel Research Journal, 7*(1), 1115.
213. Umar, A., Khan, M. A., Kumar, R., & Algarni, H. (2017). Ag-doped ZnO nanoparticles for enhanced ethanol gas sensing application. *Journal of nanoscience and nanotechnology, 18*(5), 3557–3562.

214. Gomes, D. G., Serna-Loaiza, S., Cardona, C. A., Gama, M., & Domingues, L. (2018). Insights into the economic viability of cellulases recycling on bioethanol production from recycled paper sludge. *Bioresource technology, 267,* 347–355.
215. Kim, N. J., Li, H., Jung, K., Chang, H. N., & Lee, P. C. (2011). Ethanol production from marine algal hydrolysates using Escherichia coli KO11. *Bioresource technology, 102*(16), 7466–7469.
216. Baruah, J., Nath, B. K., Sharma, R., Kumar, S., Deka, R. C., Baruah, D. C., & Kalita, E. (2018). Recent trends in the pretreatment of lignocellulosic biomass for value-added products. *Frontiers in Energy Research, 6,* 141.
217. Waghmare, P. R., Khandare, R. V., Jeon, B. H., & Govindwar, S. P. (2018). Enzymatic hydrolysis of biologically pretreated sorghum husk for bioethanol production. *Biofuel Res. J, 5*(3), 846–853.
218. Nielsen, F., Galbe, M., Zacchi, G., & Wallberg, O. (2020). The effect of mixed agricultural feedstocks on steam pretreatment, enzymatic hydrolysis, and cofermentation in the lignocellulose-to-ethanol process. *Biomass Conversion and Biorefinery, 10*(2), 253–266.
219. Nogueira, K. M. V., Mendes, V., Carraro, C. B., Taveira, I. C., Oshiquiri, L. H., Gupta, V. K., & Silva, R. N. (2020). Sugar transporters from industrial fungi: key to improving second-generation ethanol production. *Renewable and Sustainable Energy Reviews, 131,* 109991.
220. Hilares, R. T., de Almeida, G. F., Ahmed, M. A., Antunes, F. A., da Silva, S. S., Han, J. I., & Dos Santos, J. C. (2017). Hydrodynamic cavitation as an efficient pretreatment method for lignocellulosic biomass: A parametric study. *Bioresource technology, 235,* 301–308.
221. Enerting, R. M., Shonnard, D. R., Griffing, E. M., Lai, A., & Palou-Rivera, I. (2016). Life cycle assessments of ethanol production via gas fermentation: anticipated greenhouse gas emissions for cellulosic and waste gas feedstocks. *Industrial & Engineering Chemistry Research, 55*(12), 3253–3261.
222. Bondancia, T. J., Côrrea, L. J., Cruz, A. J., BADINO, A., Mattoso, L. H. C., Marconcini, J. M., & Farinas, C. S. (2017). Nanofibras de celulose via hidrólise enzimática em reator de tanque agitado. In *Embrapa Instrumentação-Artigo em anais de congresso (ALICE).* In: *Workshop Da Rede De Nanotecnologia Aplicada Ao Agronegócio, 9,* 2017, Editors: Caue Ribeiro de Oliveira, Elaine Cristina Paris, Luiz Henrique Capparelli Mattoso, Marcelo Porto Bemquerer, Maria Alice Martins, Odílio Benedito Garrido de Assis. São Carlos. Anais. São Carlos: Embrapa Instrumentação, 2017. pp. 287–290.
223. Moran, P. J., Vacek, A. T., Racelis, A. E., Pratt, P. D., & Goolsby, J. A. (2017). Impact of the arundo wasp, *Tetramesa romana* (Hymenoptera: *Eurytomidae*), on biomass of the invasive weed, *Arundo donax* (Poaceae: *Arundinoideae*), and on revegetation of riparian habitat along the Rio Grande in Texas. *Biocontrol Science and Technology, 27*(1), 96–114.
224. Patni, N., Pillai, S. G., & Dwivedi, A. H. (2013). Wheat as a promising substitute of corn for bioethanol production. *Procedia Engineering, 51,* 355–362.
225. Gupta, R., & Lee, Y. Y. (2010). Investigation of biomass degradation mechanism in pretreatment of switchgrass by aqueous ammonia and sodium hydroxide. *Bioresource technology, 101*(21), 8185–8191.
226. RFA Agricultural Outlook 2016–2020. https:// https://ethanolrfa.org/
227. EMBRAPA, 2016: Faleiro, F. G., & Junqueira, N. T. V. (2016). Maracujá: o produtor pergunta, a Embrapa responde. *Embrapa Cerrados-Livro técnico (INFOTECA-E).*
228. Pereira, J. P., Oliveira, V. B., & Pinto, A. M. F. R. (2017). Modeling of passive direct ethanol fuel cells. *Energy, 133,* 652–665.

229. Dias, M. O. S., Ferraz, A., Milagres, A. M. F., Santos, J. C., & Bonomi, A. (2017). Techno-economic impacts of varied compositional profiles of sugarcane experimental hybrids on a biorefinery producing sugar, ethanol and electricity. *Chemical Engineering Research and Design, 125,* 72–78.

230. Pereira, L. G., Dias, M. O., Junqueira, T. L., Pavanello, L. G., Chagas, M. F., Cavalett, O., ... & Bonomi, A. (2014). Butanol production in a sugarcane biorefinery using ethanol as feedstock. Part II: Integration to a second generation sugarcane distillery. *Chemical Engineering Research and Design, 92*(8), 1452–1462.

231. DeLuca, T. H., Gundale, M. J., Brimmer, R. J., & Gao, S. (2020). Pyrogenic carbon generation from fire and forest restoration treatments. *Frontiers in Forests and Global Change, 3,* 24.

232. Chen, Z., Jacoby, W. A., & Wan, C. (2019). Ternary deep eutectic solvents for effective biomass deconstruction at high solids and low enzyme loadings. *Bioresource technology, 279,* 281–286.

233. Schuette, S. E. (2017). Insights into thermophilic plant biomass hydrolysis from Caldicellulosiruptor systems biology. *Microorganisms, 8*(3), 385.

234. Gibbens, W., & Hughes, S. (2009). Integrated biorefineries with engineered microbes and high-value co-products for profitable biofuels production. *Biofuels,* 265–283.

235. Carvalho, D. J., Moretti, R. R., Colodette, J. L., & Bizzo, W. A. (2020). Assessment of the self-sustained energy generation of an integrated first and second generation ethanol production from sugarcane through the characterization of the hydrolysis process residues. *Energy Conversion and Management, 203,* 112267. https://www. conab.gov.br/.

236. Gissén, C., Prade, T., Kreuger, E., Nges, I. A., Rosenqvist, H., Svensson, S. E., & Björnsson, L. (2014). Comparing energy crops for biogas production–Yields, energy input and costs in cultivation using digestate and mineral fertilisation. *Biomass and bioenergy, 64,* 199–210.

237. Furlan, A. L., Bianucci, E., del Carmen Tordable, M., Castro, S., & Dietz, K. J. (2014). Antioxidant enzyme activities and gene expression patterns in peanut nodules during a drought and rehydration cycle. *Functional Plant Biology, 41*(7), 704–713.

238. Peterson, T., Sharma, N., Shojaeiarani, J., & Bajwa, S. G. (2014). A review of densified solid biomass for energy production. *Renewable and Sustainable Energy Reviews, 96,* 296–305.

239. Kazi Kazeema, N., Borana Pravin, H., & Ikale Vijay, H. (2021). Formulation and Evaluation of An Anti-Oxidant Product From Betel Leaf Extract.

240. Kathi, Y., An, M., Liu, K., Nagai, S., Shigematsu, T., Morimura, S., & Kida, K. (2006). Ethanol production from acid hydrolysate of wood biomass using the flocculating yeast Saccharomyces cerevisiae strain KF-7. *Process Biochemistry, 41*(4), 909–914.

241. Werner, F. (2017). Background report for the life cycle inventories of wood and wood based products for updates of ecoinvent 2.2.

242. Dias, M. O., Lima, D. R., & Mariano, A. P. (2018). Techno-economic analysis of cogeneration of heat and electricity and second-generation ethanol production from sugarcane. In *Advances in Sugarcane Biorefinery* (pp. 197–212). Elsevier.

243. Patrizi, N., Pulselli, F. M., Morandi, F., & Bastianoni, S. (2010). Evaluation of the emergy investment needed for bioethanol production in a biorefinery using residual resources and energy. *Journal of Cleaner Production, 96,* 549–556.

244. Šobotník, J., Bourguignon, T., Carrijo, T. F., Bordereau, C., Robert, A., Křížková, B., & Cancello, E. M. (2015). The nasus gland: a new gland in soldiers of *Angularitermes* (Termitidae, *Nasutitermitinae*). *Arthropod Structure and Development, 44*(5), 401–406.

245. De Lucca, M. R. Z. (2020). Análise empírica sobre os determinantes da decisão de produção de açúcar e etanol na agroindústria canavieira paulista [*Empirical analysis of the determinants of the decision to produce sugar and ethanol in the sugarcane agroindustry in São Paulo*].

246. Kativado, K. A., Rathod, A. P., Wasewar, K. L., & Labhsetwar, N. K. (2016). Comparative study of different waste biomass for energy application. *Waste Management, 47*, 40–45.

CHAPTER 11

Conversion of Sweet Sorghum Juice to Bioethanol

IOSVANY LÓPEZ-SANDIN,[1] FRANCISCO ZAVALA-GARCÍA,[1]
GUADALUPE GUTIÉRREZ-SOTO,[1] and HÉCTOR RUIZ LEZA[2]

[1]*University of Nuevo León, Faculty of Agronomy, Francisco Villa S/N Col.
Ex Hacienda El Canadá – 66415, General Escobedo, N. L., México,
E-mail: francisco.zavala.garcia@gmail.com (F. Zavala-García)*

[2]*Biorefinery Group, Food Research Department,
Faculty of Chemistry Sciences, Autonomous University of Coahuila,
Saltillo–25280, Coahuila, Mexico*

ABSTRACT

The continuous decrease in fossil fuel reserves and their impact on the environment, lead the search for new alternative energy sources. In this context, biofuels provide a viable substitute that contributes to the reduction of polluting emissions into the atmosphere and increases the conventional fuels efficiency. However, biofuels confront great challenges such as obtaining raw material, requiring biomass sources to mitigate this problem. In this sense, the sweet sorghum variety ROGER has shown potential in the production of bioethanol, which depending on the production conditions can have ethanol concentrations in the juice of 81.2 g/l. This chapter mentions aspects related to the bioethanol production from sweet sorghum juice.

11.1 INTRODUCTION

The continuous use of fossil fuels has led to the reduction of their reserves and environmental pollution because of the emission of considerable volumes of GHG and other pollutants into the atmosphere [1]. However, over the years, actions aimed at reversing the energy and environmental paradigm through the search for new renewable and eco-friendly energy sources have increased [2]. In this sense, biofuels generate a significant reduction in GHG emissions with respect to those derived from oil [3, 4].

The biofuels production faces four major challenges: 1) the raw material must not compete with food production, 2) it must have a neutral balance of polluting emissions (CO_2), 3) its production must be continuous, ensuring the supply of industrial facilities (extraction plants, biorefineries, fermenters), and 4) extraction methods must be profitable from the energy and economic point of view to compete in the long term with the price of other sources of non-renewable energy [5].

Bioenergy is the most extensively used energy source in the world, mainly in solid form and to a minor extent in liquid and gaseous form. Thus, the continuous technological advance has allowed the development of more efficient biomass conversion technologies that make it possible to obtain biofuels in their different physical states. However, most of these achievements have not been implemented in commercial facilities yet [6, 7].

With respect to liquid biofuels (bioethanol and biodiesel), they are used as mixtures or substitutes for gasoline and diesel, being used mainly in transportation and industry. According to the origin of the raw material and the technology used to obtain it, they are classified as first, second, third, and fourth generation, the latter still under development [5, 8], which depend on sources of sugars, starch, and LCB. Where the conversion to bioethanol depends on the nature of the raw material, mainly on its biochemical composition [7].

In the case of first-generation bioethanol, it is a product that derives mainly from the sugars fermentation contained in agricultural crops with a high sucrose content, such as sugar cane, beet, and sweet sorghum, or from the starch contained in the corn grain, wheat, barley, tubers (like potato) and roots (like cassava), with the disadvantage of being crops intended for food, so their use as energy material generates a food safety controversy. In addition to contributing to deforestation by eliminating large areas of primary forests for their establishment [9]. However, sweet sorghum [*Sorghum bicolor* (L.) Moench] has agronomic characteristics that allow it to supply raw material for the production of biofuels, without affecting the diet and agricultural areas [10, 11], since it is the food base of millions of people, mainly in semi-arid regions around the world [12].

Sweet sorghum is a short-cycle plant with a high degree of tolerance to biotic and abiotic stresses, which can achieve high biomass yields and accumulate considerable volumes of fermentable sugars in the stem parenchyma [13–15]. Sorghum has historically been used in the production of syrup and molasses, but its bioenergetic potential makes it a viable raw material for the manufacture of bioethanol and other products [16, 17].

The agronomic yields of sweet sorghum are related to the variety and cultivation conditions, consolidating a strong line of research aimed at determining the factors that can affect the productivity of agronomic traits associated with obtaining bioethanol [94]. Hence, the importance of studying the crop production process to identify (among other factors) the best conditions for plant development in terms of energy and production, considering agricultural activities such as soil preparation [18] and fertilization [19]. They have a high influence on the development of the crop, on the energy consumptions of the system [20] and the definition of the optimal moment of its harvest [17]. Although this would not be complete without the study of the fermentation kinetics that defines the final production of bioethanol [21]. For this reason, ethanol productivity varies between sweet sorghum genotypes, due to its characteristics such as the amount of juice they can retain in their stems and the concentration of sugars, which can influence the process of conversion to bioethanol, where the temperature and the size of the yeast inoculum play an important role. The latest has a great influence on the typical biomass conversion processes, specifically in the sugar fermentation, where one of the traditionally used yeasts is Saccharomyces cerevisiae [7].

Thus, the potential of sorghum in the bioethanol production is also determined by the energy necessary to obtain it, energy balances being necessary in order to reduce consumption and achieve high energy efficiency. In this sense, an agricultural operation is efficient when the energy obtained is greater than that used. However, the energy ratio depends on the crop, the production system and the intensity of management. The last is mainly associated with the optimal management of the resources used in the production process such as water, fuels, fertilizers, pesticides, machinery, electricity, and others [22].

In general, obtaining bioethanol is a complex process that must be continuously improved, even more so by the introduction of new raw materials. Therefore, the incorporation of a new plant material always requires the study and evaluation of its potential application, as well as the study of the sweet sorghum variety. This chapter give general aspects related to the production of bioethanol from sweet sorghum juice, considering agronomic and energetic parameters of the crop, besides of the conversion process to bioethanol, using a new sweet sorghum variety developed in the Universidad

Autónoma de Nuevo León, leaded by the research group focused in sorghum plant breeding of the Agronomy Department.

11.2   BIOLOGY OF SWEET SORGHUM

Sorghum is an herbaceous plant of type C_4 that belongs to the grass family (Poaceae) with high ecological plasticity that allows its cultivation in tropical, subtropical, temperate, and semi-arid regions. Thus, taking advantage of soils with reduced natural fertility such as sandy ones and, due to its multiple uses, it is divided into grain, broomcorn, sweet, and forage sorghum [23, 24]. The morphological and physiological variations of sorghum plants have provided the material for natural selection to act on them and accumulate individual differences in a similar or artificial way. These differences allow the different uses of cultivated sorghums, such as the production of ethanol, grain, honey, brooms, etc. In such a way, that they are currently grouped into various categories, based on their main products and uses, as well as the distinctive or genetic characteristics of the plant [24, 25].

Taxonomically, sorghum was first described by Linnaeus in 1753 under the name *Holcus*. He originally delineated several species of *Holcus*, some of which were later moved to the *Avenae* tribe, where the generic name *Holcus* now belongs. In 1794, Moench distinguished the genus Sorghum from the genus Holcus [26]. The genus *Sorgum* is divided into five subgenera: *sorghum, chaetosorghum, heterosorghum, parasorghum*, and *stiposorghum*. Within the Sorghum subgenus, the wild species, *S. bicolor, S. halepense,* and *S. propinquum* can be mentioned [27].

Sorghum exhibits several morphophysiological forms and great variation for flower morphology, resulting in the classification of several basic and intermediate races. Harlan and de Wet [96], classified *Sorghum bicolor* (L.) Moench, subspp. *bicolor* in five basic races and ten hybrids as shown in the following subsections.

11.2.1   BASIC RACES

- Race *bicolor* (B);
- Race *guinea* (G);
- Race *caudatum* (C);
- Race *kafir* (K);
- Race *durra* (D).

11.2.2 INTERMEDIATE/HYBRID RACES

- Race *guinea-bicolor* (GB);
- Race *caudatum-bicolor* (CB);
- Race *kafir-bicolor* (KB);
- Race *durra-bicolor* (DB);
- Race *guinea-caudatum* (GC);
- Race *guinea-kafir* (GK);
- Race de *guinea-durra* (GD);
- Race *kafir-caudatum* (KC);
- Race *durra-caudatum* (DC);
- Race *kafir-durra* (KD).

11.3 GLOBAL SORGHUM PRODUCTION

The different uses of sorghum and its high capacity for adaptation to different climatic conditions have maintained its productive levels, as well as the areas devoted to this crop. The global trend of sorghum production is shown in Table 11.1. According to Mundia et al. [28], there are mainly 10 factors that significantly impact sorghum production: (1) climate change, (2) population growth/economic development, (3) non-food demand, (4) agricultural inputs, (5) demand for other crops, (6) scarcity of agricultural resources, (7) biodiversity, (8) cultural influence, (9) prices, and (10) armed conflicts. Furthermore, several of these factors can simultaneously affect sorghum production and their degree of impact can be the combination of several factors, and their magnitude differs geographically.

TABLE 11.1 Global Sorghum Production by Tonnage (in millions)

Region	Year				
	2010	2012	2014	2016	2018
World	60.181	57.321	68.278	63.661	59.342
Africa	25.073	23.583	29.298	30.430	29.782
America	22.500	21.056	26.820	23.024	19.244
Asia	10.384	9.676	9.498	7.132	7.974
Europe	0.713	0.763	1.375	1.280	1.079
Oceania	1.512	2.243	1.287	1.796	1.262

Source: FAO [29], Food and Agriculture Organization of the United Nations.

11.4   SWEET SORGHUM VARIETIES

In the production of first-generation bioethanol, many sweet sorghum genotypes have been reported (Table 11.2). Some of these varieties stand out for
their high resistance and adaptation to different production conditions [30],
for example:

- **Dale:** Resistant to lodging and diseases, with approximately 120 days
  to maturity.
- **Della:** Mid-season variety, disease-resistant, approximately 114 days
  to maturity.
- **Sugar Drip:** High productivity with late sowing, susceptible to most
  sorghum diseases, with approximately 110 days to maturity (early
  maturity).
- **M81-E:** Late maturing, resistant to lodging and diseases.
- **Theis and Brandes:** Late maturing, at least 2–3 weeks later than
  Dale, resistant to lodging and disease.

TABLE 11.2   Agronomic Performance of Some Sweet Sorghum Varieties Used to Produce
Bioethanol

Varieties	Juice (m³/ha)	°Bx (%)	Ethanol (m³/ha)	References
Keller	18.9–20.8	16.5–18.5	1.487–1.544	[31, 32]
Dale	21.1–22.9	15.3–18.7	1.331–1.360	[31, 32]
Della	18.2–24.2	13.7–16.2	1.281–1.429	[31, 32]
M81-E	23.1–23.4	14.9–17.1	1.419–1.496	[31, 32]
Sugar Drip	9.7–13.8	14.4–16.3	0.704–0.842	[31, 32]

11.5   ADVANTAGES OF SWEET SORGHUM

Sweet sorghum is a subspecies of sorghum [*Sorghum bicolor* (L.) Moench]
that is characterized by its high sugar content rather than grain production
[33]. Its agronomic characteristics make it a viable source in the current
energy situation and the food conflicts generated using agricultural crops
in the production of fuels. In addition, it has considerable advantages [10,
34–36] such as:

- It does not compete for agricultural ground as it can to be cultivated
  on marginal lands that are not optimal for food production;

- It is tolerant to drought, high temperatures, floods, soil salinity, acidity toxicity, allowing its development under different agroclimatic conditions;
- Presents high productivity and efficiency in the use of solar radiation, as well as nitrogen (N)-based fertilizers compared to other crops;
- It requires less inputs, chemical reactions and energy in the bioethanol obtainment from stem juice with a minimum cost of cultivation;
- It has high concentrations of fermentable sugars in the juice of its stems, producing more ethanol per unit area than other crops;
- Provides high volumes of raw material to produce second-generation bioethanol.

11.6 SWEET SORGHUM PRODUCTION

Sweet sorghum is one of the energy crops with high potential for the bioethanol production, which can be obtained from the juice of the stems, grains, bagasse, and straw. However, the cultivars with greater interest stand out in certain agronomic traits such as: (i) high biomass yield; (ii) stems rich in juices with high sugar content; (iii) high percentage of extractable juice; (iv) stem geometry capable of resisting lodging; (v) a long period of industrial use that prolongs the harvest season; and (vi) tolerance to different production conditions [11, 30].

Compared to other plant materials with potential in the bioethanol production such as sugar cane, sugar beet, and corn, sweet sorghum has higher levels of directly fermentable reducing sugars in its stems and requires fewer inputs per production unit of biomass. In addition, it has high adaptability to adverse edaphoclimatic conditions and a greater efficiency in the use of solar radiation and nitrogen fertilizers than corn and sugar cane [34, 37].

Although sweet sorghum is a crop with a great capacity to adapt to various agroclimatic conditions, crop yield in any region is associated with plant growth, which depends on climatic, biological, edaphic factors and agronomic practices. Some of these factors are summarized below:

1. **Precipitation and Temperatures:** It adapts to a wide range of annual precipitations (from 550 to 800 mm) and temperatures (15 to 45°C), under various climatic and soil conditions. Although the optimum temperature for plant development is between 25 and 40°C and requires a day duration between 10–14 h [38]. However, these

annual differences (particularly low rainfall and low temperatures) can influence biomass production, stem juice and sugar content, affecting crop yields and consequently ethanol. Also, they limit the response of the plant to the application of N and the magnitude of its effects will depend on the growing season [17, 39].

2. **Edaphic Conditions:** The soil provides the nutrients and other elements necessary for the plant development [40]. The case of sweet sorghum can be cultivated in different types of soils and, therefore, availability of nutrients. However, knowing the nutrient and water requirements for the plant and the physical, chemical, and biological properties of the soil. In most cases, it is necessary to modify mainly the initial soil conditions to favor the growth of the crop and increase yield [41, 97]. Ramírez et al. [42] reported that in Mexico there are more than 19 million hectares with high optimal conditions under irrigation conditions, although, with the supply of water, it can be cultivated throughout the country.

3. **Seedtime:** Sweet sorghum is an annual plant with a short life cycle (approximately 4 months) that, depending mainly on the agroclimatic conditions and the genotype, can be planted two or three times a year [14, 43]. According to Han et al. [44], sweet sorghum can be planted from mid-March to early June with favorable results. Nevertheless, planting sorghum in early. May and harvested at the hard dough stage can substantially increase the fermentable sugar content. Though, a late seedtime can decrease the efficiency of the use of solar radiation and consequently the productivity of the crop [45], which may be due to the gradual decrease of the foliar area and the height of the plant as a consequence of sensitivity to photoperiod or shorter days [46]. In addition, it delays the harvest exposing the plants to the attack of pests and diseases that are predominant at the end of the crop cycle [47]. In general, the sowing date can influence the contents of sucrose, total sugar, pH, electrical conductivity of the juice and the Brix degrees [48].

4. **Sowing Density:** It influences the growth parameters of sorghum and its yield. For example, optimal plant density can increase biomass yields, since they maintain a high and stable net assimilation rate, besides a positive balance between photosynthesis and respiration processes [40, 49, 50]. Since the plants have a greater vertical angle of the leaf with respect to the main stem, increasing the capture of light and consequently a higher photosynthetic rate [51, 52]. In this

sense, it has been observed that a high planting density ($\approx$16.7 plants/m^2) increases the juice, sugar, and biomass yields. However, the stem diameter and the internodes number decrease contributing to a higher lodging rate [49, 53].

5. **Fertilization:** The constant exploitation of agricultural soils has decreased their nutrient reserves, requiring artificial sources to compensate for the deficit. Which depends largely on the properties of the soil and the type of crop to be established. For example, N, and C deficiency limits plant growth and decreases their agronomic yields [50, 54, 55]. For the sustainable production of sweet sorghum as a raw material to produce bioethanol, it depends on agronomic practices that maintain the necessary levels of N and C in the soil [56]. Thus, the physical-chemical analysis of the soil is necessary to make up for its deficiencies. In this sense, a positive response of sweet sorghum has been observed with applications of $\geq$40 kg/ha of N, increasing the biomass, juice, and sugars yields [57–60].

6. **Pests Management and Control, Diseases, and Weeds:** Undesirable plants compete for nutrients, water, and light, they can release substances through their roots and leaves that turn out to be toxic to crops. In addition, they provide a favorable habitat for the proliferation of other pests (arthropods, mites, pathogens, and others) by serving as their hosts, which can affect the harvesting process and contaminate the production obtained [61]. On the other hand, pest, and disease cause considerable crop losses; for example, the yellow aphid *Melanaphis sacchari* causes crop damage at all stages of sorghum development [62, 63]. This pest feeds on the sap of the plant, reducing its vigor and yield [64], and even causing 100% of crop losses under severe attack [65].

7. **Irrigation:** It tries to supply rainwater when it is insufficient to supply the water needs of the crop, maintaining productive soils with optimal humidity levels for growth, in addition to contributing to increased agricultural yields. Water helps plants in the absorption of nutrients from the soil and to perform other physiological functions [66]. Sweet sorghum is a drought-resistant crop that, compared to other crops (such as sugar cane), requires a lower volume of irrigation water (approximately 400 mm), which depends mainly on climatic conditions and the variety [67, 68]. Thus, there is the possibility of reducing irrigation water needs through sweet sorghum cultivars with deep roots that intercept most rainwater and in some productive environments no require irrigation [69].

8. **Harvest:** The optimal harvest time is important in reducing production costs and obtaining high-quality plant material (high concentrations of fermentable sugars and biomass) destined for the bioethanol production [70, 71]. In this sense, it has been reported that the sweet sorghum harvest may be carried out from the soft mass stage of the grain (the filling of the grains is complete and they begin to harden) and extend to physiological maturity (the grains begin to dry out), which would imply between 104 to 130 days after sowing. Although, this time will depend on the variety and environmental conditions [71–74].

Once harvested, sweet sorghum faces some challenges such as the rapid fermentation of sugars, generating an acceleration of its decomposition in a few hours, indicating that it should not be stored for long periods [70, 75]. Therefore, the rapid stem grinding and the processing of the extracted juice (4 to 5 hours) is essential considering the high content of fermentable sugars and the metabolic activity of contaminating bacteria. However, to reduce this postharvest effect, the juice may be refrigerated, or chemical preservatives can be added [76, 77]. However, it could change the chemical composition of the juice [78].

In general, the bioethanol production requires the continuous supply of sweet sorghum, the transport and storage of high volumes of plant material, as well as minimal post-harvest losses of fermentable sugars [34].

11.7 ENERGY USE IN THE SWEET SORGHUM PRODUCTION

The energy balance allows evaluating the efficiency of the crop production methods, considering the energy inputs and outputs of the production system. The energy balance differs widely between agricultural production systems and life cycle analysis is commonly used to assess the energy efficiency and environmental effect of bioethanol production. Therefore, energy demand is classified according to the energy inputs of each production system [79]. Accordingly, an efficient process requires that the energy invested in producing a unit of biofuel, including the agricultural and industrial stages, be less than the energy leaving the system.

In the sweet sorghum production, the inputs associated with agricultural operations are decisive in energy consumption. Wherein production systems with low input requirements are characterized by lower consumption and, hence, greater energy efficiency. In this sense, it has been reported that the greatest contribution to energy consumption corresponds to the diesel used to move agricultural machinery and in nitrogen fertilization [22, 80]. Thereupon, consumption of energy has a wide range of variation depending on the production conditions of sweet sorghum. For example, Jankowski et al. [80] obtained that the low input technology required between 14.9 and 15.8 GJ/ha, with an energy gain of 170 GJ/ha, although high input technology had higher energy gains with 226 GJ/ha.

On the other hand, Garofalo et al. [22] with conventional tillage and the application of 150 kg/ha of N had an energy consumption in diesel of 4,508 MJ/ha, with an energy yield of 35.2 GJ/ha. Compared to the treatment without tillage and without fertilizer with an energy consumption of 2,260 MJ/ha. For the bioethanol production, an energy production of 22.6 to 70.5 GJ/ha has been obtained from fresh sweet sorghum biomass [81]. Likewise, a net energy gain of 8.37 MJ/l of bioethanol and an energy efficiency of 1.56 have been reported [79].

11.8 CONVERSION OF SWEET SORGHUM JUICE TO BIOETHANOL

The juice of the sweet sorghum stalk has shown to be a raw material of great potential for the biofuels production which, after being transformed (Figure 11.1) into anhydrous ethanol, is mixed with gasoline and is used in motors. During the bioethanol production, the stems are crushed to extract the juice rich in fermentable sugars with an exact proportion that varies between genotypes (53–85% sucrose, 9–33% glucose and 6–21% fructose) and that can be converted directly into ethanol with high efficiency, mainly by the yeast *Saccharomyces cerevisiae*. Which is a fast-acting microorganism that tolerates high alcohol concentrations (up to 150 g/l) and shows high bioethanol yields. Furthermore, it maintains high cell viability under different fermentation conditions [30, 82, 83]. When used in sweet sorghum juice in the bioethanol production, its growth curve shows three phases: the exponential phase observed approximately in the first 18 h of incubation, the slow phase observed approximately between 18 to 24 h and the stationary phase observed approximately between 24 to 72 hours [95]. Likewise, the kinetics of bioethanol production depend on several factors such as temperature, yeast load, pH, aeration time and rate, stirring conditions, nitrogen source and initial sugar concentration. Therefore, in the bioethanol production from sweet sorghum juice, a wide range of yield between 39.2 and 127.8 g/l has been reported [83–85].

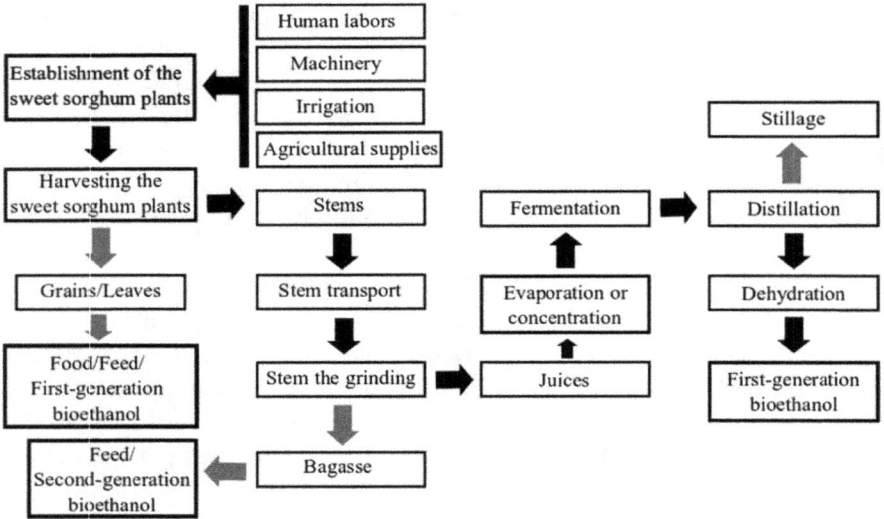

FIGURE 11.1 Conventional process diagram for obtaining first-generation bioethanol from sweet sorghum juice.

11.9 THE SWEET SORGHUM VARIETY "ROGER" IN THE PRODUCTION OF BIOETHANOL

The variety ROGER was registered under the same name and with registration number SOG-261-050315 by the Universidad Autónoma de Nuevo León in the National Catalog of Plant Varieties. This genotype is characterized by an average of 75 days to flowering and a cultivation cycle of 130 days. Depending on the production conditions, this crop have approximately a fresh biomass yield of 51.66 t/ha, juice of 22.53 m³/ha and °Bx of 16.04% [74].

ROGER was produced in the Marín experimental field of the Faculty of Agronomy belonging to the Universidad Autónoma de Nuevo León (UANL), located in the municipality of Marín, in the state of Nuevo Leon, Mexico. This is geographically located at 25° 52′ 13.5″ north latitude and 100° 02′ 22.56″ west longitude, with an altitude of 355 meters above sea level. The climate corresponds to a BS1 (h̄) hw (é), described as a dry hot steppe with rains in summer, annual rainfall that fluctuates between 250–500 mm and average annual temperature of 22°C. The type of soil is vertisols, thin, compacted, with high content of clay and calcium carbonate, low content of organic matter and pH between 7.5–8.5. ROGER, like other varieties of sweet sorghum, has a high capacity to adapt to different agro-climatic conditions. However, when these are not optimal for the efficient growth of the crop, they need to be reestablished by carrying out different agricultural operations.

11.9.1   AGRICULTURAL OPERATIONS

In one of our experiments, in the soil preparation, the traditional tillage integrated by the operations of clearing, subsoil plowing (up to a depth of 0.75 m), plowing, and harrowing was used. The sowing was carried out at a depth of 0.05 m and distance between rows of 0.8 m, using an experimental planter (Almaco® CTS EO, Nevada, Iowa, USA). Organic fertilization (chicken manure) was applied 20 days before sowing with doses of 3 t/ha (N-60 kg/ha, P-65 kg/ha and K-75 kg/ha) and 20 days after sowing thinning was performed to obtain an average density of 18 plants/m². For pest control, a dose of 0.6 l/ha of the pesticide Pounce 340 CE® was applied, 28 and 59 days after sowing and in particular for the yellow aphid a dose of 0.3 l/ha of Murralla® was applied at 32 and 81 days after sowing. For weed control, 1 l/ha of Phyto Amina 40® was applied at 35 and 64 days after sowing. Irrigation by band (superficial drip) was carried out before sowing and five auxiliary irrigation after sowing at intervals of 14 days. The harvest was carried out manually (130 days after sowing) at a cutting height of 0.03 m to 0.04 m from the stem base. However, to determine the harvest moment with the highest bioethanol yield, samples were taken at different plant physiological stages (PS) described by Vanderlip and Reeves [86]. In the first sampling, the grains showed soft mass (PS7). In the second sampling, the grains showed a stage of dough (PS8). In the third sampling, the culture was at physiological maturity (PS9).

11.9.2   ENERGY CONSUMPTION

In our study, the tillage and fertilization effect on energy consumption during the establishment of the crop was evaluated. The tillage treatments consisted of a minimum tillage system, conventional tillage and conventional tillage with breaking of the plow layer; while the fertilization treatments consisted of the use of organic and inorganic fertilizer and without fertilizer. For the above, the methodology described by De las Cuevas et al. [87] and Paneque and Sánchez [88], based on the proposals of Bridges and Smith [89] and supported by the information presented by Stout [90]. This methodology determines the energy cost in mechanized agricultural operations considering the energy required in construction, manufacturing, and transportation materials, fuel, lubricants/filters, maintenance/repair, labor, fertilizers, pesticides, and seeds. As a result, it was obtained that the energy demand varied depending on the production system used, where the systems with low inputs required less energy supply. In this sense, minimum tillage and no fertilization showed the lowest energy consumption values with 21.3 and 17.2 GJ/ha, respectively. Likewise, these systems had the highest energy efficiency values with 15.1 and 18.7, respectively. On the other hand, the highest net energy production was 340.3 and 351.4 GJ/ha, obtained with deep conventional tillage and organic fertilization, respectively. Where, the highest contribution to energy consumption was using diesel used in agricultural machinery in a range of 0.41 to 1.7 GJ/ha and in inorganic fertilization with 8.4 GJ/ha [74].

11.9.3   BIOETHANOL PRODUCTION

To evaluate the potential of the variety ROGER in the obtaining bioethanol, several tests were carried out at the laboratory level in which the effect of stirring conditions, inoculum size, phenological stage of the plant and freezing of the juice was evaluated. This was done using a standard submerged culture methodology in 250 ml Erlenmeyer flasks, incubated at 30°C for 48 h after being inoculated with the yeast *Saccharomyces cerevisiae* PE-2. To estimate the juice ethanol yield, 100 ml of the supernatants were taken and placed in a rotavapor (IKA® RV 10 Digital, China) at 30°C, 132 mbar and 50 rpm for a period of 40 min. The distillate obtained from the rotavapor was used to determine ethanol concentration using the potassium dichromate method [91, 92].

Table 11.3 shows the most representative results of the effects of inoculum, agitation, juice conditions (fresh and frozen) and the phenological stage of the plants in the bioethanol production. The highest bioethanol yield (78.1 g/l) was obtained with the juice of PS7 plants at 48 h of fermentation under stirring conditions and with 2% inoculum. Although, there was no significant difference between the phenological stages evaluated. Regarding the freezing of the juices, it had no influence on the obtaining of bioethanol, since in comparison with fresh juices and under the same fermentation conditions, there was no significant difference between them. This allows the juice to be preserved for long periods without affecting its quality and consequently the bioethanol yield.

TABLE 11.3 Bioethanol Production Conditions of with Juice of the Sweet Sorghum Variety ROGER

Treatments	Culture Conditions	Bioethanol Yield (g/l)	Time (h)
Fresh juice	Agitation 150 rpm, 2% inoculum	78.1	48
Fresh juice	Agitation 150 rpm, 3% inoculum	80.9	48
Fresh juice	Agitation 150 rpm, 1% inoculum	52.3	24
Fresh juice	Without agitation, 2% inoculum	34.7	24
Frozen juice	Agitation 150 rpm, 1% inoculum	47.3	24
Juice PS7*	Agitation 150 rpm, 2% inoculum	70.3	48
Juice PS8*		65.7	48
Juice PS9*		57.9	48

*Phenological stage of the plant described by Vanderlip and Reeves [86]. All the tests were carried out in triplicate, with a standard deviation of less than 5%.

In a test with 100 l of juice, the bioethanol production kinetic was determined. The juice was placed and sterilized in a stainless-steel tank with a capacity of 200 l and the changes in pH, °Bx, reducing sugar concentration (RS), and bioethanol concentration were determined at 0, 24, 48 and 72 h of fermentation. The pH measurement was made with a potentiometer (Corning® pH meter 430, USA). The °Bx were determined with a portable digital refractometer (ATAGO® 3810, PAL-1, Atago USA Inc., Bellevue, USA). The RS quantification was performed by the method described by Miller [93], using glucose as standard.

As a result, the bioethanol production curve showed a logarithmic phase up to 48 h, maintaining the stationary phase until 72 h (see Figure 11.2), during which the maximum bioethanol concentration was reached (81.2 g/l). The concentration of reducing sugars and °Bx showed a decrease of 64.5 and 59.1%, respectively at 72 h. However, the highest consumption rate was observed at 48 h with 59.6 and 52.3%, respectively. It is worth mentioning that the juice was not supplemented with other sources of carbon, nitrogen or other types of nutrients, nor was the pH adjusted during the fermentation process. Therefore, the behavior of the production kinetics suggests that the juice obtained from the variety does not have levels of compound that inhibit the development of yeast.

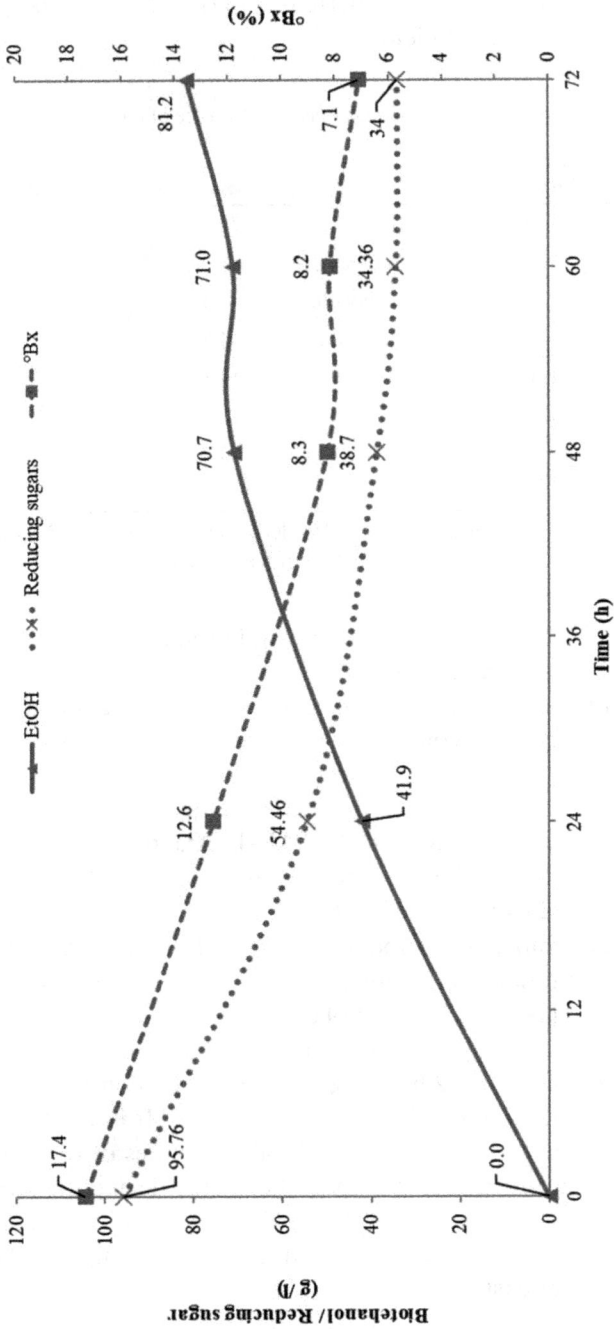

FIGURE 11.2 Bioethanol production kinetic.

11.10 CONCLUSIONS

Sweet sorghum is a low-input energy crop that has high adaptability to various production conditions and great potential in bioethanol production. In this sense, the sweet sorghum variety ROGER has potential in the first-generation bioethanol production and can be cultivated in semi-arid climates, achieving a high production of biomass, juice, and sugar. On the other hand, during the crop establishment, energy consumption oscillates depending on the production system used, with low-input systems being those with the lowest energy consumption. The higher energy input was from the consumption of diesel fuel (0.41 to 1.7 GJ/ha) and of chemical fertilizer (8.4 GJ/ha). Thus, the higher energy efficiency with values of 15.1 and 18.7 was obtained with minimum tillage and without fertilizer application, respectively. Whereas for the stem juice, it does not require supplementation or addition of nutrients for its conversion to bioethanol. However, the yields fluctuate depending on the fermentation conditions, where the maximum yield can be obtained after 72 h of fermentation (81.2 g/l).

KEYWORDS

- **bioenergy**
- **biofuels**
- **fossil fuels**
- **greenhouse gases**
- **production systems**
- **variety ROGER**

REFERENCES

1. Rezk, H., Nassef, A. M., Inayat, A., Sayed, E. T., Shahbaz, M., & Olabi, A. G., (2019). Improving the environmental impact of palm kernel shell through maximizing its production of hydrogen and syngas using advanced artificial intelligence. *Sci. Total Environ., 658*, 1150–1160.
2. Nazir, M. S., Mahdi, A. J., Bilal, M., Sohail, H. M., Ali, N., & Iqbal, H. M., (2019). Environmental impact and pollution-related challenges of renewable wind energy paradigm: A review. *Sci. Total Environ., 683*, 436–444.

3. Adler, P. R., Spatari, S., D'Ottone, F., Vazquez, D., Peterson, L., Del, G. S. J., Baethgen, W. B., & Parton, W. J., (2018). Legacy effects of individual crops affect N_2O emissions accounting within crop rotations. *GCB Bioenergy, 10,* 123–136.
4. Holmberg, K., & Erdemir, A., (2019). The impact of tribology on energy use and CO_2 emission globally and in combustion engine and electric cars. *Tribol. Int., 135,* 389–396.
5. Leyva, A., & De Gauna, G. R., (2011). Cultivos energéticos y biocombustibles. *Lychnos, 6,* 44–49.
6. Toklu, E., (2017). Biomass energy potential and utilization in Turkey. *Renewable Energy, 107,* 235–244.
7. Zabed, H., Sahu, J. N., Suely, A., Boyce, A. N., & Faruq, G., (2017). Bioethanol production from renewable sources: Current perspectives and technological progress. *Renew. Sustain. Energy Rev., 71,* 475–501.
8. Ramachandra, T. V., & Hebbale, D., (2020). Bioethanol from macroalgae: Prospects and challenges. *Renewable and Sustainable Energy Reviews, 117,* 109479.
9. Valdés-Rodríguez, O. A., & Palacios-Wassenaar, O. M., (2016). Evolution and current situation of plantations for biofuel: perspectives and challenges for México. *Agroproductividad, 9,* 33–41.
10. Dar, R. A., Dar, E. A., Kaur, A., & Phutela, U. G., (2018). Sweet sorghum-a promising alternative feedstock for biofuel production. *Renewable Sustainable Energy Rev., 82,* 4070–4090.
11. Jiang, D., Hao, M., Fu, J., Liu, K., & Yan, X., (2019). Potential bioethanol production from sweet sorghum on marginal land in China. *J. Cleaner Prod., 220,* 225–234.
12. Mengistu, G., Shimelis, H., Laing, M., Lule, D., & Mathew, I., (2020). Genetic variability among Ethiopian sorghum landrace accessions for major agro-morphological traits and anthracnose resistance. *Euphytica, 216,* 1–15.
13. Shukla, S., Felderhoff, T. J., Saballos, A., & Vermerris, W., (2017). The relationship between plant height and sugar accumulation in the stems of sweet sorghum (*Sorghum bicolor* (L.) Moench). *Field Crop. Res., 203,* 181–191.
14. Mathur, S., Umakanth, A. V., Tonapi, V. A., Sharma, R., & Sharma, M. K., (2017). Sweet sorghum as biofuel feedstock: Recent advances and available resources. *Biotechnol. Biofuels, 10,* 146.
15. Bojović, R., Popović, V. M., Ikanović, J., Živanović, L., Rakaščan, N., Popović, S., Ugrenović, V., & Simić, D., (2019). Morphological characterization of sweet sorghum genotypes across environments. *JAPS: J. Ani. Plant. Sci., 29,* 721–729.
16. Ou, M. S., Awasthi, D., Nieves, I., Wang, L., Erickson, J., Vermerris, W., Ingram, L. O., & Shanmugam, K. T., (2016). Sweet sorghum juice and bagasse as feedstocks for the production of optically pure lactic acid by native and engineered *Bacillus coagulans* strains. *BioEnergy Res., 9,* 123–131.
17. Pabendon, M. B., Efendi, R., Santoso, S. B., & Prastowo, B., (2017). Varieties of sweet sorghum super-1 and super-2 and its equipment for bioethanol in Indonesia. *IOP Conference Series: Earth and Environmental Science* (Vol. 65). International conference on biomass: technology, application, and sustainable development-2016; Bogor, Indonesia.
18. Adimassu, Z., Alemu, G., & Tamene, L., (2019). Effects of tillage and crop residue management on runoff, soil loss and crop yield in the humid highlands of Ethiopia. *Agric. Syst., 168,* 11–18.
19. Thivierge, M. N., Chantigny, M. H., Bélanger, G., Seguin, P., Bertrand, A., & Vanasse, A., (2015). Response to nitrogen of sweet pearl millet and sweet sorghum grown for ethanol in eastern Canada. *Bioenergy Res., 8,* 807–820.

20. Ding, N., Yang, Y., Cai, H., Liu, J., Ren, L., Yang, J., & Xie, G. H., (2017). Life cycle assessment of fuel ethanol produced from soluble sugar in sweet sorghum stalks in North China. *J. Clean. Prod., 161*, 335–344.

21. Phukoetphim, N., Chan-u-tit, P., Laopaiboon, P., & Laopaiboon, L., (2019). Improvement of bioethanol production from sweet sorghum juice under very high gravity fermentation: Effect of nitrogen, osmoprotectant, and aeration. *Energies, 12*, 3620.

22. Garofalo, P., Campi, P., Vonella, A. V., & Mastrorilli, M., (2018). Application of multi-metric analysis for the evaluation of energy performance and energy use efficiency of sweet sorghum in the bioethanol supply-chain: A fuzzy-based expert system approach. *App. Energy, 220*, 313–324.

23. Draghici, I., Draghici, R., Diaconu, A., Croitoru, M., Paraschiv, A. N., Dima, M., Ciuciuc, E., & Ciuciuc, D., (2019). Results on bioenergetic potential of some sweet sorghum hybrids cultivated under psamosols conditions in Southern Oltenia. In: *E3S Web of Conferences* (Vol. 112). EDP Sciences.

24. Martiwi, I. N. A., Nugroho, L. H., Daryono, B. S., & Susandarini, R., (2020). Morphological variability and taxonomic relationship of *Sorghum bicolor* (L.) moench accessions based on qualitative characters. *Annu. Res. Rev. Biol., 35*, 40–52.

25. Hilley, J. L., Weers, B. D., Truong, S. K., McCormick, R. F., Mattison, A. J., McKinley, B. A., Morishige, D. T., & Mullet, J. E., (2017). Sorghum Dw2 encodes a protein kinase regulator of stem internode length. *Sci. Rep., 7*, 1–13.

26. Lazarides, M., Hacker, J. B., & Andrew, M. H., (1991). Taxonomy, cytology and ecology of indigenous Australian sorghums (Sorghum Moench: Andropogoneae: Poaceae). *Aust. Syst. Bot., 4*, 591–635.

27. Saballos, A., (2008). Development and utilization of sorghum as a bioenergy crop. In: *Genetic Improvement of Bioenergy Crops* (pp. 211–248). Springer, New York, NY, (2008).

28. Mundia, C. W., Secchi, S., Akamani, K., & Wang, G., (2019). A regional comparison of factors affecting global sorghum production: The case of North America, Asia and Africa's Sahel. *Sustainability, 11*, 2135.

29. FAO, (2020). *Food and Agriculture Organization of the United Nations*. http://www.fao.org/faostat/es/#data/QC (accessed on 28 October 2021).

30. Appiah-Nkansah, N. B., Li, J., Rooney, W., & Wang, D., (2019). A review of sweet sorghum as a viable renewable bioenergy crop and its techno-economic analysis. *Renewable Energy, 143*, 1121–1132.

31. Rutto, L. K., Xu, Y., Brandt, M., Ren, S., & Kering, M. K., (2013). Juice, ethanol, and grain yield potential of five sweet sorghum (*Sorghum bicolor* [L.] moench) cultivars. *J. Sustain. Bioenergy Syst., 3*, 113–118.

32. Briand, C. H., Geleta, S. B., & Kratochvil, R. J., (2018). Sweet sorghum (*Sorghum bicolor* [L.] Moench) a potential biofuel feedstock: Analysis of cultivar performance in the Mid-Atlantic. *Renewable Energy, 129*, 328–333.

33. Perrin, R., Fulginiti, L. E., Bairagi, S., & Dweikat, I., (2018). Sweet Sorghum as feedstock in great plains corn ethanol plants: The role of biofuel policy. *J. Agr. Resour. Econ., 43*, 34–45.

34. Anami, S. E., Zhang, L. M., Xia, Y., Zhang, Y. M., Liu, Z. Q., & Jing, H. C., (2015). Sweet sorghum ideotypes: Genetic improvement of stress tolerance. *Food Energy Secur., 4*, 3–24.

35. Bonin, C. L., Heaton, E. A., Cogdill, T. J., & Moore, K. J., (2016). Management of sweet sorghum for biomass production. *Sugar Tech., 18*, 150–159.

36. Larnaudie, V., Rochon, E., Ferrari, M. D., & Lareo, C., (2016). Energy evaluation of fuel bioethanol production from sweet sorghum using very high gravity (VHG) conditions. *Renewable Energy, 88*, 280–287.

37. Jardim, A. M. D. R. F., Da Silva, G. Í. N., Biesdorf, E. M., Pinheiro, A. G., Da Silva, M. V., Araújo, J., Dos, S. A., et al., (2020). Production potential of *Sorghum bicolor* (L.) Moench crop in the Brazilian semiarid. *PUBVET, 14*, 1–13.

38. Umakanth, A. V., Kumar, A. A., Vermerris, W., & Tonapi, V. A., (2019). Sweet sorghum for biofuel industry. In: *Breeding Sorghum for Diverse End Uses* (pp. 255–270). Woodhead Publishing.

39. Maw, M. J., Houx, III. J. H., & Fritschi, F. B., (2016). Sweet sorghum ethanol yield component response to nitrogen fertilization. *Ind. Crops Prod., 84*, 43–49.

40. Xuan, T. D., Phuong, N. T., Khang, D. T., & Khanh, T. D., (2015). Influence of sowing times, densities, and soils to biomass and ethanol yield of sweet sorghum. *Sustainability, 7*, 11657–11678.

41. Zuo, W., Gu, C., Zhang, W., Xu, K., Wang, Y., Bai, Y., & Dai, Q., (2019). Sewage sludge amendment improved soil properties and sweet sorghum yield and quality in a newly reclaimed mudflat land. *Sci. Total. Environ., 654*, 541–549.

42. Ramírez-Jaramillo, G., (2020). Agroclimatic conditions for growing *Sorghum bicolor* L. moench, under irrigation conditions in Mexico. *Open Access Library Journal, 7*, 1–14.

43. Aguilar-Sánchez, P., Navarro-Pineda, F. S., Sacramento-Rivero, J. C., & Barahona-Pérez, L. F., (2018). Life-cycle assessment of bioethanol production from sweet sorghum stalks cultivated in the state of Yucatan, Mexico. *Clean. Technol. Environ. Policy, 20*, 1685–1696.

44. Han, K. J., Alison, M. W., Pitman, W. D., Day, D. F., Kim, M., & Madsen, L., (2012). Planting date and harvest maturity impact on biofuel feedstock productivity and quality of sweet sorghum grown under temperate Louisiana conditions. *Agron. J., 104*, 1618–1624.

45. Houx, III. J. H., & Fritschi, F. B., (2015). Influence of late planting on light interception, radiation use efficiency and biomass production of four sweet sorghum cultivars. *Ind. Crops Prod., 76*, 62–68.

46. Wolabu, T. W., Zhang, F., Niu, L., Kalve, S., Bhatnagar-Mathur, P., Muszynski, M. G., & Tadege, M., (2016). Three flowering locus T-like genes function as potential florigens and mediate photoperiod response in sorghum. *New Phytol., 210*, 946–959.

47. Almodares, A., & Hadi, M. R., (2009). Production of bioethanol from sweet sorghum: A review. *Afr. J. Agr. Res., 4*, 772–780.

48. Uchimiya, M., Knoll, J. E., Anderson, W. F., & Harris-Shultz, K. R., (2017). Chemical analysis of fermentable sugars and secondary products in 23 sweet sorghum cultivars. *J. Agric. Food Chem., 65*, 7629–7637.

49. Tang, C., Sun, C., Du, F., Chen, F., Ameen, A., Fu, T., & Xie, G. H., (2018). Effect of plant density on sweet and biomass Sorghum production on semiarid marginal land. *Sugar Tech., 20*, 312–322.

50. Sahu, H., Tomar, G. S., & Thakur, V. S., (2020). Effects of planting density and levels of nitrogen on growth and development of sweet sorghum [*Sorghum bicolor* (L.) Moench] varieties. *J. Pharmacogn. Phytochem., 9*, 1894–1898.

51. Tian, F., Bradbury, P. J., Brown, P. J., Hung, H., Sun, Q., Flint-Garcia, S., Rocheford, T. R., et al., (2011). Genome-wide association study of leaf architecture in the maize nested association mapping population. *Nature Genetics, 43*, 159–162.

52. Zhang, N., Van, W. A., He, L., Evers, J. B., Anten, N. P., & Marcelis, L. F., (2020). Light from below matters: Quantifying the consequences of responses to far-red light reflected upwards for plant performance in heterogeneous canopies. *Plant. Cell. Environ.*, 1–12.
53. Choi, Y., Han, H., Shin, S., Heo, B., Choi, K., & Kwon, S., (2019). Effect of planting density on growth and yield components of the sweet sorghum cultivar,'Chorong'. *Korean J. Crop Sci., 64*, 40–47.
54. Heitman, A. J., Castillo, M. S., Smyth, T. J., Crozier, C. R., Wang, Z., Heiniger, R. W., & Gehl, R. J., (2017). Nitrogen fertilization effects on yield and nutrient removal of biomass and sweet sorghum. *Agron. J., 109*, 1352–1358.
55. Mishra, J. S., Kumar, R., & Rao, S. S., (2017). Performance of sweet sorghum (*Sorghum bicolor*) cultivars as a source of green fodder under varying levels of nitrogen in semi-arid tropical India. *Sugar Tech., 19*, 532–538.
56. Sainju, U. M., Singh, H. P., Singh, B. P., Whitehead, W. F., Chiluwal, A., & Paudel, R., (2018). Cover crop and nitrogen fertilization influence soil carbon and nitrogen under bioenergy sweet sorghum. *Agron. J., 110*, 463–471.
57. Ameen, A., Yang, X., Chen, F., Tang, C., Du, F., Fahad, S., & Xie, G. H., (2017). Biomass yield and nutrient uptake of energy sorghum in response to nitrogen fertilizer rate on marginal land in a semi-arid region. *BioEnerg. Res., 10*, 363–376.
58. Kering, M. K., Temu, V. W., & Rutto, L. K., (2017). Nitrogen fertilizer and panicle removal in Sweet Sorghum production: Effect on biomass, juice yield and soluble sugar content. *J. Sustain Bioenergy Syst., 7*, 14–26.
59. Sowiński, J., Konieczny, M., & Jama-Rodzeńska, A., (2018). The effect of yararega fertilization on the nitrogen effectiveness and yield of sweet sorghum (*Sorghum bicolor* (L.) moench). *Acta Sci. Pol. Agric., 16*, 235–246.
60. Maw, M. J., Houx, III. J. H., & Fritschi, F. B., (2019). Nitrogen content and use efficiency of sweet sorghum grown in the lower Midwest. *Agron. J., 111*, 2920–2928.
61. Esperbent, C. E., (2015). *Malezas: El Desafío Para el Agro Que Viene.* [Online], *12*, 235–240 http://www.redalyc.org/pdf/864/86443147004.pdf (accessed on 28 October 2021).
62. Vázquez-Navarro, J. M., Carrillo-Aguilera, J. C., & Cisneros-Flores, B. A., (2016). A population study in a forage sorghum crop infested with sugarcane aphid *Melanaphis sacchari* (Zehntner, 1897) (Hemiptera: Aphididae) at the Comarca Lagunera Region. *Entomología Mexicana, 3*, 395–400.
63. Tejeda-Reyes, M. A., Díaz-Nájera, J. F., Rodríguez-Maciel, J. C., Vargas-Hernández, M., Solís-Aguilar, J. F., Ayvar-Serna, S., & Flores-Yáñez, J. A., (2017). Evaluación en campo de insecticidas sobre melanaphis sacchari (Zehntner) 1 en sorgo. *Southwest. Entomol., 42*, 545–551.
64. Bowling, R. D., Brewer, M. J., Kerns, D. L., Gordy, J., Seiter, N., Elliott, N. E., Butin, G. D., et al., (2016). Sugarcane aphid (Hemiptera: Aphididae): A new pest on sorghum in North America. *J. Integr. Pest. Manage., 7*, 12–25.
65. Rodríguez-del-Bosque, L. A., & Terán, A. P., (2015). *Melanaphis sacchari* (Hemiptera: Aphididae): A new sorghum insect pest in Mexico. *Southwest. Entomol., 40*, 433–435.
66. Dercas, N., & Liakatas, A., (2018). Sweet Sorghum Canopy Development in Relation to Radiation and Water Use. *Environ. Process, 5*, 413–425.
67. Mengistu, M. G., Steyn, J. M., Kunz, R. P., Doidge, I., Hlophe, H. B., Everson, C. S., Jewitt, G. P. W., & Clulow, A. D., (2016). A preliminary investigation of the water use efficiency of sweet sorghum for biofuel in South Africa. *Water SA, 42*, 152–160.

68. Prasad, S., Sheetal, K. R., Renjith, P. S., Kumar, A., & Kumar, S., (2019), Sweet sorghum: An excellent crop for renewable fuels production. In: Rastegari, A., Yadav, A., & Gupta, A., (eds.), *Prospects of Renewable Bioprocessing in Future Energy Systems*: *Biofuel Bior. Technol.* (Vol. 10, pp. 291–314), Springer, Cham.

69. Lopez, J. R., Erickson, J. E., Asseng, S., & Bobeda, E. L., (2017). Modification of the CERES grain sorghum model to simulate optimum sweet sorghum rooting depth for rainfed production on coarse textured soils in a sub-tropical environment. *Agric. Water Managet., 181*, 47–55.

70. Costa, G. H. G., Ciaramello, S., Giachini, J. W., Gazzola, W. C. B., Giachini, L. E., & Uribe, R. A. M., (2018). Effects of sweet sorghum harvest systems on raw material quality. *Sugar Tech., 20*, 730–733.

71. Vlachos, C. E., Pavli, O. I., Flemetakis, E., & Skaracis, G. N., (2020). Exploiting pre-and post-harvest metabolism in sweet sorghum genotypes to promote sustainable bioenergy production. *Ind. Crops. Prod., 155*, 112758.

72. Oyier, M. O., Owuoche, J. O., Oyoo, M. E., Cheruiyot, E., Mulianga, B., & Rono, J., (2017). Effect of harvesting stage on sweet sorghum (*Sorghum bicolor* L.) genotypes in western Kenya. *Sci. World. J., 2017*, 1–10.

73. Dutra, E. D., Alencar, B. R. A., Galdino, J. J., Tabosa, J. N., Menezes, R. S. C., De Araújo, F. R. N., Dário, C. P., et al., (2018). First and second generation of ethanol production for five sweet sorghum cultivars during soft dough grain. *J. Exp. Agric., 25*, 1–12.

74. López-Sandin, I., Gutiérrez-Soto, G., Gutiérrez-Díez, A., Medina-Herrera, N., Gutiérrez-Castorena, E., & Zavala-García, F., (2019). Evaluation of the use of energy in the production of sweet sorghum (Sorghum Bicolor (L.) Moench) under different production systems. *Energies, 12*, 1713.

75. Ebrahimiaqda, E., & Ogden, K. L., (2018). Evaluation and modeling of bioethanol yield efficiency from sweet sorghum juice. *BioEnergy Res., 11*, 449–455.

76. Wu, X., Staggenborg, S., Propheter, J. L., Rooney, W. L., Yu, J., & Wang, D., (2010). Features of sweet sorghum juice and their performance in ethanol fermentation. *Ind. Crops. Prod., 31*, 164–170.

77. Kumar, C. G., Rao, P. S., Gupta, S., Malapaka, J., & Kamal, A., (2015). Chemical preservatives-based storage studies and ethanol production from juice of sweet sorghum cultivar, ICSV 93046. *Sugar tech., 17*, 404–411.

78. Bridgers, E. N., Chinn, M. S., Veal, M. W., & Stikeleather, L. F., (2011). Influence of juice preparations on the fermentability of sweet sorghum. *Biol. Engineer. Trans., 4*, 57–67.

79. Wang, M., Chen, Y., Xia, X., Li, J., & Liu, J., (2014). Energy efficiency and environmental performance of bioethanol production from sweet sorghum stem based on life cycle analysis. *Bioresour. Technol., 163*, 74–81.

80. Jankowski, K. J., Sokólski, M. M., Dubis, B., Załuski, D., & Szempliński, W., (2020). Sweet sorghum-Biomass production and energy balance at different levels of agricultural inputs. A six-year field experiment in north-eastern Poland. *Eur. J. Agron., 119*, 126119.

81. Glab, L., Sowiński, J., Chmielewska, J., Prask, H., Fugol, M., & Szlachta, J., (2019). Comparison of the energy efficiency of methane and ethanol production from sweet sorghum (*Sorghum bicolor* (L.) Moench) with a variety of feedstock management technologies. *Biomass Bioenergy, 129*, 105332.

82. Argote, F. E., Cuervo, R. A., Osorio, E., Ospina, J. D., & Castillo, H. S. V., (2015). Evaluation of ethanol production from molasses with native strains of Saccharomyces cerevisiae. Biotecnol. *Sector Agropecuario Agroind, 13*, 40–48.

83. Buruiană, C. T., Vizireanu, C., & Furdui, B., (2018). Bioethanol production from sweet sorghum stalk juice by ethanol-tolerant *Saccharomyces cerevisiae* strains: An overview. *The Annals of the University Dunarea de Jos of Galati. Fascicle VI-Food Technology, 42*, 153–167.

84. Pilap, W., Thanonkeo, S., Klanrit, P., & Thanonkeo, P., (2018). The potential of the newly isolated thermotolerant *Kluyveromyces marxianus* for high-temperature ethanol production using sweet sorghum juice. *3 Biotech, 8*, 126.

85. Laopaiboon, P., Khongsay, N., Phukoetphim, N., & Laopaiboon, L., (2019). Ethanol production from sweet sorghum juice under very high gravity fermentation by *Saccharomyces cerevisiae*: Aeration strategy and its effect on yeast intracellular composition. *Chiang Mai J. Sci., 46*, 481–494.

86. Vanderlip, R. L., & Reeves, H. E., (1972). Growth stages of sorghum [*Sorghum bicolor*, (L.) moench.]. *Agron. J., 64*, 13–16.

87. De Las, C. H. R. M., Rodríguez, T. H., Paneque, P. R., & Díaz, M. A., (2011). Energy cost of the knife roller CEMA 1400 for vegetable covering. *Rev. Cienc. Téc. Agropecu., 20*, 53–56.

88. Paneque, P., & Rodríguez, Y. S., (2006). Energy cost of the rice mechanized harvest in Cuba. *Cienc. Téc. Agropecu., 15*, 19–23.

89. Bridges, T. C., & Smith, E. M., (1979). A method for determining the total energy input for agricultural practices. *Trans. ASAE., 2*, 781–0784.

90. Stout, B. A., (1990). *Handbook of Energy for World Agriculture* (1st edn.). Texas A & M University: College Station, TX, USA, pp. 50–95, ISBN: 1-85166-349-5.

91. Crowell, E. A., & Ough, C. S., (1979). A modified procedure for alcohol determination by dichromate oxidation. *Am. J. Enol. Viticult., 30*, 61–63.

92. Khalil, S. R., Abdelhafez, A. A., & Amer, E. A. M., (2015). Evaluation of bioethanol production from juice and bagasse of some sweet sorghum varieties. *Ann. Agric. Sci., 60*, 317-324.

93. Miller, G. L., (1959). Use of dinitrosalicylic acid reagent for determination of reducing sugars. *Anal. Chem., 31*, 426–428.

94. Disasa, T., Feyissa, T., & Admassu, B., (2017). Characterization of Ethiopian sweet sorghum accessions for 0 brix, morphological and grain yield traits. *Sugar Tech., 19*, 72–82.

95. Sarungallo, R. S., Melawaty, L., Djonny, M., Bulo, L., Mangera, L., Pabendon, M. B., & Sarungallo, Z. L., (2019). Fermentation juice sweet sorghum genotip 4-183A using batch system by optimizing the concentration of inoculum and substrate. The 1st international conference on education and technology (ICETECH). *IOP Conf. Series: Journal of Physics: Conf. Series 1464* (2020), 012050.

96. Harlan, J. R., & de Wet, J. M. J. (1972). A simplified classification of cultivated sorghum. *Crop Science, 12*, 172–176.

97. Bogunovic, I., Pereira, P., Kisic, I., Sajko, K., & Srakac, M. (2018). Tillage management impacts on soil compaction, erosion and crop yield in Stagnosols (Croatia). *CATENA 160*, 376–384.

CHAPTER 12

Biotechnology Development of Bioethanol from Sweet Sorghum Bagasse

DANIEL TINÔCO

Biochemical Engineering Department, School of Chemistry,
Federal University of Rio de Janeiro, Rio de Janeiro–21941909, RJ, Brazil,
E-mail: dneto@peq.coppe.ufrj.br

ABSTRACT

Sorghum is one of the most cultivated cereals in the world, with great potential for energy and fuel production. Its lignocellulosic fraction can be used for cellulosic ethanol production, after undergoing treatment steps to make sugars available. This chapter presents the most recent trends in the bioprocess for the ethanol production from sweet sorghum bagasse, highlighting the main physical, chemical, physical-chemical, and biological pretreatments used for cellulose digestibility such as acid-base, SE, ammonia fiber expansion (AFEX), organic solvents, ILs, microwave, and combined methods. Saccharification and fermentation approaches were also presented, such as simultaneous (SSF), separate (SHF), and co-fermentation (SScF) processes. Finally, a biotechnological evolution of sweet sorghum bagasse was presented from the main scientific reports in the last 12 years on its use as a raw material for the fuel cellulosic ethanol production.

12.1 INTRODUCTION

Sorghum is one of the most cultivated cereals in the world, along with wheat, rice, corn, and barley [1], being composed on average of 50–60% lignocellulosic biomass (LCB) [2]. Considered a viable energy crop for alcohol fuel production, sorghum, especially the sweet variety, has great potential for the generation of cellulose-based products such as ethanol, butanol, and wood-plastic composites [3]. In addition to the free fermentable sugars present in the stem, sweet sorghum bagasse can be processed and converted into fuel ethanol and energy, giving this biomass a multi-purpose aspect, therefore being considered as a biorefinery crop [4].

The sweet sorghum bagasse processing for the bioethanol production is characterized by an initial treatment step, followed by the cellulose saccharification, and the fermentation of released pentoses and hexoses. After harvesting, drying, and storage, sweet sorghum bagasse can be submitted to physical, chemical, physical-chemical, and biological action, responsible for preparing the LCB for the enzymatic hydrolysis. The main physical treatments are drying, grinding, sieving, granulating, extruding, steam flaking, extraction, and decortication. Chemical methods are based on dilute acids and alkalis, organic solvents, ILs, and oxidative agents. Physical-chemical treatments combine mechanical and chemical processes, being the SE, liquid hot water (LHW), supercritical CO_2, AFEX, and microwave the most used methods. Biological treatment can be combined with other methodologies, being marked by the use of enzymes and microorganisms (bacteria and fungi) capable of degrading the lignocellulosic material [5–7].

The next step is the enzymatic hydrolysis and fermentation. These processes can be conducted in different approaches: simultaneously in the same bioreactor or separately in different bioreactors [7]. SSF is generally conducted at low temperatures due to the thermal tolerance of the microorganisms used. However, this condition limits the fermentation efficiency, requiring a pre-saccharification step under optimized conditions. Once the sugars are released, the temperature can be readjusted to the ideal fermentation condition and, then, the microorganism inoculation can be completed. This process is classified as delayed simultaneous saccharification and fermentation (DSSF) [8]. The SHF has the advantage that the saccharification and fermentation steps can be conducted under ideal individualized conditions. Meanwhile, the ethanol yield and productivity, and the bioprocess economy tend to be lower than those obtained with the SSF approach. Simultaneous saccharification and co-fermentation (SScF) are similar to the SSF process, with the additional fermentation of free sugars present in the sweet sorghum stem. Therefore, the C5 and C6 sugars can be simultaneously converted to ethanol [9].

This chapter aimed to present the main trends related to the cellulosic ethanol biotechnology from sweet sorghum bagasse, highlighting the treatment methodologies and the most used saccharification and fermentation approaches in recent years, especially between 2009 and 2020. The biotechnological evolution was also presented for the cellulosic ethanol production, highlighting the most used combinations of pretreatment and sweet sorghum fermentation, as well as the microorganisms and most relevant aspects involved in the sweet sorghum ethanol bioprocess.

12.2 SORGHUM

12.2.1 WORLD PRODUCTION

Sorghum is the fifth most cultivated cereal in the world, with an expected production of 327 Mtons by 2027 [10]. The main producing regions are Africa (40.1%) and the Americas (38.1%), where sorghum is used in human food and for the alcoholic drinks and biofuels production. Next are Asia (17.3%), Oceania (3.2%), and Europe (1.4%) (Figure 12.1).

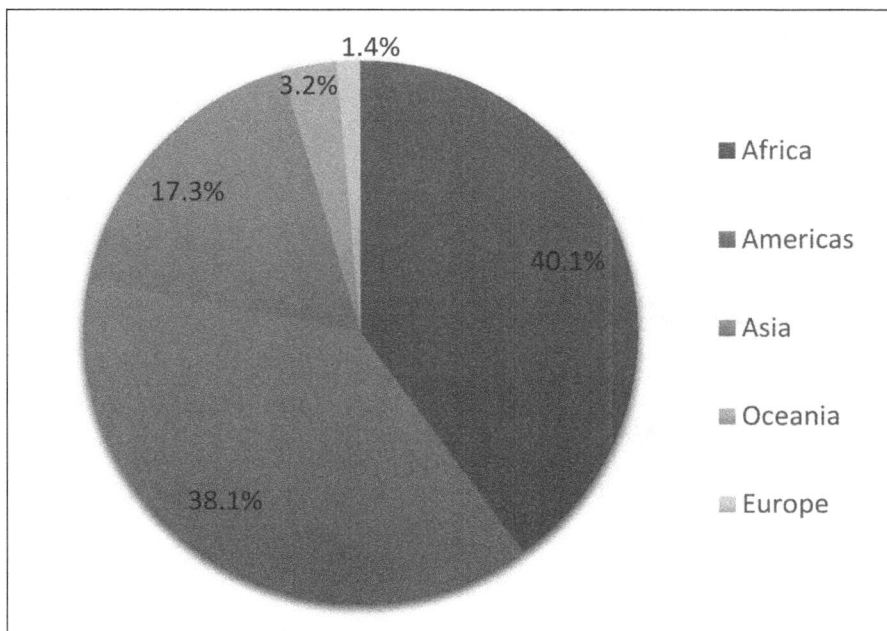

FIGURE 12.1 World sorghum production [1].

According to the Food and Agriculture Organization of the United Nations (FAOSTAT), among the main countries producing sorghum in the world between 1998 and 2018, four are African, three American, two Asian, and one from Oceania (Figure 12.2). The United States is the largest sorghum producer in the world, with an average production of 10.6 Mtons, followed by Nigeria with around 7.4 Mtons, India with 6.65 Mtons, and Mexico with 6.1 Mtons. Sudan and Sudan (former) account for a production of 4.4 and 3.8 Mtons, respectively [1]. Argentina is a reference in South American sorghum production with 3 Mtons, together with Brazil, which has widely cultivated sorghum in association with sugarcane for bioethanol production, although not among the 10 largest producers [11]. Since 1960, Ethiopia has increased its sorghum production, with a recent amount of 2.9 Mtons. In addition to India, China has stood out in Asia since the sorghum participation in its domestic market increased. China has imported large cereal quantities in the past, about 3 Mtons in 2012 and 18 Mtons in 2014 [10]. Today, China has an average production of 2.6 Mtons, being the ninth largest producer in the world and contributing, together with India, with more than 85% of Asian sorghum production [12]. In Oceania, the only main contributor is Australia, with a production of 1.9 Mtons.

The average world sorghum production and the corresponding planted area for the period 1998–2018 was 63.1 Mtons and 45 Mha, respectively, which represents a world average yield of 1.4 t/ha^1 (Figure 12.3).

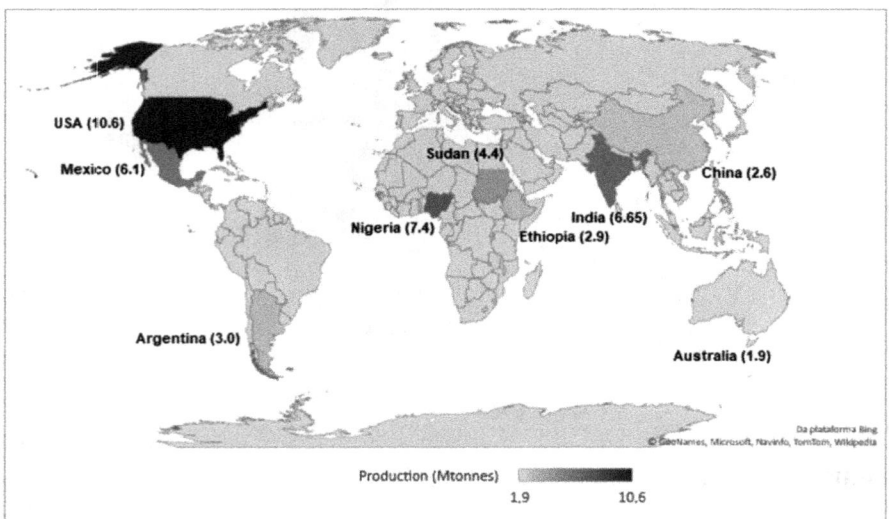

FIGURE 12.2 Main sorghum producers-average 1998–2018 [1].

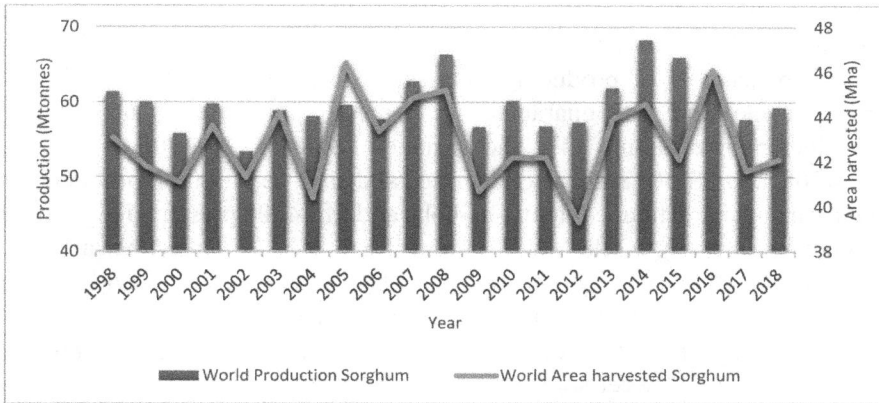

FIGURE 12.3   Area harvested and production of sorghum in the world [1].

12.2.2   SWEET SORGHUM

Sorghum (*Sorghum bicolor* (L.) Moench) is a tropical crop from an African source, classified into basic types: sweet, grain, forage, low-lignin, and biomass [7]. The sweet type has great industrial and agricultural interest due to its chemical characteristics, in particular, the composition in free fermentable sugars of its stem, which contains about 12 to 18% (w/v) sugar. The sugars present in sweet sorghum juice correspond to 53–85% sucrose, 9–33% glucose, and 6–21% fructose [4]. In addition to a sugar-rich composition, sweet sorghum has the following agricultural characteristics: high productivity, high-stress tolerance (temperature, salinity, and water), and high adaptation to the cultivation and management infrastructure [4, 12]. Compared to traditional biomasses such as sugarcane, sweet sorghum requires about 2 and 4-fold less fertilizer and water, respectively, being grown in up to 3 annual cycles [2]. The most suitable soils for the sweet sorghum cultivation are the red or black clay [4], and the average plant growth temperature ranges from 12°C to 37°C [12].

The most popular sweet sorghum varieties, identified with biofuel production potential, were: Dale, Della, M81-E, Sugar Drip, Top 76–6, Umbrella, Keller, Rio, Wray, and AFL Tcv27751. The Top 76–6 variety was considered the most suitable for the bioethanol production in a study carried out with six of these 10 varieties, as it presented higher productivity, resulting from the soluble sugars found in its stem, and a greater nutritional value of its grains, with a low concentration of polyphenols and high-quality relative protein [13]. Hybrid varieties have been investigated, such as Hybrid H8015, which are capable of producing biomass with higher yields, in a shorter growth cycle, and with guaranteed seed production, which contributes to the bioethanol economy [4]. The varieties GK-coba, Mn-1054, Ramada, Mn-4508, and SS-301 were also identified and analyzed for their productivity and their sugar and fiber content. GK-coba and Mn-4508 presented high sugar content in the stem, while Mn-1054 and Ramada had a large fiber amount. The SS-301 variety had a high sugar and fiber content, which provided juice and bagasse with high potential for ethanol production [14].

Sweet sorghum can be cultivated in tropical and temperate regions. In tropical countries, sweet sorghum is widely cultivated in the rainy seasons from June to July, and dry seasons from September to October, while in temperate countries, the planting is limited to once a year [4]. In Brazil, where sugarcane is the main feedstock for bioethanol production, sweet sorghum has been cultivated in the sugarcane off-season, between November and May, allowing the operational integration of the commercial distilleries [15]. In the USA, with their corn bioethanol, sweet sorghum has been planted between May and June, which has contributed to biomass yields of 26 to 29 t/ha, with an ethanol production of at least 14,500 L/ha [16].

12.2.3 BIOENERGY PRODUCTION

Sweet sorghum can be used in the production of solid (biochar), liquid (bioethanol, biodiesel, and bio-oil) and gas (bio-hydrogen, biogas, and synthesis gas) fuels [7], from use of all its constituent parts: grains (starch), juice (sucrose) and biomass (lignocellulosic fraction). Sweet sorghum can be also used for heat and power co-generation. Therefore, it is classified as an important biorefinery crop [4].

Sweet sorghum is capable of producing integrated first and second-generation ethanol, due to its rich saccharine juice and its high lignocellulosic composition, with 50–60% bagasse [2]. Although promising, the bioethanol production from sweet sorghum by biorefinery concept has challenges as the biological conversion, which requires a high yield of the generated bioproducts. The process implementation based on microorganisms and technologies able to simultaneously hydrolyze cellulose, overcome the lignin recalcitrance, and ferment different sugars like glucose, xylose, and cellobiose, with minimal inhibitor release, are requirements for the bioprocess success. The use of low-cost cellulolytic enzymes and its reuse also contributes to making biological conversion even more productive and economically competitive [7].

12.2.4 SWEET SORGHUM BAGASSE

Bagasse, a fibrous lignocellulosic material, represents approximately two-thirds of the sweet sorghum dry matter [17]. It consists of 27–44.6% cellulose, 25–27% hemicellulose, 11–24.7% lignin [4], and other compounds as: minerals (K, P, Mg, and Ca), nutrients (niacin, thiamine, riboflavin, and B6), amino acids (proteins), dietary fibers (soluble and insoluble), fatty constituents (saturated and unsaturated), and vitamins [12]. Its calorific value (18.3 MJ/kg) is comparable with other lignocellulosic feedstocks as switchgrass (18.4 MJ/kg) and big bluestem (18.6 MJ/kg) [4]. Due to its chemical and power characteristics, sweet sorghum bagasse can be used for the paper and animal feed production, in soil applications as a fertilizer, for the power co-generation, and as a raw material for fuel cellulosic ethanol [17].

12.3 CELLULOSIC ETHANOL PRODUCTION

Sweet sorghum bagasse requires some pretreatment steps, a long processing time, and a large investment to make the sugar present in its lignocellulosic fraction available for fermentation [7]. The main cellulosic ethanol production steps involve: biomass pretreatment, lignocellulosic material saccharification, and released sugars fermentation.

12.3.1 BIOMASS PRETREATMENT

Pretreatment is the most critical bioprocess step in the cost-efficient conversion of biomass to ethanol and other bio-based products [18], since it is responsible for preparing the cellulose for the enzymatic and biological action, therefore influencing the productivity and economy of the fermentative process. This step requires a low biomass constituents' degradation and a low inhibitors formation for being considered adequate [7]. The pretreatment aims to reduce cellulose recalcitrance by removing lignin and releasing hemicellulose, preferably, with an energy requirement as low as possible [19]. Production costs, including ethanol recovery costs, are directly affected by the pretreatment [20].

These costs normally increase with a decrease in the solids load of the treated material (dilute solution form) [19].

The main pretreatments used with lignocellulosic materials include physical, chemical, physical-chemical, and biological methods. Many of these treatments can be combined to improve cellulose accessibility and sugar yield in the enzymatic hydrolysis step. The main technologies for the sweet sorghum bagasse treatment were summarized, according to previous studies (Figure 12.4).

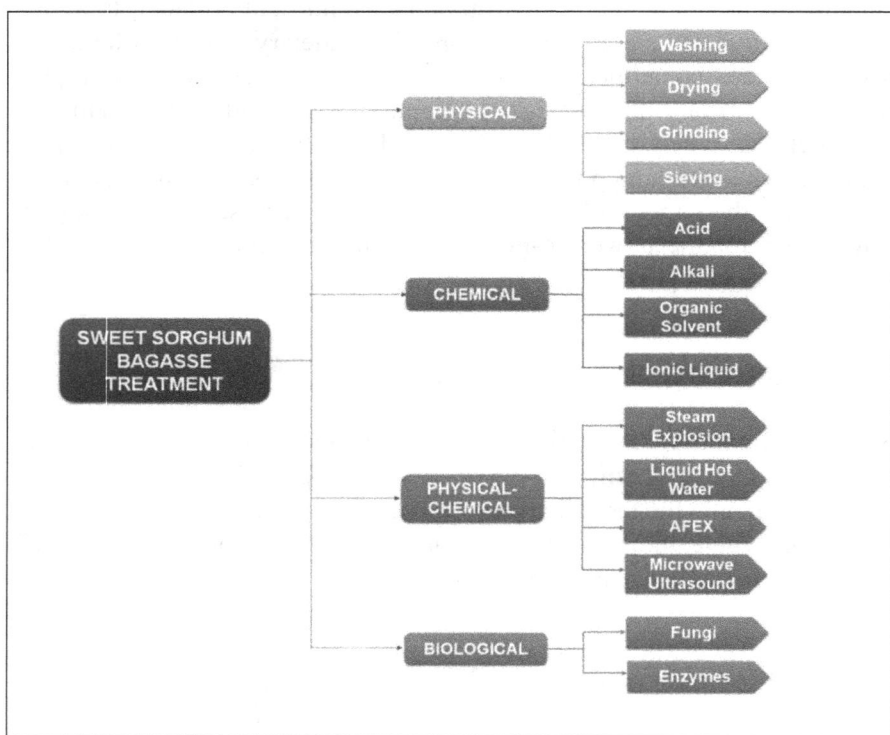

FIGURE 12.4   Main methodologies for the sweet sorghum bagasse treatment.

12.3.1.1   PHYSICAL

Physical pretreatment of sweet sorghum bagasse involved mechanical methods of preparing biomass, responsible for increasing the surface area of the material, by reducing the particle size and the cellulose crystallinity [18].

The main physical methods investigated were: washing, drying, grinding, and sieving [7].

Sweet sorghum bagasse is usually chopped, milled in hammer or rotary mills, washed, and dried at 50–60°C to reduce the moisture content to about 10 to 15% [21]. The grinding reduces the particle size by mechanical shear, which can be classified according to their diameter by the sieving process. The particle size is a parameter related to the material's saccharification efficiency, as it is inversely proportional to its surface area and, therefore, represents the contact area available to the action of the hydrolysis agents. In several studies, the size range of the sweet sorghum bagasse particles after grinding and sieving was 0.2 to 1.2 mm. Cao et al. [22] used a three-roller mill to separate broth and sweet sorghum bagasse, which was air-dried, ground, and sieved at 0.42 mm. Chen et al. [23] obtained the sweet sorghum bagasse by crushing in a roller press, drying at 20°C to maintain the final moisture content at 20%, grinding, and sieving in three fractions: 9.5–18, 4–6, and 1–2 mm. Darkwah et al. [24] used a mechanical extractor to separate the juice from the bagasse, which was air-dried and ground with a 1 mm mesh in a knife mill. Lavudi et al. [25] submitted the sorghum bagasse to oven drying to reduce the moisture content to 9–10%, being subsequently chopped, ground, and sieved at 0.6 mm.

12.3.1.2   CHEMICAL

Chemical pretreatment is based on the application of solvents and chemical compounds capable of breaking down lignocellulolytic structures with high efficiency. During this process, the production of the toxic compounds, with carbohydrate loss, can happen, being it the main limitation of the method. Furthermore, it is a high cost and environmental risk technology [12]. The main chemical pretreatment methodologies used with sweet sorghum bagasse were: acid and base treatment, organic solvent treatment, and ionic liquid treatment [7].

12.3.1.2.1   Acid and Base Treatment

Acid treatment is based on the application of organic (formic, acetic, and propionic) and inorganic acids (sulfuric, nitric, hydrochloric, and phosphoric) to remove lignin and, mainly, hydrolyze vegetable fibers, with the removal of hemicellulose. The acids can be employed in the concentrated (greater than 30% v/v) or diluted (up to 10% v/v) forms. The concentrated acid-based process typically occurs at 100°C, for 2 to 10 h, and at atmospheric pressure. It is considered slow, toxic, and corrosive, requiring corrosion-resistant equipment, which increases the capital, operational, and maintenance costs. Conversion rates for cellulose and hemicellulose can reach values above 90% [18, 26, 27]. Diluted inorganic acid is used at 100–240°C and a pressure greater than 10 atm for a few seconds or minutes [27]. The process is considered fast, with no need to recycle the acid. However, there is the formation of biomass degradation products that can compromise the fermentation process [12]. The application of different acids and pretreatment conditions were responsible for different yields of released sugar. The treatment with 0.5% (v/v) H_2SO_4 at 180°C for 5 min allowed a release of 66% glucan, with total xylan removal. Compared to untreated bagasse, the available glucan amount increased by 51% with treatment based on diluted acid [28]. In another investigation, an improvement in the enzymatic hydrolysis step was verified with the application of 1.75% (v/v) H_2SO_4 at 121°C for 40 min, which favored the release of 14.22 g/L xylose and 2.42 g/L glucose. A small amount of inhibitory compounds was also produced: 1.34 g/L acetic acid, 0.90 g/L phenol, 0.12 g/L hydroxymethylfurfural (HMF), and 0.98 g/L furfural [29]. In the treatment based on H_3PO_4 (85% v/v) at 40–85°C, it was not observed the HMF and furfural formation. After acid pretreatment at 50°C for 43 min, followed by enzymatic hydrolysis, the released glucose yield was 85% [30]. The diluted H_2SO_4 treatment was investigated at 150°C for 1 h, at high pressure in four different concentrations: 0.5; 1.0, 1.5 and 2.0% (v/v). A greater recovery of glucan and xylan, around 88% and 91%, respectively, was observed in the pre-treatment with 2% (v/v) H_2SO_4 [31].

Basic treatment is based on the application of hydroxides (NaOH, KOH, NH_4OH and Ca $(OH)_2$), Na_2CO_3 and similar alkaline compounds, usually at 140–200°C, for a residence time of minutes to hours [19], which are responsible for the saponification reactions of the ester bonds between lignin and other carbohydrates [18]. These reactions reduce the polymerization degree and the lignocellulosic material crystallinity, due to the lignin destruction [27] and the cellulose and hemicellulose swelling, with consequent degradation of their crystals [12]. Basic treatment causes lesser sugar loss than acid treatment, and it is usually carried out with diluted alkalis [27]. Different alkalis have been used to treat sweet sorghum bagasse. Treatment with NaOH at 121°C, for 1 hour at 15 psi, led to an approximate 37% lignin removal, with 88% cellulose recovery. In this method, the furfural and HMF formation was not verified [32]. The treatment based on alkaline distillation using 10% (w/w) NaOH was carried out at 100°C, for 30 min at 0.04 MPa. A recovery of approximately 94.5% glucan and 86% xylan, and a removal of about 72% lignin were obtained [33]. The alkaline hydrogen-peroxide treatment was investigated in two conditions: at 35°C, 2% (w/v) H_2O_2, for 24 h, in a dark room (mild condition); and at 121°C (autoclave), 2% (w/v) NaOH, for 1 h + 5% (w/v) H_2O_2, at 20°C, for 24 h, in a dark room (severe condition). In both conditions, more than 90% of cellulose was recovered. However, the severe treatment was more efficient in obtaining fermentable sugars, with higher concentrations in the hydrolysis step, reaching a value 2-fold higher than the mild treatment [34]. The NaOH application under the same conditions established by Yu et al. [33], but without biomass washing, increased the fermentable sugar conversion from 44% to 65% [35]. The residual alkalis use is an economical and environmental method alternative for the basic treatment of sweet sorghum bagasse. The black liquor, resulting from the basic pretreatment of the empty fruit cluster, was used at 150°C for 30 min. This residue contained about 76% NaOH and pH 13, which contributed to approximately 62% bagasse delignification [20]. The green liquor, generated in kraft pulp mills, consisting of Na_2CO_3 and NaS_2 (simulated composition), was used at 160°C for 110 min, with 18% total titratable alkaline charge and 40% sulfidity. A sugar yield of 82.6% was achieved after this treatment [36].

The comminated treatment is an efficient alternative for the treatment of the lignocellulosic material, especially between acid and base, since together, these methodologies can remove lignin and hemicellulose, making cellulose available with greater purity. The combination of 1.4% (v/v) H_2SO_4 at 120°C for 47 min, and 0.25 mol/L NaOH at 120C for 5 min, resulted in 92% hemicellulose removal 90% lignin removal, with only 2.3% cellulose lost, whose composition in the bagasse after treatment was approximately 79% [37]. The acid treatment with 1.5% (v/v) H_2SO_4 in autoclave for 33 min followed by the basic treatment with 4% (v/v) H_2O_2 for 45 h, pH 11.5, allowed almost 77% hemicellulose and lignin removal, making available about 79% cellulose for the reducing sugars generation [38].

12.3.1.2.2 Solvent-Based Treatment

The treatment using organic solvents is known as organosolv or solvolysis. This method is capable of removing lignin and hemicellulose via solubilization, characterized by the cleavage of α-O-aryl, β-O-aryl bonds, and esters of 4-O-methylglucuronic acid [18]. The most used solvents are ethanol, methanol, acetone, organic acids (formic and acetic), and ethylene glycol [19]. Although expensive, most organic solvents can be reused after recovery by evaporation and other extractive technologies, thereby reducing the adverse effects of their presence in the fermentation step [27]. When treated with 50% ethanol and 1% H_2SO_4, at 140°C for 30 min, sweet sorghum stem showed a sugar yield of 77%, 2-fold greater than untreated bagasse [39]. Acetone treatment at 180°C for 60 min, improved the enzymatic hydrolysis step by 94.2% in a process intended for the joint production of acetone, butanol, and ethanol (ABE process), allowing the removal of 143 g lignin and the release of 250 g hemicellulose for each 1 Kg bagasse treated [40]. Concerning the biorefinery concept, treatment with an aqueous solution of 60% (v/v) ethanol and 20% (v/v) isopropanol at 140°C for 30 min, with 1% sulfuric acid as catalyst, contributed to a hydrolysis yield of 90.3%, producing 24.9 g/L glucose [41].

12.3.1.2.3 *Ionic Liquids (ILs) Treatment*

Ionic liquids (ILs) are salts formed by a large volume organic cation and an inorganic anion of different size, usually smaller than the cation, which are liquids at room temperature [19]. ILs are considered special solvent types, being classified as green solvents, as they do not release toxic or flammable substances during the cellulose treatment. These compounds are capable of dissolving lignin and other carbohydrates by forming hydrogen bonds with the hydroxyls of the sugars present in the lignocellulosic material [42]. The sweet sorghum bagasse treated with 1-Butyl-3-methylimiazolium chloride ([BMIM] Cl) at 110°C for 120 min, showed a reduction in the hemicellulose composition from 26.3% in untreated bagasse to 16.7% after treatment [43]. ILs such as 1-allyl-3-methylimdazolium acetate ([AMIM] OAc), 1-ethyl-3-methylimidazolium formate ([EMIM] Fmt) and 1-ethyl-3-methy-limidazolium acetate ([EMIM] OAc) showed high capacity dissolving the lignocellulosic fraction of sweet sorghum, without causing damage to the cellulase enzymatic action. A sugar yield 7.5-fold higher was verified with the treatment using [EMIM] Fmt, compared to untreated bagasse, which also favored a bacterial nanocellulose productivity of 0.71 g/(L.d) [44].

12.3.1.3 *PHYSICAL-CHEMICAL*

Physical-chemical treatment is a special type of combined treatment, resulting from the union between mechanical and chemical methods of treating lignocellulosic material. The physical-chemical methods most used with sweet sorghum bagasse have been: SE, LHW, AFEX, and microwave- or ultrasound-assisted technologies [7].

12.3.1.3.1 Steam Explosion (SE) Treatment

Steam explosion (SE) is a hydrothermal pretreatment capable of hydrolyzing the acetyl group present in hemicellulose, and promoting structural changes in cellulose by the hydrogen bonds rupture, increasing its digestibility [27]. The biomass is rapidly heated by saturated steam at high temperatures and pressures [45], often in the range of 160–270°C and 20–50 bar, respectively, for a few seconds or minutes [12], without the addition of a chemical reagent. Therefore, it is a lower environmental impact method, responsible for the complete sugar recovery when compared to other pretreatments [27]. SE can be associated with water and SO_2 impregnations to increase the lignocellulosic fiber fracture efficiency. The application of the SE treatment increased in 2.5-fold the maximum cellulose conversion and the glucose release from sweet sorghum bagasse treated in a reactor with 0.25 t/h high pressure, at 160°C, for 5 min [43]. Sweet sorghum bagasse samples impregnated with SO_2 gas were treated with steam at 190C for 5 min. Approximately 54% glucan, 10% xylan, and 26% lignin were obtained after treatment [46]. The sweet sorghum bagasse submitted to steam at 200°C for 5 min released leachate, which had its liquid and solid fractions separated by a cyclone. A composition of 52% cellulose, 9% hemicellulose, and 25% lignin was verified for the released solid fraction [47]. Impregnation with water was also investigated in the SE treatment of dry sweet sorghum bagasse. After being subjected to steam at 215°C for 2 min, about 37.3% of total sugars could be recovered from the biomass treated [48].

12.3.1.3.2 Liquid Hot Water (LHW) Treatment

Liquid hot water (LHW) or hot-compressed water acts as a solvent and reaction medium capable of breaking down cellulose, by combining retention time, temperature, and pressure [26]. Usually, the method is carried out at 200–230°C, and at high pressure, so that the water can penetrate the biomass and, thus, favor the cellulose hydrolysis, in addition to the removal of the hemicellulose [27]. Sweet sorghum bagasse, when hydrolyzed with water at 60°C for 75 min, showed a sugar yield of almost 68% [21]. When it was treated at 200°C for 6.5 min, hemicellulose solubilization of 85% was achieved, being greater than the 74% yield obtained with the treatment at 190°C for 20 min [49].

12.3.1.3.3   *Ammonia Fiber Expansion (AFEX) Treatment*

Ammonia fiber expansion (AFEX) is a pretreatment based on the application of liquid anhydrous or gaseous ammonia at elevated temperatures (90–100°C) and high pressures (1–5.2 MPa) [27]. With the pressure release, ammonia is quickly evaporated, which promotes the cellulose crystallinity reduction, hemicellulose depolymerization, and deacetylation of acetyl groups [45]. In AFEX treatment, the fermentation inhibitors formation and a liquid current are not verified being, therefore, considered a dry-to-dry process [18]. Ammonia can be reused in this method, reducing process costs, and avoiding environmental problems with its disposal [27]. The optimization of AFEX treatment conditions was defined as 140°C, 30 min, 2:1 for ammonia/biomass ratio, and a mixture content of 120%. Under these conditions, the glucan and xylan conversion reached 90% for sweet sorghum bagasse. Free sugars destruction was observed during the AFEX treatment, which was solved with the previous biomass washing [50].

12.3.1.3.4   *Microwave- or Ultrasound-Assisted Treatment*

Microwave and ultrasound irradiation methods are considered energetical efficient, simple to operate, and capable of heating the lignocellulosic material quickly [27]. These treatments can accelerate the chemicals released and the biological and physical processes action, through collisions of polar and ionic molecules promoted by heat [45]. Although microwave treatment improves the cellulose accessibility to hydrolytic enzymes, its investment cost (equipment cost) is higher than other technologies [27]. Microwave treatment assisted by ammonia heated to 130°C for 1 h promoted the release of 42 g glucose for each 100 g dry sweet sorghum biomass [23]. The association of microwaves and ultrasound treatments with other methods contributes to the increase in cellulose digestibility. The optimization of the microwave-assisted alkali pretreatment conducted at 1,000 W as 0.1 g lime, 10 ml water content, and exposure time of 4 min resulted in a sugar yield of 52.6%. In the lime absence and maintaining the water content and exposure time, this yield increased to 65.1% [51]. For the microwave treatment of sweet sorghum bagasse at

180 W for 20 min, equivalent to a power of 43.2 kJ/g of dry biomass, in the sulfuric acid solution (50 g/Kg of bagasse) presence at 82°C, a yield of 820 g sugar for each 1 Kg treated bagasse was obtained after the process [52]. The microwave-assisted acid treatment at 180 W and microwave-assisted alkali at 300 W employing two different microorganisms in a co-fermentation process at different cell ratios were investigated. Better ethanol yields were achieved through microwave-assisted acid treatment, regardless of the cell ratio used. Microwave-assisted acid pretreatment is more efficient, requiring less energy [53]. A sugar yield of 57% was achieved by treating sweet sorghum bagasse with diluted aqueous ammonia and assisted by 90 W ultrasound [54].

12.3.1.4 BIOLOGICAL

Biological treatments are based on the action of enzymes and microorganisms such as bacteria and fungi, capable of carrying out the fermentation in the solid-state and producing several lignocellulolytic enzymes, responsible for degrading the sweet sorghum biomass [7]. Microbial consortia, use of fungal species, and enzymatic technologies are examples of biological treatments [45]. The biological method is effective in lignin degradation, having as advantages the use of mild conditions, low energy consumption, and low environmental impact. However, it is limited by long-term cell cycles and consumption of cellulose and hemicellulose for the growth of some fungi, high degradation conditions for bacteria, and low activity of lignocellulolytic enzymes, which limits its industrial application [27]. *Coriolus versicolor* fungus is widely used in the sweet sorghum bagasse treatment due to its selective lignin degradation capacity. In an investigation using this fungus with $CuSO_4$-syringic acid supplements, the synergistic effect of $CuSO_4$ and serum acid was evaluated. An approximate degradation of 36% lignin was achieved, which resulted in a sugar production 2.2-fold greater in the enzymatic hydrolysis step [55]. In another study using the *C. versicolor*, supplements of serum acid and gallic acid were added to the biological treatment. The synergistic action of these acids resulted in the degradation of about 31% lignin, contributing again to a sugar production 2.2-fold greater in the enzymatic hydrolysis step [56]. *C. versicolor* was also used in a solid-state fermentation in a Mesh Tray Bioreactor. In this investigation, an increase in the production of the lignocellulolytic enzymes laccase and xylanase was observed, due to the proper conduction of the fermentation in the tray mesh, and in the ideal airflow in the bioreactor. The lignin degradation was approximately 46%, with an almost 8% cellulose loss. Enzymatic hydrolysis produced about 2.5-fold more fermentable sugars than the control assays using untreated sweet sorghum bagasse in a meshless bioreactor and without a humid airflow [57].

12.3.2 SACCHARIFICATION AND FERMENTATION

After initial treatment, sweet sorghum bagasse is submitted to enzymatic and microbial action to convert cellulose and hemicellulose into ethanol. Treated biomass saccharification can be carried out simultaneously with the fermentation process in the same bioreactor (SSF), or in the different bioreactors (SHF). When the SSF process steps take place at relatively large intervals, the process is classified as DSSF. Treated biomass can also be saccharified and fermented together with sweet sorghum juice in a process classified as SSCF (Figure 12.5).

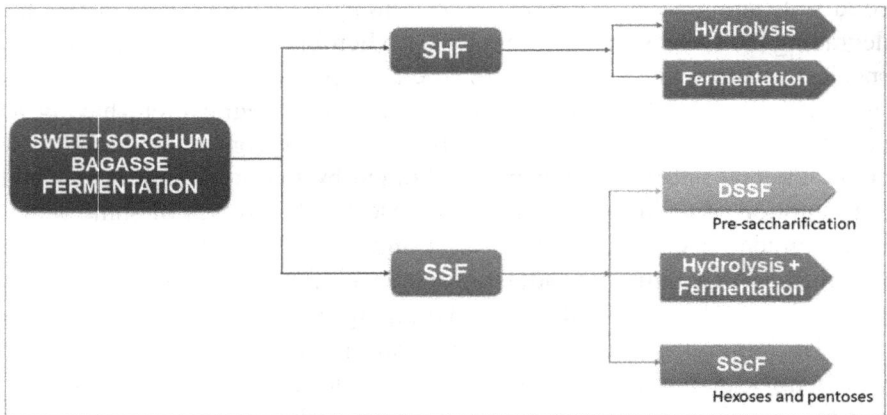

FIGURE 12.5 Saccharifications and fermentation approaches for the cellulosic ethanol production from sweet sorghum bagasse. [*Abbreviations:* SHF: separate hydrolysis and fermentation; SSF: simultaneous saccharification and fermentation; DSSF: delayed simultaneous saccharification and fermentation; SScF: simultaneous saccharification and co-fermentation].

12.3.2.1 SIMULTANEOUS SACCHARIFICATION AND FERMENTATION (SSF)

Simultaneous saccharification and fermentation (SSF) process contribute to the reduction of operating and capital costs, avoiding the cellular inhibition caused by the sugar-released excess after lignocellulosic material enzymatic hydrolysis. Although advantageous to ethanol yield and low biomass loss, the SSF process is limited by the recovery difficulty and reuse of the micro-organisms and enzymes, also requiring distinct favorable conditions for each step such as optimal temperature and pH [58]. Treated sweet sorghum bagasse can be submitted to the SSF process, using different enzymatic loads and microorganisms, for the cellulosic ethanol production. The sweet sorghum bagasse treated with diluted H_2SO_4 was fermented in two steps. Firstly, *Neurospora crassa* DSM 1129 fungus was used to produce lignocel-lulolytic enzymes. Posteriorly, the SSF process was performed using these enzymes, with cellulase and β-glucosidase supplementation at 6 FPU/g. About 27.6 g/L ethanol was produced, with a yield of 84.7% [59]. The SSF process optimization was performed to the efficient cellulosic ethanol production. At 37°C, 25 FPU/g enzymatic load, and 1.4 g/L *Saccharomyces cerevisiae* ATCC 24858, about 38 g/L ethanol, with 89.4% yield and 1.28 g/L/h productivity, were produced from biomass treated with dilute H_2SO_4 [28]. Approximately 85 g/L ethanol was achieved in 21 h fermentation by *S. cerevisiae* JP1 at 37°C, after 6 h enzymatic hydrolysis at 50°C, with 32.8 FPU/g enzymatic load, and combined acid-base pretreatment. An ethanol yield of 63% and 4 g/L/h productivity were obtained [37]. At severe alkaline treatment using NaOH and H_2O_2, about 19.3 g/L ethanol was obtained by *S. cerevisiae* CICC1308 at 36°C, after the pre-saccharification at 60 FPU/g cellulase and 80 IU/g β-glucosidase, at 50°C for 12 h. Ethanol yield reached more than 88% in 48 h of SSF [34]. The combined acid-base treatment followed by hydrolysis at 20 IU/g, at 60°C for 58 h, and fermentation at 35°C by *Pichia kudriavzevii* HOP-1 resulted in 26.8 g/L ethanol in 48 h. *P. kudriavzevii* HOP-1 was considered advantageous for the SSF process due to its thermotolerant capacity, which confers economic advantages to the commercial cellulosic ethanol production [25]. The black liquor application for 30 min followed by enzymatic hydrolysis with 30 FPU/g enzymatic load and fermentation at 32°C by commercial *S. cerevisiae* resulted in 45.06 g/L ethanol in 72 h [20]. The alkaline treatment based on Na_2CO_3, resulting from the reaction between NaOH from the bagasse basic treatment and the CO_2 released from the sweet sorghum juice fermentation, was associated with saccharification and fermentation at 32°C for 72 h. A theoretical ethanol yield of almost 82% glucan by *S. cerevisiae* was obtained [60]. The DSSF process was also investigated after hydrothermal treatment associated with liquefaction-saccharification at 50°C, with 10 FPU/g enzymatic load, for 24 h, and addition of extra enzymes. About 48.3 g/L ethanol by baker's yeast at 30°C was reached. The ethanol yield and productivity under these conditions were equal to 71.2% and 2.2 g/L/h, respectively [61].

SSF process can be performed by a fed-batch approach, as verified in a study using sweet sorghum bagasse treated with acetic acid. This process was carried out at 37°C for 96 h, resulting in 53.1 g/L ethanol by *S. cerevisiae* ATCC 24858, which was 2-fold higher than the batch SSF production, equal to 25.7 g/L. Fed-batch SSF improved the final ethanol production and contributed to reduce the cell inhibition caused by the high fermentation medium viscosity containing a high solid load content [24].

SSF process can also be combined with biological pretreatment by solid-state fermentation of sweet sorghum bagasse, in which filamentous fungi are used to degrade the lignocellulosic fraction. In a study using *Mucor indicus* CCUG 22424, the solid-state fermentation product was submitted to the SSF process, with 15 FPU/g cellulase and 30 IU/g β-glucosidase, and fermentation at 37°C for 48 h. Compared to the SSF process without biological pretreatment, the solid-state fermentation allowed a 4.3-fold higher ethanol yield, equal to 85.6% [62].

12.3.2.2   SEPARATE HYDROLYSIS AND FERMENTATION (SHF)

The separate hydrolysis and fermentation (SHF) process is widely used in the ethanol production from sweet sorghum bagasse, although its yield achieved is slightly lower than the SSF process yield [58]. The separation of these steps allows each process to be conducted under optimal temperature and pH conditions, thus contributing to an efficient conversion of cellulose and a high ethanol production [8]. SHF process also allows the cellular biomass and hydrolysate recovery, being widely used with filamentous fungi. In a study using *M. hiemalis* CCUG 16148, the sweet sorghum bagasse treated by the ultrasound-assisted NaOH was submitted to the SHF process at 32°C for 24 h, and 81% yield and 0.70 g/L/h ethanol productivity have been achieved [63]. In another investigation, the sweet sorghum bagasse treated by microwave-assisted diluted ammonia was submitted to 60 FPU/g cellulase and 64 CBU/g hemicellulase, at 55°C for 24 h. After 48 h fermentation by *S. cerevisiae* (D5A) at 30°C, about 21 g ethanol/100 g treated biomass were produced [23]. Cellulosic ethanol was also obtained by *Dekkera bruxellensis* GDB 248, yeast capable of assimilating cellobiose and glucose. After the alkaline treatment using H_2O_2 and enzymatic hydrolysis at 50°C, with 20 FPU/g commercial cellulase, for 48 h, and ethanol yield of 0.44 g/g was achieved in 7 h fermentation at 32°C [64]. The acid treatment based on H_2SO_4 was optimized and used for the sweet sorghum bagasse hydrolysis, whose hemicellulosic fraction generated was submitted to fermentation at 30°C by *Scheffersomyces stipitis*

ATCC 58376. About 22 g/L ethanol, 0.40 g/g yield, and 0.34 g/L/h productivity were obtained after 55 h [29]. The cellulosic ethanol production was also investigated without pre-treatment. For an enzymatic hydrolysis with 8.32 FPU/g load and a fermentation at 30°C by *Trichosporon fermentans* CBS 439.83, approximately 20.7 g/L ethanol were obtained. The external nitrogen supplementation increased ethanol production to 23.5 g/L, with 0.118 g/g yield and 0.196 g/L/h productivity in 120 h [65].

Enzymatic hydrolysis step is usually performed in batch approaches. However, some studies suggest that the fed-batch approaches can improve the glucose release during the saccharification process. The sweet sorghum bagasse treated with hot liquid water at 180°C, for 20 min at 4 MPa, was submitted to saccharification in fed-batch, with supplemented Tween80. The solid loads were fed in 24 h and 48 h, together with 20 FPU/g and 30 FPU/g cellulase at 50°C, to reach a final load of 20% and 30%, respectively. Approximately 89 g/L glucose was released after 120 h, and, then, submitted to fermentation by *S. cerevisiae* Y2034 at 30°C for 72 h. About 43.4 g/L ethanol was produced [66].

SHF processes are generally compared to SSF processes, considering the same biomass pretreatment, to identify and differentiate their advantages and limitations. The SSF and SHF processes were used after sweet sorghum bagasse steam-treated to produce cellulosic ethanol by *S. cerevisiae* (Tembec 1). In the SSF process, the hydrolysis step was conducted at 50°C for 12 h, and the fermentation step at 37°C for 120 h. About 23.3 g/L ethanol was produced, with a yield of almost 64%. In the SHF process, the hydrolysis step was conducted at 50°C and the fermentation step at 30°C. Approximately 21.2 g/L ethanol was obtained, with a yield of almost 58% [46]. The ethanol produced by *P. kudriavzevii* HOP-1 was investigated in the SSF and SHF processes, after the acid-base pretreatment of sweet sorghum bagasse. Both methods resulted in just over 26 g/L ethanol, however, the SSF productivity (0.56 g/L/h) was 40% higher than the SHF productivity (0.40 g/L/h), due to the time required for the maximum ethanol production, which was 48 h for SSF and 66 h for SHF [25]. *S. cerevisiae* TISTR 5606 was used to convert sweet sorghum bagasse treated by H_2O_2 and NaOH into cellulosic ethanol. The saccharification and fermentation of the hydrolysate were carried out by SSF, SHF, and DSSF. In the SSF process, the substrate, inoculum, and cellulase were mixed in the same bioreactor, at 37°C, for 72 h, which resulted in 22.3 g/L ethanol. In the DSSF process, the hydrolysis step was carried out at 50°C for 6 h, followed by fermentation at 37°C for 66 h. About 28.3 g/L ethanol was produced by DSSF. The DSSF productivity was 26% higher than the SSF and SHF productivities of 0.31 g/L/h [8].

Combined fermentation using different microorganisms was also verified in the cellulosic ethanol production from sweet sorghum bagasse. The microbial consortium is applied to ensure that 5- and 6-carbon sugars are converted to ethanol. The treated hemicellulosic fraction of sweet sorghum by diluted H_2SO_4, self-hydrolyzed by SE, and detoxified by the over-liming process, was converted into 38.7 g/L ethanol, with 82.5% yield, by the synergic action of *P. stipitis* NCIM 3497 and *Debaryomyces hansenii* sp. at 30°C. About 92 g/L sugars, mainly xylose and glucose, were assimilated by both microorganisms [67]. *Zymomonas mobilis* ATCC 31821 and commercial *S. cerevisiae* were used in the sweet sorghum bagasse assimilation, after it was treated by microwave. The combined fermentation was performed at 32°C for 24 h, which resulted in 94% conversion efficiency and 480 g ethanol/Kg treated biomass. Concerning only lignocellulosic fraction, about 33 m³/ha ethanol can be produced from sweet sorghum [52]. *Z. mobilis* ATCC 31821 and commercial *S. cerevisiae* were also used for the sorghum hydrolysate fermentation, treated by microwave-assisted acid/base. For 10 g/L *Z. mobilis* and 5 g/L *S. cerevisiae*, a yield of 0.50 g/g ethanol was obtained after 24 h fermentation at 32°C [53].

12.3.2.3   SIMULTANEOUS SACCHARIFICATION AND CO-FERMENTATION (SSCF)

All sweet sorghum components (grain, juice, and bagasse) can be used to increase ethanol production from its lignocellulosic fraction, since the low concentration of fermentative products is the main limitation to the industrial production of the cellulosic ethanol [49]. The bagasse saccharification and fermentation together with the juice fermentation can improve ethanol yield from different feedstocks, making the bioprocess economically advantageous. A previously treated and dehydrated sweet sorghum juice and bagasse was responsible for 53 g/L ethanol production in 168 h. The nutrients present in the sweet sorghum juice were beneficial to the fermentation process, improving in 92% the final ethanol, in lesser time of 72 h [49]. The fermentative integration was also performed for the sweet sorghum bagasse obtained from the solid-state process. This biomass was treated by NaOH and submitted to SScF by *Z. mobilis* TSH-ZM-01 at 32°C. Under optimized conditions, about 69.5% yield was achieved, which can reduce the capital costs and energy consumption, making the ethanol bioprocess commercially viable on a large scale [33]. The bagasse, treated by SE-assisted dilute phosphoric acid, and submitted to liquefaction, followed by SScF at 37°C using an engineered ethanologenic *E. coli* strain, was converted into 27.5 g/L ethanol in 96 h. Concerning to juice and bagasse parts, about 10,600 L/ha ethanol has been achieved from sweet sorghum [68]. The SScF integrated process was investigated after H_2SO_4 treatment. Under optimized conditions of 2% H_2SO_4, 20 FPU/g cellulase, and fermentation at 37°C using *S. cerevisiae* M-HT 3013, about 120.4 g/L ethanol was produced in 216 h. This production corresponded to a yield of 67.75 g ethanol/Kg treated biomass [31]. The SE treatment followed by the fermentation of the free fermentable and the lignocellulosic sugars resulted in more than 90% ethanol yield, with low inhibitor formation such as 5-HMF, furfural, and levulinic acid, by a wild-strain *S. cerevisiae*. The detoxification step was not necessary, which reduces the production costs [69]. The co-fermentation of juice and bagasse from sweet sorghum was also investigated by the cellular consortium fermentation. Firstly, the lignocellulosic fraction was treated by H_2SO_4 (98% v/v) at 120°C for 1 h. Next, the treated biomass and the juice were fermented at 30°C for 96 h, using *S. cerevisiae* ATCC 7754 and *Z. mobilis* ATCC 29191. For each 1 Kg treated sweet sorghum (variety SS-301), about 160 mL ethanol was produced [14].

12.4 BIOTECHNOLOGY EVOLUTION

Cellulosic ethanol production from residual sweet sorghum has been extensively studied in recent years, as a promising alternative to sugarcane ethanol. In 2009–2020, several investigations were carried out, using different treatment technologies and saccharification and fermentation approaches. Approximately 32% of scientific articles investigated physical, chemical, physical-chemical, and biological treatments to make cellulose digestibility more efficient and to available the sweet sorghum bagasse sugars for fermentation. Of the 68% scientific articles addressing the fermentation stage, about 54.3% corresponded to the SHF process, 30.4% to the SSF process, and 15.2% to the SScF process (Figure 12.6). Although lesser efficient than the SSF process, the SHF process has been the most used method in recent scientific research for converting sweet sorghum bagasse into ethanol. Possibly, the conducting of saccharification and fermentation steps under respective optimal conditions can explain this predominance. In turn, the promising SScF process was the least reported method for sweet sorghum processing since it is still a recent technology, starting its investigations in 2017.

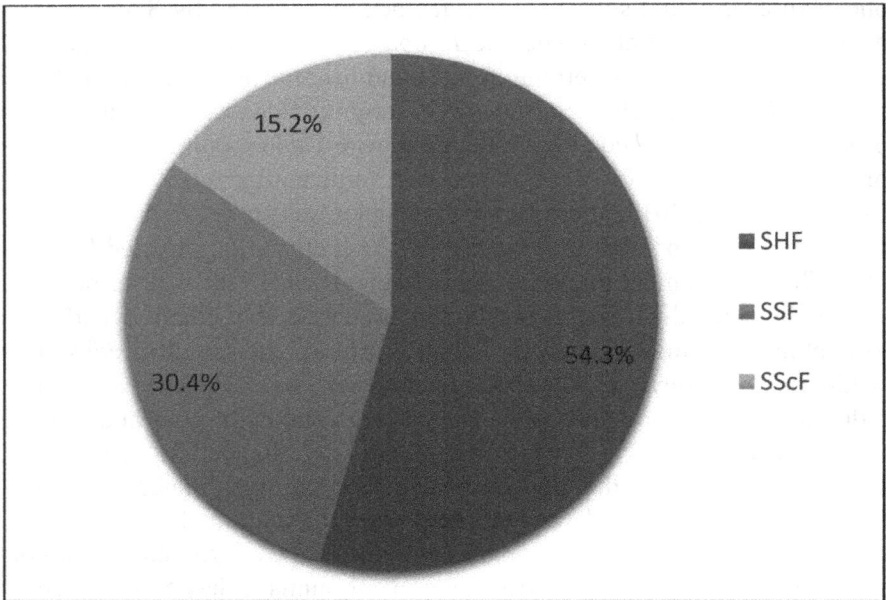

FIGURE 12.6   Main saccharifications and fermentation approaches reported by scientific articles in 2009–2020.

The three fermentation approaches reported were preceded by the treatment of sweet sorghum biomass. For the SSF process, the acid pretreatment was the most used technology, followed by the combined treatment, in 2009–2016 [24, 28, 37, 59, 70, 71]. The main used acid was the diluted H_2SO_4, and its combination with NaOH was considered one of the most efficient pretreatments for the cellulose release. In 2017–2020, the alkaline treatment gained prominence, accounting for 60% of reported articles employing the SSF process [8, 20, 60]. Again, NaOH was the most used alkali for processing the lignocellulosic fraction of sweet sorghum (Figure 12.7).

In 2009–2012, the following treatments were investigated in 60% of articles using the SHF process: alkaline, SE, and combined. In addition to NaOH, $Ca(OH)_2$ [32] and H_2O_2 [22] were used in basic treatments as alkaline compounds. The SE treatment was conducted independently [46, 72] or combined with H_2SO_4 [67], and it did not require an enzymatic hydrolysis step, since it was able to release glucose efficiently. The microwave-assisted NaOH [63] and microwave-assisted diluted NH_3 treatments [23] were considered important combined treatments without the saccharification step. In 2013–2016, the alkaline and combined treatments predominated, being used NaOH and H_2O_2 as alkalis, with, and without NaOH washing [35, 73]. The association between NaOH, H_2O_2, and H_2SO_4 [74], and the SE-associated diluted H_2SO_4 [52] were the main combined treatments used in the period. In 2017–2020, the acid, the alkaline, and the acid-base treatments were also wildly investigated [8, 25, 29] (Figure 12.8).

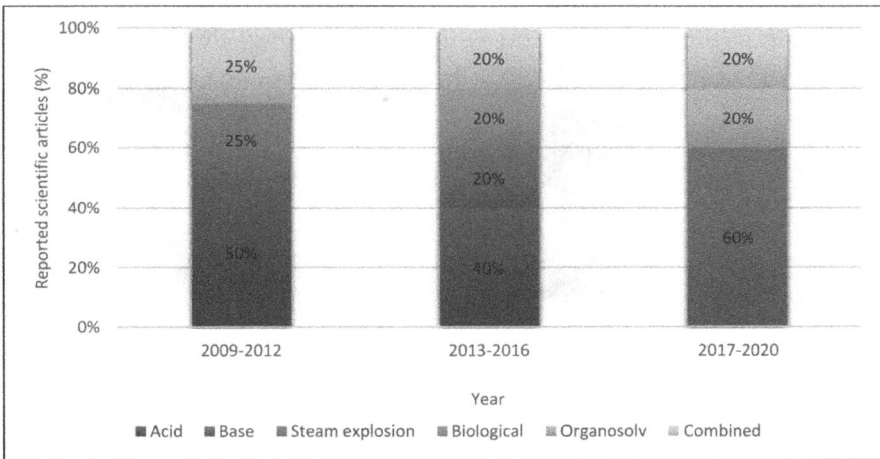

FIGURE 12.7 Main methods for the sweet sorghum treatment using the SSF approach.

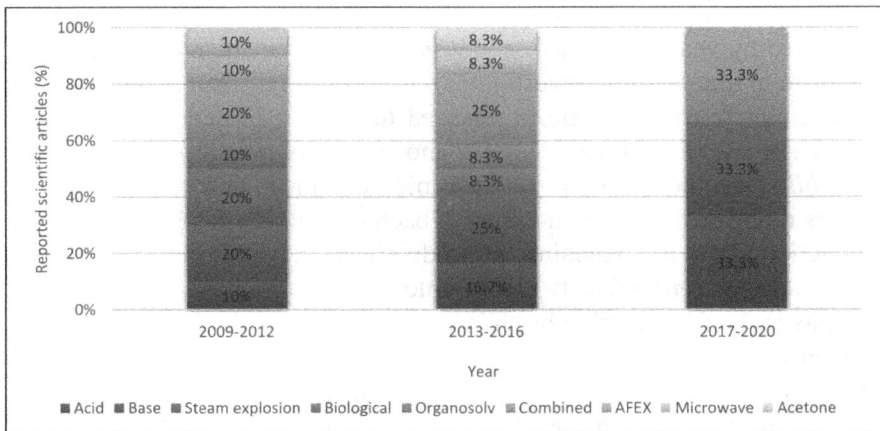

FIGURE 12.8 Main methods for the sweet sorghum treatment using the SHF approach.

SScF has been recently investigated, begging in 2013. In 2013–2016, the alkaline treatment using NaOH [33], the hot liquid water [49], and the biological-assisted NaOH treatments [75], were the most methods investigated. However, in 2017–2020, 50% of scientific articles employed the SE independently [48, 69] or assisted by diluted acid [68] to release glucose to the fermentation step (Figure 12.9).

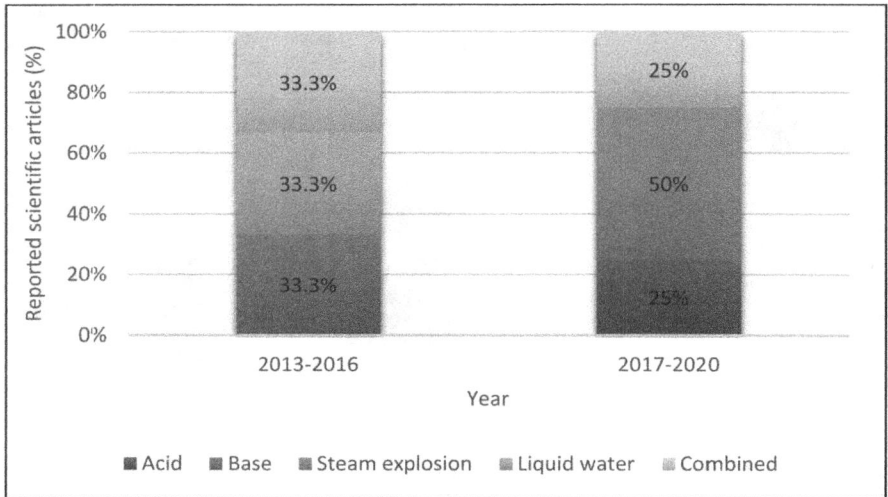

FIGURE 12.9   Main methods for the sweet sorghum treatment using the SScF approach.

In all fermentative processes, the fermentations have been mainly performed by yeasts such as *S. cerevisiae*, which was the most used microorganism in the scientific investigations developed between 2009 and 2020. Approximately 65% of articles reported this yeast, in cellular consortium or isolated one, for the cellulosic ethanol production from sweet sorghum. *Z. mobilis* was the second most used microorganism, according to 14% of articles reported. It is a gram-negative bacterium that has a high bioethanol production capacity, reaching a productivity 2.5-fold higher than that obtained by *S. cerevisiae*. It is also able to tolerate up to 120 g/L ethanol presents in the fermentation broth [53] (Figure 12.10).

Concerning the sweet sorghum bagasse treatment, without the fermentation step, there was an increase in the use of combined treatments, and a decrease in the acid and the SE treatments over the years (Figure 12.11). While the acid and the SE treatments are reported in the main articles, going from 25% (2009–2012) to 20% (2013–2016), and from 25% (2009–2012) to 9% (2017–2020), respectively, the combined treatment grew from 25% (2009–2012) to 55% (2017–2020). The association of different treatments can assist the cellulose digestibility since they integrate the advantages observed in each separate technology in a single process. Therefore, an increase in the saccharification step efficiency can be reached. The operational combination can also reduce some technological problems such as environmental pollution, energy consumption, and reaction time, associated with a single treatment methodology [27], justifying, thus, its large application in recent years.

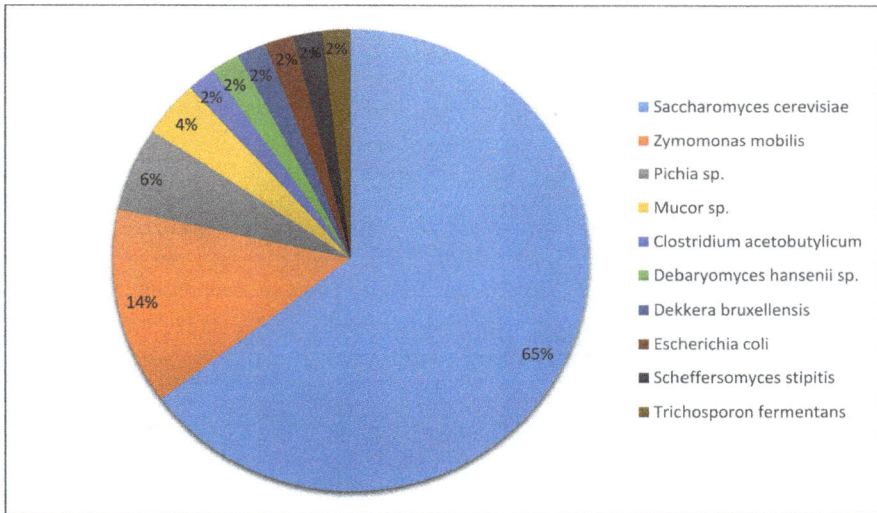

FIGURE 12.10 Main producer microorganism of cellulosic ethanol from sweet sorghum bagasse.

12.5 CONCLUSIONS

The bioprocess for the sweet sorghum bagasse ethanol production has changed in recent years, especially due to the development of more efficient, cheaper, and eco-friendlier treatment technologies and fermentation approaches. The combination of different treatments will be increasingly used, since it can take advantage of the separated methodologies benefits to improve the ethanol conversion and yield. Furthermore, the simultaneous saccharifications and fermentation processes using cellular consortia and the free sugar co-fermentation can contribute to an optimized and economically feasible large-scale implementation. Therefore, the full sweet sorghum bagasse potential should be used with biotechnology evolution, consolidating this biomass as an important and economical biorefinery crop.

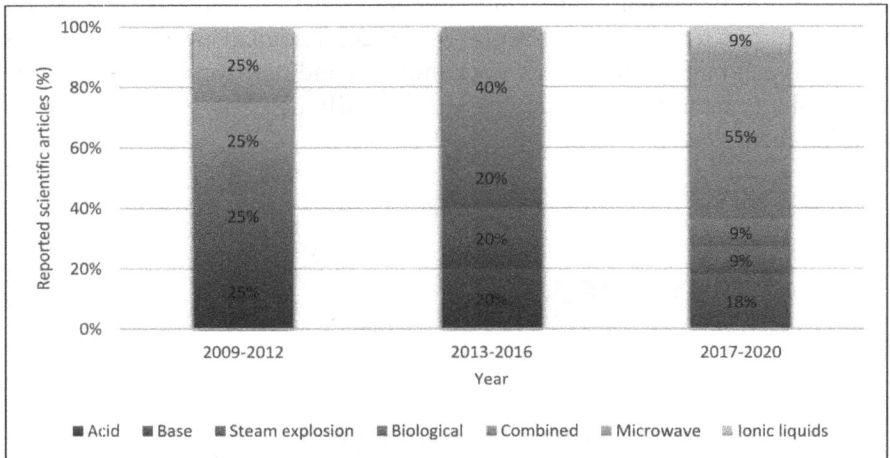

FIGURE 12.11   Evolution of the main methods for the sweet sorghum treatment.

KEYWORDS

- biotechnology evolution
- cellulosic ethanol
- fermentation approaches
- pretreatment methodologies
- saccharification and fermentation
- sweet sorghum bagasse

REFERENCES

1. Food and Agriculture Organization of the United Nations. (2018). *FAOSTAT Crops.* http://www.fao.org/faostat/en/#data/QC/visualize (accessed on 28 October 2021).
2. Cutz, L., & Santana, D., (2014). Techno-economic analysis of integrating sweet sorghum into sugar mills: The central American case. *Biomass and Bioenergy, 68*, 195–214. https://doi.org/10.1016/j.biombioe.2014.06.011.

3. Yu, J., Zhang, T., Zhong, J., Zhang, X., & Tan, T., (2012). Biorefinery of sweet sorghum stem. *Biotechnol. Adv., 30*(4), 811–816. https://doi.org/10.1016/j.biotechadv.2012.01.014.

4. Appiah-Nkansah, N. B., Li, J., Rooney, W., & Wang, D., (2019). A review of sweet sorghum as a viable renewable bioenergy crop and its techno-economic analysis. *Renew. Energy, 143*, 1121–1132. https://doi.org/10.1016/j.renene.2019.05.066.

5. Akhtar, N., Gupta, K., Goyal, D., & Goyal, A., (2016). *Recent Advances in Pretreatment Technologies for Efficient Hydrolysis of Lignocellulosic Biomass, 35*(2), 489–511. https://doi.org/10.1002/ep.

6. Lima, I., Bigner, R., & Wright, M., (2017). Conversion of sweet sorghum bagasse into value-added biochar. *Sugar Tech., 19*(5), 553–561. https://doi.org/10.1007/s12355-017-0508-8.

7. Stamenković, O. S., Siliveru, K., Veljković, V. B., Banković-Ilić, I. B., Tasić, M. B., Ciampitti, I. A., Đalović, I. G., et al., (2020). Production of biofuels from sorghum. *Renew. Sustain. Energy Rev., 124*. https://doi.org/10.1016/j.rser.2020.109769.

8. Thanapimmetha, A., Saisriyoot, M., Khomlaem, C., Chisti, Y., & Srinophakun, P., (2019). A comparison of methods of ethanol production from sweet sorghum bagasse. *Biochem. Eng. J., 151*, 107352. https://doi.org/10.1016/j.bej.2019.107352.

9. Chandel, A. K., Garlapati, V. K., Singh, A. K., Antunes, F. A. F., & Da Silva, S. S., (2018). The path forward for lignocellulose biorefineries: Bottlenecks, solutions, and perspective on commercialization. *Bioresour. Technol., 264*, 370–381. https://doi.org/10.1016/j.biortech.2018.06.004.

10. OECD-FAO, (2018). Chapter 3. *Cereals Agricultural Outlook 2018–2027* (pp. 109–126). OCED/FAO.

11. Mejila, D., & Lewis, B., (1999). Sorghum: Post-harvest operations. *INPhO - Post-harvest Compendium, 33*.

12. Velmurugan, B., Narra, M., Rudakiya, D. M., & Madamwar, D., (2020). Sweet sorghum: A potential resource for bioenergy production. In: *Refining Biomass Residues for Sustainable Energy and Bioproducts* (pp. 215–242). Elsevier. https://doi.org/10.1016/B978-0-12-818996-2.00010-7.

13. Cifuentes, R., Bressani, R., & Rolz, C., (2014). Energy for sustainable development the potential of sweet sorghum as a source of ethanol and protein. *Energy Sustain. Dev., 21*, 13–19. https://doi.org/10.1016/j.esd.2014.04.002.

14. Khalil, S. R. A., Abdelhafez, A. A., & Amer, E. A. M., (2015). Evaluation of bioethanol production from juice and bagasse of some sweet sorghum varieties. *Ann. Agric. Sci., 60*(2), 317–324. https://doi.org/10.1016/j.aoas.2015.10.005.

15. Khawaja, C., Janssen, R., Rutz, D., Luquet, D., Trouche, G., Reddy, B., Rao, P.S., Basavaraj, G., Schaffert, R., Damasceno, C., et al. (2014). *A Handbook of Energy Sorghum: An alternative energy crop* (WIP–Renewable Energies Sylvensteintr). Available at: http://oar.icrisat.org/9049/1/Sweetfuel%20Handbook%20English%20version.pdf (accessed on 23 July 2020).

16. Bonin, C. L., Heaton, E. A., Cogdill, T. J., & Moore, K. J., (2016). Management of sweet sorghum for biomass production. *Sugar Tech, 18*(2), 150–159. https://doi.org/10.1007/s12355-015-0377-y.

17. Regassa, T. H., & Wortmann, C. S., (2014). Sweet sorghum as a bioenergy crop: Literature review. *Biomass and Bioenergy, 64*, 348–355. https://doi.org/10.1016/j.biombioe.2014.03.052.

18. Kumar, B., Bhardwaj, N., Agrawal, K., Chaturvedi, V., & Verma, P., (2020). *Current Perspective on Pretreatment Technologies Using Lignocellulosic Biomass: An Emerging Biorefinery Concept, 199.* https://doi.org/10.1016/j.fuproc.2019.106244.

19. Galbe, M., & Wallberg, O., (2019). Pretreatment for biorefineries: A review of common methods for efficient utilization of lignocellulosic materials. *Biotechnol. Biofuels, 12*(1), 294. https://doi.org/10.1186/s13068-019-1634-1.

20. Muryanto, & Sari, A. A., (2018). Pretreatment of sweet sorghum bagasse using EFB-based black liquor for ethanol production. In: *Sustainable Future for Human Security* (pp. 85–95). Springer Singapore: Singapore. https://doi.org/10.1007/978-981-10-5430-3_8.

21. Heredia-Olea, E., Pérez-Carrillo, E., & Serna-Saldívar, S. O., (2013). Production of ethanol from sweet sorghum bagasse pretreated with different chemical and physical processes and saccharified with fiber degrading enzymes. *Bioresour. Technol., 134*, 386–390. https://doi.org/10.1016/j.biortech.2013.01.162.

22. Cao, W., Sun, C., Liu, R., Yin, R., & Wu, X., (2012). Comparison of the effects of five pretreatment methods on enhancing the enzymatic digestibility and ethanol production from sweet sorghum bagasse. *Bioresour. Technol., 111*, 215–221. https://doi.org/10.1016/j.biortech.2012.02.034.

23. Chen, C., Boldor, D., Aita, G., & Walker, M., (2012). Ethanol production from sorghum by a microwave-assisted dilute ammonia pretreatment. *Bioresour. Technol., 110*, 190–197. https://doi.org/10.1016/j.biortech.2012.01.021.

24. Darkwah, K., Wang, L., & Shahbazi, A., (2016). Simultaneous saccharification and fermentation of sweet sorghum after acid pretreatment. *Energy Sources, Part A Recover. Util. Environ. Eff., 38*(10), 1485–1492. https://doi.org/10.1080/15567036.2012.724146.

25. Lavudi, S., Oberoi, H. S., & Mangamoori, L. N., (2017). Ethanol production from sweet sorghum bagasse through process optimization using response surface methodology. *3 Biotech, 7*(4), 233. https://doi.org/10.1007/s13205-017-0863-x.

26. Rabemanolontsoa, H., & Saka, S., (2016). Various pretreatments of lignocellulosics. *Bioresour. Technol., 199*, 83–91. https://doi.org/10.1016/j.biortech.2015.08.029.

27. Chen, H., Liu, J., Chang, X., Chen, D., Xue, Y., Liu, P., Lin, H., & Han, S., (2017). A Review on the pretreatment of lignocellulose for high-value chemicals. *Fuel Process. Technol., 160*, 196–206. https://doi.org/10.1016/j.fuproc.2016.12.007.

28. Wang, L., Luo, Z., & Shahbazi, A., (2013). Optimization of simultaneous saccharification and fermentation for the production of ethanol from sweet sorghum (*Sorghum bicolor*) bagasse using response surface methodology. *Ind. Crop. Prod., 42*, 280–291. https://doi.org/10.1016/j.indcrop.2012.06.005.

29. Camargo, D., Sydney, E. B., Leonel, L. V., Pintro, T. C., & Sene, L., (2019). Dilute acid hydrolysis of sweet sorghum bagasse and fermentability of the hemicellulosic hydrolysate. *Brazilian J. Chem. Eng., 36*(1), 143–156. https://doi.org/10.1590/0104-6632.20190361s20170643.

30. Lo, E., Brabo-Catala, L., Dogaris, I., Ammar, E. M., & Philippidis, G. P., (2020). Biochemical conversion of sweet sorghum bagasse to succinic acid. *J. Biosci. Bioeng., 129*(1), 104–109. https://doi.org/10.1016/j.jbiosc.2019.07.003.

31. Zhang, C., Wen, H., Chen, C., Cai, D., Fu, C., Li, P., Qin, P., & Tan, T., (2019). Simultaneous saccharification and juice co-fermentation for high-titer ethanol production using sweet sorghum stalk. *Renew. Energy, 134*, 44–53. https://doi.org/10.1016/j.renene.2018.11.005.

32. Kim, M., Han, K., Jeong, Y., & Day, D. F., (2012). *Utilization of Whole Sweet Sorghum Containing Juice, Leaves, and Bagasse for Bio-Ethanol Production, 21*(4), 1075–1080. https://doi.org/10.1007/s10068-012-0139-5.

33. Yu, M., Li, J., Li, S., Du, R., Jiang, Y., Fan, G., Zhao, G., & Chang, S., (2014). A cost-effective integrated process to convert solid-state fermented sweet sorghum bagasse into cellulosic ethanol. *Appl. Energy, 115*, 331–336. https://doi.org/10.1016/j.apenergy.2013.11.020.

34. Cao, W., Sun, C., Qiu, J., Li, X., Liu, R., & Zhang, L., (2016). Pretreatment of sweet sorghum bagasse by alkaline hydrogen peroxide for enhancing ethanol production. *Korean J. Chem. Eng., 33*(3), 873–879. https://doi.org/10.1007/s11814-015-0217-5.

35. Yu, M., Li, J., Chang, S., Zhang, L., Mao, Y., Cui, T., Yan, Z., et al., (2016). Bioethanol production using the sodium hydroxide pretreated sweet sorghum bagasse without washing. *Fuel, 175*, 20–25. https://doi.org/10.1016/j.fuel.2016.02.012.

36. Pham, H. T. T., Nghiem, N. P., & Kim, T. H., (2018). Near theoretical saccharification of sweet sorghum bagasse using simulated green liquor pretreatment and enzymatic hydrolysis. *Energy, 157*, 894–903. https://doi.org/10.1016/j.energy.2018.06.005.

37. Barcelos, C. A., Maeda, R. N., Santa, A. L. M. M., & Pereira, N., (2016). Sweet sorghum as a whole-crop feedstock for ethanol production. *Biomass and Bioenergy, 94*, 46–56. https://doi.org/10.1016/j.biombioe.2016.08.012.

38. Guarneros-flores, J., Aguilar-uscanga, M. G., Morales-martínez, J. L., & López-zamora, L., (2019). *Maximization of Fermentable Sugar Production from Sweet Sorghum Bagasse (Dry and Wet Bases) Using Response Surface Methodology (RSM)*, 633–639.

39. Ostovareh, S., Karimi, K., & Zamani, A., (2015). Efficient conversion of sweet sorghum stalks to biogas and ethanol using organosolv pretreatment. *Ind. Crop. Prod., 66*, 170–177. https://doi.org/10.1016/j.indcrop.2014.12.023.

40. Jafari, Y., Amiri, H., & Karimi, K., (2016). Acetone pretreatment for improvement of acetone, butanol, and ethanol production from sweet sorghum bagasse. *Appl. Energy, 168*, 216–225. https://doi.org/10.1016/j.apenergy.2016.01.090.

41. Nozari, B., Mirmohamadsadeghi, S., & Karimi, K. (2018). Bioenergy production from sweet sorghum stalks via a biorefinery perspective. *Appl. Microbiol. Biotechnol. 102*(7), 3425–3438. https://doi.org/10.1007/s00253–018–8833–8.

42. Chen, D., Gao, D., Capareda, S. C., E, S., Jia, F., & Wang, Y., (2020). Influences of hydrochloric acid washing on the thermal decomposition behavior and thermodynamic parameters of sweet sorghum stalk. *Renew. Energy, 148*, 1244–1255. https://doi.org/10.1016/j.renene.2019.10.064.

43. Zhang, J., Ma, X., Yu, J., Zhang, X., & Tan, T., (2011). The effects of four different pretreatments on enzymatic hydrolysis of sweet sorghum bagasse. *Bioresour. Technol., 102*(6), 4585–4589. https://doi.org/10.1016/j.biortech.2010.12.093.

44. Chen, G., Chen, L., Wang, W., & Hong, F. F., (2018). Evaluation of six ionic liquids and application in pretreatment of sweet sorghum bagasse for bacterial nanocellulose production. *J. Chem. Technol. Biotechnol., 93*(12), 3452–3461. https://doi.org/10.1002/jctb.5703.

45. Batista Meneses, D., Montes de Oca-Vásquez, G., Vega-Baudrit, J. R., Rojas-Álvarez, M., Corrales-Castillo, J., & Murillo-Araya, L. C. (2020). Pretreatment methods of lignocellulosic wastes into value-added products: recent advances and possibilities. *Biomass Conv. Bioref.* https://doi.org/10.1007/s13399–020–00722–0

46. Shen, F., Hu, J., Zhong, Y., Liu, M. L. Y., & Saddler, J. N., (2012). Ethanol production from steam-pretreated sweet sorghum bagasse with high substrate consistency enzymatic hydrolysis. *JBB, 41*, 157–164. https://doi.org/10.1016/j.biombioe.2012.02.022.

47. Pengilly, C., Diedericks, D., Brienzo, M., & Görgens, J. F., (2015). Enzymatic hydrolysis of steam-pretreated sweet sorghum bagasse by combinations of cellulase and endoxylanase. *Fuel, 154*, 352–360. https://doi.org/10.1016/j.fuel.2015.03.072.

48. Damay, J., Duret, X., Ghislain, T., Lalonde, O., & Lavoie, J., (2018). Steam explosion of sweet sorghum stems: Optimization of the production of sugars by response surface methodology combined with the severity factor. *Ind. Crops Prod., 111, 482*–493. https://doi.org/10.1016/j.indcrop.2017.11.006.

49. Rohowsky, B., Häßler, T., Gladis, A., Remmele, E., Schieder, D., & Faulstich, M., (2013). Feasibility of simultaneous saccharification and juice co-fermentation on hydrothermal pretreated sweet sorghum bagasse for ethanol production. *Appl. Energy, 102*, 211–219. https://doi.org/10.1016/j.apenergy.2012.03.039.

50. Li, B., Balan, V., Yuan, Y., & Dale, B. E., (2010). Process optimization to convert forage and sweet sorghum bagasse to ethanol based on ammonia fiber expansion (AFEX) pretreatment. *Bioresour. Technol., 101*(4), 1285–1292. https://doi.org/10.1016/j.biortech.2009.09.044.

51. Choudhary, R., Loku, A., Liang, Y., & Siddaramu, T., (2012). Microwave pretreatment for enzymatic saccharification of sweet sorghum bagasse. *Biomass and Bioenergy, 39*, 218–226. https://doi.org/10.1016/j.biombioe.2012.01.006.

52. Marx, S., Ndaba, B., Chiyanzu, I., & Schabort, C., (2014). Fuel ethanol production from sweet sorghum bagasse using microwave irradiation. *Biomass and Bioenergy, 65*, 145–150. https://doi.org/10.1016/j.biombioe.2013.11.019.

53. Ndaba, B., Chiyanzu, I., Marx, S., & Obiero, G., (2014). Effect of *saccharomyces cerevisiae* and *Zymomonas mobilis* on the co-fermentation of sweet sorghum bagasse hydrolysates pretreated under varying conditions. *Biomass and Bioenergy, 71*, 350–356. https://doi.org/10.1016/j.biombioe.2014.09.022.

54. Xu, Q., Zhao, M., Yu, Z., Yin, J., Li, G., Zhen, M., & Zhang, Q., (2017). Enhancing enzymatic hydrolysis of corn cob, corn stover and sorghum stalk by dilute aqueous ammonia combined with ultrasonic pretreatment. *Ind. Crops Prod., 109*, 220–226. https://doi.org/10.1016/j.indcrop.2017.08.038.

55. Mishra, V., Jana, A. K., Jana, M. M., & Gupta, A., (2017). Improvement of selective lignin degradation in fungal pretreatment of sweet sorghum bagasse using synergistic $CuSO_4$-syringic acid supplements. *J. Environ. Manage., 193*, 558–566. https://doi.org/10.1016/j.jenvman.2017.02.057.

56. Mishra, V., Jana, A. K., Jana, M. M., & Gupta, A., (2017). Synergistic Effect of Syringic Acid and Gallic Acid Supplements in Fungal Pretreatment of Sweet Sorghum Bagasse for Improved Lignin Degradation and Enzymatic Saccharification. *Process Biochem.* https://doi.org/10.1016/j.procbio.2017.02.011.

57. Mishra, V., & Jana, A. K., (2019). Sweet sorghum bagasse pretreatment by coriolus versicolor in mesh tray bioreactor for selective delignification and improved saccharification. *Waste and Biomass Valorization, 10*(9), 2689–2702. https://doi.org/ 10.1007/s12649-018-0276-z.

58. Olofsson, K., Bertilsson, M., & Lidén, G., (2008). A short review on SSF – an interesting process option for ethanol production from lignocellulosic feedstocks. *Biotechnol. Biofuels, 1*(1), 7. https://doi.org/10.1186/1754-6834-1-7.

59. Dogaris, I., Gkounta, O., & Mamma, D., (2012). *Bioconversion of Dilute-Acid Pretreated Sorghum Bagasse to Ethanol by Neurospora Crassa,* 541–550. https://doi.org/10.1007/s00253-012-4113-1.

60. Nghiem, N. P., & Toht, M. J., (2019). Pretreatment of sweet sorghum bagasse for ethanol production using Na_2CO_3 obtained by NaOH absorption of CO_2 generated in sweet sorghum juice ethanol fermentation. *Fermentation, 5*(4), *91.* https://doi.org/10.3390/fermentation5040091.

61. Matsakas, L., & Christakopoulos, P., (2013). Fermentation of liquefacted hydrothermally pretreated sweet sorghum bagasse to ethanol at high-solids content. *Bioresour. Technol., 127,* 202–208. https://doi.org/10.1016/j.biortech.2012.09.107.

62. Molaverdi, M., Karimi, K., Khanahmadi, M., & Goshadrou, A., (2013). Enhanced sweet sorghum stalk to ethanol by fungus mucor indicus using solid-state fermentation followed by simultaneous saccharification and fermentation. *Ind. Crops Prod., 49,* 580–585. https://doi.org/10.1016/j.indcrop.2013.06.024.

63. Goshadrou, A., Karimi, K., & Taherzadeh, M. J., (2011). Bioethanol production from sweet sorghum bagasse by mucor hiemalis. *Ind. Crops Prod., 34*(1), 1219–1225. https://doi.org/10.1016/j.indcrop.2011.04.018.

64. Reis, A. L. S., Damilano, E. D., Menezes, R. S. C., & De Morais, Jr. M. A., (2016). Second-generation ethanol from sugarcane and sweet sorghum bagasse using the yeast Dekkera bruxellensis. *Ind. Crops Prod., 92,* 255–262. https://doi.org/10.1016/j.indcrop.2016.08.007.

65. Antonopoulou, I., Spanopoulos, A., & Matsakas, L., (2020). Single-cell oil and ethanol production by the oleaginous yeast *Trichosporon fermentans* utilizing dried sweet sorghum stalks. *Renew. Energy, 146,* 1609–1617. https://doi.org/10.1016/j.renene.2019.07.107.

66. Wang, W., Zhuang, X., Yuan, Z., Yu, Q., Qi, W., Wang, Q., & Tan, X., (2012). High consistency enzymatic saccharification of sweet sorghum bagasse pretreated with liquid hot water. *Bioresour. Technol., 108,* 252–257. https://doi.org/10.1016/j.biortech.2011.12.092.

67. Kurian, J. K., Minu, K. A., Banerji, A., & Kishore, V. V., (2010). Bioconversion of hemicellulose hydrolysate of sweet sorghum bagasse to ethanol by using *Pichia stipitis* NCIM 3497 and *Debaryomyces hansenii* sp. *BioResources, 5,* 2404–2416.

68. Castro, E., Nieves, I. U., Rondón, V., Sagues, W. J., Fernández-Sandoval, M. T., Yomano, L. P., York, S. W., et al., (2017). Potential for ethanol production from different sorghum cultivars. *Ind. Crops Prod., 109,* 367–373. https://doi.org/10.1016/j.indcrop.2017.08.050.

69. Damay, J., Boboescu, I. Z., Duret, X., Lalonde, O., & Lavoie, J. M., (2018). A novel hybrid first and second generation hemicellulosic bioethanol production process through steam treatment of dried sorghum biomass. *Bioresour. Technol., 263,* 103–111. https://doi.org/10.1016/j.biortech.2018.04.045.

70. Yu, J., Zhong, J., Zhang, X., & Tan, T., (2010). *Ethanol Production from H_2SO_3-Steam-Pretreated Fresh Sweet Sorghum Stem by Simultaneous Saccharification and Fermentation,* 401–409. https://doi.org/10.1007/s12010-008-8333-x.

71. Massoud, M. I., & El-razek, A. M. A., (2011). Suitability of *Sorghum bicolor* L. stalks and grains for bioproduction of ethanol. *Ann. Agric. Sci., 56*(2), 83–87. https://doi.org/10.1016/j.aoas.2011.07.004.

72. Sipos, B., Réczey, J., & Somorai, Z., (2009). *Sweet Sorghum as Feedstock for Ethanol Production: Enzymatic Hydrolysis of Steam-Pretreated Bagasse,* 151–162. https://doi.org/10.1007/s12010-008-8423-9.

73. Li, J., Li, S., Han, B., Yu, M., Li, G., & Jiang, Y., (2013). A novel cost-effective technology to convert sucrose and homocelluloses in sweet sorghum stalks into ethanol. *Biotechnol. Biofuels, 6*(174), 1–12. https://doi.org/http://www.biotechnologyforbiofuels.com/content/6/1/174.

74. Kaur, P., Uppal, S. K., Dhir, C., Sharma, P., & Kaur, R., (2015). Comparative study of chemical pretreatments and acid saccharification of bagasse of sugar crops for ethanol production. *Sugar Tech., 17*(4), 412–417. https://doi.org/10.1007/s12355-014-0346-x.

75. Yu, M., Li, J., Chang, S., Du, R., Li, S., Zhang, L., Fan, G., et al., (2014). Optimization of ethanol production from NaOH-pretreated solid-state fermented sweet sorghum bagasse. *Energies, 7*(7), 4054–4067. https://doi.org/10.3390/en7074054.

CHAPTER 13

Bioethanol Production Using *Agave americana* L. Leaves Wastes from Coahuila

CÉSAR D. PINALES-MÁRQUEZ, SHIVA, ROHIT SAXENA,
ROSA M. RODRÍGUEZ-JASSO, and HÉCTOR RUIZ LEZA

*Biorefinery Group, Food Research Department, Faculty of Chemistry Sciences, Autonomous University of Coahuila, Saltillo–25280, Coahuila, Mexico, Phone: (+52)-1-844-416-12-38,
E-mail: rrodriguezjasso@uadec.edu.mx (R. M. Rodríguez-Jasso),
hector_ruiz_leza@uadec.edu.mx (H. A. Ruiz)*

ABSTRACT

Currently, there is a tendency to mitigate through advances in science and technology the environmental problems caused by the abuse of burning fossil fuels, and this resulted in accelerated climate change that has negatively impacted the quality of human life and countless species. The residues from the Agave industry in Mexico represent an excellent opportunity to adapt large amounts of LCB into the biorefinery sustainable processing model to produce various bioproducts, including bioethanol, a fuel that can improve the substantivity of combustible compared to those produced by hydrocarbons. In this study, residues from *Agave americana* L leaves were used as biomass for the production of bioethanol under the following processing line: drying, milling, hydrothermal pretreatment, enzymatic hydrolysis, and fermentation. Also, a rich pretreated solid of cellulose (48.21%) was obtained after hydrothermal pretreatment (190C/50 min). In this study, pre-saccharification, and fermentation strategy was applied in the bioethanol production, 33.78 g/L of sugars were produced at 16 h of hydrolysis and 13.58 g/L of ethanol. Therefore, the development of this process allowed the use of a promising raw material in the production of biofuels and high added-value compounds in terms of biorefinery in Mexico.

13.1 INTRODUCTION

In recent decades, civilization has had an accelerated growth, resulting in different environmental problems, one of them being a large number of emissions into the atmosphere, causing high levels of air pollution, resulting in one of the most critical issues in the public health worldwide [1]. The World Health Organization released a statement in which nine out of 10 people are exposed to environments with high levels of pollutants in the air and it is resulting in the death of 4.2 million people worldwide per year due to stroke, heart disease, lung cancer, and chronic respiratory diseases [2].

Air pollution is derived from a complex mixture of particle matters (PM), vapors, and gases that can be emitted in different ways, both natural and synthetic. PM is generally classified into particles of 10 (PM_{10}) and 2.5 ($PM_{2.5}$) micrometers and is formed by the conglomerate of suspended solids, humidity, and atmospheric conditions, PM_{10} particles tend to accumulate in the nasal concavities and the respiratory tract superior. At the same time, the finer particles ($< PM_{2.5}$) can also be absorbed in the lower respiratory tract, causing diseases such as lung cancer, ischemic heart disease, respiratory tract infections, allergies, and type 2 diabetes due to saturation of suspended solid particles in the environment [1, 3]. On the other hand, gases released into the atmosphere have produced anormal amounts of CO_2, CO, $NO_{(x)}$, CH_4, which in addition to decreasing air quality and deteriorating human health, is the cause of climate change because these gases cause the greenhouse effect, which is the main reason of the increase in the temperature of the earth. The increase in air contamination is directly related to the excessive use of fossil fuels and industrial growth [1, 4].

Based on this problem, efforts have been made to try to mitigate or control climate change. One of these efforts is The Paris Agreement of the United Nations, whose main objective is to regulate polluting emissions that cause the greenhouse effect and to keep the increase in world temperature below 2°C, limiting this increase to 1.5°C and this in a time frame that allows ecosystems to adapt to climate change and enable sustainable economic progress [5].

The biorefinery concept fits as an emerging solution to this problem because it seeks the substitution of hydrocarbon-produced energy compounds for biofuels produced from renewable sources such as biomass, some of the proposed fuels are bioethanol, biogas, biohydrogen, biodiesel, bio-oil, vegetable oils, biosynthetic gas, bio-char among others, increasing the substantivity of energy sources, which are also produced through increasingly sustainable and efficient strategies [6].

The operation of a biorefinery is similar to a conventional refinery which the raw material is oil, since different products, including energy compounds, are produced from a petroleum complex mixture through various stages, however, given the non-renewable nature of this material and the environmental problems that result from the excessive use of these products, make the concept of biorefinery an attractive alternative, since using different types of biomass, which can come from various sources and have meager costs due to its high availability, and result in a wide variety of compounds with novel applications in addition to bioenergetics [7].

In general, biomass is defined as all material from a living organism, and its use in biorefineries has generated a classification as generations, depending on the biomass used. Currently, there are four generations of biomass, the first generation (1G) comprises those of biorefineries that use food crops as fuels, due to their high content in sugars, starch, and natural oils, resulting in the first-generation biofuels, this generation represents a significant concern because it uses edible resources creating direct competition between food and fuel production [8]. The second-generation biorefinery (2G) includes biorefineries whose raw material is LCB, that is, those that are composed of hemicellulose, cellulose, and lignin. These biorefineries present significant advantages since the raw material is found in large quantities and is usually the residue of agricultural, forestry, and food industries, so it does not compete with the crops generated for the food sector, and due to the main components of the matrix it is possible to produce a wide variety of byproducts and biofuels [9]. The third-generation (3G) is characterized by the use of aquatic biomass as raw material, specifically microalgae and macroalgae, this alternative is very attractive since it does not generate competition with arable land, it also has a high carbohydrate content and it does not have the strong union of terrestrial biomasses, so there is a great opportunity to produce a wide variety of different products and biofuels [10]. The fourth-generation (4G) consists of microorganisms and genetically modified plants to have high amounts of carbon for the production of fuels and various biochemicals [8].

Finally, the use of biomass has become a sustainable alternative to the future for the supply of renewable products. It generates an alternative to non-renewable energy sources, promoting the growth of a biologically based economy, of a sustainable nature, also affected circular bioeconomy [8].

13.2   LIGNOCELLULOSIC MATERIALS: A NOVEL PROPOSAL FOR RAW MATERIAL IN BIOREFINERIES

13.2.1   *OVERVIEW OF LIGNOCELLULOSIC BIOMASS (LCB) FOR BIOREFINERIES*

In general, lignocellulosic materials are mainly composed of cellulose (30–50%), hemicellulose (20–40%), and lignin (20–30%) [11]. Cellulose is made up of a polymer of linked β-1,4 glucose units, Hemicellulose by the polymer of pentoses, hexoses, and uronic acids, where its main component is xylan, and lignin is constituted by polymers cross-linked phenolic which function as the main structuring agent of plants [12, 13]. This matrix of components is an opportunity for the creation of different compounds from a single raw material, through different chemical, physical, biological, and enzymatic processes [12]. Finally, annual worldwide production of LCB of 200×10^9 tons per year is estimated, making abundant feedstocks available for the production of biofuels and biochemicals [8, 14].

Some examples of possible products from the second generation biorefineries are formic acid, ethylene glycol, acetic acid, lactic acid, glycerol, glycolic acid from cellulose, but the primary designated use for this polymer is for the production of bioethanol under the model of a biorefinery. In the case of lignin, the transformation of this polymer can lead to quinones, phenol benzene, syringaldehyde, pyruvate, and different lipids. And finally, hemicellulose can be transformed into furfural, hydroxymethylfurfural (HMF), levulinic acid, pentane, among others [15]. However, numerous investigations have been directed towards the use of xylan from hemicellulose, because the compounds derived from this polymer have various food applications such as xylitol and xylose (Low-calorie sweetener substitutes) [13, 16], but with a particular interest in the production of oligomers such as xylooligosaccharides (XOs), which have multiple uses in food and pharmaceutical technology due to their prebiotic, antioxidant, and cytoprotective activity [17], especially with those of lower molecular weight or with a lower degree of polymerization (DP = 2–4, oligomers made up of 2 to 4 xylose units) [13, 16] and those with higher molecular weight, such as biopolymer with (DP > 5–10) which have a wide variety of applications such as its use in the development of food covers [18]. Finally, the design of operational chains that promote an integral use of biomass, through the combination of the production of biofuels and high added value co-products, because all these biochemicals promote the economic projections of the second generation biorefineries. For this reason, it is essential to highlight the importance of the great variety of products that can be produced from LCB and its economic and social contribution [19, 20].

One of the common obstacles when it comes to LCB is its resistance to fractionation, this factor is known as recalcitrance, and it also worsens stages such as the saccharification of the constituent polymers of lignocellulosic materials such as glucan and xylan. Nevertheless, there are processing models that improve the availability of the material using mechanisms such as autohydrolysis in treatments such as LHW and SE, which use only water as a catalytic medium to promote the separation of the lignocellulosic matrix and improve the later stages [21].

Nowadays, biorefineries around the world mainly produce biodiesel and bioethanol using first-generation raw materials, due to its high content of sugars, starches, or natural oils [22]. As reported by the Renewable Fuels Association, 15,776 million bioethanol gallons were produced in 2019 using primarily grains like corn and sorghum [23]. However, although first-generation biofuels diversify the raw materials with which biofuels are produced, they are surrounded by environmental and social contradictions such as environmental degradation due to the change in land use derived from the expansion of agriculture, affecting indirectly or directly biodiversity and the amount of CO_2 produced, and also the creation of direct competition with agriculture for food generation [22].

In addition, the environmental problem tends to worsen, because as the population grows, there will be a higher demand for resources, specifically energy and food, by 2030 it is expected that there will be 8.5 billion people in the world, at a growth rate 83 million people each year, which projects an increase of 70 million ha to meet global food demand by 2050 [24]. All these factors drive towards a transition that first generates a renewable alternative for the production of fuels and does not compromise the use of food resources and the environment, taking into account the raw material and the sustainability of the process. Finally, this promotes the concept of second-generation biorefineries, whose main objective is to generate bioeconomic circular routes (Figure 13.1) where non-edible lignocellulosic waste is used with increasingly optimal and sustainable processes [22, 24].

13.2.2   SECOND GENERATION BIOREFINERY DEVELOPMENT OPPORTUNITY IN MEXICO

Mexico is one of the most promising developing countries, due to its geographical position, this country has a great variety of favorable climates for agricultural production, and also is a sunny country, which also has various seas, rivers, and lakes. For these reasons, Mexico has enormous potential to develop more sustainable alternative energies such as solar, wind, hydraulic, and thermal energy, but also with large amounts of biomass available due to its vast agricultural activity [25].

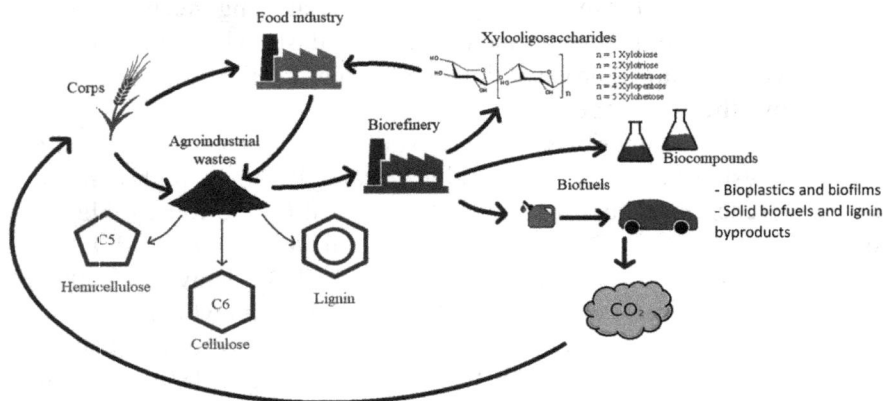

FIGURE 13.1   Overview of second-generation biorefinery and its applications.

According to the National Institute of Statistics and Geography (INEGI), in 2017, there was a total of 18.900,000 ha for agricultural activity in Mexico, of this available land 83% was destined for annual crops and the rest for perennial plants. The five main crops in the country during that year were: sugar cane (56,354,945 ton), white grain corn (23,142,203 ton), yellow grain corn (8,071,840 ton), wheat grain (3,214,047 ton), red tomato (3,008,036 ton) [26]. As a result, the availability of lignocellulosic residues in Mexico is high, given the level and the great variety of agricultural products in the country.

According to the Mexican National Energy Secretariat (SENER) and the National Energy Balance databases, the national energy consumption is 9,236,858 PJ, of which the energy consumption for fuels corresponds to 5,283,705 PJ or 57.2%. Only 6.01% of use for fuels was produced by renewable sources in 2018, where firewood (249,084 PJ or 4.71%) and cane bagasse (55,716 PJ or 1.05%) are the primary sources of energy generation from biomass. On the other hand, petroleum products corresponding to 57.07% (3,015,637 PJ) plus the contribution derived from coal of 3.54% (186,931 PJ) [26]. This demonstrates that the current generation of fuels in

Mexico is still highly inclined towards the generation of energy through non-renewable sources and where the burning of fossil fuels has a great presence in the country [25].

Today, despite political limitations and the legal framework for the use of biomass for the production of biofuels is still not defined, there has been a growing interest in the development of renewable energy in the country. In 2016 the Mexican Center for Innovation in Bioenergy (CEMIE-Bio) was created to carry out research and development of technologies for the sustainable production of fuels from biomass, said the group works in 5 specialized groups for each type of biofuel: solid biofuels, bioalcohol, biodiesel, biogas, and bioturbosin. The results of these research groups are expected to increase the use of biomass in Mexico and contribute to the generation of at least 35% of clean energy by 2024 [27].

13.2.3   AGAVE WASTE AS RAW MATERIAL UNDER BIOREFINERY

The use of agave has a strong presence in Mexico since it is used in various artisanal and industrial activities, this generates large amounts of solid waste with enormous potential for its use [28]. Various of agave species contribute to the economy of Mexico via the production of contrasting products such as in northern Mexico, where agave is used for the production of textile fibers, while, in the southern region tradition Mexico developing tequila, mezcal, and pulque [28]. For example, Agave *tequila weber* is distinguished for the production of tequila in central Mexico through cooking and crushing this plant to create juices with high sugar content, which are fermented and distilled to create tequila; however, this process leaves large amounts of solid residue that can be used for the creation of biofuels and high added-value compounds [29, 30].

Through the biorefinery model this solid bagasse can be used to produce the biofuels and biproducts. Agave bagasse contains the cellulose, hemicellulose, and lignin in (31–43%), (11–22%), and (11–20%) w/w respectively. Biofuel refineries using the agave bagasse with the maximum concentration of amorphous cellulose and hemicellulose, but the recalcitrance property of lignin make it difficult to extract which act as a barrier [31].

In addition to the presence of the Agave *tequilana weber*, the climate of Mexico allows different species of Agave to grow such as *Agave americana*,

Agave *angustifolia*, Agave *fourcroydes* and Agave *sisalana*, and these also form in the economic activity of the country [32]. Agave species reflects the low rate of transpiration, have crassulacean acid metabolism capability which upgrade water using efficiency in semiarid region [33].

13.2.3.1   AGAVE AMERICANA L.

The *Agave americana* has broad leaves, and its color is green with slightly gray with a whitish color, and also, is commonly used as an ornamental plant in gardening or for the extraction of mead also known as "Aguamiel." These plants have very resistant fibers, used to make handcraft textile items. *Agave americana* has an average size of 1.0 m to 1.4 m, with leaves 80 to 120 cm long and 15 to 20 cm wide, and its leaves are fleshy and smooth to the touch [32].

The origin of this plant is Southern Africa and it is reported that the size of this plant could reach a height of up to 2 meters. In Mexico, *the Agave americana* L. is used in various activities, which makes a plant with great importance for the economy and culture of the country. In Mexico, generates the sustenance of 38,000 families, which also maintains a product range in the country of 412,900 tons to 998,400 tons and wherein 2008 there was a maximum in the production of this input with 1,125,100 tons, which shows that it is a company that promises and has remained within Mexican culture [34].

The Industrial Scientific Research Council (CSIR) studied about the processing of *Agave americana* L. and highlighted the enormous potential for fiber production and paper manufacturing, and also conducted important market research on the global demand for inulin as a by-product which could be derived from respected processes [34]. All this to carry out the development of the technology to establish an industry based on the *Agave americana* L. in Southern Africa, although studies have also been carried out in our country to use in a more optimal way this plant and that can also adapt to the process of manufacturing textile fibers since according to the CSIR, the contribution of fiber in this plant is 1.5% resulting in the rest of the plant having no use and also with these conditions prevents the exclusive processing from this raw material to be unreachable. For the production of mead *Agave americana*, L. is cut and generates a hole at the center of the plant, and allow to collect the liquid of the plant also from the rain later it could be supplied to tequila industries for the purification and sterilization of the mead [34].

13.3 PRELIMINARY STAGES FOR THE CONVERSION PROCESS OF AGAVE WASTES TO BIOMASS

13.3.1 DRYING AND GRINDING OF BIOMASS AS A PRELIMINARY STAGE

The first stage contemplated for the biomass transformation process consists of a series of mechanical treatments, which can be drying and crushing. This process aims to improve the disposition of the material to the later stages of the process. Generally, the dry and crushed materials have higher surface areas and lower crystallinity, which can promote, for example, hydrolysis or mass transfer phenomena within the biomass. Mechanical pretreatment as the only pretreatment is not very useful for LCB es because it cannot fractionate hemicellulose, cellulose, and lignin, so a series of subsequent steps are necessary which can separate the biomass. However, mechanical pretreatment is required to achieve essential characteristics in the process, such as the appropriate particle size for later stages of the process, making this an indispensable step for biorefineries [8].

The mechanical pretreatment of *Agave Americana* L consisted of a primary stage where the leaves were cut into thin strips to facilitate drying. Subsequently, sunlight was used to dry the biomass partially, and then the strips were cut in 1 cm^2 pieces to be dried in a laboratory oven at 75°C until reaching an internal humidity of 8% (w/w). Finally, a blade mill (Thomas Wiley, Swedesboro, NJ, USA) was used to achieve a particle size ≈ (1–0.3 mm) [36].

13.3.2 CHEMICAL CHARACTERIZATION OF BIOMASS

Only with the biomass, which had a particle size of (0.5–0.3 mm) the characterization of the raw material was carried out. The analysis considered were: a humidity determination at 121°C, ash determination, protein content (N × 5.67) by Kjeldahl method, and solvent and aqueous extractives (through a Soxhlet type extraction with acetone and water respectively). Also, it was necessary to make a physicochemical characterization to determine the cellulose, hemicellulose, and lignin content, which was made according to the standard analytical procedures of the National Renewable Energy Laboratory (NREL/TP-510-42618), through the quantification of monomers and acid by high-performance liquid chromatography (HPLC) (Section 13.3.2.1). Finally, Kalson lignin was quantified by the gravimetry method (Table 13.1) [36, 37].

TABLE 13.1 Characterization of *Agave americana* L. Leaves (% on Total Dry Weight) [38, 36]

Component	Agave americana L.	Agave tequilana Leaves [38]	Agave tequilana Bagasse [36]
Cellulose	29.89 ± 3.80	34.81	20.85
Hemicellulose	14.61 ± 0.10	17.98	17.31
Xylan	13.65 ± 0.02	16.49	–
Arabinan	1.31 ± 0.54	1.5	–
Klason lignin	13.65 ± 0.13	9.89	17.31
Ashes	10.59 ± 0.51	–	7.67
Protein	6.018 ± 0.0	8.35	–
Solvent extractives	2.55 ± 0.10	–	1.53
Aqueous extractives	43.55 ± 1.21	12.48	8.36

Although the composition of *Agave americana* L and Agave *Tequilana* is similar, there is a clear difference in the amount of soluble aqueous extractives. However, some reports indicate that the soluble extracts in *Agave Americana* L are approximately 55.5% (w/w) [39]. Due to the natural morphology of the leaf, The Agave *Tequilana* leaf is quite thin. In contrast, the *Agave Americana* L leaf is much thicker, resulting in much more soluble material inside.

13.3.2.1 ANALYTICAL METHOD (HPLC)

For the physicochemical characterization and the reading of monomers in the various stages of the process, they were quantified under the same conditions, using an Agilent 1260 Infinity II HPLC with a refractive index for glucose, xylose, arabinose, acetic acid, ethanol, and degradation compounds. All these compounds were detected and quantified using calibration curves made with pure agents. Besides, a MetaCarb 87 H column (300 × 7.8 mm, Agilent) was used for the analyzes; the column temperature was 60°C and a mobile phase with a concentration of 0.005 mol/L of sulfuric acid, with a flow rate of 0.7 mL/min and an injection volume of 20 μL [36].

13.3.3 GENERAL ASPECTS OF AVAILABLE PRETREATMENTS FOR BIOMASS CONVERSION

Because drying and grinding treatments are not capable of breaking the complex union that exists between the components of lignocellulosic material, it is necessary to add a stage to the process that has this task as the main function. This stage is known as pretreatment. This operation is crucial since it directly influences the subsequent steps because it also changes the adaptability of the raw material, such as for enzymatic hydrolysis, fermentation, inhibitor production, filtration of flows, and the treatment of the waste to be produced during the process. For this reason, it is estimated that the choice of pretreatment constitutes 40% of the total cost of the process [40].

In addition to mechanical treatments, there are physical treatments that can modify the structure of the biomass. Such is the case of microwaves, ultrasound, high-energy electron radiation, pretreatment, and high-temperature pyrolysis. There are also chemical pretreatments such as acid and alkaline treatments, oxidative treatments, ionic liquid, Organosolv, among others. Another branch of pretreatments is made up of biological processes carried out by microorganisms to structurally change biomass and, finally, physicochemical treatments, in which there are processes such as steam or CO_2 explosion and finally hydrothermal pretreatments [41].

13.4 EFFECTS OF HYDROTHERMAL PROCESSING ON THE BIOMASS

13.4.1 FUNDAMENTALS OF HYDROTHERMAL PRETREATMENT

Hydrothermal processing or also known as liquid-hot water pretreatment is a process widely used in lignocellulosic materials. This stage is applied to a wide variety of operational conditions and strategies. Usually, the operating parameters are 150–230°C in isothermal regime (maintaining the set-point for 10–50 min) and non-isothermal, at pressures ranging from 4.9 to 20 bars [21, 42].

In this range of temperatures, the hydrogen bond of the water begins to weaken, which results in the autoionization of the water in hydronium ion (H_3O^+) which acts as a catalyst due to its acid potential, and the hydroxide ions are formed equally (OH^-) whose potential is naturally basic. Thanks to the hydrogenation of the acetyl groups contained in the hemicellulose present in the biomass, it is how the ions formed by water act, causing that at the end of the reaction between 40 or 60% of the total mass on dry base have been lost, diluting the aqueous compounds and hemicellulose [43].

13.4.2 AN OVERVIEW OF THE USES OF HYDROTHERMAL PRETREATMENT (LIQUID HOT WATER (LHW) AND STEAM EXPLOSION (SE) PROCESS)

Nowadays, the hydrothermal process is a favorable process among the researchers which could hydrolysis the polysaccharides into monomers and oligomers. During the hydrothermal pretreatment, hydronium ions act as a catalyst and are able to eradicate the rigidity of the biomass as well as prevents the corrosion in the equipment which results in a reduction of the capital cost of the end product. Pretreatment by water supports the reduction in the capital cost of chemicals likewise the cost of expensive non-corrosive metals. The Autohydrolysis technique reflects the economic feasibility because the process does not require the acid, neutralizing agents, expensive corrosive metals for the support of degradation of biomass, and prevention from corrosion respectively [21, 42, 44].

The solid-liquid ratio of biomass and water composes the medium complex and hike the processing scale which requires a greater amount of the power consumption to heat the slurry mixture. Recent studies use the heat exchanger to recover these amounts of energy which could enhance the overall cost of the process because of the use of heat exchanger [45, 46]. Autohydrolysis technique adequate to improve the quality of solid also removal of the metal component from the lignocellulosic feedstock. Mainly removal of metal from the lignocellulosic raw material plays a significant role during the production of bio-oil [47]. Studies state the preference of the autohydrolysis technique reduces the chances of development of inhibitors that would not disrupt the downstream processing and complement its accessibility for efficient saccharification using hydrolyzes enzymes [45, 48].

13.4.2.1 SEVERITY FACTOR AS PARAMETER

The effectiveness of the hydrothermal treatment can be related to the severity factor (R_0). This parameter describes the relation between temperature and reaction time; this relationship is described with the following Eqns. (1) and (2) [49].

$$Log(R_0) = \int_0^t exp\left[\frac{T-100}{w}\right] \tag{1}$$

where; T is the process temperature (°C); t is the reaction time (min); 100 is the reference temperature in degree Celsius (°C); and w is an empirical parameter related to the activation energy, assuming kinetics of first-order (w=14.74) [50]. However, this equation in isothermal regime considers the contributors to heating, steady-state, and cooling during the process, resulting in the following Eqns. (2) and (3).

$$Log(R_0) = [Heating\ (R_0)] + [Isothermal\ (R_0)] + [Cooling\ (R_0)] \tag{2}$$

$$LogRo = \left[\int_0^{t_1} \left[\frac{T(t)-100}{w} \right] + \int_{t_1}^{t_2} exp\left[\frac{T(t)-100}{w} \right] + \int_{t_2}^{t_f} \left[\frac{T(t)-100}{w} \right] \right] dt \tag{3}$$

When the hydrothermal pretreatment is in an isothermal regime, it is necessary to make a distinction in the stages of the temperature profile. In the case of the second equation, the first term refers to the heating stage, so the value of t_1 is the time in which the temperature setpoint is reached. The isothermal period begins, which is described by the second term of the equation and is evaluated from t_1 to t_2. With this temperature t_2, the cooling stage begins and ends with the final time of the process t_f [36].

13.4.3 HYDROTHERMAL PRETREATMENT OF AGAVE AMERICANA L. LEAVES

13.4.3.1 DESCRIPTION OF THE LIQUID HOT WATER (LHW) PRETREATMENT REACTOR AND ITS OPERATION

The autohydrolysis of the *Agave americana* L biomass was carried out in a stainless-steel batch reactor with a stirring propeller and a maximum capacity of 750 mL (design by biorefinery group: http://www.biorefinerygroup.com). The pre-treatment was carried out with the biomass whose particle size was ≈ (1–0.3 mm), the particle size distribution in was 0% > 2 mm, 16% > 1 mm, 54% > 0.5 mm and 28% 0.3 mm and finally 2% fines (% w/w). The liquid/solid ratio was 10:1 (% w/v), with a maximum working volume of 300 mL, with a maximum temperature of 190°C and a minimum of 150°C, a maximum isotherm duration of 50 min and a minimum of 10 min, and a stirring of 120 rpm. The reactor had a PID controller to regulate the temperature during the isotherm with the intensity of the heating jacket and a stainless-steel thermocouple and the rotations with a sensor in the propeller shaft located outside the reactor [51].

13.4.3.2   EXPERIMENTAL DESIGN FOR THE HYDROTHERMAL PRETREATMENT OF BIOMASS

A central composite was used to identify the conditions in which the highest cellulose content was found in the pre-treated solid phase. To determine the number of conditions was calculated using the following equation [36, 51].

$$N = 2^k + 2 * k + 1 \tag{4}$$

where; k is equal to the number of variables to manipulate during the experiment ($k = 2$, time, and temperature). Resulting in a total of 9 conditions with 3 extra repetitions of the central value (170°C–30 min) [36, 51].

13.4.3.3   CHANGES IN BIOMASS DUE TO HYDROTHERMAL PRETREATMENT

Finally, after the operation, 2 streams were generated, the first a solid phase with a high concentration of cellulose and lignin with which the ethanol will be produced, and a liquid one where the presence of XOs and monomers will be checked during the treatment with the HPLC (Section 13.3.2.1) [51].

During the pretreatment stage, the heating time, isothermal, and cooling time was captured each 10C (Figure 13.2). With this information, it was possible to make a temperature profile and calculate the area under the curve. The severity factor was calculated with the software Polymath v6.0., between 100°C of heating and 100°C of the cooling stage, for each pretreatment to obtain the severity factor ($Log(R_0)$) (Figure 13.3).

From biomass pretreatment and the calculation of the severity factor, it was possible to measure the contribution of the operation in the transformation of the biomass and to relate this change to the increase in the severity factor, as the pretreatment conditions were more intense. This factor is an interesting relationship between the parameters of the pretreatment of LHW, serving this same as a guideline to relate the behavior of biomass in different reactors or the scaling-up of the processes (Table 13.2).

With the results of the previous table, we can see how hemicellulose dilutes as the severity of the treatment increases. Furthermore, it is possible to observe an increase in the percentage of cellulose and lignin in the solid phase, starting from 36.67% to 50.66% (w/w) in the case of cellulose, and for lignin from 18.38% to a 33.06% (w/w).

Besides, it is possible to see the reduction of factors such as pH, checking the acidification of the medium by acetylation of the medium. XOs are an exciting factor since it is possible to increase their production in treatments where the severity factor is not as high, such as 170°C for 10 min, 190°C for 10 min, approximately with severity factors of ≈ 3.8. Finally, monomers and XOs in the liquid phase tend to degrade under severe treatment conditions, resulting in some acids, such as acetic acid and degradation compounds such as (HMF) or furfural [52].

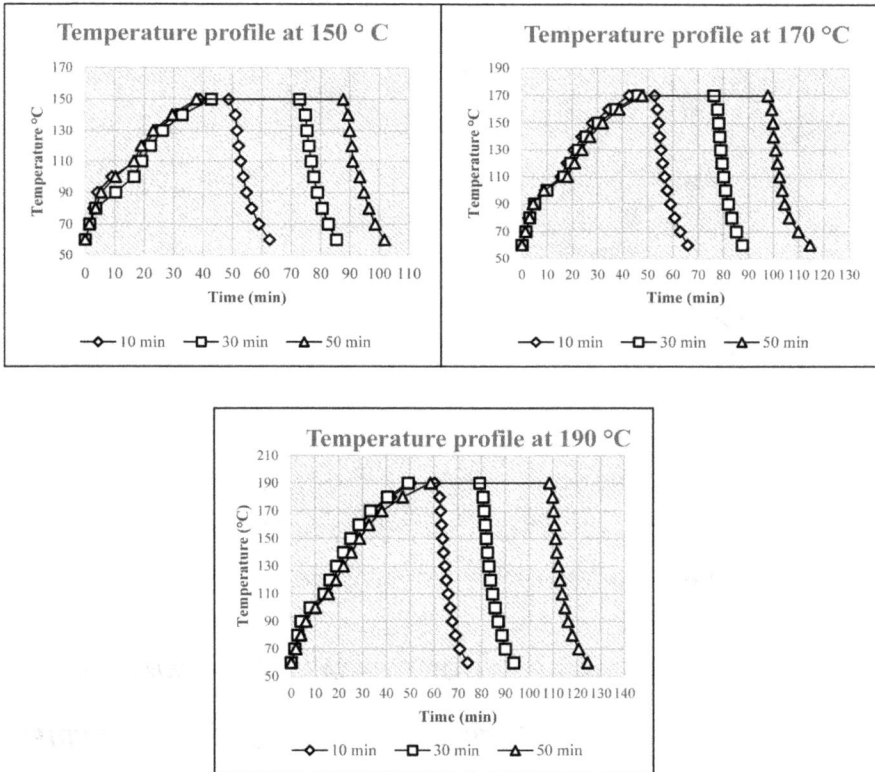

FIGURE 13.2 Temperature profiles of hydrothermal pretreatments.

It was determined that the best treatment for ethanol production, after data analysis, is the condition with higher cellulose production and lower hemicellulose content. For this reason, the state of 190°C and 50 min was directly chosen for enzymatic hydrolysis, due to avoid inhibition of the enzyme by traces of compounds derived from hemicellulose.

This behavior can be compared to that obtained by Aguirre-Fierro et al. [53] where they used agave bagasse for ethanol production, which was pretreated by high-pressure CO_2-H_2O. In this study, it was found that at temperatures from 150–190°C (10–50 min), they produced 110.5 g/L of reducing sugars.

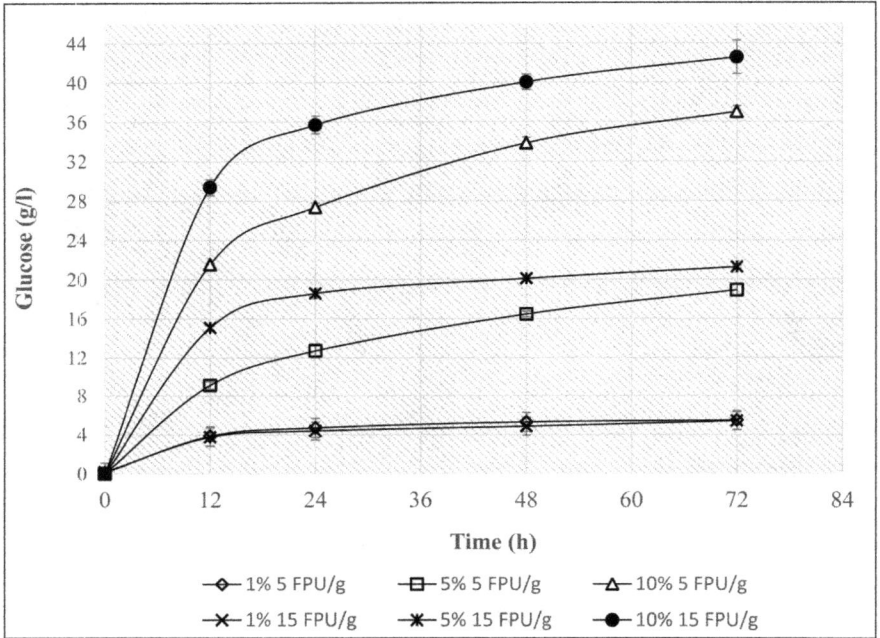

FIGURE 13.3   Glucose production from enzymatic hydrolysis condition.

13.5   ENZYME SACCHARIFICATION OF *AGAVE AMERICANA* L. WASTE

13.5.1   *USE OF ENZYMES IN SECOND-GENERATION BIOREFINERIES*

Enzymes used as biological catalysts derives from their versatility, accepting a great variety of complex molecules, including synthetic structures and found in nature. Enzymes stand out for their selectivity, and factor of great application within the chemical industry. Consequently, enzymes have gained relevance within different sectors and processes, are usually characterized as a sustainable technology that promotes a transition towards processes with higher environmental principles. Finally, science has directed the use of enzymes as a substitute for chemical processes, with low efficiency or where there is a high generation of waste [54].

TABLE 13.2 Operational Conditions for Hydrothermal Pretreatment and Chemical Composition of Pretreated Solid and Liquid Phase

Temperature (°C)	150			170			190		
Time (min)	10	30	50	10	30	50	10	30	50
$Log(R_0)$	3.76	3.93	4.00	3.85	4.03	4.22	3.96	4.10	4.17
pH	5.64	5.01	4.88	4.70	4.56	4.46	4.40	4.32	4.28
Heating Rate	2.31	2.10	2.38	2.57	2.37	2.29	2.15	2.63	2.21
Solid Phase Composition (% on Total Dry Weight)									
Cellulose	36.67±0.92	38.085±1.90	40.026±1.77	37.692±0.99	43.935±1.03	48.215±0.84	37.755±0.95	44.933±0.90	50.660±0.82
Lignin	18.38±0.42	22.57±1.72	25.590±0.55	27.472±0.38	29.270±0.38	31.568±0.52	28.642±0.52	32.069±0.23	33.066±0.09
Hemicellulose	11.22±0.00	9.3953±0.33	8.745±0.472	7.514±0.472	3.088±0.177	2.587±0.14	2.273±0.13	2.273±0.13	00.000±0.00
Liquid Phase Composition (g/L)									
Glucose	4.438±0.05	4.161±0.01	4.132±0.14	3.912±0.23	2.472±0.36	1.940±0.27	1.698±0.01	1.403±0.00	1.117±0.01
Xylose	4.602±0.00	4.754±0.01	5.196±0.14	6.648±0.00	6.543±0.01	4.105±0.02	6.147±0.01	4.366±0.01	3.234±0.11
Arabinose	1.166±0.04	1.140±0.00	1.193±0.21	1.387±0.00	1.082±0.00	0.000±0.00	0.000±0.00	0.000±0.00	0.000±0.00
XOs	2.581	2.787	3.221	4.349	3.938	1.313	4.17	2.409	1.475
Acetic acid	0.117	0.129	0.144	0.124	0.126	0.128	0.157	0.185	0.215
Levulinic acid	0.000	0.000	0.000	0.000	0.000	0.000	0.000	0.000	0.000
Formic acid	0.000	0.000	0.000	0.000	0.000	0.025	0.253	0.034	0.035
Hydroxy-methylfurfural	0.000	0.000	0.000	0.000	0.000	0.001	0.023	0.045	0.054

In second-generation biorefineries, the depolymerization of the polymer constituents of biomass to monomers or sugars represents a technological barrier. The way enzymes can be produced from various sources, such as fungal taxa, bacteria, and archaea. In the case of lignin, there are non-hydrolytic lignase enzymes (laccases, lignin-peroxidases, and manganese peroxidases (MnPs)) that act through the generation of highly reactive free radicals that break the bonds between carbon-carbon and ether units in the structure of the lignin. In the case of hemicellulose, due to its heterogeneous nature, there are a great variety of hydrolytic and non-hydrolytic enzymes with different specificities that work cooperatively, which attack the main chains of hemicellulose polymers, working as debranching enzymes, such as α-arabinofuranosidase, ferulic acid esterase, acetyl xylan esterase, and α-glucuronidase. Finally, the most common enzymes for cellulose degradation are endo-glucanases, exo-glucanases, cellobiohydrolases, and β-glucosidases; these enzymes can be found as a set for the hydrolysis of cellulose to glucose [55].

13.5.2 ENZYMATIC HYDROLYSIS IN AGAVE AMERICANA L.

With the biomass pretreated at 190°C and 50 min, enzymatic hydrolysis was carried out for the saccharification of glucan. The enzyme consisted of 125 mL shake flasks at 50°C under 150 rpm agitation for 72 h with enzyme concentration of 5 and 15 FPU/g of glucan of Cellic CTec2 enzyme cocktail, with a 50 mM citrate buffer to maintain the reaction at a pH of 4.8. sodium azide was added at a concentration of 0.2% (w/v) to prevent microbial growth [56]. The operating volume was 50 mL. The sampling was performed at 0, 12, 24, 48, and 72 h with a sample size of 2 mL. The amount of glucose produced was determined using the methodology described in Section 13.3.2.1. The experiments were carried out in triplicate, with the use of blanks to compare that there is no conversion of glucan to glucose without the enzyme [36, 57].

13.5.2.1 CELLIC CTec2 COMMERCIAL ENZYME COCKTAIL

The enzymatic cocktail used for the conversion of *Agave americana* L to biomass was the cellulase Cellic® CTec2 by Novozymes (North America, USA). This cocktail is composed of Exo-1,4-β-D-glucanase, Endo-1,4-β-D-glucanase, and

1,4-β-D-glucosidases. This mixture of enzymes manages to break the cellulose chains through different mechanisms (Figure 13.4). Finally, the enzyme activity was determined using 0.5 mL of a diluted enzyme solution, 50 mg of a Whatman No. 1 filter paper (1 cm × 6 cm) and 1 mL of 50 mM citrate buffer at pH 4.8. Finally, reducing sugars were quantified with Miller's method (DNS), resulting in 123 FPU/mL [52, 56, 58].

13.5.2.2 YIELD CALCULATION FOR ENZYMATIC HYDROLYSIS

Through the following Eqn. (5), the yield of enzymatic hydrolysis is calculated [56].

$$Enzymatic\ saccharification\ yield\ (\%) = \frac{[Glucose] + 1.053[Cellobiose]}{f \times [Biomass] \times 1.111} * 100 \quad (5)$$

where; [Glucose] is glucose concentration (g/L); 1.053 is the conversion factor of cellobiose to glucose; [Cellobiose] is the concentration of cellobiose (g/L); f is the cellulose fraction in dry biomass (g/g); [Biomass] is dry biomass concentration at the beginning of the enzymatic saccharification (g/L); and 1.111 is the conversion factor of cellulose to Glucose.

13.5.2.3 ENZYMATIC HYDROLYSIS OF PRETREATED SOLID BIOMASS

The quantification of the transformation of the pretreated biomass of Agave America L was measured and analyzed using three main parameters: Production of glucose from glucan (g/L), the yield of the reaction (%), and the initial reaction rates of each treatment. Each of these measures was analyzed individually, to choose which of the conditions would be the best candidate for the production of bioethanol. Within the analysis, it was a priority to maximize the amount of glucose produced (Figure 13.4), as well as a reasonable reaction yield. Still, the factor of the initial reaction rate was an indispensable factor in choosing the condition since the ethanol produced during the reaction would greatly influence fermentation.

At this point, the condition of 10% of solid loading and 15 FPU/g glucan is the treatment where there was a higher conversion of glucan to glucose. However, it was also necessary to check if there was a significant difference appropriate for the choice of this condition (Table 13.3).

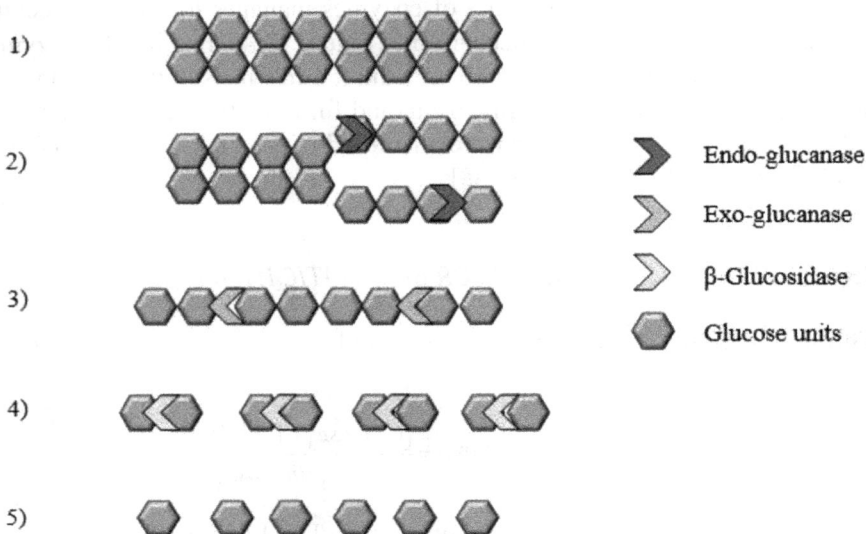

FIGURE 13.4　Graphical representation of the enzyme cocktail.

TABLE 13.3　Single-Factor Analysis of Variance for Yield and Initial Reaction Rate

Effect	Degrees of Freedom	Sum of Square	Mean Sum of Square	F-Value	Significant (p<0.005)
ANOVA-Enzymatic Saccharification Yield					
Treatment	5	2257.729	451.546	206.153	$F_{5,6}$ = 4.39
Error	6	13.142	2.19	–	–
Total	**11**	**2270.871**	–	–	–
ANOVA-Initial Reaction Rates					
Treatment	5	7.53	1.506	408.436	$F_{5,6}$ = 4.39
Error	6	0.022	0.004	–	–
Total	**11**	**7.55**	–	–	–

When this statistical difference was verified, a comparison of means was made for the reaction speed and yield (Table 13.4).

By analyzing the parameters, the idea of using the treatment of 10% pretreated solids loading and 15 FPU/g of glucan was reinforced, since it has the second-best reaction yield of all the conditions, the highest amount of glucose produced, and finally, was the condition with the highest glucose conversion rate. Finally, the results of the blanks showed that it was not possible to convert glucan to glucose without enzymes since no trace of glucose was detected in these blank tests.

TABLE 13.4 Enzymatic Conversion Yield, Initial Rates and Mean Comparison of the Hydrolysis

Hydrolysis Conditions	Enzymatic Saccharification Yield (%)	Comparison of Means (95%)	Initial Rate of Reaction (g L–1 h–1)	Comparison of Means (95%)
1%–5 FPU/g	96.24±2.83	a	0.327±0.011	e
5%–5 FPU/g	65.97±0.33	c	0.7405±0.015	e
10%–5 FPU/g	60.88±0.38	d	1.783±0.145	d
1%–10 FPU/g	94.98±0.65	a	0.302±0.008	c
5%–10 FPU/g	73.67±1.67	b	1.259±0.017	b
10%–10 FPU/g	70.46±1.27	b	2.487±0.0103	a

These results can be compared with the information reported by Láinez et al. [59], who from *Agave Salmiana* leaves residues produced bioethanol using the model of a second-generation biorefinery. Through an acid-alkaline pretreatment, where the hemicellulose was removed, and enzymatic hydrolysis with a solid load was 5.62% and 10 FPU/g of glucan, it resulted in 49.04 g/L of glucose released and also a behavior similar to leading in the enzymatic hydrolysis of *Agave Americana* L. The enzyme used during this experiment was Celluclast 1.5L (Novozymes), an enzymatic complex of cellulases and β-glucosidase.

13.6 BIOETHANOL FERMENTATION OF THE RESIDUES

13.6.1 *BIOETHANOL FERMENTATION*

Fermentation is a biological process in which microorganisms, mainly yeasts, and bacteria, convert fermentable monomeric sugars into acids, gases, and ethanol. Saccharomyces cerevisiae yeast (baker's yeast) is the most widely used microorganism in these processes for the production of alcohol, due to its high productivity and performance with different raw materials. The stoichiometric equations characterize the fermentation of hexose ($C_6H_{12}O_6$) and pentose ($C_5H_{10}O_5$) to ethanol with yields close to (51.1%) since CO_2 is an integral part of the reaction [35].

13.6.2 *EXPLORATORY FERMENTATION OF AGAVE AMERICANA L. RESIDUES*

According to the results obtained from the two previous stages, the isothermal hydrothermal pretreatment of 190°C for 50 min, and the enzymatic hydrolysis of 10% solid-loading and 15 FPU/g of glucan, an exploratory fermentation was carried out as to demonstrate the conversion of glucose produced during hydrolysis to ethanol. The pre-saccharification and fermentation strategy (PSSF) experimentation was carried out in 125 mL flasks with a 25 mL working volume. Each flask contained 10% of pretreated solids loading, the enzyme concentration was maintained at 15 FPU/g glucan, 5% (v/v) citrate buffer to maintain the pH of the reaction at 4.8. This flask was placed in an incubator with a shaking of 150 rpm at 50°C for 16 h so that the pre-hydrolysis step was carried out. Subsequently, the temperature of the incubator would be lowered to 32°C; at this temperature, 1.5 mL of suspended yeast *Saccharomyces cerevisiae* PE-2 was added. From this point, samples were taken every 12, 24, 36, and 48 hours to monitor fermentation. Each of the tests was carried out in triplicate [52]. The conversion of glucose to ethanol was measured as described by (Section 13.3.2.1) using HPLC.

13.6.2.1   *SACCHAROMYCES CEREVISIAE PE-2 INOCULUM*

The yeast strain PE-2 was grown during this experiment in a 500 mL flask, with a total working volume of 125 mL. The medium contained 10 g/L yeast extract, 10 g/L peptone, 10 g/L agar, 10 g/L dextrose, and 0.5 g/L of $(NH_4)_2HPO_4$ and $MgSO_4 \cdot 7H_2O$. The glucose was dissolved in an aqueous solution and sterilized separately. The flask with the medium and the yeast was incubated at 30°C, 150 rpm for 12 h. The yeast suspension was made from the pre-inoculum centrifugation at 5,000 rpm at 4°C for 10 min. Finally, it was suspended in a 0.4% NaCl solution [35].

13.6.2.2   *FERMENTATION CONVERSION YIELD*

According to the following equation, the ethanol production yield is calculated [57].

$$Ethanol\ yield\,(\%) = \frac{\left[EtOH_f\right]}{0.51 \times f \times [Biomass] \times 1.111} \times 100 \qquad (6)$$

where; [EtOH] is the final concentration of ethanol (g/L); 0.51 is the theoretical conversion of glucose to ethanol based on stoichiometric biochemistry of yeast; [Biomass] is dry biomass concentration at the beginning of the enzymatic saccharification (g/L); f is the cellulose fraction in dry biomass (g/g); and 1.111 is the conversion factor of cellulose to glucose.

13.6.2.3 PRODUCTION OF BIOETHANOL FROM PRETREATED AND HYDROLYZED BIOMASS OF AGAVE AMERICANA L.

From the operation route outlined during this article, it was possible to produce bioethanol from the biomass produced by the waste of *Agave Americana* L leaf, within the concept of a second-generation biorefinery. The maximum glucose concentration reached in the pre-hydrolysis stage was ≈33.78 g/L at 16 h. Later the yeast was added, and a maximum level of ethanol was reached ≈13.53 g/L after 12 h (Figure 13.5), which is equivalent to an ethanol yield of 43.15±0.48%. A decrease in ethanol concentration can also be observed after 12 hours; this may be due to the consumption of ethanol by microorganisms as a carbon source.

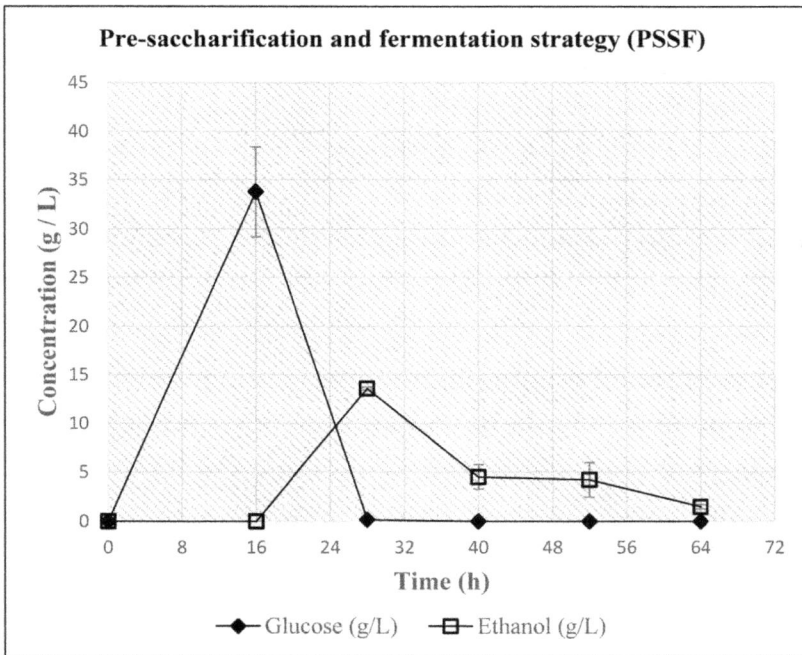

FIGURE 13.5 Conversion of glucose to ethanol from hydrolyzed biomass.

13.7 CONCLUSIONS AND PERSPECTIVES

The search for different sources of energy leads to the development of emerging technologies that solve the environmental problems that the use of fossil fuels implies. From this study, it was observed that the remnants of different processes, such as the use of *Agave americana* L, are an option for the production of biofuels such as bioethanol, revaluing the residues of this economic activity. Also, the use of waste from desert biomass implies generating lignocellulosic materials with less use of water and cares in their production. For this and other reasons, it is necessary to continue in the search for new technological alternatives that reduce the cost and increase the production volumes of fuels such as bioethanol as well as it is also necessary to generate options that integrate sustainable processes such as the use of enzymes and hydroelectric pretreatment. Finally, the use of the residues of *Agave americana* L, reinforces the concept of adaptability of biorefineries and gives rise to a new area of application of resources that avoid competition with food crops, an essential factor in developing countries such as Mexico.

ACKNOWLEDGMENTS

This project was funded by the Secretary of Public Education of Mexico-Mexican Science and Technology Council (SEP-CONACYT) with the Basic Science Project-2015-01 (Ref. 254808). We gratefully acknowledge support for this research by the Energy Sustainability Fund 2014-05 (CONACYT-SENER), Mexican Center for Innovation in Bioenergy (Cemie-Bio), Cluster of Bioalcohols (Ref. 249564). César D. Pinales-Márquez, Shiva, and Rohit Saxena thank the National Council for Science and Technology (CONACYT, Mexico) for theirs Master and PhD Fellowship support, respectively, (grant number: 1001882).

KEYWORDS

- **biofuels**
- **biomass**
- **enzymatic hydrolysis**
- **fermentation**
- **hydrothermal processing**
- **lignocellulosic material**
- **pretreatment**
- **second-generation biorefinery**

REFERENCES

1. Duan, R., Hao, K., & Yang, T., (2020). Air pollution and chronic obstructive pulmonary disease. *Chronic Dis. Transl. Med.* https://doi.org/10.1016/j.cdtm.2020.05.004.
2. World Health Organization. (2020). *Air Pollution.* https://www.who.int/health-topics/air-pollution#tab=tab_1 (accessed on 28 October 2021).
3. Özener, O., & Özkan, M., (2020). Fuel consumption and emission evaluation of a rapid bus transport system at different operating conditions. *Fuel, 265,* 117016. https://doi.org/10.1016/j.fuel.2020.117016.
4. Lozhkin, V., Lozhkina, O., & Dobromirov, V., (2018). A study of air pollution by exhaust gases from cars in well courtyards of Saint Petersburg. *Transp. Res. Procedia, 36,* 453–458. https://doi.org/10.1016/j.trpro.2018.12.124.
5. Gao, Y., Gao, X., & Zhang, X., (2017). The 2°C global temperature target and the evolution of the long-term goal of addressing climate change—from the United Nations framework convention on climate change to the Paris agreement. *Engineering, 3*(2), 272–278. https://doi.org/10.1016/J.ENG.2017.01.022.
6. Carrillo-Nieves, D., Rostro, A. M. J., De La Cruz, Q. R., Ruiz, H. A., Iqbal, H. M. N., & Parra-Saldívar, R., (2019). Current status and future trends of bioethanol production from agro-industrial wastes in Mexico. *Renew. Sustain. Energy Rev., 102,* 63–74. https://doi.org/10.1016/j.rser.2018.11.031.
7. Ruiz, H. A., Thomsen, M. H., & Trajano, H. L., (2017). *Hydrothermal Processing in Biorefineries.* Springer International Publishing, Cham, Switzerland. https://www.springer.com/gp/book/9783319564562 (accessed on 28 October 2021).
8. Kumar, B., Bhardwaj, N., Agrawal, K., Chaturvedi, V., & Verma, P., (2020). Current perspective on pretreatment technologies using lignocellulosic biomass: An emerging biorefinery concept. *Fuel Process. Technol., 199,* 106244. https://doi.org/10.1016/j.fuproc.2019.106244.
9. Ajao, O., Marinova, M., Savadogo, O., & Paris, J., (2018). Hemicellulose based integrated forest biorefineries: Implementation strategies. *Ind. Crops Prod., 126,* 250–260. https://doi.org/10.1016/j.indcrop.2018.10.025.
10. Lara, A., Rodríguez-Jasso, R. M., Loredo-Treviño, A., Aguilar, C. N., Meyer, A. S., & Ruiz, H. A., (2020). Enzymes in the third generation biorefinery for macroalgae biomass. In: *Biomass, Biofuels, Biochemicals* (pp. 363–396). Elsevier. https://doi.org/10.1016/B978-0-12-819820-9.00017-X.
11. Singh, A., Rodríguez, J. R. M., Gonzalez-Gloria, K. D., Rosales, M., Belmares, C. R., Aguilar, C. N., Singhania, R. R., & Ruiz, H. A., (2019). The enzyme biorefinery platform for advanced biofuels production. *Bioresour. Technol. Reports, 7,* 100257. https://doi.org/10.1016/j.biteb.2019.100257.
12. Kupski, L., Telles, A. C., Gonçalves, L. M., Nora, N. S., & Furlong, E. B., (2018). Recovery of functional compounds from lignocellulosic material: An innovative enzymatic approach. *Food Biosci., 22,* 26–31. https://doi.org/10.1016/j.fbio.2018.01.001.
13. Banerjee, S., Patti, A. F., Ranganathan, V., & Arora, A., (2019). Hemicellulose based biorefinery from pineapple peel waste: Xylan extraction and its conversion into xylooligosaccharides. *Food Bioprod. Process, 117,* 38–50. https://doi.org/10.1016/j.fbp.2019.06.012.
14. Haykiri-Acma, H., Yaman, S., & Kucukbayrak, S., (2010). Comparison of the thermal reactivities of isolated lignin and holocellulose during pyrolysis. *Fuel Process. Technol., 91*(7), 759–764. https://doi.org/https://doi.org/10.1016/B978-0-12-818178-2.00005-5.

15. Islam, M. K., Wang, H., Rehman, S., Dong, C., Hsu, H. Y., Lin, C. S. K., & Leu, S. Y., (2020). Sustainability metrics of pretreatment processes in a waste derived lignocellulosic biomass biorefinery. *Bioresour. Technol., 298*, 122558. https://doi.org/10.1016/j.biortech.2019.122558.

16. Singh, R. D., Talekar, S., Muir, J., & Arora, A., (2019). Low degree of polymerization xylooligosaccharides production from almond shell using immobilized nano-bio catalyst. *Enzyme Microb. Technol., 130*, 109368. https://doi.org/10.1016/j.enzmictec.2019.109368.

17. Samanta, A. K., Jayapal, N., Jayaram, C., Roy, S., Kolte, A. P., Senani, S., & Sridhar, M., (2015). Xylooligosaccharides as prebiotics from agricultural by-products: Production and applications. *Bioact. Carbohydrates Diet. Fibre, 5*(1), 62–71. https://doi.org/10.1016/j.bcdf.2014.12.003.

18. Ruiz, H. A., Rodríguez-Jasso, R. M., Fernandes, B. D., Vicente, A. A., & Teixeira, J. A., (2013). Hydrothermal processing, as an alternative for upgrading agriculture residues and marine biomass according to the biorefinery concept: A review. *Renew. Sustain. Energy Rev., 21*, 35–51. https://doi.org/10.1016/j.rser.2012.11.069.

19. Galanopoulos, C., Giuliano, A., Barletta, D., & Zondervan, E., (2020). An integrated methodology for the economic and environmental assessment of a biorefinery supply chain. *Chem. Eng. Res. Des., 160*, 199–215. https://doi.org/10.1016/j.cherd.2020.05.016.

20. Ubando, A. T., Felix, C. B., & Chen, W. H., (2020). Biorefineries in circular bioeconomy: A comprehensive review. *Bioresour. Technol., 299*. https://doi.org/10.1016/j.biortech.2019.122585.

21. Ruiz, H. A., Vicente, A. A., & Teixeira, J. A., (2012). Kinetic modeling of enzymatic saccharification using wheat straw pretreated under autohydrolysis and organosolv process. *Ind. Crops Prod., 36*, 100–107. https://www.sciencedirect.com/science/article/pii/S0926669011003682 (accessed on 28 October 2021).

22. Correa, D. F., Beyer, H. L., Possingham, H. P., Thomas-Hall, S. R., & Schenk, P. M., (2017). Biodiversity impacts of bioenergy production: Microalgae vs. first-generation biofuels. *Renew. Sustain. Energy Rev., 74*, 1131–1146. https://doi.org/10.1016/j.rser.2017.02.068.

23. Renewable Fuels Association. https://ethanolrfa.org/markets-and-statistics/view-all-markets-and-statistics (accessed on 28 October 2021).

24. Hassan, S. S., Williams, G. A., & Jaiswal, A. K., (2019). Moving towards the Second generation of lignocellulosic biorefineries in the EU: Drivers, challenges, and opportunities. *Renew. Sustain. Energy Rev., 101*, 590–599. https://doi.org/10.1016/j.rser.2018.11.041.

25. Ruiz, H. A., Martínez, A., & Vermerris, W., (2016). Bioenergy potential, energy crops, and biofuel production in Mexico. *BioEnergy Res., 9*(4), 981–984. https://doi.org/10.1007/s12155-016-9802-7.

26. Secretary of Energy (SENER). (2020). *National Energy Balance: Total Final Energy Consumption by Fuel.* https://sie.energia.gob.mx/bdiController.do?action=cuadro&cvecua=IE7C01 (accessed on 28 October 2021).

27. Honorato-Salazar, J. A., & Sadhukhan, J., (2020). Annual biomass variation of agriculture crops and forestry residues, and seasonality of crop residues for energy production in Mexico. *Food Bioprod. Process, 119*, 1–19. https://doi.org/10.1016/j.fbp.2019.10.005.

28. Ponce-Ortega, J. M., & Santibañez-Aguilar, J. E., (2019). Optimization of the supply chain associated to the production of bioethanol from residues of agave from the tequila process in Mexico. In: *Strategic Planning for the Sustainable Production of Biofuels* (pp. 115–146). Elsevier. https://doi.org/10.1016/B978-0-12-818178-2.00005-5.

29. Mielenz, J. R., (2001). Ethanol production from biomass: Technology and commercialization status. *Curr. Opin. Microbiol., 4*(3), 324–329. https://doi.org/10.1016/S1369-5274 (00)00211-3.

30. Perez-Pimienta, J. A., Flores-Gómez, C. A., Ruiz, H. A., Sathitsuksanoh, N., Balan, V., da Costa, S. L., Dale, B. E., Singh, S., & Simmons, B. A., (2016). Evaluation of agave bagasse recalcitrance using AFEXTM, autohydrolysis, and ionic liquid pretreatments. *Bioresour. Technol., 211*, 216–223.

31. Palomo-Briones, R., López-Gutiérrez, I., Islas-Lugo, F., Galindo-Hernández, K. L., Munguía-Aguilar, D., Rincón-Pérez, J. A., Cortés-Carmona, M. Á., et al., (2018). Agave bagasse biorefinery: Processing and perspectives. *Clean Technol. Environ. Policy, 20*(7), 1423–1441. https://doi.org/10.1007/s10098-017-1421-2.

32. Parra, N. L. A., Del, V. Q. P., & Prieto, R. A., (2010). Extraction of agave fibers to make paper and crafts. *Acta Univ., 20*, 77–83. https://doi.org/10.15174/au.2010.63.

33. Valdez-Vazquez, I., Alatriste-Mondragón, F., Arreola-Vargas, J., Buitrón, G., Carrillo-Reyes, J., León-Becerril, E., Mendez-Acosta, H. O., et al., (2020). A comparison of biological, enzymatic, chemical and hydrothermal pretreatments for producing biomethane from agave bagasse. *Ind. Crops Prod., 145*, 112160. https://doi.org/10.1016/j.indcrop.2020.112160.

34. Ganduri, L., Van, D. M. A. F., & Matope, S., (2015). Economic model for the production of spirit, inulin and syrup from the locally eco-friendly agave americana. *Procedia CIRP, 28*, 173–178. https://doi.org/10.1016/j.procir.2015.04.030.

35. Pereira, F. B., Romaní, A., Ruiz, H. A., Teixeira, J. A., & Domingues, L., (2014). Industrial robust yeast isolates with great potential for fermentation of lignocellulosic biomass. *Bioresour. Technol., 161*, 192–199. https://doi.org/10.1016/j.biortech.2014.03.043.

36. Pino, M. S., Rodríguez-Jasso, R. M., Michelin, M., & Ruiz, H. A., (2019). Enhancement and modeling of enzymatic hydrolysis on cellulose from agave bagasse hydrothermally pretreated in a horizontal bioreactor. *Carbohydr. Polym., 211*, 349–359. https://doi.org/10.1016/j.carbpol.2019.01.111.

37. Devi, R., & Sit, N., (2019). Effect of single and dual steps annealing in combination with hydroxypropylation on physicochemical, functional and rheological properties of barley starch. *Int. J. Biol. Macromol., 129*, 1006–1014. https://doi.org/10.1016/j.ijbiomac.2019.02.104.

38. Rijal, D., Vancov, T., McIntosh, S., Ashwath, N., & Stanley, G. A., (2016). Process options for conversion of *Agave tequilana* leaves into bioethanol. *Ind. Crops Prod., 84*, 263–272. https://doi.org/10.1016/j.indcrop.2016.02.011.

39. Corbin, K. R., Byrt, C. S., Bauer, S., DeBolt, S., Chambers, D., Holtum, J. A. M., Karem, G., et al., (2015). Prospecting for energy-rich renewable raw materials: Agave leaf case study. *PLoS One, 10*(8), e0135382. https://doi.org/10.1371/journal.pone.0135382.

40. Sindhu, R., Binod, P., & Pandey, A., (2016). Biological pretreatment of lignocellulosic biomass: An overview. *Bioresour. Technol., 199*, 76–82. https://doi.org/10.1016/j.biortech.2015.08.030.

41. Chen, H., Liu, J., Chang, X., Chen, D., Xue, Y., Liu, P., Lin, H., & Han, S., (2017). A review on the pretreatment of lignocellulose for high-value chemicals. *Fuel Process. Technol., 160*, 196–206. https://doi.org/10.1016/j.fuproc.2016.12.007.

42. Lara-Flores, A. A., Araújo, R. G., Rodríguez-Jasso, R. M., Aguedo, M., Aguilar, C. N., Trajano, H. L., & Ruiz, H. A., (2018). *Bioeconomy and Biorefinery: Valorization of*

Hemicellulose from Lignocellulosic Biomass and Potential Use of Avocado Residues as a Promising Resource of Bioproducts, pp. 141–170. https://doi.org/10.1039/D0RA04031B.

43. Pino, M. S., Rodríguez-Jasso, R. M., Michelin, M., Flores-Gallegos, A. C., Morales-Rodríguez, R., Teixeira, J. A., & Ruiz, H. A., (2018). Bioreactor design for enzymatic hydrolysis of biomass under the biorefinery concept. *Chem. Eng. J., 347*, 119–136. https://www.sciencedirect.com/science/article/pii/S1385894718306235 (accessed on 28 October 2021).

44. Ruiz, H. A., Conrad, M., Sun, S. N., Sanchez, A., Rocha, G. J. M., Romaní, A., Castro, E., et al., (2020). Engineering aspects of hydrothermal pretreatment: From batch to continuous operation, scale-up and pilot reactor under biorefinery concept. *Bioresour. Technol., 299*, 122685. https://doi.org/10.1016/j.biortech.2019.122685.

45. Fujimoto, S., Inoue, S., & Yoshida, M., (2018). High solid concentrations during the hydrothermal pretreatment of eucalyptus accelerate hemicellulose decomposition and subsequent enzymatic glucose production. *Bioresour. Technol. Reports, 4*, 16–20. https://doi.org/10.1016/j.biteb.2018.09.006.

46. Liu, W., Wu, R., Hu, Y., Ren, Q., Hou, Q., & Ni, Y., (2020). Improving enzymatic hydrolysis of mechanically refined poplar branches with assistance of hydrothermal and Fenton pretreatment. *Bioresour. Technol., 316*, 123920. https://doi.org/10.1016/j.biortech.2020.123920.

47. Ge, J., Wu, Y., Han, Y., Qin, C., Nie, S., Liu, S., Wang, S., & Yao, S., (2020). Effect of hydrothermal pretreatment on the demineralization and thermal degradation behavior of eucalyptus. *Bioresour. Technol., 307*, 123246. https://doi.org/10.1016/j.biortech.2020.123246.

48. Mariano, A. P. B., Unpaprom, Y., & Ramaraj, R., (2020). Hydrothermal pretreatment and acid hydrolysis of coconut pulp residue for fermentable sugar production. *Food Bioprod. Process, 122*, 31–40. https://doi.org/10.1016/j.fbp.2020.04.003.

49. Overend, R. P., Chornet, E., & Gascoigne, J. A., (1987). Fractionation of lignocellulosics by steam-aqueous pretreatments. *Philos. Trans. R. Soc. London. Ser. A, Math. Phys. Sci., 321*(1561), 523–536. https://doi.org/10.1098/rsta.1987.0029.

50. Ruiz, H. A., Cerqueira, M. A., Silva, H. D., Rodríguez-Jasso, R. M., Vicente, A. A., & Teixeira, J. A., (2013). Biorefinery valorization of autohydrolysis wheat straw hemicellulose to be applied in a polymer-blend film. *Carbohydr. Polym., 92*(2), 2154–2162. https://doi.org/10.1016/j.carbpol.2012.11.054.

51. Ruiz, H. A., Silva, D. P., Ruzene, D. S., Lima, L. F., Vicente, A. A., & Teixeira, J. A., (2012). Bioethanol production from hydrothermal pretreated wheat straw by a flocculating *Saccharomyces Cerevisiae* strain - effect of process conditions. *Fuel, 95*, 528–536. https://doi.org/10.1016/j.fuel.2011.10.060.

52. Aguilar, D. L., Rodríguez-Jasso, R. M., Zanuso, E., De Rodríguez, D. J., Amaya-Delgado, L., Sanchez, A., & Ruiz, H. A., (2018). Scale-up and evaluation of hydrothermal pretreatment in isothermal and non-isothermal regimen for bioethanol production using agave bagasse. *Bioresour. Technol., 263*, 112–119. https://doi.org/10.1016/j.biortech.2018.04.100.

53. Aguirre-Fierro, A., Ruiz, H. A., Cerqueira, M. A., Ramos-González, R., Rodríguez-Jasso, R. M., Marques, S., & Lukasik, R. M., (2020). Sustainable approach of high-pressure agave bagasse pretreatment for ethanol production. *Renew. Energy., 155*, 1347–1354. https://www.sciencedirect.com/science/article/pii/S0960148120305875 (accessed on 28 October 2021).

54. Pellis, A., Cantone, S., Ebert, C., & Gardossi, L., (2018). Evolving biocatalysis to meet bioeconomy challenges and opportunities. *N. Biotechnol., 40*, 154–169. https://doi.org/10.1016/j.nbt.2017.07.005.

55. Champreda, V., Mhuantong, W., Lekakarn, H., Bunterngsook, B., Kanokratana, P., Zhao, X. Q., Zhang, F., et al., (2019). Designing cellulolytic enzyme systems for biorefinery: From nature to application. *J. Biosci. Bioeng., 128*(6), 637–654. https://doi.org/10.1016/j.jbiosc.2019.05.007.

56. Adney, B., & Nrel, J. B., (2008). *Measurement of Cellulase Activities.* NERL/TP-510–42628. National Renewable Energy Laboratory, Golden, CO.

57. Dowe, N., & Mcmillan, J., (2008). *SSF Experimental Protocols--Lignocellulosic Biomass Hydrolysis and Fermentation: Laboratory Analytical Procedure (LAP).* Issue Date: 10/30/2001.

58. Rodrigues, A. C., Haven, M. Ø., Lindedam, J., Felby, C., & Gama, M., (2015). Celluclast and cellic® CTec2: Saccharification/fermentation of wheat straw, solid-liquid partition and potential of enzyme recycling by alkaline washing. *Enzyme Microb. Technol., 79, 80*, 70–77. https://doi.org/10.1016/j.enzmictec.2015.06.019.

59. Láinez, M., Ruiz, H. A., Arellano-Plaza, M., & Martínez-Hernández, S., (2019). Bioethanol production from enzymatic hydrolysates of *Agave salmiana* leaves comparing *S. cerevisiae* and *K. Marxianus. Renew. Energy, 138*, 1127–1133. https://doi.org/10.1016/j.renene. 2019.02.058.

CHAPTER 14

Sustainable Ethanol Production from Lignocellulosic Biomass: A Water Footprint Analysis over Pretreatment Technologies

OZNUR YILDIRIM, DOGUKAN TUNAY, BESTAMI OZKAYA, and
AHMET DEMIR

*Department of Environmental Engineering, Yildiz Technical University,
Turkey, Davutpasa Campus, Istanbul–34220, Turkey*

ABSTRACT

Lignocellulosic biomass (LCB) has a complex body consisting of hetero-polymers such as lignin, cellulose, and hemicellulose with high tolerancse to biodegradation, which is abundant and cheap in nature, and significantly reduces CO_2 emissions in the atmosphere when used in ethanol production. With the increase of global warming (GW), the concept of lignocellulosic biorefinery has become a critical focus. There are many pretreatment methods which have been developed to disrupt this structure even they are durable to biodegradation. The most important parameters affecting the sustainability and economy of the biorefinery industry are the amount of water and chemical consumption. The pretreatment process is the most critical step in the production of valuable products from LCB in terms of water and chemical consumption. This chapter aims to contribute to reducing costs during the production of high value-added products by examining all available pretreatment technologies in terms of water, chemical, and energy consumption.

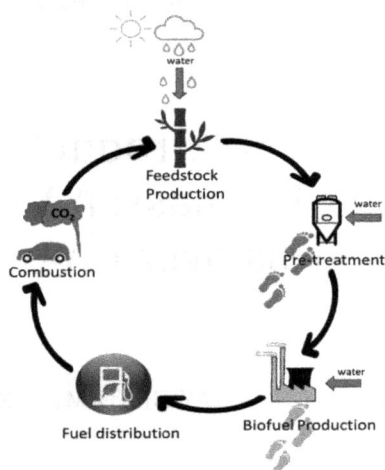

Graphical abstract

14.1 INTRODUCTION

In recent years, ethanol production from LCB has gained great momentum. LCB mostly consists of agricultural wastes and forestry residues [1]. Energy production from LCB is more significant in the economic, social, and environmental aspects for all over the world. Agricultural wastes are substantial focus area for renewable energy production because of containing high amounts of carbohydrates. Special pretreatment applications should be applied to make LCB open to biological activities by breaking the rigid structure of the lignocellulosic matrix [2, 3].

Cellulose, a kind of polymer, found abundantly in nature, formed as a result of glucose monomers combined by glucosidic bonds [4]. Cellulose polymers of different lengths and structures bond to each other with weak hydrogen bonds to form amorphous or crystalline cellulose fibrils [5]. As the number of cellulose chains formed increases, the strength of the cellulose increases, and an indigestible crystal structure is formed, which makes it arduous to break down. Similarly, as the Glucose number of cellulose increases, it is assumed that the DP increases which is proportional to the glucose number [6]. The most significant limiting factor affecting the efficiency of hydrolysis is the cellulose crystallinity. The amorphous cellulose chains can decompose faster than those with a crystalline structure [7]. Hemicellulose, which is formed by the combination of different compounds such as pentoses, hexoses, and uronic acids, attaches to cellulose fibers and binds cellulose to lignin and gives the biomass hardness [8]. This situation also prevents enzymes from accessing cellulose during enzymatic hydrolysis. The use of some chemicals during the pretreatment of hemicelluloses can cause the formation of inhibitors such as furfurals and hydroxymethylfurfural, which may adversely affect microorganisms during fermentation [9–11]. Lignin, a heteropolysaccharide consisting of three separate phenyl propane units and having an irregular structure, enables plants to be resistant, impermeable, and resistant to microbial attacks [12]. There is no order between the units in the lignin structure formed by phenyl propane units and ether bonds and carbon-carbon bonds [13].

With the use of lignocellulosic ethanol in the transportation sector, the net CO_2 emissions released into the atmosphere are decreasing [14]. However, it is no longer sufficient to produce the maximum amount of ethanol by increasing lignocellulosic ethanol production efficiency with different pretreatment processes. Due to the rapid reduction of water resources, water usage during the process and its reduction have become one of the most important parameters. It is known that most water usage in the lignocellulose-based ethanol production occurs during the pretreatment stage [15]. Water usage should be reduced in the pretreatment stage to make this process more sustainable. It is anticipated that if the water resources in the world continue to be consumed at the current rate, it will be able to manage for about 30 more years. Water demand is expected to increase by 40% by 2050, and 25% of people will not have adequate access to freshwater [16].

For the production of ethanol, water is essential, especially in the grinding, liquefaction, and fermentation processes. In recent years, water consumption per gallon of ethanol has decreased significantly. According to the plants that were operated in 2002, older plants consuming much more water. For instance, water usage of older plants were more than the 15-Gal water/Gal etOH. However, it was declined from 1 gallon to 11 gallons and on average 4.7 gallons for production of 1 gallon of ethanol from the 2002 data (Figure 14.1) [17].

There are two types of fermentation processes that is used in the ethanol industry, which are continuous and batch. 27% of ethanol was produced with the continuous fermentation process, and the remaining was produced with the batch process according to the 2002 survey of ethanol plants. When the ethanol plants were analyzed due to their process selection, it can be said that the continuous fermentation system was more preferred in large plants. While 4 of these large plants were operated continuous, 17 out of 21 plants were operated with batch system [17].

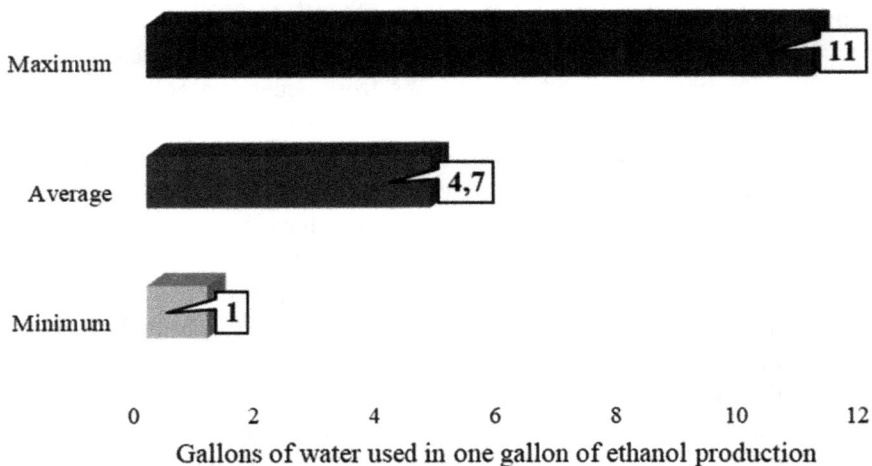

FIGURE 14.1   Gallons of water usage per gallon of ethanol produced for the 21 ethanol plants in 2002 [17].

Although lignocellulosic ethanol production is important in terms of slowing the effects of GW by reducing CO_2 emissions, if a large amount of fresh water is used during the process, a sustainable fuel production cannot be obtained when the whole process is examined. Therefore, it is increasingly important to conduct water footprint studies to compare the water consumption data of different pretreatment processes. This section details the water footprint data of different pretreatment types in detail and introduces more sustainable methods.

14.2   WATER FOOTPRINT

Rise in water consumption due to population increase and industrialization have a significant effect on the environment and ecological life. Even though almost 70% of the Earth's surface is covered by the water, only 2.5% of this water is fresh and can be directly be used [18]. It is an undeniable fact that these limited water resources will be run out one day in the future unless precautions are not taken. For this reason, water consumption should be reduced and the water cycle should be maintained by recycling the used water. Water footprint is a measurement of water required for the production of any tools, products, goods, or services that takes a place for daily life. There are three main components of water footprint which are called blue, green, and gray water footprints. Green water footprint can be explained as the water that is deposited, transformed, evaporated by plants, or consumption of water from precipitation that is held by the roots of the soil. In short, everything is incorporated with plant, agriculture, and forestry taking charge of water cycle called the green water footprint. On the other hand, surface and groundwater sources that are used for domestic, industrial, and irrigation water for agricultural purposes are related to the blue water footprint. Bluewater consumption estimations need detailed analyzes and study since there could be many reasons for the water usage directly or indirectly. For instance, there is a study regarding the water footprint for coffee and tea consumption in the Netherlands. If the virtual and real water consumptions were compared, there will be a huge difference. For example, while the real water consumptions were 0.125 L and 0.250 L water for the coffee and tea, respectively. On the other hand, virtual water consumptions rise to approximately 140 L and 34 L for coffee and tea [19]. Greywater footprint is the other component that is considered the freshwater usage for the assimilation of pollutants according to the water quality standards. It is considered as a point source pollution discharge through a pipeline, runoff, or impervious surfaces.

14.2.1 BLUEWATER FOOTPRINT FOR BIOREFINERY PURPOSES FOR LIGNOCELLULOSIC BIOMASS (LCB)

Industrialization, energy need, and increasing living standards had significantly affected to studies related with the biorefinery in last decades [20–22]. However, water scarcity is another vital issue that should be considered because of the limited freshwater sources. It is estimated from a national research council (NRC) that water consumption for the ethanol production from corn kernels and cellulosic feedstocks are 4- and 9.5-Gal water/Gal etOH [23]. Based on these predictions, it can be said that approximately 256 billion gallons of fresh water are needed in a year for the required 36 billion gallons of ethanol in 2022 [24]. For the production of biofuels from woody biomass, there are different approaches. The first approach supports the production of the crops for direct fuel production such as (biogas, bioethanol, biohydrogen, biochar, etc.) [25, 26]. In the first approach, the problem is the usage of feedstocks that is already used as a food for the bioethanol production. The use of agricultural land for fuel production can also be of concern for the future. Therefore, the requirement of fuel production should be carefully analyzed compared to other demands. In the second discernment, the biofuel production can be carried out depending on the usage of a lignocellulosic waste or biomass that is not to be used as a food or food products. Nevertheless, these feedstocks could need more water to break down lignin, and biofuel production yields can be dropped down [6, 27].

Process water which is used in the pretreatment step is evaluated as blue water. Besides, LCB holds some amount of water in their fabrics which is interpreted as green water. Green water can decrease the water usage for the preparation operations; however, it is also part of a water cycle and it should be considered in terms of process water. Figure 14.2 shows the components of the water footprint during biofuel production from LCB. Chiu et al. [8] investigated that water consumption according to bioethanol production stage and irrigation need for different locations of the United States. Bioethanol production consumes 0.5–28% of the total water which includes process water and irrigation water. It shows that growing crops for bioethanol production has a much higher water footprint due to the lignocellulosic waste or uncontrolled growing herbs or plants usage.

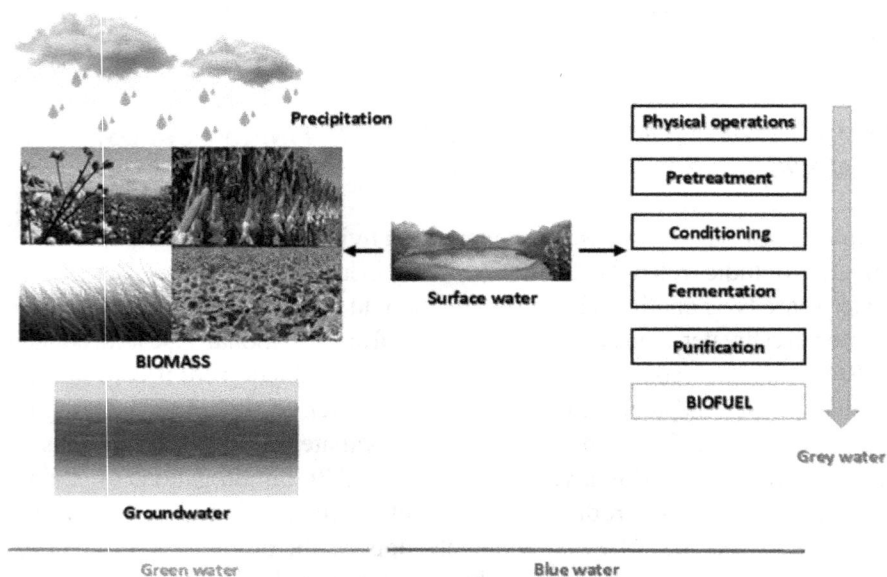

FIGURE 14.2 The components of water footprint during biofuel production from lignocellulosic biomass.

Irrigation should be considered as green water is considered as a kind of green water source in the water that plants hold in themselves, even if the water used directly from freshwater sources (process water) is gained, it is included in the current water consumption within the water cycle. The water used in the pretreatment process is used as process water for fuel production and is therefore considered blue water. Therefore, in this section, studies on the amount of water consumed during the pretreatment, in other words, the determination of the blue water footprint for pretreatment will be examined in current studies.

14.3 PRETREATMENT METHODS

The carbohydrate polymers in lignocellulose are tightly bonded to lignin with hydrogen bonds and covalent bonds, creating a complex and crystalline structure that prevents the carbohydrate in the raw material from being used by microorganisms and enzymes [28]. The object of the pretreatment is to decrease the crystallinity of lignocellulose, break their resistance to enzymatic depolymerization, and disrupt the heterogeneous structure of lignocellulosic materials. For obtaining valuable products from lignocellulosic materials, their complex structure must be disrupted and broken down into smaller carbohydrate components. Pretreatment types are basically classified as physical, chemical, physicochemical, and biological.

For a conventional pretreatment process to be efficient; It would be expected to increase the accessible cellulose surface area, break the lignin barrier and cellulose crystallinity, prevent the formation of toxic by-products that will cause inhibition in the fermentation stages, to reduce the loss of reducing sugar resulting from the process and to be economically efficient [29]. However, new issues to be considered in addition to these items have become a current issue. Now, in addition to the requirements such as high sugar conversion rate of the pretreatment process, being economical, and providing low inhibitor production, lower water usage has become one of the most important parameters. Due to the depletion of water resources and GW, efforts to develop new methods that reduce water consumption during the process and wastewater production after the process to make environmentally friendly waste-to-fuel production processes such as biofuel production from lignocellulosic wastes more sustainable.

The literature includes different approaches for pretreatment. Several methods of pretreatment were tested by applying them to different lignocellulosic wastes. In this chapter, studies that mostly conduct water footprint studies and indicate how much water is consumed per waste are included.

14.4   WATER CONSUMPTION IN PHYSICAL PRETREATMENT

In physical pretreatment, the crystallinity of lignocellulosic material is broken, so the surface area accessible by microorganisms is increased by reducing the size of wastes by cutting, grinding, and milling [30]. Methods such as mechanical extrusion, grinding, and microwave are in the physical pretreatment category. In addition, there are methods such as ultrasound and pulsed electric field, which have recently been applied.

In milling and chipping methods, which are the most preferred types of physical treatment, size reduction is performed with the help of grinders and cutting tools without any water usage. However, it may be necessary to use water in some methods such as microwave. One of the physical processes that are used for biofuel production from lignocellulosic feedstocks is the cooling necessity. Wu and Sawyer [31] found that 65% of the total process of water consumption for renewable diesel blendstock (RDB) is caused by cooling. Water necessity for the pretreatment and enzymatic hydrolysis were found 0.16 L water/L RDB and 0.18 L water/L RDB, respectively.

14.5   WATER CONSUMPTION IN CHEMICAL PRE-TREATMENT

Chemical pretreatment includes acid, alkaline, ozone, and ionic liquid methods. At the end of the chemical pretreatment process, the hemicellulose, cellulose, and lignin matrix split from each other and becomes explicit to the attacks of enzymes [32]. Large cellulose and hemicellulose chains are divided into shorter pieces due to chemical pretreatment after lignin hemicellulose and cellulose are segregated from each other and enzymes can act. However, the use of chemicals in pretreatment processes is an important disadvantage in terms of the cost-effectiveness. Furthermore, usage of chemicals also increases water consumption during application. The most preferred methods are acid and alkaline pretreatment. Acid and alkaline applications are examined under two subtitles as concentrated and diluted [33]. Dilute acid and alkaline applications are more preferred because the use of concentrated acid causes more chemical consumption. However, a considerable amount of clean water is spent even to purify only the pretreated wastes from acid or base residue [34]. After pretreatment with acid or base, some inhibitory by-products can be formed due to harsh pretreatment conditions. If these inhibitors remain on the pre-treated fermentation raw material, it may damage the fermentation microorganisms. In addition, as a result of pretreatment with acid and base, the washing is a significant step to bring the pH to a neutral level. Approximately 10 L of water is consumed during the 1 L of ethanol production process [35].

In a study [36], a greenhouse gas (GHG) and water use inventory were created in case of use of cholinium lysate ([Ch] [Lys]) chemical known as ionic liquids (ILs) in the lignocellulosic biofuel production process. They compared two different methods, the traditional water washing (WW) method and the integrated high gravity (iHG) method, which does not require water use. As a result of the study, they observed that pretreatment of corn stover using [Ch] [Lys] resulted in inadmissible high GHG in the WW process, while the iGH process could reduce GHG by 45% as per gasoline. When compared in terms of water consumption, WW and iHG processes have been observed to have almost similar water consumption (about 0.30–0.40 L/MJ). The reason the water usage values are similar for the WW process and the iHG process is that all the water used in the WW stage is eventually recovered and recycled through highly efficient evaporative dehydration process.

According to the National Renewable Energy Laboratory Technical Report, Aden et al. [37] investigated the ethanol process design and economical approaches for corn stover biomass. Water balances were derived from the ASPEN simulations. It was found that 0.252 kg ethanol can be produced from 1 kg of corn stover with the 0.224 kg of water consumption for the pre-hydrolysis (0.113 kg of water/kg ethanol produced) and saccharification (0.111 kg of water/kg ethanol produced). According to ASPEN simulations, it was proven the pretreatment applied during ethanol generation from the corn stove affects the water usage by 2.62%.

14.6   WATER CONSUMPTION IN PHYSICOCHEMICAL PRE-TREATMENT

The processes where physical and chemical pretreatment processes are used together are defined as a physicochemical pretreatment method [2]. In the methods applied in this type of treatment, treatment is done in harsh conditions such as high temperature and pressure. LHW, wet oxidation, ammonia fiber explosion (AFEX), and CO_2 explosion are the most used techniques. As the name suggests, methods such as LHW and wet oxidation are methods that cause too much water consumption. In terms of process efficiency, they have less sugar conversion efficiency compared to acid and base applications under high temperature and pressure.

Dong et al. [38], produced ethanol using sulfite pretreated Monterey pine and completed the water and energy footprint analysis of the whole process. As a new pretreatment in the study, the process of breaking down oven-dried (OD) wood chips with sulfuric acid and sodium bisulfite was applied in a rotary electric heated digester. They proved that the newly pretreatment process they performed when they examined the water footprint of the process provides 3.65 tons/ton of dry biomass water consumption and this value provides 25–51% less water consumption than pretreatment processes such as steam explosion (SE). They also claimed in their report that the highest water consumption of the SE, diluted acid, and organosolv pretreatment types where organosolv > dilute acid > steam explosion, respectively. Organosolv pretreatment leads not only to the highest water consumption, but also to energy consumption. Pretreatment with dilute acid consumes less energy than steam putting process. It is also stated that the acid pretreatment process mostly results in higher sugar conversion efficiency than SE [39]. In this case, if both the efficiencies of energy, water consumption, and pretreatment are compared, the most preferred pretreatment form, dilute acid, tends to be more advantageous.

In another study [40], while investigating the effect of reactor filling and solid-liquid ratio on ethanol yield during the production of ethanol from corn stover, water, and steam consumption amounts were investigated. At the end of the process, it was observed that the steam consumption was significantly reduced and no acidic waste containing acid was produced. in pretreatment conditions where the highest sugar content and minimum water consumption occur; 190°C; 3.00% sulfuric acid; 3 min, water, and steam consumption, respectively, 97.3 g water and 44 g steam per g dry waste were used.

Pan et al. [41], after dilute acid pretreatment for corn stover based ethanol production, water footprint analysis of various conditioning processes were carried out. These methods were; overliming, ammonia addition, two-stage treatment, and membrane separations. In overliming with conditioning application, 4.94 kg/kg dry biomass direct process water was used in total and approximately 35% of this was used during pretreatment. In conditioning with ammonia and two-stage conditioning application, a total of 4.58 kg and 4.25 kg of water were used respectively, and the most water was used in the pH adjustment stage. In the air conditioning application with membrane, the total water usage (2.54 kg) has the lowest value compared to other processes. The results showed that the amount of steam and water used in the dilute acid pretreatment stage ranged from 14% to 74% for the four methods applied. The 4.94 L of water per kg-DB was used in conventional overliming. With membrane separations of approximately 2.54 L per kg-DB, water use was found to be the lowest application, and almost half the use of water compared to overliming. Also, a considerable amount of water was used for the hydrolyzate neutralization. Membrane applications as a conditioning method could be offered as a solution to accomplish water-efficient both dilute acid removal and pH adjustment. In addition to the blue WF, which is defined as direct/indirect water use in the study, gray WF was determined by considering process discharge in all scenarios. Among the four different treatment processes, both blue and gray WF values were the lowest application membrane separations.

14.7 WATER CONSUMPTION IN BIOLOGICAL PRE-TREATMENT

Biological pretreatment, also known as microbial pretreatment, is considered as the most environmentally friendly method since it is performed without using chemicals. Microorganisms and rotting fungi are used in biological pretreatment. Microorganisms often secrete extracellular enzymes to disrupt lignin and hemicellulose. Enzymes from different species such as Aspergillus and Streptomyces are used for biodegradation of lignocellulosic materials [42]. The degradation of lignin by fungi occurs by the effect of lignin-destructive enzymes, such as peroxidases and laccase. Rotting fungi are examined in two groups as white and brown. While brown rot fungi can only degrade cellulose, white-rot fungi can break down both cellulose and lignin [43]. Biological pretreatment is suitable for lignocellulosic material pretreatment due to its low energy usage and the absence of toxic by-product production as in chemical pretreatment. However, biological pretreatment has also some drawbacks such as consuming some of the raw material and long process time.

In a study [44], a new procedure was developed to increase hydrolysis efficiency and reduce the water footprint. In the newly developed method, both solvent and water are used to separate the confectionery and lignocellulose components. By adding trace amounts of inorganic acid, lignin is dissolved and hemicellulose is hydrolyzed under the appropriate reaction temperature and pressure. In the conventional method, it is known to wash with plenty of water to remove the solid phase solvent after pretreatment. In the newly developed method, after hydrolysis with acid and solvent, drying is carried out with a relatively low-temperature air suction using the relative volatility difference between solvent and water to remove solvents in the solid phase.

Thus, while achieving high efficiency in enzymatic hydrolysis, water consumption and wastewater generation are reduced. As a result of the pretreatment proposed by the present invention, the lignin removal rate can be increased to 90% and the enzymatic hydrolysis efficiency of the cellulose solid residue to 80–90%. In addition, the amount of water used is saved about 10 to 20 times the weight of the cellulose solid residue, which significantly reduces the water used in production, resulting in an important decrease in the water footprint. As in the mentioned study, the original efficiency of the process is preserved and even increased, while water consumption can be reduced by new methods (Table 14.1).

As a result, there are many pretreatment methods used for lignocellulose-based ethanol production. The sugar conversion efficiency of the pretreatment method used varies according to the type, content, location, and harvesting method. In order to figure out which pretreatment process would result in maximum efficiency, different parameters such as the amount of water consumption and output in the production of ethanol need to be examined. Table 14.1 shows the pretreatment type applied, the LCB used, the amount of water needed for ethanol production, and the ethanol production efficiency of some studies.

14.8 CONCLUSION

Traditional ethanol production, which accelerated because of depletion of fossil fuels and the damage it caused to the environment, became uncontrollable, and was replaced by lignocellulosic ethanol production as a result of the anticipation of the depletion of food sources due to the increase in the world population. In order to make the process more sustainable, it is of great importance to conduct studies to examine and reduce the amount of water consumed during the process. Using LCB as a raw material, the amount of water consumed in biofuel production is larger (10 L water/L ethanol) compared to the use of corn as a raw material [59]. When thermochemically biofuel production and crude oil production processes are evaluated, it is seen that water consumption is 2 L/L and 2.5 L/L, respectively [60]. From a different perspective, the incineration and landfilling of agricultural wastes are not accepted in the circular economy. Ethanol production from lignocellulosic wastes not only provides sustainable management of agricultural wastes but also reduces the amount of net CO_2 emission released into the atmosphere compared to gasoline. As a result, lignocellulosic ethanol production draws attention as an environmentally friendly method in

TABLE 14.1 Water Demands for the Process of Ethanol Production from Different Lignocellulosic Biomass

Pretreatment Methods		Biomass	Water Demand (m³ Water/t Lignocellulosic Biomass)	Water/t Ethanol Yield (t/t Dry Matter)	References
Physical Pretreatment	Liquid hot water	N/A	8.55	0.174	[45, 46]
	Liquid hot water	Palm-oil residues	2.62–7.62	0.09–0.136	[47, 48]
	Steam explosion	Corn stover	1.5	N/A	[49]
Chemical Pre-Treatment	Diluted acid (0.5–2%)	N/A	4.76	0.216	[45, 46]
	SO$_2$ stem explosion	N/A	4.80	–	[45, 46]
	Acid	Olive tree biomass	2.13–2.42	0.097–0.11	[47, 50]
	Lime	N/A	11.39	0.202	[45, 46]
	Ozonolysis	Oil palm	0.5	N/A	[51, 52]
	Ionic liquid	Palm biomass	15	N/A	[51, 53]
	Mild NaOH	Raw cogon grass	46	0.134–0.174	[47, 54]
Physicochemical Pretreatment	Ammonia fiber explosion	N/A	2	0.220	[45, 46]
	Ammonia fiber explosion	Corn stover	1	N/A	[51, 55]
	Soaking in aqueous ammonia (SAA)	N/A	5.53	0.148	[45, 46]
	Biochemical conversion (diluted acid 1.1% + enzymatic hydrolysis)	Corn stover	8.55	0.252	[56]
Biological Pretreatment	*Irpex lacteus CD2*	Corn stalks	2.5	N/A	[57]
	Ganoderma boninense	Water hyacinth	1.43	N/A	[58]

all aspects. By developing new methods that will reduce the high freshwater consumption, which is one of the handicaps of this process, the process will be made more preferred. More novel technology development and techno-economic analysis of these methods are required for lignocellulosic ethanol production.

KEYWORDS

- **biorefinery**
- **ethanol**
- **ionic liquids**
- **lignocellulosic biomass**
- **pretreatment**
- **water footprint**

REFERENCES

1. Anwar, Z., Gulfraz, M., & Irshad, M., (2014). Agro-industrial lignocellulosic biomass a key to unlock the future bio-energy: A brief review. *J. Radiat. Res. Appl. Sci., 7*(2), 163–173.

2. Batista, M. D., De Oca-Vásquez, G., Vega-Baudrit, J. R., Rojas-Álvarez, M., Corrales-Castillo, J., & Murillo-Araya, L. C., (2020). Pretreatment methods of lignocellulosic wastes into value-added products: Recent advances and possibilities. *Biomass Convers. Biorefinery,* 1–18.

3. Mancini, G., Papirio, S., Lens, P. N. L., & Esposito, G., (2016). Solvent pretreatments of lignocellulosic materials to enhance biogas production: A review. *Energy & Fuels, 30*(3), 1892–1903.

4. Wang, B. T., Hu, S., Yu, X. Y., Jin, L., Zhu, Y. J., & Jin, F. J., (2020). Studies of cellulose and starch utilization and the regulatory mechanisms of related enzymes in Fungi. *Polymers (Basel), 12*(3), 1–17.

5. Chundawat, S. P. S., et al., (2011). Restructuring the crystalline cellulose hydrogen bond network enhances its depolymerization rate. *J. Am. Chem. Soc., 133*(29), 11163–11174.

6. Zoghlami, A., & Paës, G., (2019). Lignocellulosic biomass: Understanding recalcitrance and predicting hydrolysis. *Front. Chem.,* 7.

7. Yu, Y., et al., (2010). Significant differences in the hydrolysis behavior of amorphous and crystalline portions within microcrystalline cellulose in hot-compressed water. *Ind. Eng. Chem. Res., 49*(8), 3902–3909.

8. Angell, S., Norris, F. W., & Resch, C. E., (1936). The analysis of carbohydrates of the cell wall of plants: The determination of pentoses as single substances and in mixtures containing uronic acids and hexoses. *Biochem. J., 30*(12), 2146–2154.

9. Liu, K., et al., (2015). Butanol production from hydrothermolysis-pretreated switchgrass: Quantification of inhibitors and detoxification of hydrolyzate. *Bioresour. Technol., 189,* 292–301.

10. Ahmed, B., et al., (2019). Improvement of anaerobic digestion of lignocellulosic biomass by hydrothermal pretreatment. *Appl. Sci., 9*(18), 1–17.

11. Da Conceição, G. A., Rodrigues, M. I., De França P. D., Machado De C. A., Maria, M. S. A. L., & Pereira, N., (2019). Acetone-butanol-ethanol fermentation from sugarcane bagasse hydrolysates: Utilization of C5 and C6 sugars. *Electron. J. Biotechnol.*

12. Zabed, H., Sahu, J. N., Boyce, A. N., & Faruq, G., (2016). Fuel ethanol production from lignocellulosic biomass: An overview on feedstocks and technological approaches. *Renew. Sustain. Energy Rev., 66,* 751–774.

13. Gellerstedt, G., & Henriksson, G., (2008). Lignins: Major sources, structure and properties. In: *Monomers, Polymers and Composites from Renewable Resources* (pp. 201–224). Elsevier.

14. ICF, (2017). A life-cycle analysis of the greenhouse gas emissions of corn-based ethanol. *Ind. Biotechnol., 13*(1), 19–22.

15. Karimi, K., (2015). *Lignocellulose-Based Bioproducts.* Switzerland: Springer International Publishing. (p. 270).

16. Shaikh, A., (2017). *The Bad News? The World will Begin Running Out of Water By 2050.* The good news? It's not 2050 yet. United Nations News & Commentary Global News.

17. Shapouri, H., Gallagher, P., & Graboski, M. S. (2002). USDA's 1998 ethanol cost-of-production survey (No. 1473–2021–024).

18. Gleick, P. H., et al., (1993). World freshwater resources. *Water Cris. a Guide. to World's Freshwater Resour.*

19. Chapagain, A. K., & Hoekstra, A. Y., (2007). The water footprint of coffee and tea consumption in the Netherlands. *Ecol. Econ., 64*(1), 109–118.

20. De Jong, E., & Jungmeier, G., (2015). *Biorefinery Concepts in Comparison to Petrochemical Refineries.* Elsevier B.V.

21. Hingsamer, M., & Jungmeier, G., (2018). *Biorefineries.* Elsevier Inc.

22. Gnansounou, E., & Pandey, A., (2017). *Classification of Biorefineries Taking into Account Sustainability Potentials and Flexibility.* Elsevier B.V.

23. Council, N. R., et al., (2008). *Water Implications of Biofuels Production in the United States.* National Academies Press.

24. Lingaraju, B. P., Lee, J. Y., & Yang, Y. J., (2013). Process and utility water requirements for cellulosic ethanol production processes via fermentation pathway. *Environ. Prog. Sustain. Energy, 32*(2), 396–405.

25. Singh, O. V., & Harvey, S. P., (2010). Sustainable biotechnology: Sources of renewable energy. *Sustain. Biotechnol. Sources Renew. Energy,* pp. 1–323.

26. Putro, J. N., Soetaredjo, F. E., Lin, S. Y., Ju, Y. H., & Ismadji, S., (2016). Pretreatment and conversion of lignocellulose biomass into valuable chemicals. *RSC Adv., 6*(52), 46834–46852.

27. Welker, C. M., Balasubramanian, V. K., Petti, C., Rai, K. M., De Bolt, S., & Mendu, V., (2015). Engineering plant biomass lignin content and composition for biofuels and bioproducts. *Energies, 8*(8), 7654–7676.

28. Shishir, P. S. C., Gregg T, B., Michael E, H., & Bruce E, D. (2011). Deconstruction of Lignocellulosic Biomass to Fuels and Chemicals. *Annual Review of Chemical and Biomolecular Engineering, 2*(1).

29. Kuila, A., & Sharma, V. (Eds.). (2017). *Lignocellulosic Biomass Production and Industrial Applications.* John Wiley & Sons.

30. Kumar, S. J., Reetu, S., & Lakshmi, T., (2015). *Lignocellulosic Agriculture Wastes as Biomass Feedstocks for Second-generation Bioethanol Production: Concepts and Recent Developments,* 337–353.

31. Wu, M. M., & Sawyer, B. M. (2016). Estimating Water Footprint and Managing Biorefinery Wastewater in the Production of Bio-based Renewable Diesel Blendstock (No. ANL/ESD-17/2). Argonne National Lab.(ANL), Argonne, IL (United States).

32. Behera, S., Arora, R., Nandhagopal, N., & Kumar, S., (2014). Importance of chemical pretreatment for bioconversion of lignocellulosic biomass. *Renew. Sustain. Energy Rev., 36,* 91–106.

33. Chen, H., et al., (2017). *A Review on the Pretreatment of Lignocellulose for High-Value Chemicals, 160,* 196–206.

34. Guerrero, A. B., Ballesteros, I., & Ballesteros, M., (2017). Optimal conditions of acid-catalyzed steam explosion pretreatment of banana lignocellulosic biomass for fermentable sugar production. *J. Chem. Technol. Biotechnol., 92*(9), 2351–2359.

35. Wu, M., Mintz, M., Wang, M., & Arora, S., (2009). Water consumption in the production of ethanol and petroleum gasoline. *Environ. Manage., 44*(5), 981.

36. Neupane, B., Konda, N. V. S. N. M., Singh, S., Simmons, B. A., & Scown, C. D., (2017). *Life-Cycle Greenhouse Gas and Water Intensity of Cellulosic Biofuel Production Using Cholinium Lysinate Ionic Liquid Pretreatment,* 10176–10185.

37. Aden, A., Ruth, M., Ibsen, K., Jechura, J., Neeves, K., Sheehan, J. & Lukas, J. (2002). Lignocellulosic biomass to ethanol process design and economics utilizing co-current dilute acid prehydrolysis and enzymatic hydrolysis for corn stover (No. NREL/ TP-510–32438). National Renewable Energy Lab., Golden, CO. (US).

38. Dong, C., Wang, Y., Chan, K. L., Bhatia, A., & Leu, S. Y., (2018). Temperature profiling to maximize energy yield with reduced water input in a lignocellulosic ethanol biorefinery. *Appl. Energy, 214,* 63–72.

39. Singh, J., Suhag, M., & Dhaka, A., (2015). Augmented digestion of lignocellulose by steam explosion, acid and alkaline pretreatment methods: A review. *Carbohydr. Polym., 117,* 624–631.

40. Zhang, J., Wang, X., Chu, D., He, Y., & Bao, J., (2011). Dry pretreatment of lignocellulose with extremely low steam and water usage for bioethanol production. *Bioresour. Technol., 102*(6), 4480–4488.

41. Pan, S. Y., Lin, Y. J., Snyder, S. W., Ma, H. W., & Chiang, P. C., (2016). Assessing the environmental impacts and water consumption of pretreatment and conditioning processes of corn stover hydrolysate liquor in biorefineries. *Energy, 116,* 436–444.

42. Binod, P., Gnansounou, E., Sindhu, R., & Pandey, A., (2018). Enzymes for second-generation biofuels: Recent developments and future perspectives. *Bioresour. Technol. Reports.*

43. Sun, Y., & Cheng, J., (2002). Hydrolysis of lignocellulosic materials for ethanol production: A review. *Bioresour. Technol., 83*(1), 1–11.

44. Yen, F. Y., Chen, H. H., Guo, G. L., Jang, M. F., & Chen, W. H., (2016). *Method of Pretreating Lignocellulose Using Solvent with Little Water Footprint* (Vol. 1). United States Pat. Appl. Publ.

45. Zhuang, X., et al., (2016). Liquid hot water pretreatment of lignocellulosic biomass for bioethanol production accompanying with high valuable products. *Bioresour. Technol., 199,* 68–75.

46. Tao, L., et al., (2011). Process and techno-economic analysis of leading pretreatment technologies for lignocellulosic ethanol production using switchgrass. *Bioresour. Technol., 102*(24), 11105–11114.

47. Cheah, W. Y., et al., (2020). Pretreatment methods for lignocellulosic biofuels production: Current advances, challenges and future prospects. *Biofuel Res. J., 7*(1), 1115–1127.

48. Cardona, E., Llano, B., Peñuela, M., Peña, J., & Rios, L. A., (2018). Liquid-hot-water pretreatment of palm-oil residues for ethanol production: An economic approach to the selection of the processing conditions. *Energy, 160*, 441–451.

49. Zimbardi, F., Viola, E., Nanna, F., Larocca, E., Cardinale, M., & Barisano, D., (2007). Acid impregnation and steam explosion of corn stover in batch processes. *Ind. Crops Prod., 26*(2), 195–206.

50. Solarte-Toro, J. C., Romero-García, J. M., Martínez-Patiño, J. C., Ruiz-Ramos, E., Castro-Galiano, E., & Cardona-Alzate, C. A., (2019). Acid pretreatment of lignocellulosic biomass for energy vectors production: A review focused on operational conditions and techno-economic assessment for bioethanol production. *Renew. Sustain. Energy Rev., 107*, 587–601.

51. Abdul, R. A. Z., & Saidina, A. N. A., (2019). Environmental sustainability assessment of palm lignocellulosic biomass pretreatment methods. *Chem. Eng. Trans., 72*, 13–18.

52. Omar, W. N. N. W., & Amin, N. A. S., (2016). Multi response optimization of oil palm frond pretreatment by ozonolysis. *Ind. Crops Prod., 85*, 389–402.

53. Tan, H. T., Lee, K. T., & Mohamed, A. R., (2011). Pretreatment of lignocellulosic palm biomass using a solvent-ionic liquid [BMIM] Cl for glucose recovery: An optimization study using response surface methodology. *Carbohydr. Polym., 83*(4), 1862–1868.

54. Goshadrou, A., (2019). Bioethanol production from Cogon grass by sequential recycling of black liquor and wastewater in a mild-alkali pretreatment. *Fuel, 258*, 116141.

55. Da Costa, S. L., et al., (2016). Next-generation ammonia pretreatment enhances cellulosic biofuel production. *Energy Environ. Sci., 9*(4), 1215–1223.

56. Aden, A., et al., (2002a). *Lignocellulosic Biomass to Ethanol Process Design and Economics Utilizing Co-Current Dilute Acid Prehydrolysis and Enzymatic Hydrolysis for Corn Stover*. (No. NREL/TP-510–32438). National Renewable Energy Lab., Golden, CO.(US).

57. Du, W., et al., (2011). The promoting effect of byproducts from *Irpex lacteus* on subsequent enzymatic hydrolysis of bio-pretreated cornstalks. *Biotechnol. Biofuels, 4*, 1–8.

58. Amriani, F., Salim, F. A., Iskandinata, I., Khumsupan, D., & Barta, Z., (2016). Physical and biophysical pretreatment of water hyacinth biomass for cellulase enzyme production. *Chem. Biochem. Eng. Q., 30*(2), 237–244.

59. Aden, A., & others, (2007). Water usage for current and future ethanol production. *Southwest Hydrol., 6*(5), 22, 23.

60. Gu, Y. M., Kim, H., Sang, B. I., & Lee, J. H., (2018). Effects of water content on ball milling pretreatment and the enzymatic digestibility of corn stover. *Water-Energy Nexus, 1*(1), 61–65.

CHAPTER 15

Biodegradation and Biocatalysis Aspects of Direct Bioethanol Production by Fungi in a Single Step Named Consolidated Bioprocessing

LUIS FERNANDO AMADOR CASTRO and DANAY CARRILLO NIEVES

Tecnologico de Monterrey, Escuela de Ingenieria y Ciencias,
Av. General Ramon Corona No. 2514, Zapopan–45201, Jal., Mexico
E-mail: danay.carrillo@tec.mx (Danay Carrillo Nieves)

ABSTRACT

The increasing industrialization and population growth are coupled with rising energy demand. However, using fossil fuels to satisfy this demand is not a feasible strategy as climate change has intensified due to anthropogenic-related emissions. The previous triggered the search for cleaner energies, and among the options, ethanol was presented as a potential alternative fuel; but its success was clouded as its production was considered a threat for food security. To assess this problem, the use of lignocellulosic residues and non-edible crops has been suggested, but its production process is more expensive. Consolidated bioprocessing (CBP) recently surged as a solution as it reduces the number of steps involved in ethanol production, thus reducing the cost of the final product. However, the search for suitable microorganisms that can be employed in this type of process continues, following that they need to be able to degrade the recalcitrant structure of the cell walls and ferment sugars. Fungi are promising microorganisms, as they can produce lignocellulosic enzymes and can be engineered for efficient fermentation. This chapter aims to give the reader some aspects regarding the use of lignocellulosic materials for ethanol production.

15.1 INTRODUCTION

Currently, climate change poses a risk to modern societies. It has been estimated that a global temperature increase by 1.5°C from pre-industrial levels will result in negative effects on several aspects, including human health, economic growth, food security, biodiversity, and natural ecosystems [1]. Anthropogenic-related emissions are the greatest contributor to climate change, being carbon dioxide (CO_2) emissions the most abundant [2]. Given that around 90% of CO_2 emissions proceed from fossil fuel usage and industrial sources [3], the search for alternative energy sources stands as a viable strategy for mitigating the increasing effects of climate change [4, 5]. The use of biomass has been proposed as one of the alternatives to fossil fuels being lignocellulose of particular importance as it is known to be the most abundant biomass on Earth [6].

Lignocellulose is an organic material that constitutes the plant cell walls. It can be readily obtained from feedstocks and forestry as well as from their wastes [7]. Biofuels that can be obtained from this material include ethanol, butanol, and biodiesel; among the previous ethanol has the highest relevance [8]. Ethanol can be readily mixed with gasoline reducing hydrocarbon and carbon monoxide emissions [9]. Ethanol-gasoline blends can also contribute to decreasing the effects of climate change given that feedstocks used for ethanol production sequester CO_2 for its growth [10]. Additionally, the use of lignocellulosic wastes and non-edible crops for ethanol production can favor rural economic development and at the same time does not comprise food security [11]. However, the implementation of this technology faces significant challenges.

Serving as a defense mechanism against pathogens, lignocellulose is highly recalcitrant, and its hydrolysis requires the employment of chemical or enzymatic procedures. Furthermore, to improve the ethanol yield biomass should be pretreated either by physical or chemical means [12]. The requirement of a multiple-step process for ethanol production from LCB makes the process not economically competitive when compared to currently existing fuels [11]. Aiming towards cost reduction, multiple strategies have been proposed, including the elimination of the pretreatment procedures, the increase in cellulose hydrolysis yield by optimizing the enzymes involved in the process and improving the ethanol yields [13]. However, one of the most promising strategies is to perform the conversion of lignocellulose into ethanol by implementing a single-step process known as CBP [14].

CBP possesses multiple advantages when compared to separate hydrolysis and fermentation (SHF) processes. But this technology is still under development and finding or engineering an appropriate microorganism for the process has resulted to be not an easy task. Therefore, we aim to provide the reader with some insights regarding the structure of the plant cell walls and the enzymes that are involved in their synthesis and degradation. Further, we explain the concepts, advantages, and challenges around CBP; and finally, we review some fungi that have been used or have the potential to be used in this type of process as they are known to be the most efficient lignocellulosic enzymes producers.

15.2 LIGNOCELLULOSIC MATERIALS, THE STRUCTURE, AND RECALCITRANCE OF CELL WALL

The cell wall is a critical component of plant cells. Apart from encasing cell's protoplasm, it participates in biological processes such as differentiation, intracellular communication, water, and nutrient transport and as a defense mechanism against pathogens [15]. Cell wall composition varies greatly among different plant species and within cells from the same plant depending on the physiological role they exert. Furthermore, there are some plant cells that contain two cell walls, primary, and secondary, which composition is also different [15]. Plant cell walls have a complex structure consisting of a mixture of carbohydrates, proteins, and aromatic compounds. Carbohydrates that form the plant cell walls include cellulose, hemicellulose, and pectin and can account for as much as 90% of the content of the primary cell wall [16]. Aromatic compounds like lignin which are present in the walls of many plant cells provide plants with further structural support. Proteins present on the cell walls can serve as structural molecules or signaling receptors which are essential for plant development. This section will briefly describe the main components of the plant cell walls.

15.2.1 LIGNIN

Lignin is a complex aromatic polymer mainly present in secondary thickened plant cell walls. It is covalently bound to hemicellulose and its function is to provide the cell wall with strength and rigidity, allowing plants to grow upward and serve as a vascular system for transport of water and solutes. This highly cross-linked polymer is commonly synthesized from the oxidative coupling of p-hydroxycinnamyl alcohol monomers, often referred to as monolignols [17]. The three most common monolignols are p-coumaryl alcohol, coniferyl alcohol, and sinapyl alcohol, molecules from which the hydroxyphenyl (H), guaiacyl (G), and syringyl (S) lignin subunits, respectively derive [18]. Recent evidence indicates that apart from hydroxycinnamyl alcohols, lignin can be synthesized to a less extent from other phenolic monomers, further increasing the complexity of this polymer [19]. Still, all currently known lignin building blocks derived from the general phenylpropanoid pathway in which phenylalanine serves as the substrate for this pathway in plants and tyrosine as an additional substrate in grasses. The amount of lignin as well as its composition vary significantly among and within plant species cell types and even cell walls; being influenced by diverse environmental factors [19]. Generally, lignin polymers from dicotyledonous angiosperms are composed mostly of G and S subunits having only traces of H subunits. Gymnosperm lignins lack S subunits and are composed as much as 90% from G subunits. Lignins from grasses contain G and S subunits at similar levels and a lower amount of H [20]. In lignin, these subunits are typically connected by carbon-oxygen ether linkages and carbon-carbon linkages being β-O-4 is the most common linkage. Other molecular bonds include α-O-4, β-5, 5-5, 4-O-5, β-1, and β-β linkages [20]. The interest in the study of lignin arises as this polymer, which accounts for up to 30% of LCB, can greatly hinder the process of the bioconversion of LCB into ethanol. Therefore, reducing its content or changing its composition can represent significant improvements regarding the use of lignocellulosic materials in multiple industries.

15.2.2   CELLULOSE

Cellulose is the most abundant polymer of the plant cell being present in both primary and secondary walls. It provides structural support allowing plants to withstand physical stress and act as a barrier against other organisms. Cellulose is not exclusive to plants as certain bacteria, fungi, and animals can synthesize it [21]. This polymer consists of chains of repeating D-glucopyranose molecules which are linked by β-1,4-glycosidic bonds. Chains of cellulose comprise a non-reducing end, which is a glucose molecule with its original C4-OH group and a reducing end where C1-OH is present; the rest of the polymer contains anhydroglucose molecules containing three hydroxyl groups. Each anhydroglucose molecule is rotated 180° in the plane, being two adjacent molecules, the cellulose monomer called cellobiose [22].

In nature, cellulose can take more complex supramolecular structures which include crystalline and amorphous areas. Crystalline cellulose, which is denser than amorphous cellulose, can account for approximately 50–90% of the total cellulose and not readily accessible to water [23]. Structures include elementary fibrils, microfibrils, and microfibrillar bands [22]. Cellulose microfibrils constitute the main component of plant cell walls. Microfibrils are inelastic and surround cells providing resistance to osmotic pressures. These cable-like structures are regularly composed of about 36 chains containing from 500 to 14,000 β-1,4-linked glucose molecules [24]. Molecular weight and DP, which is the number of glucose residues in the chain, vary depending on its location within the cell. Primary cell wall cellulose chains have a lower molecular weight and less DP when compared to those of the secondary cell wall [21]. Being the main polymer present in LCB, cellulose regularly accounts for 30–50% of its dry weight. Further, glucose molecules proceeding from cellulose degradation can undergo fermentation, thus being an important substrate for biofuel production.

15.2.3 HEMICELLULOSE

Hemicellulose is a diverse group of polysaccharides that accounts for about one-third of the total dry weight of LCB. Its structure and amount, as in the case of previously mentioned polymers vary depending on the plant species and within the same plant depending on the tissue type, developmental stage, and its physical location (primary or secondary cell wall) [25]. The term hemicellulose is not as clearly established as it is the case of cellulose or lignin and some polysaccharides like galactans, arabinans are sometimes included within this group. However, here we consider hemicellulose as a group of polysaccharides which include xyloglucan, xylans, mannans, glucomannans, and mixed linkage glucans [26]. Xyloglucan is present on all land plant species, being the most common hemicellulose in dicots primary cell wall. Xyloglucan comprises a β-1,4-linked glucan backbone partially substituted with side chains at the O-6 positions [25]. Xylans are a group of polysaccharides that share the feature of having a backbone of β-1,4-linked xylose residues. In grasses, xylans, following cellulose, represent the second most abundant polysaccharides of the primary cell wall with content around 20% [26]. Mannans and glucomannans are also β-1,4-linked polysaccharides. Its backbone can contain solely mannose, as in the case of mannans and galactomannans, or consist of a combination of mannose and glucose in a non-repeating pattern like it can be observed for glucomannans and galactoglucomannans. They are the main hemicellulose in charophytes and galactoglucomannans are the main components of the secondary walls of gymnosperms [26]. In higher plants, mixed linkage glucans are limited to the cell walls of grasses. Hemicelluloses often interact with the other cell wall components, thus adding more complexity to the plant cell walls.

15.2.4   PECTIN

Pectin is considered the most structurally complex family of plant polysaccharides [27]. It can be found in the cell walls of plants and certain algae, constituting about 35% of primary walls in non-graminaceous monocots and dicots, 2–10% of grass primary walls, and as much as 5% of woody tissue walls. As previous cell wall constituents, pectin is involved in vital plant functions including growth processes, morphology, development, and defense against environment and pathogens. In the cell walls, it also provides strength and flexibility [28]. Pectin polysaccharides include homogalacturonan, rhamnogalacturonan I (RGI), rhamnogalacturonan II (RGII), and xylogalacturonan (XGA). The total pectin content and the relative proportion of the polysaccharides that constitute it varies depending on environmental factors, from plant to plant and within same the plant. Homogalacturonan is the most common polysaccharide, it can account for up to 65% of pectin; RGI constitutes from 20% to 35%. XGA and RGII are present in minor amounts, each accounting for less than 10% of pectin [27]. All previous polysaccharides contain galacturonic acid linked at the O-1 and the O-4 position. Homogalacturonan is a linear polymer consisting only of α-1,4-linked galacturonic acid molecules. The common length of the chain is of about 100 galacturonic acid molecules, however, shorter chains can be found. RGII structure is far more complex, it consists of a backbone of homogalacturonan that contains branches formed from different sugars. XGA is composed of a homogalacturonan chain substituted at O-3 with a β-linked xylose. Finally, RGI consists of a backbone of the disaccharide repeats of rhamnose and galacturonic acid, that can be substituted with side chains containing L-galactose and L-arabinose residues [27].

15.2.5   EXTRACTIVES AND ASHES

Apart from the cited polysaccharides and lignin, plant cell walls contain other minor components that may be taken into consideration when designing a process where LCB is the raw material. Plants contain a variety of minerals, sometimes referred to as ashes, which are important for their development. The presence of high contents of ashes can have negative effects on enzymatic hydrolysis and fermentation processes [7]. Plants also contain pigments such as flavonoids and anthocyanins that, when extracted, can affect the bioconversion of lignocellulose into certain products, therefore pretreatment processes should be considered depending on the final use of the biomass. A variety of proteins can also be found on the plant cell wall, these proteins exert mainly structural functions, although some are involved in morphogenesis having signaling functions. Most cell wall proteins are glycosylated and contain hydroxyproline except for glycine-rich proteins [29]. The presence of proteins adds to further increase the complexity of the plant cell walls, which degradation will involve the participation of multiple enzymes as we well address in further section.

15.3 LIGNOCELLULOLYTIC ENZYMES PRODUCED BY FUNGI ACT AS DEGRADERS AND BIOCATALYST OF THE MAIN POLYMERS OF THE CELL WALL

The recalcitrance of the cell wall is one of the first challenges for the bioprocessing of lignocellulosic materials for their use in biofuel production. As aforementioned, plant cell wall is composed mainly of lignin, cellulose, hemicellulose, and pectin; therefore, its complete enzymatic degradation entails the participation of several lignocellulosic enzymes. However, nature can give us the solution, as lignocellulosic microorganisms like fungi possess different enzymes that can make feasible the degradation of this structure. Lignin recalcitrance to degradation is due to its complex structure and linkage heterogeneity [30]. Lignin-degrading enzymes can be classified mainly into two groups namely heme peroxidases and laccases [31, 32]. Ligninolytic heme peroxidases comprise lignin, manganese, and versatile and dye decolorizing peroxidases (DyP) [31]. Laccases assist lignin degradation and are part of the multicopper oxidase superfamily [30]. Additional enzymes that produce hydrogen peroxide (H_2O_2), which is required for peroxidase-based lignin degradation, include glyoxal oxidase, pyranose-2 oxidase, and aryl alcohol oxidase [33]. Cellulose and hemicellulose contain most of the sugars of the plant cell walls, but to make these sugars available for fermentation, the complex structures of these polymers need to be degraded. The hydrolysis of these cell walls' components requires the synergistic action of multiple cellulases and hemicellulases. Pectin degradation is performed by a set of pectinases, enzymes that have found applications in different industries.

Besides directing their attention to the enzymatic cell wall degradation, some scientists have also focused on engineering feedstocks that can be easily degraded by previously cited lignocellulosic enzymes [34]. Pursuing this strategy requires extensive knowledge of the processes involved in the formation of the cell wall; an area that, despite extensive investigations remains with many questions. This section will briefly describe some of the enzymes that are involved in plant cell wall degradation and give some insights into cell wall formation and fermentation processes.

15.3.1 LIGNIN DEGRADING PEROXIDASES

Peroxidases are widely distributed among plants, animals, and microorganisms. These enzymes catalyze the oxidation of different substrates, using H_2O_2 or other peroxides as electron acceptors [35]. Most of the peroxidases are heme enzymes that contain an iron protoporphyrin IX prosthetic group, although there are many nonheme peroxidases [36]. Heme peroxidases can be mainly classified into two superfamilies animal and non-animal (plant) peroxidases [37, 38]. Non-animal peroxidases are further subdivided into three classes. Class I are intracellular peroxidases that have been found in bacteria, fungi, plants, and some protists, examples include cytochrome c, catalase, and ascorbate peroxidases. Class II refers to extracellular fungal peroxidases in which lignin, manganese, and versatile peroxidases (VPs) are included. Class III are secreted peroxidases only found in plants [39]. Animal peroxidases have also been subclassified as in the case of plant peroxidases, however, its subclassification is more complex [37]. Some heme peroxidases such as DyPs do not fit within the previously mentioned superfamily classification and constitute its family [40]. Different peroxidase phylogenetic classifications have also been proposed [41, 42].

15.3.1.1 LIGNIN PEROXIDASE (LiP)

Lignin peroxidase (LiP) (EC 1.11.1.14) can catalyze lignin depolymerization using H_2O_2 as an electron acceptor. These enzymes have been found on different white-rot and brown-rot fungi as well as in some types of bacteria [43]. Various LiPs isoenzymes have been encountered in some fungi, and up to 16 different forms of LiPs have been identified in *Trametes versicolor* [44]. LiPs are monomeric hemoproteins that have a molecular mass of about 40 kDa, an isoelectric point around 3.3 to 4.7, and an optimal acidic pH of approximately 3 [45–47]. LiPs are capable of oxidizing different non-phenolic lignin compounds for example arylglycerol β-aryl ethers [31]. The oxidation of β-O-4 linked compounds requires the formation of a radical cation through one-electron oxidation, the process is followed by cleavage of the side chain, demethylation, intramolecular addition, and some molecular rearrangements [47]. Additional to the oxidation of non-phenolic molecules, LiPs can also oxidase different phenolic compounds such as ring- and N-substituted anilines [48]. The catalytic cycle of LiPs is similar to that of other peroxidases such as the horseradish peroxidase [49]. The cycle can be divided into three steps. First, the native ferric enzyme [Fe (III)] undergoes two-electron oxidation to form an intermediate iron porphyrin radical cation [Fe (IV)] using H_2O_2 as the electron acceptor. In the second step, the intermediate, known as compound I, is reduced by a substrate which translocases one electron to it to form another intermediate referred to as compound II. The final step compound II receives another electron from the reduced substrate and consequently, the enzyme returns to its original oxidation state [50].

15.3.1.2 MANGANESE PEROXIDASE (MnP)

Manganese peroxidase (MnPs) (EC 1.11.1.13) is another heme extracellular fungal peroxidase capable of degrading lignin as well as other phenolic substrates by oxidizing Mn^{2+} to Mn^{3+} in a peroxide-dependent manner [51]. MnP is one of the major lignin-degrading enzymes produced by certain basidiomycetes, particularly white-rot fungi [35, 52]. Same as with of LiPs, different isoforms of MnP exists among fungi and up to 11 different forms of MnPs have been described in *Ceriporiopsis subvermispora* [53]. MnPs are monomeric hemoproteins with masses between 32 and 62.5 kDa with an optimum pH of 4–7 and a temperature of 40–60°C [43]. The catalytic cycle of MnP is like that of LiPs. First, the native ferric enzyme reacts with H_2O_2 to yield an intermediate Fe^{4+}-oxo-porphyrin radical complex, that receives the name of complex I. Further Mn^{2+} oxidation into Mn^{3+} allows the one-electron reduction of compound I to form compound II. In the last step, Mn^{2+} oxidation is used for the final reduction of compound II to return the enzyme into its native oxidation state [54]. Chelated Mn^{3+} ions derived from previous reactions act as diffusible redox mediators that allow the oxidation of lignin phenolic structures [51]. For the oxidation of non-phenolic substrates by Mn^{3+} reactive radicals like superoxide (O_2^-) or peroxyl radicals must be formed in the presence of a second mediator such as oxalate and malonate [51, 55].

15.3.1.3   VERSATILE PEROXIDASE (VP)

Versatile peroxidases (VPs) (EC 1.11.1.16) are ligninolytic heme peroxidases that are not only specific to Mn^{2+} as in the case of MnPs, but also can oxidase phenolic and non-phenolic substrates that are commonly employed by LiPs without the presence of manganese [56]. VPs have mainly been reported to be produced natively by fungi species of the genera *Bjerkandera* and *Pleurotus* [56, 57]. Similarly, to lignin and MnPs, multiple isoforms of VPs exists [58]. The molecular structure of VP possesses characteristics of both LiPs and MnP, however, it is more like LiP than MnP [59]. VPs have a mass around 37 kDa, an isoelectric point of ~3.5, and an acidic optimum pH of about 4.5 [60]. This type of peroxidases can oxidize substrates in a broad range of potentials, including low and high redox potentials compounds. The catalytic cycle of VP is like that of previously described lignin and manganese heme peroxidases. It involves the first two-electron oxidation followed by two reduction reactions in which intermediate compounds are formed to return the enzyme to its native oxidation state. The main difference is that these enzymes have the versatility of oxidizing either Mn^{2+} or other substrates to carry out their reduction reactions [59].

15.3.1.4   DYE DECOLORIZING PEROXIDASE (DyP)

Dye decolorizing peroxidases (DyPs) (EC 1.11.1.19) are heme-containing peroxidases that comprise a peroxidase family that differ from other plant peroxidases due to their primary and tertiary structures as well as their specific reaction characteristics [40]. DyPs are produced mainly by some species of bacteria, although they have been reported to be present in fungi and some archaea [61, 62]. DyPs have a mass between 40–60 kDa, an isoelectric point of around 3.8, and an acidic optimum pH around 2–3 depending on the employed substrate [62, 63]. DyP has been named after certain enzymes that belong to this family were identified as capable of dye degradation. Apart from dyes DyPs have been shown to oxidize a variety of substrates. DyP from fungi *Auricularia auricula-judae* has been observed to degrade the nonphenolic β-O-4 lignin model compound and oxidize veratryl alcohol at low pH [64]. However, the physiological role of these enzymes is not completely elucidated, but it has been associated with the oxidation of nonphenolic substrates like lignin and toxic aromatic products or as a defense mechanism against oxidative stress [62, 63].

15.3.2 LACCASES

Laccases (EC 1.10.3.2, benzenediol: oxygen oxidoreductase) are multi-copper enzymes produced by some species of fungi, plants, bacteria, and insects [65]. Laccases are involved in numerous biological processes; in plants, they contribute to the radical-based mechanisms for the formation of the lignin polymer. In fungi, there have been related to immunity, pathogenesis, morphogenesis, and lignin degradation [65, 66]. These enzymes are of biotechnological importance due to their capacity of oxidizing a wide variety of substrates by only using molecular oxygen (O_2) and producing water as a by-product [67]. In contrast with ligninolytic peroxidases which are only extracellular, laccases can be localized extra or intracellularly. The localization of these enzymes may be related to the physiological function they exert, and this localization also determines the variety of substrates that can be used by them [66]. Generally, fungal laccases have masses around 60–70 kDa, an isoelectric point about 4, and an acidic optimum pH between 2.5–4 [66].

The catalytic cycle of laccases reduces one molecule of O_2 to two molecules of water with the concomitant oxidation of four molecules of a substrate to produce four radicals. This redox process is aided by a cluster of four copper atoms that serve as the catalytic core of the enzyme [67]. Laccases can catalyze direct oxidation of phenolic compounds and certain amines [68]. Nevertheless, oxidization of high redox potential compounds such as non-phenolic model compounds and β-1 lignin dimers requires the presence of a mediator. Some commonly used mediators are 3-hydroxyanthranilic acid (HAA), N-hydroxybenzotriazole, and 2,20-azino-bis-(3-ethyl-benzothiazoline-6-sulphonic acid) (ABTS) [67]. However, the use of these mediators increases the overall cost of the process.

15.3.3 DEGRADATION OF CELLULOSE AND HEMICELLULOSE

As described earlier, cellulose and hemicellulose are major constituents of the plant cell walls. Like in the case of lignin, the degradation of these compounds is challenging due to their complex structure, linkage strength, and high molecular weight [69]. However, microorganisms have developed a set of enzymes called cellulases and hemicellulases that can degrade these compounds into simpler carbohydrate monomers. Their degradation capacity has made these enzymes of biotechnological importance in various sectors including textile, wine, and brewery, food, pulp, and paper, agriculture, and biofuel industries [70, 71].

15.3.3.1 CELLULASES

Cellulases are a family of three groups of enzymes, endoglucanases (EC 3.2.1.4), exoglucanases (EC 3.2.1.91), and β-glucosidases (EC 3.2.1.21) [72]. Cellulases are expressed natively by different microorganisms including fungi and bacteria. However, most commercially available cellulases are produced from aerobic cellulolytic fungi due to the high yields that some engineered strains have in producing these enzymes [73]. In bacteria, cellulases are present as extracellular aggregated structures attached to the cells called cellulosomes [74]. Fungal cellulases have a relatively simple molecular structure containing a catalytic domain and a cellulose-binding domain. Extensive reviews describing the structure and characteristics of these enzymes are available on refs [75, 76]. The catalytic domain is linked to the cellulose-binding domain by a linker peptide. The anchoring of the cellulose-binding domain to cellulose permits the enzyme to optimally perform its catalytic function [75]. Cellulases degrade cellulose using different synergistic approaches. Endoglucanases, also called non-processive cellulases, can cleave the internal *O*-glycosidic bonds producing polysaccharide chains of variable lengths. Exoglucanases or processive cellulases act by coupling to the cellulose chain ends and hydrolyzing cellulose into cellobiose. β-glucosidases can then cleave cellobiose into glucose that can be used by the microorganism [76].

15.3.3.2 HEMICELLULASES

Hemicelluloses are a diverse group of enzymes capable of degrading hemicellulose. Due to the variety of monomers that conform to hemicellulose, its complete degradation requires the synergistic interaction from different enzymes [77]. Hemicellulases can be classified according to their catalytic modules in either glycoside hydrolases that can hydrolyze glycosidic bonds or carbohydrate esterases (CEs), which are enzymes capable of hydrolyzing ester linkages of acetate or ferulic acid side groups [76]. Among hemicellulose degrading enzymes we can find xylanases (EC 3.2.1.8), β-β-mannanases (EC 3.2.1.78), α-L-arabinofuranosidases (EC 3.2.1.55) and α-L-arabinanases (EC 3.2.1.99), α-glucuronidases (EC 3.2.1.139), β-xylosidases (EC 3.2.1.37) and acetyl xylan esterases (EC 3.1.1.72) [78, 79]. Many of these hemicellulolytic enzymes have been found on various types of microorganisms, including fungi and bacteria [78]. Describing the catalytic mechanisms and structure of the previously mentioned enzymes is beyond the scope of this chapter.

15.3.4 PECTINASES

Pectin, as formerly mentioned, is a complex polysaccharide of the cell walls that contributes to their strength and flexibility. It contains acidic sugars that are involved in some cell wall characteristics including adhesion, extensibility, porosity, and electrostatic potential [80]. Because of its contribution to porosity and charge, pectin has been a target for studies aiming at cell wall degradation [81]. Pectinases are a group of enzymes capable of degrading pectin that has found application in the fruit, paper, and pulp, textile, and biofuel industries [81, 82]. Pectinases are natively produced by a variety of organisms including fungi, bacteria, insects, protozoa, and plants [83]. However, for the industrial production of these enzymes, fungi, and bacteria are the most commonly employed organisms [84]. Different classifications for pectinases have been proposed that are based on their preferred substrate, their mode of action, and cleavage characteristics [85]. Microorganisms can produce different forms of pectinases that have been encountered to work in a wide range of pH and temperature values [82]. Optimum pH values may vary between 4–10 and optimal temperature from 30 to 70°C [82]. Such variability can be related to the natural environmental factors to which the microorganism is exposed [83]. Because of the previous recombinant pectinases are produced from different sources so that they can be applied in processes that involve harsh temperature and pH conditions [84].

15.3.5 PLANT CELL WALL BIOSYNTHESIS

Even though lignocellulose is considered the most abundant biomass on Earth [86], some of the processes that are related to its biosynthesis and the construction of the plant cell walls remain to be elucidated [87, 88]. Cell wall synthesis is a complex process that involves numerous enzymes and processes such as intracellular trafficking of cell wall precursors and proteins, the extracellular assembly of cell wall polymers, the remodeling of these polymers, and the recycling of some cellular sugars [87]. The assembly of the nucleotide sugars which are required for cellulose and hemicellulose synthesis is catalyzed by several glycosyltransferases, enzymes capable of forming glycosidic bonds by transferring sugar moieties between the donor and accepting molecules [87]. On the other hand, the building blocks for lignin biosynthesis proceed from the general phenylpropanoid pathway, being phenylalanine and sometimes tyrosine the substrates for this pathway [19]. Furthermore, laccases, and peroxidases, enzymes that are involved in the lignin degradation process, are also involved in lignin biosynthesis by catalyzing the oxidation of phenolic monolignols into phenolic monomers, a process essential for the initiation of *in vivo* polymerization of lignin [30].

The study regarding cell wall synthesis is of great interest due as one strategy to overcome its recalcitrance may be to develop feedstock varieties with more labile cell walls that apart from maintaining is agricultural yields can be useful for biofuel production [89]. Some of the enzymes that participate in the degradation of the main components of the plant cell wall have also been identified to be involved in its biosynthesis. This is the case of the membrane-anchored endoglucanase Korrigan which appears to be related to cellulose biosynthesis in the primary and secondary cell walls of *Arabidopsis thaliana*. Mutant genotypes have exhibited defects in cell wall formation, including reduced cellulose content and increased pectin synthesis deriving in abnormal plant morphology [90, 91]. Comprehensive reviews can provide the reader with a complete picture of the current understanding of the biosynthesis of the main components of the cell wall [19, 25, 80, 92].

15.3.6   INSIGHTS INTO ALCOHOLIC FERMENTATION

Once that lignocellulosic materials have been pretreated and lignocellulolytic enzymes have degraded the main components of the cell wall, sugars are fermented to produce ethanol [93]. Sugars derived from cellulose and hemicellulose degradation include hexoses such as glucose, mannose, fructose, and galactose, as well as pentoses, namely xylose and arabinose [94]. Some microorganisms like yeast, fungi, and bacteria possess the mechanisms for fermenting these sugars into ethanol; being yeasts the most studied and industrially-employed organisms [95]. *Saccharomyces cerevisiae* can natively ferment hexoses; fermentation of glucose, mannose, and fructose occurs via the Embden-Meyerhof pathway of glycolysis, and galactose fermentation via the Leloir pathway [93]. The Entner-Doudoroff (ED) pathway is an additional pathway that allows glucose fermentation bacteria such as *Zymomonas* [96]. However, the use of these species of bacteria may not be suitable for industrial bioethanol production as it can only ferment glucose, fructose, and sucrose [96]. Wildtype *S. cerevisiae* is not capable of fermenting neither xylose nor arabinose. But some species of *Candida* have been identified to ferment these substrates into ethanol [93]. Rhamnose, 6-deoxy sugar, can also be available for fermentation as it is a constituent of the rhamnogalacturonan part of pectin and hemicellulose. However, as its alcoholic fermentation pathway is not natively expressed in most organisms, metabolic engineering will be required to express a route for its fermentation in microorganisms that are currently used for industrial fermentation [93]. The use of genetic engineering can greatly increase the yield of bioethanol from LCB by allowing the engineered organism to ferment most of the available sugars. Furthermore, the integration of lignocellulosic enzymes will make able to develop a consolidated bioprocess to simultaneously perform the saccharification and fermentation processes.

15.4 CONSOLIDATED BIOPROCESSING (CBP)

Ethanol production from lignocellulosic materials is a multiple-step process that includes biomass collection and storage, pretreatment, saccharification, fermentation, and ethanol recovery [13]. The need for multiple steps is translated into a higher cost of production when compared to traditional ethanol production from starch- or sugar-based crops [97]. Conventionally bioethanol production process involves that saccharification and fermentation processes are performed separately, which results in a time consuming and costly process [98]. Apart from the previous disadvantage, SHF processes have the drawback that cellulases activity during the hydrolysis process is inhibited by their product, limiting the efficiency of the hydrolysis process [98].

The necessity to reduce costs and optimize the process has led to the development of simultaneous saccharification and fermentation (SSF) processes [13]. In SSF processes biomass hydrolysis and fermentation are performed within the same bioreactor; this allows to reduce costs and obtain higher ethanol yields as sugars are fermented as they appear in the solution [99]. However, the SSF process is focused on fermenting only hexoses leaving potentially fermentable sugars unused. To overcome this problem, simultaneous saccharification and co-fermentation (SSCF) processes have been developed in which fermentation of hexoses and pentoses occur in a single step [100]. Nevertheless, the SSCF process still requires the addition of lignocellulolytic enzymes into the vessel, which represents a disadvantage as the cost of the biomass, its pretreatment, and the enzymes used for hydrolysis account for about two-thirds of the ethanol production cost, being the cost of the enzyme the most significant [101]. All the previous has led to the development of CBP, where enzyme production along with saccharification and fermentation are produced simultaneously within the same vessel [101]. For CBP to be feasible, it is required that either multiple or a single microorganism be able to produce both lignocellulosic and fermentative enzymes [102].

Despite being a promising strategy to optimize the process for ethanol production, CBP remains in its early development stages. One known limiting factor is that the optimal temperature for saccharification (45–50°C) differs from that required for efficient fermentation which is around 28–35°C [94, 96]. Another limitation is that ethanol has the capacity of inhibiting the growth of its producing microorganisms once a certain concentration has been reached; thus, discontinuing the fermentation process which may result in low ethanol yield [99]. These factors have led to the quest of finding microorganisms that are optimal for this type of bioprocess. However, microbes capable of overcoming previous difficulties and that possess the enzymatic machinery to perform both saccharification and fermentation efficiently have not yet been encountered in nature [101, 102]. Therefore, different strategies have been proposed to develop these ideal microorganisms. Mainly there is a focus on two strategies, the native strategy, and the recombinant strategy. The first one aims to genetically engineer a microorganism natively capable of producing lignocellulosic enzymes to make it ethanologenic. The second is targeted to engineer an ethanol-producing microorganism to make it capable of degrading lignocellulose [97, 102].

Following the native or recombinant strategies has some challenges. By engineering microorganisms according to the native strategy, scientists need to use the currently available genetic engineering tools to produce a strain capable of yielding high ethanol concentrations and that is robust enough for its industrial application [102]. Most promising candidate microorganisms for following this strategy are lignocellulosic fungi and bacteria, as microorganisms like bacterium *Clostridium thermocellum* and a variety of white-rot fungi have been already used in ethanol production via CBP [103, 104]. On the other hand, the production of organisms by the recombinant strategy faces the challenge that the selected organisms need to produce enough quantities of a wide variety of enzymes to degrade the main components of the plant cell walls [97]. The recombinant strategy has focused mainly on modifying yeasts and bacteria, with most of the advances being performed in *S. cerevisiae* [102, 105].

Besides previously mentioned challenges, other aspects limit the full potential of CBP. Genetically engineered microorganisms may present problems regarding their genetic stability, and productivity [106]. Ethanol yields are lower, and fermentation is longer (taking between 3 to 12 days) when using organisms like bacteria and fungi [107]. Further, lignocellulose degradation rates of some modified non-lignocellulolytic microorganisms remain notably lower than those exhibited by lignocellulolytic fungi [106]. Additionally, the simultaneous expression of multiple cellulose-degrading genes at the appropriate levels is difficult and is still under investigation [106]. Lastly, it is important to consider that biomass pretreatments can release toxic compounds that may reduce the process performance including, furan derivatives, pentose, and hexoses, weak acids, and lignin-derived phenolic compounds [95]. Therefore, the ideal microorganisms would need to be engineered to overcome these difficulties. In the following section, we will present some fungi that are good candidates or that already have been employed in the CBP of ethanol production from LCB.

15.5 ROT FUNGI FOR DIRECT ETHANOL PRODUCTION FROM LIGNOCELLULOSE MATERIALS

At the hand of the abundant presence of biomass and the pressing need for different fuel sources, the finding of fungi with the potential to produce lignocellulolytic enzymes is a viable solution, regarding the problem. The fungi are grouped in relation to the type and characteristics of produced rot, in spite of the classification system is not accurately aimed. Table 15.1 depicts the following groups of fungi-rot.

TABLE 15.1 Examples of Fungi Belonging to Different Groups

Fungi-Rot	Examples
White rot fungi	*Agrocybe aegerita, Auricularia auricula-judae, Bjerkandera adusta, Coprinellus radians, Ceriporiopsis subvermispora, Heterobasidium annosum, Irpex lacteus, Mycena haematopus, Phanerochaete chrysosporium, Pleurotus eryngii, Phlebia radiata, Pleurotus ostreatus, Pycnoporus cinnabarinus, Trametes versicolor, Stropharia rugosoannulata,* and *Xylaria hypoxylon* [108–112]
Soft rot fungi	*Alternaria alternata, Chaetomium globosum, Ceratocystis, Kretzschmaria deusta, Paecilomyces spp, Thielavia terrestres,* and *Ustulina deusta* [108, 113, 114]
Brown rot fungi	*Coniophora puteana, Laetiporus sulphureus, Gloeophyllum trabeum, Piptoporus betulinus, Phaeolus schweinitzii Postia placenta,* and *Serpula lacriman* [95, 114]
Stain fungi	*Ophiostoma, Ceratocystis spp., Aureobasidium pullulans, Phialophora spp.,* and *Trichoderma spp* [113]

15.5.1 WHITE ROT FUNGI

White-rot fungi produce a multi enzymatic system allowing for break down the three principal polymers of the natural lignocellulosic substrates: lignin, hemicellulose, and cellulose in progressive form. This complex system may work separately or cooperatively. In particular, the ligninolytic enzymes provided by white-rot fungi have high specificity to the substrate and they are LiP, VPs, MnPs, and copper-containing phenol oxidases (laccase) [108–112].

15.5.2 SOFT ROT FUNGI

Ascomycetes and Deuteromycetes belong to soft-rot fungi and they are the noblest fungi in terms of the level of decomposition they cause. Even, they emerge before the rotting process carried out by white and brown fungi. Despite the low degradation, soft-rot fungi degrading wood in severe environmental conditions. These kinds of fungi require fix nitrogen to synthesize cellulase enzymes, in charge of depolymerizing the cellulose of the secondary cell walls, making huge cavities [108, 113, 114].

15.5.3 BROWN ROT FUNGI

Numerous fungi cause white rot compared to brown rot. However, they play an important role in the degradation of deadwood, especially, in coniferous ecosystems. The brown rot fungi firstly attack the lignin through demethoxylation, occurring a lignin partial modification, without produce lignin-degrading enzymes. Posteriorly, the cellulose tackle is carried out by the transformation of methanol to peroxide with a methanol oxidase over-expression via Fenton chemistry. Thanks to the accelerated fractionation of cellulose, the wood loses strength rapidly [95, 114].

15.5.4 STAIN FUNGI

Stain fungi are a small group of Ascomycetes and Deuteromycetes, colonizing more favorably ray parenchyma cells and resin canals with a mild degradation. The degradation impacts on the water-soluble components and extractives components as fatty acids, diterpenoids resin acids, and triglycerides of the cell wall [113].

KEYWORDS

- **consolidated bioprocessing**
- **hydrogen peroxide**
- **lignin peroxidase**
- **manganese peroxidase**
- **rhamnogalacturonan I**
- **separate hydrolysis and fermentation**

REFERENCES

1. Intergovernmental Panel on Climate Change, (2018). *Global Warming of 1.5°C*. World Meteorological Organization: Geneva.
2. Intergovernmental Panel on Climate Change, (2014). In: Stocker, T., (ed.). *Climate Change 2013: The Physical Science Basis: Working Group I Contribution to the Fifth Assessment Report of the Intergovernmental Panel on Climate Change*. Cambridge University Press: New York.
3. Jackson, R. B., Quéré, C. L., Andrew, R. M., Canadell, J. G., Peters, G. P., Roy, J., & Wu, L., (2017). Warning signs for stabilizing global CO_2 emissions. *Environ. Res. Lett., 12*(11), 110202. https://doi.org/10.1088/1748-9326/aa9662.
4. Dresselhaus, M. S., & Thomas, I. L., (2001). Alternative energy technologies. *Nature, 414*(6861), 332–337. https://doi.org/10.1038/35104599.
5. Whitesides, G. M., & Crabtree, G. W., (2007). Don't forget long-term fundamental research in energy. *Science, 315*(5813), 796–798. https://doi.org/10.1126/science.1140362.
6. Zhang, Y. H. P., (2008). Reviving the carbohydrate economy via multi-product lignocellulose biorefineries. *J. Ind. Microbiol. Biotechnol., 35*(5), 367–375. https://doi.org/10.1007/s10295-007-0293-6.
7. Chen, H., (2014). *Biotechnology of Lignocellulose: Theory and Practice*. Springer Netherlands: Dordrecht. https://doi.org/10.1007/978-94-007-6898-7.
8. Jönsson, L. J., Alriksson, B., & Nilvebrant, N. O., (2013). Bioconversion of lignocellulose: Inhibitors and detoxification. *Biotechnol. Biofuels, 6*(1), 16. https://doi.org/10.1186/1754-6834-6-16.
9. Najafi, G., Ghobadian, B., Tavakoli, T., Buttsworth, D. R., Yusaf, T. F., & Faizollahnejad, M., (2009). Performance and exhaust emissions of a gasoline engine with ethanol-blended gasoline fuels using artificial neural network. *Appl. Energy, 86*(5), 630–639. https://doi.org/10.1016/j.apenergy.2008.09.017.
10. Tilman, D., Hill, J., & Lehman, C., (2006). Carbon-negative biofuels from low-input high-diversity grassland biomass. *Science, 314*(5805), 1598–1600. https://doi.org/10.1126/science.1133306.
11. Lynd, L. R., (2017). The grand challenge of cellulosic biofuels. *Nat. Biotechnol., 35*(10), 912–915. https://doi.org/10.1038/nbt.3976.

12. Canilha, L., Chandel, A. K., Suzane, D. S. M. T., Antunes, F. A. F., Luiz Da, C. F. W., Das, G. A. F. M., & Da Silva, S. S., (2012). Bioconversion of sugarcane biomass into ethanol: An overview about composition, pretreatment methods, detoxification of hydrolysates, enzymatic saccharification, and ethanol fermentation. *J. Biomed. Biotechnol., 2012,* 1–15. https://doi.org/10.1155/2012/989572.

13. Xu, Q., Singh, A., & Himmel, M. E., (2009). Perspectives and new directions for the production of bioethanol using consolidated bioprocessing of lignocellulose. *Curr. Opin. Biotechnol., 20*(3), 364–371. https://doi.org/10.1016/j.copbio.2009.05.006.

14. Salvachúa, D., Karp, E. M., Nimlos, C. T., Vardon, D. R., & Beckham, G. T., (2015). Towards lignin consolidated bioprocessing: Simultaneous lignin depolymerization and product generation by bacteria. *Green Chem., 17*(11), 4951–4967. https://doi.org/ 10.1039/ C5GC01165E.

15. Cosgrove, D. J., (2005). Growth of the plant cell wall. *Nat. Rev. Mol. Cell Biol., 6*(11), 850–861. https://doi.org/10.1038/nrm1746.

16. Caffall, K. H., & Mohnen, D., (2009). The structure, function, and biosynthesis of plant cell wall pectic polysaccharides. *Carbohydr. Res., 344*(14), 1879–1900. https://doi. org/10.1016/j.carres.2009.05.021.

17. Vanholme, R., Morreel, K., Ralph, J., & Boerjan, W., (2008). Lignin engineering. *Curr. Opin. Plant Biol., 11*(3), 278–285. https://doi.org/10.1016/j.pbi.2008.03.005.

18. Ragauskas, A. J., Beckham, G. T., Biddy, M. J., Chandra, R., Chen, F., Davis, M. F., Davison, B. H., et al., (2014). Lignin valorization: Improving lignin processing in the biorefinery. *Science, 344*(6185), 1246843–1246843. https://doi.org/10.1126/science.1246843.

19. Vanholme, R., De Meester, B., Ralph, J., & Boerjan, W., (2019). Lignin biosynthesis and its integration into metabolism. *Curr. Opin. Biotechnol., 56,* 230–239. https://doi. org/10.1016/j.copbio.2019.02.018.

20. Boerjan, W., Ralph, J., & Baucher, M., (2003). Lignin biosynthesis. *Annu. Rev. Plant Biol., 54*(1), 519–546. https://doi.org/10.1146/annurev.arplant.54.031902.134938.

21. Brett, C. T., (2000). Cellulose microfibrils in plants: Biosynthesis, deposition, and integration into the cell wall. In: *International Review of Cytology* (Vol. 199, pp. 161--199). Elsevier. https://doi.org/10.1016/S0074-7696(00)99004-1.

22. Klemm, D., Heublein, B., Fink, H. P., & Bohn, A., (2005). Cellulose: Fascinating biopolymer and sustainable raw material. *Angew. Chem. Int. Ed., 44*(22), 3358–3393. https://doi.org/10.1002/anie.200460587.

23. Foyle, T., Jennings, L., & Mulcahy, P., (2007). Compositional analysis of lignocellulosic materials: Evaluation of methods used for sugar analysis of waste paper and straw. *Bioresour. Technol., 98*(16), 3026–3036. https://doi.org/10.1016/j.biortech.2006.10.013.

24. Somerville, C., (2006). Cellulose synthesis in higher plants. *Annu. Rev. Cell Dev. Biol., 22*(1), 53–78. https://doi.org/10.1146/annurev.cellbio.22.022206.160206.

25. Pauly, M., Gille, S., Liu, L., Mansoori, N., De Souza, A., Schultink, A., & Xiong, G., (2013). Hemicellulose biosynthesis. *Planta, 238*(4), 627–642. https://doi.org/10.1007/ s00425-013-1921-1.

26. Scheller, H. V., & Ulvskov, P., (2010). Hemicelluloses. *Annu. Rev. Plant Biol., 61*(1), 263--289. https://doi.org/10.1146/annurev-arplant-042809-112315.

27. Mohnen, D., (2008). Pectin structure and biosynthesis. *Curr. Opin. Plant Biol., 11*(3), 266--277. https://doi.org/10.1016/j.pbi.2008.03.006.

28. Harholt, J., Suttangkakul, A., & Scheller, H. V., (2010). Biosynthesis of pectin. *Plant Physiol., 153*(2), 384–395. https://doi.org/10.1104/pp.110.156588.

29. Cassab, G. I., (1998). Plant cell wall proteins. *Annu. Rev. Plant Physiol. Plant Mol. Biol., 49*(1), 281–309. https://doi.org/10.1146/annurev.arplant.49.1.281.

30. De Gonzalo, G., Colpa, D. I., Habib, M. H. M., & Fraaije, M. W., (2016). Bacterial enzymes involved in lignin degradation. *J. Biotechnol., 236,* 110–119. https://doi.org/10.1016/j.jbiotec.2016.08.011.

31. Abdel-Hamid, A. M., Solbiati, J. O., & Cann, I. K. O., (2013). Insights into lignin degradation and its potential industrial applications. In: Sariaslani, S., & Gadd, G. M., (eds.), *Advances in Applied Microbiology* (Vol. 82, pp. 1–28). Academic Press. https://doi.org/10.1016/B978-0-12-407679-2.00001-6.

32. Kinnunen, A., Maijala, P., Jarvinen, P., & Hatakka, A., (2017). Improved efficiency in screening for lignin-modifying peroxidases and laccases of basidiomycetes. *Curr. Biotechnol., 6*(2), 105–115. https://doi.org/10.2174/2211550105666160330205138.

33. Tuomela, M., & Hatakka, A., (2019). Oxidative fungal enzymes for bioremediation. In: Moo-Young, M., (ed.), *Comprehensive Biotechnology* (3rd edn., pp. 224–239); Pergamon: Oxford. https://doi.org/10.1016/B978-0-444-64046-8.00349-9.

34. Shen, H., Poovaiah, C. R., Ziebell, A., Tschaplinski, T. J., Pattathil, S., Gjersing, E., Engle, N. L., et al., (2013). Enhanced characteristics of genetically modified switchgrass (Panicum Virgatum L.) for high biofuel production. *Biotechnol. Biofuels, 6*(1), 71. https://doi.org/10.1186/1754-6834-6-71.

35. Hamid, M., & Khalil-Ur-Rehman, (2009). Potential applications of peroxidases. *Food Chem., 115*(4), 1177–1186. https://doi.org/10.1016/j.foodchem.2009.02.035.

36. Hofrichter, M., Ullrich, R., Pecyna, M. J., Liers, C., & Lundell, T., (2010). New and classic families of secreted fungal heme peroxidases. *Appl. Microbiol. Biotechnol., 87*(3), 871–897. https://doi.org/10.1007/s00253-010-2633-0.

37. Passardi, F., Theiler, G., Zamocky, M., Cosio, C., Rouhier, N., Teixera, F., Margis-Pinheiro, M., et al., (2007). PeroxiBase: The peroxidase database. *Phytochemistry, 68*(12), 1605–1611. https://doi.org/10.1016/j.phytochem.2007.04.005.

38. Savelli, B., Li, Q., Webber, M., Jemmat, A. M., Robitaille, A., Zamocky, M., Mathé, C., & Dunand, C., (2019). RedoxiBase: A database for ROS homeostasis regulated proteins. *Redox Biol., 26,* 101247. https://doi.org/10.1016/j.redox.2019.101247.

39. Passardi, F., Bakalovic, N., Teixeira, F. K., Margis-Pinheiro, M., Penel, C., & Dunand, C., (2007). Prokaryotic origins of the non-animal peroxidase superfamily and organelle-mediated transmission to eukaryotes. *Genomics, 89*(5), 567–579. https://doi.org/10.1016/j.ygeno.2007.01.006.

40. Sugano, Y., (2009). DyP-type peroxidases comprise a novel heme peroxidase family. *Cell. Mol. Life Sci., 66*(8), 1387–1403. https://doi.org/10.1007/s00018-008-8651-8.

41. Zámocký, M., & Obinger, C., (2010). Molecular phylogeny of heme peroxidases. In: Torres, E., & Ayala, M., (eds.), *Biocatalysis Based on Heme Peroxidases: Peroxidases as Potential Industrial Biocatalysts* (pp. 7–35). Springer: Berlin, Heidelberg. https://doi.org/10.1007/978-3-642-12627-7_2.

42. Zámocký, M., Hofbauer, S., Schaffner, I., Gasselhuber, B., Nicolussi, A., Soudi, M., Pirker, K. F., et al., (2015). independent evolution of four heme peroxidase superfamilies. *Arch. Biochem. Biophys., 574,* 108–119. https://doi.org/10.1016/j.abb.2014.12.025.

43. Paliwal, R., Rawat, A. P., Rawat, M., & Rai, J. P. N., (2012). Bioligninolysis: Recent updates for biotechnological solution. *Appl. Biochem. Biotechnol., 167*(7), 1865–1889. https://doi.org/10.1007/s12010-012-9735-3.

44. Johansson, T., & Nyman, P. O., (1993). Isozymes of lignin peroxidase and manganese (II) peroxidase from the white-rot basidiomycete *Trametes versicolor*. *Arch. Biochem. Biophys., 300*(1), 49–56. https://doi.org/10.1006/abbi.1993.1007.
45. Tien, M., & Kirk, T. K., (1983). Lignin-degrading enzyme from the hymenomycete *Phanerochaete chrysosporium* burds. *Science, 221*(4611), 661–663. https://doi.org/10.1126/science.221.4611.661.
46. Hirai, H., Sugiura, M., Kawai, S., & Nishida, T., (2005). Characteristics of novel lignin peroxidases produced by white-rot fungus *Phanerochaete sordida* YK-624. *FEMS Microbiol. Lett., 246*(1), 19–24. https://doi.org/10.1016/j.femsle.2005.03.032.
47. Kirk, T. K., Tien, M., Kersten, P. J., Mozuch, M. D., & Kalyanaraman, B., (1986). Ligninase of *Phanerochaete chrysosporium*. Mechanism of its degradation of the non-phenolic arylglycerol beta-aryl ether substructure of lignin. *Biochem. J., 236*(1), 279–287.
48. Baciocchi, E., Gerini, M. F., Lanzalunga, O., Lapi, A., Lo Piparo, M. G., & Mancinelli, S., (2001). Isotope-effect profiles in the oxidative N-demethylation of N,N-dimethylanilines catalyzed by lignin peroxidase and a chemical model. *Eur. J. Org. Chem., 2001*(12), 2305–2310. https://doi.org/10.1002/1099-0690(200106)2001:12<2305::AID-EJOC2305>3.0.CO;2-E.
49. Wong, D. W. S., (1995). *Food Enzymes.* Springer US: Boston, MA. https://doi.org/10.1007/978-1-4757-2349-6.
50. Falade, A. O., Nwodo, U. U., Iweriebor, B. C., Green, E., Mabinya, L. V., & Okoh, A. I., (2017). Lignin peroxidase functionalities and prospective applications. *Microbiology Open, 6*(1), e00394. https://doi.org/10.1002/mbo3.394.
51. Dashtban, M., Schraft, H., Syed, T. A., & Qin, W., (2010). Fungal biodegradation and enzymatic modification of lignin. *Int. J. Biochem. Mol. Biol., 1*(1), 36–50.
52. Zhang, H., Zhang, J., Zhang, X., & Geng, A., (2018). Purification and characterization of a novel manganese peroxidase from white-rot fungus *Cerrena unicolor* BBP6 and its application in dye decolorization and denim bleaching. *Process Biochem., 66*, 222–229. https://doi.org/10.1016/j.procbio.2017.12.011.
53. Lobos, S., Larraín, J., Salas, L., Cullen, D., & Vicuña, R., (1994). Isoenzymes of manganese-dependent peroxidase and laccase produced by the lignin-degrading basidiomycete *Ceriporiopsis subvermispora*. *Microbiol. Read. Engl., 140*(Pt 10), 2691–2698. https://doi.org/10.1099/00221287-140-10-2691.
54. Chowdhary, P., Shukla, G., Raj, G., Ferreira, L. F. R., & Bharagava, R. N., (2018). Microbial manganese peroxidase: A ligninolytic enzyme and its ample opportunities in research. *SN Appl. Sci., 1*(1), 45. https://doi.org/10.1007/s42452-018-0046-3.
55. Reddy, G. V. B., Sridhar, M., & Gold, M. H., (2003). Cleavage of nonphenolic β-1 diarylpropane lignin model dimers by manganese peroxidase from *Phanerochaete chrysosporium*. *Eur. J. Biochem., 270*(2), 284–292. https://doi.org/10.1046/j.1432-1033. 2003.03386.x.
56. Moreira, P. R., Duez, C., Dehareng, D., Antunes, A., Almeida-Vara, E., Frère, J. M., Malcata, F. X., & Duarte, J. C., (2005). Molecular characterization of a versatile peroxidase from a *Bjerkandera* strain. *J. Biotechnol., 118*(4), 339–352. https://doi.org/10.1016/j.jbiotec.2005.05.014.
57. Kamitsuji, H., Honda, Y., Watanabe, T., & Kuwahara, M., (2005). Mn^{2+} is dispensable for the production of active MnP2 by *Pleurotus ostreatus*. *Biochem. Biophys. Res. Commun., 327*(3), 871–876. https://doi.org/10.1016/j.bbrc.2004.12.084.
58. Ruiz-Dueñas, F. J., Martínez, M. J., & Martínez, A. T., (1999). Molecular characterization of a novel peroxidase isolated from the ligninolytic fungus *Pleurotus eryngii*. *Mol. Microbiol., 31*(1), 223–235. https://doi.org/10.1046/j.1365-2958.1999.01164.x.

59. Wong, D. W. S., (2009). Structure and action mechanism of ligninolytic enzymes. *Appl. Biochem. Biotechnol., 157*(2), 174–209. https://doi.org/10.1007/s12010-008-8279-z.

60. Verdín, J., Pogni, R., Baeza, A., Baratto, M. C., Basosi, R., & Vázquez-Duhalt, R., (2006). Mechanism of versatile peroxidase inactivation by Ca^{2+} depletion. *Biophys. Chem., 121*(3), 163–170. https://doi.org/10.1016/j.bpc.2006.01.007.

61. Yoshida, T., & Sugano, Y., (2015). A structural and functional perspective of DyP-type peroxidase family. *Arch. Biochem. Biophys., 574*, 49–55. https://doi.org/10.1016/j.abb.2015.01.022.

62. Colpa, D. I., Fraaije, M. W., & Van, B. E., (2014). DyP-type peroxidases: A promising and versatile class of enzymes. *J. Ind. Microbiol. Biotechnol., 41*(1), 1–7. https://doi.org/10.1007/s10295-013-1371-6.

63. Salvachúa, D., Prieto, A., Martínez, Á. T., & Martínez, M. J., (2013). Characterization of a novel dye-decolorizing peroxidase (DyP)-type enzyme from *Irpex Lacteus* and its application in enzymatic hydrolysis of wheat straw. *Appl. Environ. Microbiol., 79*(14), 4316–4324. https://doi.org/10.1128/AEM.00699-13.

64. Liers, C., Bobeth, C., Pecyna, M., Ullrich, R., & Hofrichter, M., (2010). DyP-like peroxidases of the jelly fungus *Auricularia Auricula-Judae* oxidize nonphenolic lignin model compounds and high-redox potential dyes. *Appl. Microbiol. Biotechnol., 85*(6), 1869–1879. https://doi.org/10.1007/s00253-009-2173-7.

65. Claus, H., (2004). Laccases: Structure, reactions, distribution. *Micron, 35*(1), 93–96. https://doi.org/10.1016/j.micron.2003.10.029.

66. Baldrian, P., (2006). Fungal laccases - occurrence and properties. *FEMS Microbiol. Rev., 30*(2), 215–242. https://doi.org/10.1111/j.1574-4976.2005.00010.x.

67. Riva, S., (2006). Laccases: Blue enzymes for green chemistry. *Trends Biotechnol., 24*(5), 219–226. https://doi.org/10.1016/j.tibtech.2006.03.006.

68. Giardina, P., Faraco, V., Pezzella, C., Piscitelli, A., Vanhulle, S., & Sannia, G., (2010). Laccases: A never-ending story. *Cell. Mol. Life Sci., 67*(3), 369–385. https://doi.org/10.1007/s00018-009-0169-1.

69. Teeri, T. T., (1997). Crystalline cellulose degradation: New insight into the function of cellobiohydrolases. *Trends Biotechnol., 15*(5), 160–167. https://doi.org/10.1016/S0167-7799(97)01032-9.

70. Li, X., Chang, S. H., & Liu, R., (2018). Industrial applications of cellulases and hemicellulases. In: Fang, X., & Qu, Y., (eds.), *Fungal Cellulolytic Enzymes* (pp. 267–282). Springer Singapore: Singapore. https://doi.org/10.1007/978-981-13-0749-2_15.

71. Bhat, M. K., (2000). Cellulases and related enzymes in biotechnology. *Biotechnol. Adv., 18*(5), 355–383. https://doi.org/10.1016/S0734-9750(00)00041-0.

72. Kuhad, R. C., Gupta, R., & Singh, A., (2011). Microbial cellulases and their industrial applications. *Enzyme Res., 2011*, 1–10. https://doi.org/10.4061/2011/280696.

73. Wilson, D. B., (2009). Cellulases and biofuels. *Curr. Opin. Biotechnol., 20*(3), 295–299. https://doi.org/10.1016/j.copbio.2009.05.007.

74. Doi, R. H., & Kosugi, A., (2004). Cellulosomes: Plant-cell-wall-degrading enzyme complexes. *Nat. Rev. Microbiol., 2*(7), 541–551. https://doi.org/10.1038/nrmicro925.

75. Juturu, V., & Wu, J. C., (2014). Microbial cellulases: Engineering, production and applications. *Renew. Sustain. Energy Rev., 33*, 188–203. https://doi.org/10.1016/j.rser.2014.01.077.

76. Payne, C. M., Knott, B. C., Mayes, H. B., Hansson, H., Himmel, M. E., Sandgren, M., Ståhlberg, J., & Beckham, G. T., (2015). Fungal cellulases. *Chem. Rev., 115*(3), 1308–1448. https://doi.org/10.1021/cr500351c.

77. Gao, D., Uppugundla, N., Chundawat, S. P., Yu, X., Hermanson, S., Gowda, K., Brumm, P., et al., (2011). Hemicellulases and auxiliary enzymes for improved conversion of lignocellulosic biomass to monosaccharides. *Biotechnol. Biofuels, 4*(1), 5. https://doi.org/10.1186/1754-6834-4-5.

78. Shallom, D., & Shoham, Y., (2003). Microbial hemicellulases. *Curr. Opin. Microbiol., 6*(3), 219–228. https://doi.org/10.1016/S1369-5274(03)00056-0.

79. Houfani, A. A., Anders, N., Spiess, A. C., Baldrian, P., & Benallaoua, S., (2020). Insights from enzymatic degradation of cellulose and hemicellulose to fermentable sugars: A review. *Biomass Bioenergy, 134*, 105481. https://doi.org/10.1016/j.biombioe.2020.105481.

80. Anderson, C. T., (2016). We be jammin': An update on pectin biosynthesis, trafficking and dynamics. *J. Exp. Bot., 67*(2), 495–502. https://doi.org/10.1093/jxb/erv501.

81. Xiao, C., & Anderson, C. T., (2013). Roles of pectin in biomass yield and processing for biofuels. *Front. Plant Sci., 4*. https://doi.org/10.3389/fpls.2013.00067.

82. Gummadi, S. N., & Panda, T., (2003). Purification and biochemical properties of microbial pectinases: A review. *Process Biochem., 38*(7), 987–996. https://doi.org/10.1016/S0032-9592(02)00203-0.

83. Pedrolli, D. B., Monteiro, A. C., Gomes, E., & Carmona, E. C., (2009). Pectin and pectinases: Production, characterization and industrial application of microbial pectinolytic enzymes. *Open Biotechnol. J., 3*(1), 9–18. https://doi.org/10.2174/1874070700903010009.

84. Singh, R. S., Singh, T., & Pandey, A., (2019). Microbial enzymes—An overview. In *Advances in Enzyme Technology* (pp. 1–40). Elsevier. https://doi.org/10.1016/B978-0-444-64114-4.00001-7.

85. Kashyap, D. R., Vohra, P. K., Chopra, S., & Tewari, R., (2001). Applications of pectinases in the commercial sector: A review. *Bioresour. Technol., 77*(3), 215–227. https://doi.org/10.1016/S0960-8524(00)00118-8.

86. Zhou, C. H., Xia, X., Lin, C. X., Tong, D. S., & Beltramini, J., (2011). Catalytic conversion of lignocellulosic biomass to fine chemicals and fuels. *Chem. Soc. Rev., 40*(11), 5588–5617. https://doi.org/10.1039/C1CS15124J.

87. Verbančič, J., Lunn, J. E., Stitt, M., & Persson, S., (2018). Carbon supply and the regulation of cell wall synthesis. *Mol. Plant, 11*(1), 75–94. https://doi.org/10.1016/j.molp.2017.10.004.

88. O'Leary, B. M., (2020). Another brick in the plant cell wall: Characterization of *Arabidopsis* CSLD3 function in cell wall synthesis. *Plant Cell, 32*(5), 1359–1360. https://doi.org/10.1105/tpc.20.00190.

89. Biswal, A. K., Atmodjo, M. A., Li, M., Baxter, H. L., Yoo, C. G., Pu, Y., Lee, Y. C., et al., (2018). Sugar release and growth of biofuel crops are improved by downregulation of pectin biosynthesis. *Nat. Biotechnol., 36*(3), 249–257. https://doi.org/10.1038/nbt.4067.

90. Bhandari, S., Fujino, T., Thammanagowda, S., Zhang, D., Xu, F., & Joshi, C. P., (2006). Xylem-specific and tension stress-responsive co-expression of Korrigan endoglucanase and three secondary wall-associated cellulose synthase genes in aspen trees. *Planta, 224*(4), 828–837. https://doi.org/10.1007/s00425-006-0269-1.

91. Szyjanowicz, P. M. J., McKinnon, I., Taylor, N. G., Gardiner, J., Jarvis, M. C., & Turner, S. R., (2004). The irregular xylem 2 mutant is an allele of Korrigan that affects the secondary cell wall of *Arabidopsis thaliana*. *Plant J., 37*(5), 730–740.

92. McFarlane, H. E., Döring, A., & Persson, S., (2014). The cell biology of cellulose synthesis. *Annu. Rev. Plant Biol., 65*(1), 69–94. https://doi.org/10.1146/annurev-arplant-050213-040240.

93. Van, M. A. J. A., Abbott, D. A., Bellissimi, E., Van, D. B. J., Kuyper, M., Luttik, M. A. H., Wisselink, H. W., et al., (2006). Alcoholic fermentation of carbon sources in biomass hydrolysates by *Saccharomyces cerevisiae*: Current status. *Antonie Van Leeuwenhoek, 90*(4), 391–418. https://doi.org/10.1007/s10482-006-9085-7.

94. Schuster, B. G., & Chinn, M. S., (2013). Consolidated bioprocessing of lignocellulosic feedstocks for ethanol fuel production. *BioEnergy Res., 6*(2), 416–435. https://doi.org/10.1007/s12155-012-9278-z.

95. Plácido, J., & Capareda, S., (2015). Ligninolytic enzymes: A biotechnological alternative for bioethanol production. *Bioresour. Bioprocess, 2*(1), 23. https://doi.org/10.1186/s40643-015-0049-5.

96. Lin, Y., & Tanaka, S., (2006). Ethanol fermentation from biomass resources: Current state and prospects. *Appl. Microbiol. Biotechnol., 69*(6), 627–642. https://doi.org/10.1007/s00253-005-0229-x.

97. Amore, A., & Faraco, V., (2012). Potential of fungi as category I consolidated bioprocessing organisms for cellulosic ethanol production. *Renew. Sustain. Energy Rev., 16*(5), 3286–3301. https://doi.org/10.1016/j.rser.2012.02.050.

98. Ishola, M., Jahandideh, A., Haidarian, B., Brandberg, T., & Taherzadeh, M., (2013). Simultaneous saccharification, filtration and fermentation (SSFF): A novel method for bioethanol production from lignocellulosic biomass. *Bioresour. Technol., 133*, 68–73. https://doi.org/10.1016/j.biortech.2013.01.130.

99. Salehi, J. G., & Taherzadeh, M. J., (2015). Advances in consolidated bioprocessing systems for bioethanol and butanol production from biomass: A comprehensive review. *Biofuel Res. J., 2*(1), 152–195. https://doi.org/10.18331/BRJ2015.2.1.4.

100. Zhang, J., & Lynd, L. R., (2010). Ethanol production from paper sludge by simultaneous saccharification and co-fermentation using recombinant xylose-fermenting microorganisms. *Biotechnol. Bioeng., 107*(2), 235–244. https://doi.org/10.1002/bit.22811.

101. Sivasubramanian, V., (2018). *Bioprocess Engineering for a Green Environment*. CRC Press.

102. Olson, D. G., McBride, J. E., Joe, S. A., & Lynd, L. R., (2012). Recent progress in consolidated bioprocessing. *Curr. Opin. Biotechnol., 23*(3), 396–405. https://doi.org/10.1016/j.copbio.2011.11.026.

103. Kumagai, A., Kawamura, S., Lee, S. H., Endo, T., Rodriguez, M., & Mielenz, J. R., (2014). Simultaneous saccharification and fermentation and a consolidated bioprocessing for hinoki cypress and eucalyptus after fibrillation by steam and subsequent wet-disk milling. *Bioresour. Technol., 162*, 89–95. https://doi.org/10.1016/j.biortech.2014.03.110.

104. Kamei, I., Hirota, Y., Mori, T., Hirai, H., Meguro, S., & Kondo, R., (2012). Direct ethanol production from cellulosic materials by the hypersaline-tolerant white-rot fungus *Phlebia Sp*. MG-60. *Bioresour. Technol., 112*, 137–142. https://doi.org/10.1016/j.biortech.2012.02.109.

105. Hasunuma, T., & Kondo, A., (2012). Consolidated bioprocessing and simultaneous saccharification and fermentation of lignocellulose to ethanol with thermotolerant yeast strains. *Process Biochem., 47*(9), 1287–1294. https://doi.org/10.1016/j.procbio.2012.05.004.

106. Den, H. R., Van, R. E., Rose, S. H., Görgens, J. F., & Van, Z. W. H., (2015). Progress and challenges in the engineering of non-cellulolytic microorganisms for consolidated bioprocessing. *Curr. Opin. Biotechnol., 33*, 32–38. https://doi.org/10.1016/j.copbio.2014.10.003.

107. Sarkar, N., Ghosh, S. K., Bannerjee, S., & Aikat, K., (2012). Bioethanol production from agricultural wastes: An overview. *Renew. Energy, 37*(1), 19–27. https://doi.org/10.1016/j.renene.2011.06.045.

108. Liers, C., Arnstadt, T., Ullrich, R., & Hofrichter, M., (2011). Patterns of lignin degradation and oxidative enzyme secretion by different wood- and litter-colonizing basidiomycetes and ascomycetes grown on beech-wood. *FEMS Microbiol. Ecol., 78*(1), 91–102. https://doi.org/10.1111/j.1574-6941.2011.01144.x.

109. Martínez, A. T., Ruiz-Dueñas, F. J., Camarero, S., Serrano, A., Linde, D., Lund, H., Vind, J., et al., (2017). Oxidoreductases on their way to industrial biotransformations. *Biotechnol. Adv., 35*(6), 815–831. https://doi.org/10.1016/j.biotechadv.2017.06.003.

110. Linde, D., Ruiz-Dueñas, F. J., Fernández-Fueyo, E., Guallar, V., Hammel, K. E., Pogni, R., & Martínez, A. T., (2015). Basidiomycete DyPs: Genomic diversity, structural-functional aspects, reaction mechanism and environmental significance. *Arch. Biochem. Biophys., 574*, 66–74. https://doi.org/10.1016/j.abb.2015.01.018.

111. Fernández-Fueyo, E., Castanera, R., Ruiz-Dueñas, F. J., López-Lucendo, M. F., Ramírez, L., Pisabarro, A. G., & Martínez, A. T., (2014). Ligninolytic peroxidase gene expression by *Pleurotus ostreatus*: Differential regulation in lignocellulose medium and effect of temperature and PH. *Fungal Genet. Biol., 72*, 150–161. https://doi.org/10.1016/j.fgb.2014.02.003.

112. Kumar, A., & Chandra, R., (2020). Ligninolytic enzymes and its mechanisms for degradation of lignocellulosic waste in the environment. *Heliyon, 6*(2), e03170. https://doi.org/10.1016/j.heliyon.2020.e03170.

113. Martínez, Á. T., Speranza, M., Ruiz-Dueñas, F. J., Ferreira, P., Camarero, S., Guillén, F., Martínez, M. J., et al., (2005). Biodegradation of lignocellulosics: Microbial, chemical, and enzymatic aspects of the fungal attack of lignin. *Int. Microbiol., 8*(3), 195–204. https://doi.org/10.2436/im.v8i3.9526.

114. Martínez, Á. T., Ruiz-Dueñas, F. J., Martínez, M. J., Del, R. J. C., & Gutiérrez, A., (2009). Enzymatic delignification of plant cell wall: From nature to mill. *Curr. Opin. Biotechnol., 20*(3), 348–357. https://doi.org/10.1016/j.copbio.2009.05.002.

CHAPTER 16

Bioelectrosynthesis of Ethanol via Bioelectrochemical Systems: Future Perspectives

SILVIA YUDITH MARTÍNEZ AMADOR,[1]
LEOPOLDO JAVIER RÍOS GONZÁLEZ,[2]
MIGUEL ANGEL MEDINA MORALES,[2]
MÓNICA MARGARITA RODRÍGUEZ GARZA,[2]
ILEANA MAYELA MARÍA MORENO DÁVILA,[2]
THELMA KARINA MORALES MARTÍNEZ,[3] and
JOSÉ ANTONIO RODRÍGUEZ DE LA GARZA[2]

[1]Botany Department, Agronomic Division, Antonio Narro Autonomous Agrarian University, Mexico

[2]Department of Biotechnology, School of Chemistry, Autonomous University of Coahuila, Mexico, Phone: +52 (844) 415-5752–Ext 120, E-mail: antonio.rodriguez@uadec.edu.mx (J. A. R. D. L. Garza)

[3]Bioprocess and Microbial Biochemistry Group, School of Chemistry, Autonomous University of Coahuila, Mexico

ABSTRACT

Microbial electrosynthesis has gained much attention in the last decade in the scientific community due to that this novel process allows the production of commodity chemicals sustainably. MES (Microbial electrosynthesis cell), as commonly referred to as this technology, is a bioelectrochemical system that utilizes the reducing power generated from the anodic oxidation to produce high value-added products in the cathode. When coupled with a renewable source of electricity and organic or inorganic substances to produce biofuels, such as alcohols (ethanol, butanol, etc.), significant advantages stand out, such as the avoidance of fossil fuels to provide the energy that drives the process and food crops that serve as feedstock. The present chapter will give an insight into the novel perspectives of bioethanol production through bioelectrosynthesis as well as the more recent and relevant developments in MES technology for this purpose.

16.1   INTRODUCTION

2020 has been a year that has not only have cause several health issues all around the world but also has shown the fragility of the global economy [1] and its dependence on fossil fuel. This dependence has been more of a reminder of the need to search, and develop sustainable fuel production processes to fulfill our energy needs and, additionally, options that have higher market stability than oil prices [2–5]. In this regard, biofuels were considered an alternative to gradually replace fossil fuel over the next 30 years; however, many concerns have arisen mainly due to the feedstocks used to produce biofuels, generating food security issues all around the globe. In the past decade, it has been clear that future energy demand will be met by a combination of renewable energy (solar, hydro, thermal, biofuels, and eolic) and fossil fuels [6–8].

As stated initially, this year (2020) has presented several challenges; however, also it has opened new opportunities. The Contraction in global oil consumption has opened the windows for the renewable energy market; additionally, the energy market is being subjected to pressure to lower its gas emissions to fulfill the objectives of reverting or at least tackling the climate change issue [2–5].

The use of renewable energy, combined with bioenergy, will allow us to be more independent in energy matters and fight climate change. Among bioenergy, bio-alcohols as high value-added products from agro-industrial waste as an organic carbon source are well established in the energy market. Among bioalcohols, bioethanol is by far, the most produced biofuel in the globe, being Brazil and the United States the dominant players in bioethanol production, representing almost 80% of the global output. However, as mentioned earlier, conflicts over food supply and land use have made its production and utilization a controversial issue. To overcome this issue, in the last decade, second-generation bioethanol production technology has focused on the use of non-edible (human consumption) biomass sources, using a wide range of feedstocks, such as LCB, energy crops, cellulosic residues, and, particularly, wastes [9–11]. The quality of bioethanol produced is majorly dependent on the production routes. Bioethanol production, in general, consists of several sequential procedures, namely pre-treatment, hydrolysis, fermentation, and distillation. Each stage can have more than one possible route, and each alternative route can render different results in ethanol quality, as well as overall production cost [12].

Bioelectrosynthesis is a novel hybrid biobased process that involves electrochemical approaches to utilize bioelectrochemical systems to convert dissolved CO_2 into value-added organic compounds. In this regard, bioethanol can be produced by microbial reduction of acetate as the main intermediate of anaerobic digestion (AD) with hydrogen as an electron donor [13–15]. Bioethanol production via microbial bioeletroshyntesis (MES) from CO_2-waste streams, could alleviate the concern of food security-related issues mentioned before. Although ethanol can be produced directly from CO_2, its production from acetate is thermodynamically and energetically more favorable, providing that the undissociated acid exists in a slightly acidic medium [13–15]. Still, ethanol production through MES is far from becoming a feasible process, yet as titers and efficiencies are relatively low. The following book chapter will be focused on describing briefly on describing the current development on BES and its multiple application, followed by exploring the use of BES for ethanol production by means of MES. It will be discussed how can BES compete, in terms of energy usage, with conventional bioethanol production technologies by exploiting the energy content of some of the chemicals present in the waste biomass and CO_2 waste streams.

16.2 BIOELECTROCHEMICAL SYSTEMS OVERVIEW

In the early XX century, Potter [16] reported the production of electricity associated with microbial catabolism, for the first time experimentally demonstrating the capability of microorganisms to transport electrons extra-cellularly when degrading organic compounds [17–20]. The first microbial fuel cell (MFC) was built by Cohen [21]. However, the importance of the work was not recognized until the energy crisis got worse globally in the late 1990s [4, 5], re-stimulating the concerns in biofuel cells; since then, the research activities with bioelectrochemical systems as topics began to blossom resulting in an exponential increment in the number of publications, as shown in Figure 16.1 [22]. The concept of microbial electrolysis cell (MEC) was originated in 2005. Two different research groups independently noticed the production of hydrogen gas in an electrolysis-type process modified from MFC nearly at the same time [23, 24]. This process has previously been named as "electrochemically assisted microbial reactor (BEAMR)," then "biocatalyzed electrolysis." Finally, it was defined as "electrohydrogenesis" or "microbial electrolysis" to distinguish the hydrogen evolution in an MEC from dark fermentation [25–30]. Shortly afterwards, the function of MEC for CO_2 bioconversion to methane was discovered as well, this is, "electromethanogenesis" that we now know [15, 31–37]. Based on these initial efforts, a great deal of MEC configurations with different purposes have been designed and proposed very recently, such as microbial electrodialysis cell (MEDC), microbial reverse electrodialysis electrolysis cell (MREC), microbial desalination cell (MDC), etc., substantially extending the scope of MEC applications. From then onwards, a new era belonging to MEC is approaching (Figure 16.2) [29, 30].

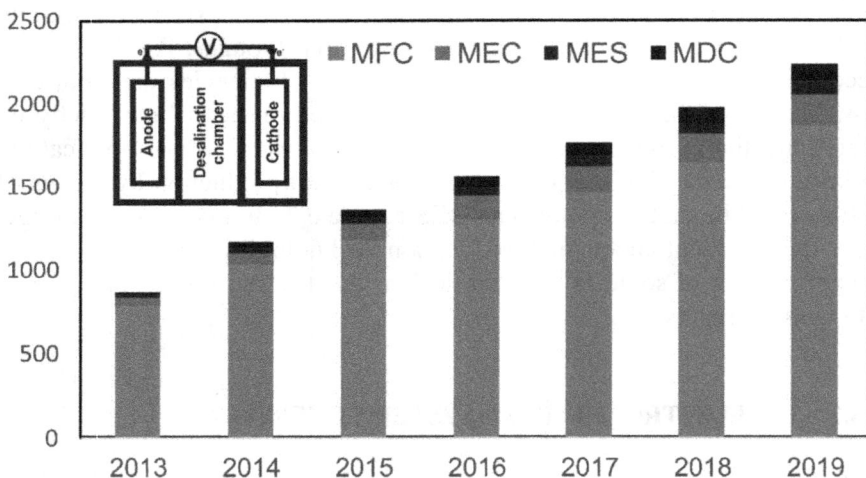

FIGURE 16.1 Number of published journal articles on MES containing the phrases "microbial fuel cell;" "microbial electrolysis cell;" "microbial electrosynthesis;" or "microbial desalination cell."

Source: Adapted from: 2013–2019 PubMed Central.
Source: Based on figure presented by Wang and Ren [22].

16.3   APPLICATIONS OF BIOELECTROCHEMICAL SYSTEMS

The different types of BES can distinguish from one another by the main function and the output of the BES. For example, a microbial fuel cell (MFC) is the most characteristic type of BES, whose main function is the generation of direct electricity. When an external power supply is added to an MFC reactor to reduce the potential of the cathode, the system becomes a microbial electrolysis cell (MEC), where hydrogen gas and other products are generated [22]. Additionally, the potential input required in a MEC can be provided by a MFC so that hydrogen can be produced directly from biomass and waste sources [38]. If the primary objective of the BES is the cathode to reduce oxidized contaminants, such as uranium, perchlorate or chlorinated solvents, it can be called a microbial remediation cell. If the main objective of the system is to synthesize value-added chemicals through cathodic reduction by microorganisms, the system can be named microbial electrosynthesis (MES). Another application for BES is water desalination, in this type of BES an additional chamber between the anode and cathode is required and this type of BES has been named as microbial desalination cell or MDS [22]. Nutrient removal or recovery can also be carried out by a BES, in the case of nitrogen, it can be removed through bioelectrochemical denitrification or recovered vía ammonium migration driven by electricity generation (Figure 16.3) [20, 39, 40].

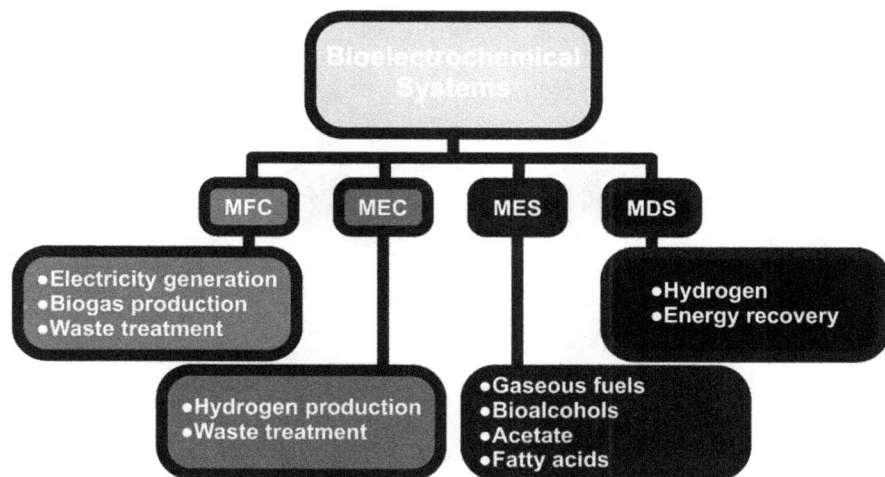

FIGURE 16.2 Overview of various types of bioelectrochemical systems and main products.

16.4 MICROBIAL FUEL CELLS (MFCS)

A MFC is a bioreactor that focuses on the production of electricity from biodegradable materials through microbial catalytic reactions [29, 42–44]. Electricity is produced by the oxidation of organic carbon in the anode under anaerobic conditions, generating electrons, protons, and products such as volatile fatty acids, biomass, and carbon dioxide (CO_2) [45]. This type of BES has the potential of a dual application, treating different types of waste streams, such as wastewater, and during this process, an electric current will be produced. In this approach, it generates an alternative energy source from wastes [46, 47].

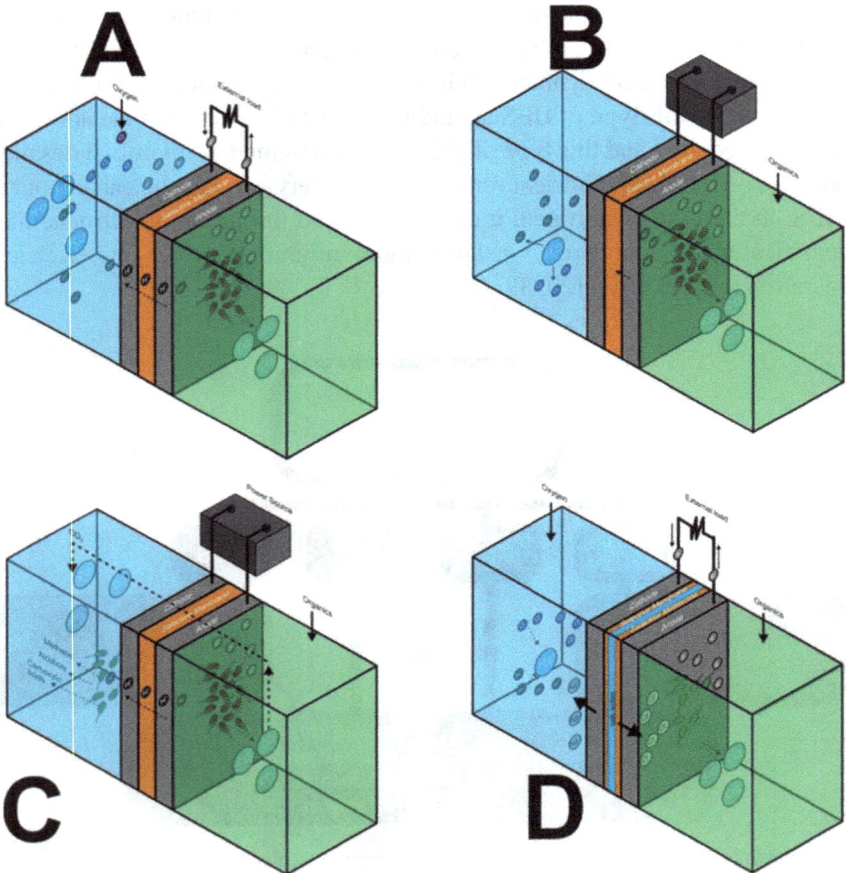

FIGURE 16.3   Schematic overview of various types of bioelectrochemical systems [41].

16.5   MICROBIAL ELECTROLYSIS CELLS (MECS)

A MEC takes advantage of the property of bacteria to transform the chemical energy stored of compounds, and with the addition of an external power applied onto the electrical circuit of MEC, this will drive electrons from anode to the cathode. In the presence of a catalyst, electrons, and protons are chemically reduced at the cathode to form hydrogen gas. Contrary to the MFC, the cathode of MEC operates under anaerobic conditions that facilitate hydrogen production. However, the anoxic environment in MEC, along with high concentrations of hydrogen production, can also promote methane production once CO_2 and methanogens are available [15, 34, 48–53]. A few of the methods to mitigate methane production includes the aeration of the cathode chamber between batches, lowering of the pH, operation at short retention times and giving a heat shock to the inoculum, or adding chemicals that inhibit the growth of methanogens [37, 53–56]. Higher electric currents are typically observed in MEC when compared to MFC, which is due to the additional applied voltage that helps overcome the cathode limitations. The energy required for MEC operation can also be provided by another separate MFC as a power source [28, 57–60]. In a MEC with bioanode and biocathode, expensive metals like platinum are not required as catalyst, and preferably, the enrichment of microbes on the carbon cathode decreases the start-up time and produces comparable current densities to those of bioanode [61–67]. Furthermore, the hydrogen synthesized in MEC can also drive the biochemical production of other high value-added chemicals [22, 63, 68, 67].

16.6 MICROBIAL ELECTROSYNTHESIS (MES)

MES, also known as bioelectrosynthesis, is a new-fangled perspective of BES which utilizes the reducing power generated from the anodic oxidation to produce value-added products at the cathode. Cathodic biocatalysts (with attached cathodic biofilms) reduce the available terminal electron acceptor to produce value-added products [8, 29, 69]. The bioelectrosynthesis process can be highly specific, depending on the biocatalyst catalyzing the redox reaction and the terminal electron acceptor involved in the process, along with the electrochemically active redox mediators or suitable reducing equivalents. Biocathodes are the key components of microbial electrosynthesis, where the electrode oxidizing microorganisms are involved in the formation of reduced value-added products such as acetate, ethanol, butyrate [8, 65, 69]. The production process of high added-value products in MES it through an electrochemical cell by electricity-driven CO_2 reduction as well as reduction/ oxidation of other organic feedstocks using microbes as biocatalyst [31, 65, 67, 69–73]. The feasibility of using MES as a technology into multi-carbon organic compounds using electrical current at the cathode has been reported widely. Additionally, MES technology also has the potential to harvest electrical energy in covalent chemical bonds of organic products. Mes technology has been extensively studied recently, highlighting the different aspects of that make it unique, such as microbiology, technology, and economics, and other practical consideration [29, 32, 34, 63, 67, 74–77].

16.7   ETHANOL PRODUCTION IN BES

Ethanol via the Wood-Ljungdahl pathway can be produced either directly from CO_2 reduction or by acetate reduction or from both. CO_2 reduction via WL pathway is highly energetically efficient, in where more than 95% of carbon and electron flow is channeled to the production of extracellular end products rather than to the microbial growth and biomass production [15, 33, 49]. The thermodynamics of ethanol production by CO_2 or acetate reduction indicates that lower pH and excess H_2 can shift the production to ethanol. The selectivity of the products of microbial CO_2 reduction has been shown to be limited by micro-elements of trace metal and vitamin components due to their effect on the activity of NADPH/NADP and Acetyl-CoA enzymes [15, 32, 78, 79]. By limiting micronutrient and trace elements (namely pantothenic acid and cobalt), higher concentrations of ethanol can be produced rather than acetate, suggesting that larger multicarbon compounds can be produced by means of applying strategies like those use in the gas fermentation process in where gaseous feedstocks such as CO, CO_2, syngas, CH_4 or biogas, into platform chemicals, fuels, polymers, etc. Thermodynamically, the reduction of acetate to ethanol is more favorable than the reduction of bicarbonate to ethanol and this has been reported when acetate accumulated and goes through further reduction (Figure 16.4). However, ethanol concentrations achieved are quite low [14, 34, 49, 56, 73, 80, 81].

One aspect that still needs further study is the mechanism of bacterial interaction with solid electrodes for CO_2 reduction. In this regard, the overpotential of the electrode will create conditions that much likely require a lower potential in the cathode. The necessary overpotential so that CO_2 reduction can take place in the cathode will require much lower cathode potential than the hydrogen evolution reaction (HER) [82]. Additionally, in the case of using a mixed culture, most likely, the HER will take place first, followed by CO_2 reduction mediated by H_2 [6, 34, 67, 69, 72]. On the other hand, when direct electron transfer takes place, it could be possible to achieve higher production rates of the targeted compound. Cytochrome-c for extracellular electron transfer on homoacetogenic bacteria has shown the potential of different of using multiple mechanisms of electron transfer, and this will depend of the biocatalyst chosen (Figure 16.5). In most of the literature, the bioelectrochemical reduction of CO_2 has been reported to take place at lower cathode potentials than the theoretical HER potential (0.61 V when reference electrode is Ag/AgCl). MES experiments showed a considerable production of acetate only when the potentials were more negative than 0.8 V (using as reference electrode Ag/AgCl) and whenever the potentials were above or equal to 0.8 V, acetate, and butyrate were degraded instead of CO_2 reduction. In this regard, acetate production occurs only below 0.9 V, in this case H_2 should mediate the CO_2 reduction reaction. Bioelectrocatalsys combined with the HER can take place when higher current densities (up to 10 ampers per m^2) are obtained at a potential of 0.9 to 1 V and when abiotic cathode is being used. However, when the biocathode is a mixed culture, it would not be clear what kind of mechanisms for electron transfer will be taking place [32, 33, 49, 52, 62, 76, 78, 81, 83–90].

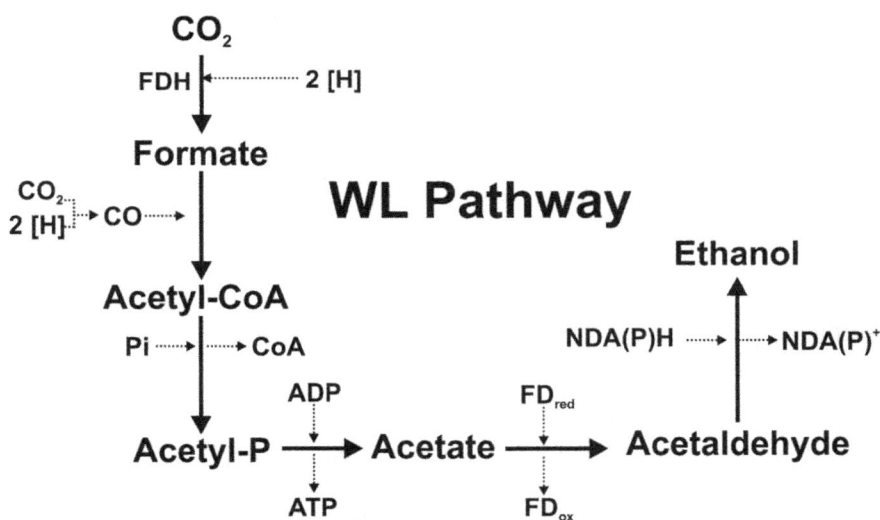

FIGURE 16.4 Schematic of the Wood-Ljungdahl pathway for CO_2 reduction for ethanol production.

16.8 CONCLUSIONS

MES can offer a wide range of opportunities that go from, energy production from wastes, to the production of high added value products from inorganic sources. MES technology offers new approaches that other heavily

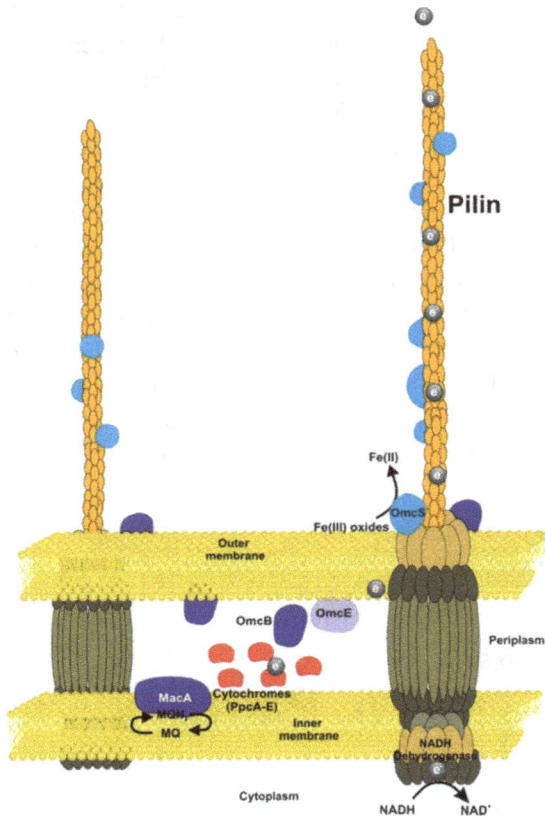

FIGURE 16.5   Extracellular electron transfer mechanism of Geobacter sulfurreducens [41].

biomass-relaying technologies that require large extension of land to obtain its feedstocks, on the other hand, MES land-use needs are low. Additionally, MES technologies, coupled with existing photovoltaic technologies can offer an almost non-polluting alternative to obtain a wide range of chemicals and energy. Operating MES system with existing photovoltaic technologies can offer higher efficiencies per surface area than conventional technologies. Additionally, fresh water and nutrients requirement can be lowered substantially in the bioproduction of chemicals compared to agricultural production. These features make MES technology very attractive, not just from an economic point of view, but also for being an almost a non-polluting technology as mentioned earlier. In these regard, MES can operate entirely without of the high input of hazardous and non-renewable raw materials in the unsustainable manner (high temperature/pressure, acid/alkaline solutions, etc.). Inexpensive and abundant carbon sources including CO_2 and waste streams can be reutilized for bioproduction with the input of electrical energy only (renewable energy). In addition, MES offers unique possibilities to promote cell growth and the controlling of redox state in bioprocesses by steering the metabolic pathways electrically. Therefore, this is a promising technology that provides prospects of renewable energy-driven waste to fuel/chemical to establish a circular economy. Still in development, this technology will require further understanding that will allow to compete in the near future with the existing commercial technologies.

KEYWORDS

- **hydrogen evolution reaction**
- **microbial desalination cell**
- **microbial electrodialysis cell**
- **microbial electrolysis cell**
- **microbial electrosynthesis**
- **microbial fuel cell**

REFERENCES

1. Atalan, A., (2020). Is the lockdown important to prevent the COVID-9 pandemic? Effects on psychology, environment and economy-perspective. *Annals of Medicine and Surgery, 56*, 38–42. Elsevier. doi: 10.1016/j.amsu.2020.06.010.
2. Prest, B. C., (2018). Explanations for the 2014 oil price decline: Supply or demand?. *Energy Economics, 74*, 63–75. doi: 10.1016/j.eneco.2018.05.029.
3. Singh, V. K., Nishant, S., & Kumar, P., (2018). Dynamic and directional network connectedness of crude oil and currencies: Evidence from implied volatility. *Energy Economics, 76*, 48–63. Elsevier B.V. doi: 10.1016/j.eneco.2018.09.018.
4. Wu, X. F., & Chen, G. Q., (2019). Global overview of crude oil use: From source to sink through inter-regional trade. *Energy Policy, 128*, 476–486. doi: 10.1016/j.enpol. 2019.01.022.
5. Go, Y. H., & Lau, W. Y., (2020). The impact of global financial crisis on informational efficiency: Evidence from price-volume relation in crude palm oil futures market. *Journal of Commodity Markets, 17*, 100081. Elsevier Ltd. doi: 10.1016/j.jcomm.2018.10.003.
6. Braden, D. J., (2005). *Fuel Cell Grade Hydrogen Production from Steam Reforming of Bio-Ethanol Over Co-based Catalyst: An Investigation of Reaction Networks and Active Sites*. Metallum. Com. Br, The Ohio S.

7. Jadhav, D. A., Ghosh, R. S., & Ghangrekar, M. M., (2017). Third generation in bio-electrochemical system research - a systematic review on mechanisms for recovery of valuable by-products from wastewater. *Renewable and Sustainable Energy Reviews, 76,* 1022–1031. doi: 10.1016/j.rser.2017.03.096.

8. Jung, S., et al., (2020). Bioelectrochemical systems for a circular bioeconomy. *Bioresource Technology,* 122748. Elsevier Ltd. doi: 10.1016/j.biortech.2020.122748.

9. Ayodele, B. V., Alsaffar, M. A., & Mustapa, S. I., (2020). An overview of integration opportunities for sustainable bioethanol production from first- and second-generation sugar-based feedstocks. *Journal of Cleaner Production, 245.* Elsevier B.V. doi: 10.1016/j. jclepro.2019.118857.

10. Sharma, B., Larroche, C., & Dussap, C. G., (2020). Comprehensive assessment of 2G bioethanol production. *Bioresource Technology, 313,* 123630. Elsevier. doi: 10.1016/j. biortech.2020.123630.

11. Zhao, J., et al., (2020). Bioconversion of industrial hemp biomass for bioethanol production: A review. *Fuel, 281.* doi: 10.1016/j.fuel.2020.118725.

12. Aditiya, H. B., et al., (2016). Second-generation bioethanol production: A critical review. *Renewable and Sustainable Energy Reviews, 66,* 631–653. Elsevier. doi: 10.1016/j.rser. 2016.07.015.

13. Warriner, K., et al., (2002). Modified microelectrode interfaces for in-line electrochemical monitoring of ethanol in fermentation processes. *Sensors and Actuators, B: Chemical, 84*(2, 3), 200–207. doi: 10.1016/S0925-4005(02)00025-4.

14. Pagnoncelli, K. C., et al., (2018). Ethanol generation, oxidation and energy production in a cooperative bioelectrochemical system. *Bioelectrochemistry, 122,* 11–25. Elsevier B.V. doi: 10.1016/j.bioelechem.2018.02.007.

15. Prévoteau, A., et al., (2020). Microbial electrosynthesis from CO_2: Forever a promise?. *Current Opinion in Biotechnology, 62,* 48–57. doi: 10.1016/j.copbio.2019.08.014.

16. Potter, M. C., (1910). *On the Difference of Potential Due to the Vital Activity of Microorganisms.* Proc Univ Durham Phil Soc.

17. Morris, J. M., et al., (2009). *Microbial Fuel Cell in Enhancing Anaerobic Biodegradation of Diesel. 146,* 161–167. doi: 10.1016/j.cej.2008.05.028.

18. Morris, J. M., & Jin, S., (2012). Enhanced biodegradation of hydrocarbon-contaminated sediments using microbial fuel cells. *Journal of Hazardous Materials, 213, 214,* 474–477. Elsevier B.V. doi: 10.1016/j.jhazmat.2012.02.029.

19. Kronenberg, M., et al., (2017). Biodegradation of polycyclic aromatic hydrocarbons: Using microbial bioelectrochemical systems to overcome an impasse. *Environmental Pollution, 231,* 509–523. Elsevier Ltd. doi: 10.1016/j.envpol.2017.08.048.

20. Nancharaiah, Y. V., Mohan, S. V., & Lens, P. N. L., (2019). Removal and recovery of metals and nutrients from wastewater using bioelectrochemical systems. *Microbial Electrochemical Technology.* Elsevier B.V. doi: 10.1016/b978-0-444-64052-9.00028-5.

21. Cohen, B., (1931). The bacterial culture as an electrical half-cell. *Journal of Bacteriology, 21,* 18–19.

22. Wang, H., & Ren, Z. J., (2013). A comprehensive review of microbial electrochemical systems as a platform technology. *Biotechnol. Adv., 31,* 1796–1807.

23. Liu, H., Grot, S., & Logan, B. E., (2005). Electrochemically assisted microbial production of hydrogen from acetate. *Environmental Science and Technology, 39*(11), 4317–4320. doi: 10.1021/es050244p.

24. Kadier, A., et al., (2016). A comprehensive review of microbial electrolysis cells (MEC) reactor designs and configurations for sustainable hydrogen gas production. *Alexandria*

Engineering Journal. Faculty of Engineering, Alexandria University, 55(1), 427–443. doi: 10.1016/j.aej.2015.10.008.

25. Oh, S. E., & Logan, B. E., (2005). Hydrogen and electricity production from a food processing wastewater using fermentation and microbial fuel cell technologies. *Water Research, 39*(19), 4673–4682. doi: 10.1016/j.watres.2005.09.019.

26. Sun, M., et al., (2010). Hydrogen production from propionate in a biocatalyzed system with in-situ utilization of the electricity generated from a microbial fuel cell. *International Biodeterioration and Biodegradation, 64*(5), 378–382. Elsevier Ltd. doi: 10.1016/j.ibiod.2010.04.004.

27. Sharma, Y., Parnas, R., & Li, B., (2011). Bioenergy production from glycerol in hydrogen-producing bioreactors (HPBs) and microbial fuel cells (MFCs). *International Journal of Hydrogen Energy, 36*(6), 3853–3861. Elsevier Ltd. doi: 10.1016/j.ijhydene.2010.12.040.

28. Montpart, P. N., (2014). *Hydrogen Production from Wastewater in Single Chamber Microbial Electrolysis Cells: Studies Towards its Scaling-up*, 192. Available at: https://www.tdx.cat/bitstream/handle/10803/283897/nmip.pdf;jsessionid=EB7C604404BCA BEE1D791B0298D3CA2D?sequence=1 (accessed on 08 December 2021).

29. Zhang, Y., et al., (2019). Microbial fuel cell hybrid systems for wastewater treatment and bioenergy production: Synergistic effects, mechanisms and challenges. *Renewable and Sustainable Energy Reviews, 103*, 13–29. Elsevier Ltd. doi: 10.1016/j.rser.2018.12.027.

30. Chu, N., et al., (2020). Microbial electrochemical platform for the production of renewable fuels and chemicals. *Biosensors and Bioelectronics, 150*, 111922. Elsevier B.V. doi: 10.1016/j.bios.2019.111922.

31. Thygesen, A., et al., (2010). Integration of microbial electrolysis cells (MECs) in the biorefinery for production of ethanol, H_2 and phenolics. *Waste and Biomass Valorization, 1*(1), 9–20. doi: 10.1007/s12649-009-9007-9.

32. Arends, J. B. A., et al., (2017). Continuous long-term electricity-driven bioproduction of carboxylates and isopropanol from CO_2 with a mixed microbial community. *Journal of CO_2 Utilization, 20*, 141–149. Elsevier. doi: 10.1016/j.jcou.2017.04.014.

33. Bajracharya, S., Van, D. B. B., et al., (2017). In situ acetate separation in microbial electrosynthesis from CO_2 using ion-exchange resin. *Electrochimica Acta, 237*, 267–275. Elsevier Ltd. doi: 10.1016/j.electacta.2017.03.209.

34. Anwer, A. H., et al., (2019). Development of novel MnO_2 coated carbon felt cathode for microbial electroreduction of CO_2 to biofuels. *Journal of Environmental Management, 249*, 109376. Elsevier. doi: 10.1016/j.jenvman.2019.109376.

35. Bian, B., et al., (2020). Microbial electrosynthesis from CO_2: Challenges, opportunities and perspectives in the context of circular bioeconomy. *Bioresource Technology, 302*, 122863. Elsevier Ltd. doi: 10.1016/j.biortech.2020.122863.

36. Singh, S., Noori, M. T., & Verma, N., (2020). Efficient bio-electroreduction of CO_2 to formate on a iron phthalocyanine-dispersed CDC in microbial electrolysis system. *Electrochimica Acta, 338*, 135887. Elsevier Ltd. doi: 10.1016/j.electacta.2020.135887.

37. Xiao, S., et al., (2020). Hybrid microbial photoelectrochemical system reduces CO_2 to CH_4 with 1.28% solar energy conversion efficiency. *Chemical Engineering Journal, 390*, 124530. Elsevier B.V. doi: 10.1016/j.cej.2020.124530.

38. Pant, D., Singh, A., Van-Bogaert, G., Alvarez-Gallego, Y., Diels, L., & Vanbroekhoven, K., (2010). An introduction to the life cycle assessment (LCA) of bioelectrochemical systems (BES) for sustainable energy and product generation: Relevance and key aspects. *Energy Rev.*, 1305–13.

39. Zhang, F., Li, J., & He, Z., (2014). A new method for nutrient removal and recovery from wastewater using a bioelectrochemical system. *Bioresour. Technol., 166*, 630–634.

40. Zhang, G., Zhou, Y., & Yang, F., (2019). Hydrogen production from microbial fuel cells-ammonia electrolysis cell coupled system fed with landfill leachate using Mo_2 C/N-doped graphene nanocomposite as HER catalyst. *Electrochimica Acta, 299*, 672–681. Elsevier Ltd. doi: 10.1016/j.electacta.2019.01.055.

41. Martínez-Amador, S. Y., Ríos-Gonzalez, L. J., Rodríguez-Garza, M. M., Moreno-Dávila, I. M. M., Morales-Martinez, T. K., Medina-Morales, M. A., & Rodríguez-De La, G. J. A., (2021). Bioelectrochemical systems for wastewater treatment and energy recovery. *Handbook of Research on Bioenergy and Biomaterials: Consolidated and Green Processes.*

42. Logan, B. E., (2009). Exoelectrogenic bacteria that power microbial fuel cells. *Nature Reviews Microbiology, 7*(5), 375–381. doi: 10.1038/nrmicro2113.

43. Rahimnejad, M., et al., (2014). *A Review on the Effect of Proton Exchange Membranes in Microbial Fuel Cells, 1*, 7–15.

44. Santoro, C., et al., (2017). Microbial fuel cells: From fundamentals to applications. A review. *Journal of Power Sources, 356*, 225–244. Elsevier B.V. doi: 10.1016/j.jpowsour. 2017.03.109.

45. Velasquez-Orta, S. B., Head, I. M., Curtis, T. P., & Scott, K., (2011). Factors affecting current production in microbial fuel cells using different industrial wastewaters. *Bioresource Technol.*, 5105–5012.

46. Zhou, M., He, H., & Jin, T., & Wang, H., (2012). Power generation enhancement in novel microbial carbon capture cells with immobilized Chlorella vulgaris. *J. Power Sources*, 216–19.

47. Venkata-Mohan, S., Velvizhi, G., Vamshi-Krishna, K., & Lenin-Babu, M. (2014). Microbial catalyzed electrochemical systems: A biofactory with multi-facet applications. *Bioresource Technol.*, 355–364.

48. Eerten-Jansen, M. C. A. A. V, et al., (2013). Microbial community analysis of a methane-producing biocathode in a bioelectrochemical system. *Archaea, 2013*. doi: 10.1155/2013/481784.

49. Gildemyn, S., et al., (2015). Integrated production, extraction, and concentration of acetic acid from CO_2 through microbial electrosynthesis. *Environmental Science and Technology Letters, 2*(11), 325–328. doi: 10.1021/acs.estlett.5b00212.

50. Sadhukhan, J., et al., (2016). A critical review of integration analysis of microbial electrosynthesis (MES) systems with waste biorefineries for the production of biofuel and chemical from reuse of CO_2. *Renewable and Sustainable Energy Reviews, 56*, 116–132. Elsevier. doi: 10.1016/j.rser.2015.11.015.

51. Bajracharya, S., Srikanth, S., et al., (2017a). Biotransformation of carbon dioxide in bioelectrochemical systems: State of the art and future prospects. *Journal of Power Sources, 356*, 256–273. Elsevier B.V. doi: 10.1016/j.jpowsour.2017.04.024.

52. Bajracharya, S., Vanbroekhoven, K., et al., (2017). Bioelectrochemical conversion of CO_2 to chemicals: CO_2 as a next-generation feedstock for electricity-driven bioproduction in batch and continuous modes. *Faraday Discussions, 202*, 433–449. doi: 10.1039/ c7fd00050b.

53. Nelabhotla, A. B. T., Bakke, R., & Dinamarca, C., (2019). Performance analysis of biocathode in bioelectrochemical CO_2 reduction. *Catalysts, 9*(8). doi: 10.3390/ catal9080683.

54. Schlager, S., et al., (2017). Bio-electrocatalytic application of microorganisms for carbon dioxide reduction to methane. *ChemSusChem, 10*(1), 226–233. doi: 10.1002/cssc.201600963.
55. Huang, Y. X., & Hu, Z., (2018). An integrated electrochemical and biochemical system for sequential reduction of CO_2 to methane. *Fuel, 220*, 8–13. Elsevier. doi: 10.1016/j.fuel.2018.01.141.
56. Kokkoli, A., Zhang, Y., & Angelidaki, I., (2018). Microbial electrochemical separation of CO_2 for biogas upgrading. *Bioresource Technology, 247*, 380–386. Elsevier Ltd. doi: 10.1016/j.biortech.2017.09.097.
57. Montpart, N., et al., (2016). Low-cost fuel-cell based sensor of hydrogen production in lab-scale microbial electrolysis cells. *International Journal of Hydrogen Energy, 41*(45), 20465–20472. doi: 10.1016/j.ijhydene.2016.09.169.
58. Fischer, F., (2018). Photoelectrode, photovoltaic and photosynthetic microbial fuel cells. *Renewable and Sustainable Energy Reviews, 90*, 16–27. Elsevier Ltd. doi: 10.1016/j.rser.2018.03.053.
59. Huarachi-Olivera, R., et al., (2018). Bioelectrogenesis with microbial fuel cells (MFCs) using the microalga *Chlorella vulgaris* and bacterial communities. *Electronic Journal of Biotechnology, 31*, 34–43. Elsevier España, S.L.U. doi: 10.1016/j.ejbt.2017.10.013.
60. Zhang, J., Yuan, H., Abu-Reesh, I. M., He, Z., & Yuan, C., (2019). Life cycle environmental impact comparison of bioelectrochemical systems for wastewater treatment. *Procedia CIRP, 80*, 382–388.
61. Butti, S. K., et al., (2016). Microbial electrochemical technologies with the perspective of harnessing bioenergy: Maneuvering towards upscaling. *Renewable and Sustainable Energy Reviews, 53*, 462–476. Elsevier. doi: 10.1016/j.rser.2015.08.058.
62. Bajracharya, S., Yuliasni, R., et al., (2017). Long-term operation of microbial electrosynthesis cell reducing CO_2 to multi-carbon chemicals with a mixed culture avoiding methanogenesis. *Bioelectrochemistry, 113*, 26–34. Elsevier B.V. doi: 10.1016/j.bioelechem.2016.09.001.
63. Kumar, G., et al., (2017). A review on bio-electrochemical systems (BESs) for the syngas and value-added biochemicals production. *Chemosphere, 177*, 84–92. Elsevier Ltd. doi: 10.1016/j.chemosphere.2017.02.135.
64. Li, J., et al., (2017). Uneven biofilm and current distribution in three-dimensional macroporous anodes of bio-electrochemical systems composed of graphite electrode arrays. *Bioresource Technology, 228*, 25–30. Elsevier Ltd. doi: 10.1016/j.biortech.2016.12.092.
65. Zou, S., & He, Z., (2018). Efficiently "pumping out" value-added resources from wastewater by bioelectrochemical systems: A review from energy perspectives. *Water Research, 131*, 62–73. Elsevier Ltd. doi: 10.1016/j.watres.2017.12.026.
66. Yang, E., et al., (2019). Critical review of bioelectrochemical systems integrated with membrane-based technologies for desalination, energy self-sufficiency, and high-efficiency water and wastewater treatment. *Desalination, 452*, 40–67. Elsevier. doi: 10.1016/j.desal.2018.11.007.
67. Luo, S., & He, Z., (2020). *Resource Recovery from Wastewater by Bioelectrochemical Systems*, 183–200. Elsevier Inc. doi: 10.1016/B978-0-12-816204-0.00009-6.
68. Bajracharya, S., et al., (2016). An overview on emerging bioelectrochemical systems (BESs): Technology for sustainable electricity, waste remediation, resource recovery,

chemical production and beyond. *Renewable Energy, 98*, 153–170. Elsevier Ltd. doi: 10.1016/j.renene.2016.03.002.

69. Ramírez-Vargas, C. A., et al., (2018). Microbial electrochemical technologies for wastewater treatment: Principles and evolution from microbial fuel cells to bioelectrochemical-based constructed wetlands. *Water (Switzerland), 10*(9), 1–29. doi: 10.3390/w10091128.

70. Lovley, D. R., (2006). Bug juice: Harvesting electricity with microorganisms. *Nat. Rev. Microbiol., 4*, 497–508.

71. González, A., et al., (2013). Microbial fuel cell with an algae-assisted cathode: A preliminary assessment. *Journal of Power Sources, 242*, 638–645. Elsevier B.V. doi: 10.1016/j.jpowsour.2013.05.110.

72. Jiang, Y., et al., (2019). Carbon dioxide and organic waste valorization by microbial electrosynthesis and electro-fermentation. *Water Research, 149*, 42–55. doi: 10.1016/j.watres.2018.10.092.

73. Schievano, A., Pant, D., & Puig, S., (2019). Editorial: Microbial synthesis, gas-fermentation and bioelectroconversion of CO_2 and other gaseous streams. *Frontiers in Energy Research, 7*, 1–4. doi: 10.3389/fenrg.2019.00110.

74. Jin, S., & Fallgren, P. H., (2014). Feasibility of using bioelectrochemical systems for bioremediation. *Microbial Biodegradation and Bioremediation*, 389–405. Elsevier. doi: 10.1016/B978-0-12-800021-2.00016-9.

75. Li, X., Angelidaki, I., & Zhang, Y., (2018). Salinity-gradient energy-driven microbial electrosynthesis of value-added chemicals from CO_2 reduction. *Water Research, 142*, 396–404. Elsevier Ltd. doi: 10.1016/j.watres.2018.06.013.

76. Irfan, M., et al., (2019). Direct microbial transformation of carbon dioxide to value-added chemicals: A comprehensive analysis and application potentials. *Bioresource Technology, 288*, 121401. Elsevier. doi: 10.1016/j.biortech.2019.121401.

77. Srivastava, R. K., (2019). Bio-energy production by contribution of effective and suitable microbial system. *Materials Science for Energy Technologies, 2*(2), 308–318. KeAi Communications Co., Ltd. doi: 10.1016/j.mset.2018.12.007.

78. Ragsdale, S. W., & Pierce, E., (2008). Acetogenesis and the wood-ljungdahl pathway of CO_2 fixation. *Biochimica et Biophysica Acta - Proteins and Proteomics, 1784*(12), 1873–1898. doi: 10.1016/j.bbapap.2008.08.012.

79. Awate, B., et al., (2017). Stimulation of electro-fermentation in single-chamber microbial electrolysis cells driven by genetically engineered anode biofilms. *Journal of Power Sources, 356*, 510–518. Elsevier B.V. doi: 10.1016/j.jpowsour.2017.02.053.

80. Speers, A. M., Young, J. M., & Reguera, G., (2014). Fermentation of glycerol into ethanol in a microbial electrolysis cell driven by a customized consortium. *Environmental Science and Technology, 48*(11), 6350–6358. doi: 10.1021/es500690a.

81. Gavilanes, J., Reddy, C. N., & Min, B., (2019). Microbial electrosynthesis of bioalcohols through reduction of high concentrations of volatile fatty acids. *Energy and Fuels, 33*(5), 4264–4271. doi: 10.1021/acs.energy fuels.8b04215.

82. Doğan, H. Ö., (2019). Ethanol electro-oxidation in alkaline media on Pd/electrodeposited reduced graphene oxide nanocomposite modified nickel foam electrode. *Solid-State Sciences, 98*. doi: 10.1016/j.solidstatesciences.2019.106029.

83. Steinbusch, K. J. J., et al., (2010). Bioelectrochemical ethanol production through mediated acetate reduction by mixed cultures. *Environmental Science and Technology, 44*(1), 513–517. doi: 10.1021/es902371e.

84. Andersen, S. J., et al., (2014). Electrolytic membrane extraction enables the production of fine chemicals from biorefinery side streams. *Environmental Science and Technology, 48*(12), 7135–7142. doi: 10.1021/es500483w.

85. Ammam, F., et al., (2016). Effect of tungstate on acetate and ethanol production by the electrosynthesis bacterium *Sporomusa ovata. Biotechnology for Biofuels. BioMed Central, 9*(1), 1–10. doi: 10.1186/s13068-016-0576-0.

86. May, H. D., Evans, P. J., & LaBelle, E. V., (2016). The electrosynthesis of acetate. *Current Opinion in Biotechnology, 42,* 225–233. Elsevier Ltd. doi: 10.1016/j.copbio. 2016.09.004.

87. Mohanakrishna, G., Vanbroekhoven, K., & Pant, D., (2016). Imperative role of applied potential and inorganic carbon source on acetate production through microbial electrosynthesis. *Journal of CO_2 Utilization, 15,* 57–64. Elsevier Ltd. doi: 10.1016/j. jcou.2016.03.003.

88. Paiano, P., et al., (2019). Electro-fermentation and redox mediators enhance glucose conversion into butyric acid with mixed microbial cultures. *Bioelectrochemistry, 130,* 107333. Elsevier B.V. doi: 10.1016/j.bioelechem.2019.107333.

89. Im, H. S., et al., (2019). Isolation of novel CO converting microorganism using zero-valent iron for a bioelectrochemical system (BES). *Biotechnology and Bioprocess Engineering, 24*(1), 232–239. doi: 10.1007/s12257-018-0373-7.

90. Haavisto, J. M., et al., (2020). The effect of start-up on energy recovery and compositional changes in brewery wastewater in bioelectrochemical systems. *Bioelectrochemistry, 132,* 107402. Elsevier LTD. doi: 10.1016/j.bioelechem.2019.107402.

CHAPTER 17

Challenges and Perspectives on Application of Biofuels in the Transport Sector

F. G. BARBOSA,[1] S. SÁNCHEZ-MUÑOZ,[1] E. MIER-ALBA,[1]
M. J. CASTRO-ALONSO,[1] R. T. HILARES,[2] P. R. F. MARCELINO,[1]
C. A. PRADO,[1] M. M. CAMPOS,[1] A. S. CARDOSO,[1] J. C. SANTOS,[3]
and S. S. DA SILVA[1]

[1]*Bioprocesses and Sustainable Products Laboratory,
Department of Biotechnology, Engineering School of Lorena,
University of São Paulo (EEL-USP), Lorena–12.602.810, SP, Brazil,
E-mails: salvador.sanchez@usp.br;
sanchezmunoz.ssm@gmail.com (S. Sánchez-Muñoz)*

[2]*Material Laboratory, Catholic University of Santa Maria (UCSM),
Yanahuara, AR–04013, Peru*

[3]*Bioprocesses, Biopolymers, Simulation, and Modeling Laboratory,
Department of Biotechnology, Engineering School of Lorena,
University of São Paulo (EEL-USP), Lorena–12.602.810, SP, Brazil*

ABSTRACT

Environmental and economic impacts caused by the use of fossil fuels have been demanding the development of renewable energy technologies based on low carbon fuels. In the last times, large-scale investments in the production of biofuels have increased in many countries intending to achieve the Sustainable Development Scenario. However, policies, economic competitors, regional feedstocks availability, technological challenges of biomass conversion, and bottlenecks for current and future biofuels productions, represent key issues to be considered for their incorporation into the transport sector.

17.1 INTRODUCTION

In the transport sector, there is the dominance of the use of fossil fuels that are derived from oil sources. With global economic development, the demand for fuels is growing proportionately. The increasing use of these fuels has been pointed to as responsible for the occurrence of climate changes due to an increased concentration of CO_2, the main greenhouse gas (GHG), in the atmosphere. Reports inform us that the transport sector consumes 30% of global energy and is responsible for 21% of the annual GHG emissions [1].

As a consequence of environmental concerns, the search for sustainable fuels (biofuels) production with low CO_2 emission and minimal climate changes has grown. In the world, the countries with the highest production of biofuels are the USA, Brazil, and China, and it is expected that these countries will double or triple their consumption of biofuels by 2030 [2, 3].

Worldwide, the biofuels most commonly used in motor vehicles, marine vessels, and aviation are bioethanol, biodiesel, and biogas [4, 5]. Bioethanol and biodiesel are mainly produced by USA and Brazil, while for biogas, the largest production is observed in Europe. Even though, policies and technologies need to be improved to allow and encourage the application of these biofuels in transportation. For this, government projects and programs are developed to intensify the production and use of these sustainable fuels. Also, financial subsidies are provided to manufacturers aiming for the production of sustainable supplies from available biomass to support bioeconomy development [6]. Due to these financial subsidies and government programs, biofuels classified as the first-generation dominate the sector of renewable fuels applied to transportation. Their production is at an advanced stage, from the processing of the biomasses involved until the industrial infrastructure. However, aspects related to land availability and the competition of feedstocks with the food sector, make second and third-generation technologies gaining attention worldwide [7, 8].

In this context, the perspectives and challenges faced for the production and application of biofuels in the transport sector will be discussed in this chapter.

17.2 BIOFUELS AS ALTERNATIVE TRANSPORT FUEL: WORLDWIDE CURRENT STATUS AND THE USE OF BIOETHANOL

Biofuels can replace conventional fuels derived from oil, such as gasoline and diesel in the transportation sector [187]. The obtaining processes of those renewable forms of fuel and their use must be environmentally friendly [9–11]. Available biofuels are bioethanol, biodiesel, biogas (biomethane), and biohydrogen [12, 13].

Biofuels are classified as first-generation (1G), second-generation (2G), and third-generation (3G) [14–19, 188]. This classification relates to the type of feedstocks used in their production, for example, 1G biofuels have as feedstock energy crops like rice, wheat, corn, sugarcane, and also vegetable oils, among others [20]. On the other hand, 2G biofuels are obtained from wood residues, inedible vegetable oil, non-food crops, and crop byproducts such as wheat straw, corn straw, sugarcane bagasse [21]. In the 3G biofuels, the biomass of organisms is used as raw material, such as microalgae and seaweed [22, 23]. Currently, industrial global production is led by 1G biofuels from traditional biomass, and it has been estimated that this technology will still be responsible for the development of biofuels for years to come. For example, bioethanol production is expected to use 14% and 24% of global corn and sugarcane production by 2028 [24].

In 2018, about 14% of the energy in the world came from renewable sources, and the other 86% corresponds to non-renewable energy, including coal, crude oil, natural gas [25]. From whole renewable energy, approximately 3.7% was directed to the transport sector [26]. Thus, biofuels contribute only with a small part of the energy supply in transportation, but their use has been increasing in parallel with the growth of the vehicle fleet [4, 27].

In 2019, the world production of biofuels was approximately 154.5 billion liters, and it is expected to increase by 25% during the 2019–2024 period, reaching about 190 billion liters of biofuels [28]. USA and Brazil were the largest producers, generating around 66.8 and 37 billion liters, respectively [25, 29]. Among the biofuels available in top producer countries for transportation, here will be highlighted biogas, biodiesel, and bioethanol [4].

Biogas corresponds to a mixture of combustible gases that can be produced by AD of manure, farm wastes, wastewater, and other residues [30–32]. It is used for the domestic generation of energy/heat, but also can be used as a motor fuel in traditional vehicle designs [33, 34]. Europe, Asia, America, and Oceania are the main biogas producers, as shown in Figure 17.1 [35]. This biofuel is mostly used in energy conversion into electricity, but its use in vehicles for transportation still has limitations [32, 36]. Although, when biogas is upgraded to biomethane, its use as a transport fuel in vehicle engines for natural gas has been successful. Due to the low emission of toxic gases, the use of biomethane in vehicles is considered one of the best renewable fuel options in transportation [36, 37].

Nowadays, biomethane is a common transportation fuel in gas filling stations in Europe, especially Germany and Sweden, either as 100% methane (CBG100) or blended with natural gas CBGXX (where XX is the percentage by volume of biomethane present in the mixture, ex. CBG10 and CBG50) [38, 39]. In Sweden, it is estimated that more than 4,000 light vehicles and fleets of public transport busses are moved by biomethane [32, 40]. On the other hand, biodiesel and bioethanol are the biofuels most used in transport vehicles in other countries. Biodiesel is a fuel formed by mono-alkyl esters of long-chain fatty acids, produced from vegetable oils [41]. It can be used pure (B100) or in diesel fuel mixtures, following the requirements of standards allowed by the policies of each country (e.g., ASTM D6751) [5, 32, 42].

USA and Brazil are one of the main producing countries of biodiesel in the world. In 2019, these countries produced around 6.5 and 5.9 billion liters of biodiesel, behind Indonesia, responsible for the production of 7.9 billion liters [43] (Figure 17.1). Currently, the biodiesel industry in the USA is the fastest-growing in the world, and it is expected to obtain favorable investments and taxes in the coming years to encourage its expansion [33]. This biofuel is also widely used in the transport sector by the European Union (EU), which in 2015 used approximately 12 billion liters [37, 44].

Among biofuels, bioethanol is the most promising to be used in the worldwide transport sector in the short and medium-term (approximately 66% of the bioethanol produced in the world is used for transportation) [45, 46]. Bioethanol of first-generation is produced by a fermentative process of feedstocks such as sugarcane, present in tropical and subtropical areas such as in Brazil, India, and Colombia, or from corn, wheat, and tubers as occurs in the USA, EU, and China [33, 47–50]. The bioethanol obtained from sugarcane has approximately 70% lower GHG emissions than fuels formed by conventional hydrocarbons, considering the life cycle balance [46].

In 2019, the USA produced the highest quantity of bioethanol in the world, generating 59.73 billion of liters, followed by Brazil (32.48 billion liters), EU (5.45 billion liters), China (3.41 billion liters), Canada (1.89 billion liters) (Figure 17.1) [51]. Although the USA is the largest producer of bioethanol in the world, it still uses gasoline as its main transport fuel. In 2019, petroleum fuels such as gasoline and diesel accounted for about 91% of the total energy use of the transportation sector in the USA [52]. In this country, it is common to find a mixture of gasoline with 10% and 15% bioethanol (E10–E15), the exact amount added varying by region [49, 53].

In this context, the current way of using biofuels is as additives for gasoline and diesel, commonly without requirements of modification in the vehicle engines. In its composition, bioethanol contains oxygen and using it as an additive to gasoline turns possible to reduce gas emissions by 25% compared to the use of pure gasoline, being beneficial for the environment [54, 55]. Recent research of Liu et al. [56], used life cycle assessment (LCA) tools to investigate three new LCB refinery systems using bioethanol mixtures E10 and E85, compared with fossil gasoline. Fuels combined with bioethanol performed better compared to gasoline, in terms of depletion of fossil fuels, with E10 and E85 were 6% and 64% to 70% lower, respectively.

To increase the use of bioethanol as an automotive biofuel, industries have been working on the development of vehicles with a Flex system. In Brazil, since 2003, automobiles with Flex engines have been sold, which have a dual supply system for fuel, bioethanol, and/or gasoline. Those vehicles are capable of using 100% bioethanol (E100), or gasoline, or any mixing rate between them [29, 57]. Vehicles with this technology have become widely used in Brazil, where about 95% of new cars sold have a Flex system [46, 58].

Although biofuels have been produced in many countries, the rate of their adoption is still low. According to information from the International Sustainable Development Strategy Group, in order to reduce the emission of GHGs and fight against climate change, it is necessary to have sustainable growth in the production of biofuel of 10% per year until 2030. Thus, political support and innovation are needed to favor the use of available feedstocks, to reduce industrial production costs, and to increase the advanced consumption of biofuels [2, 189]. The current environmental impact caused by the use of fossil fuels and the production of biofuels for the near future will be discussed below.

17.3 PERSPECTIVES IN BIOFUELS PRODUCTION: CURRENT ENVIRONMENTAL IMPACT, PROJECTIONS, AND SUSTAINABILITY

The strong economic development is the direct reason for the increasing use of fossil fuels, which causes the increase in global warming (GW) of the planet and the harmful effects of natural destruction, such as the accelerated melting of glaciers, floods, and long-term rains [34, 59, 60]. In addition, the concentration of CO_2 in the Earth's atmosphere has grown 1.48 times from 280 ppm in 1,750 to 415 ppm in 2019, representing an increase of 0.9% per year [34, 61].

To decarbonize the energy sector, several strategies are under development. One of the established strategies is the utilization of renewable energy, which production is expected to increase by 24% in the next decade [62]. However, the dynamics of production of renewable energy and CO_2 emissions in the

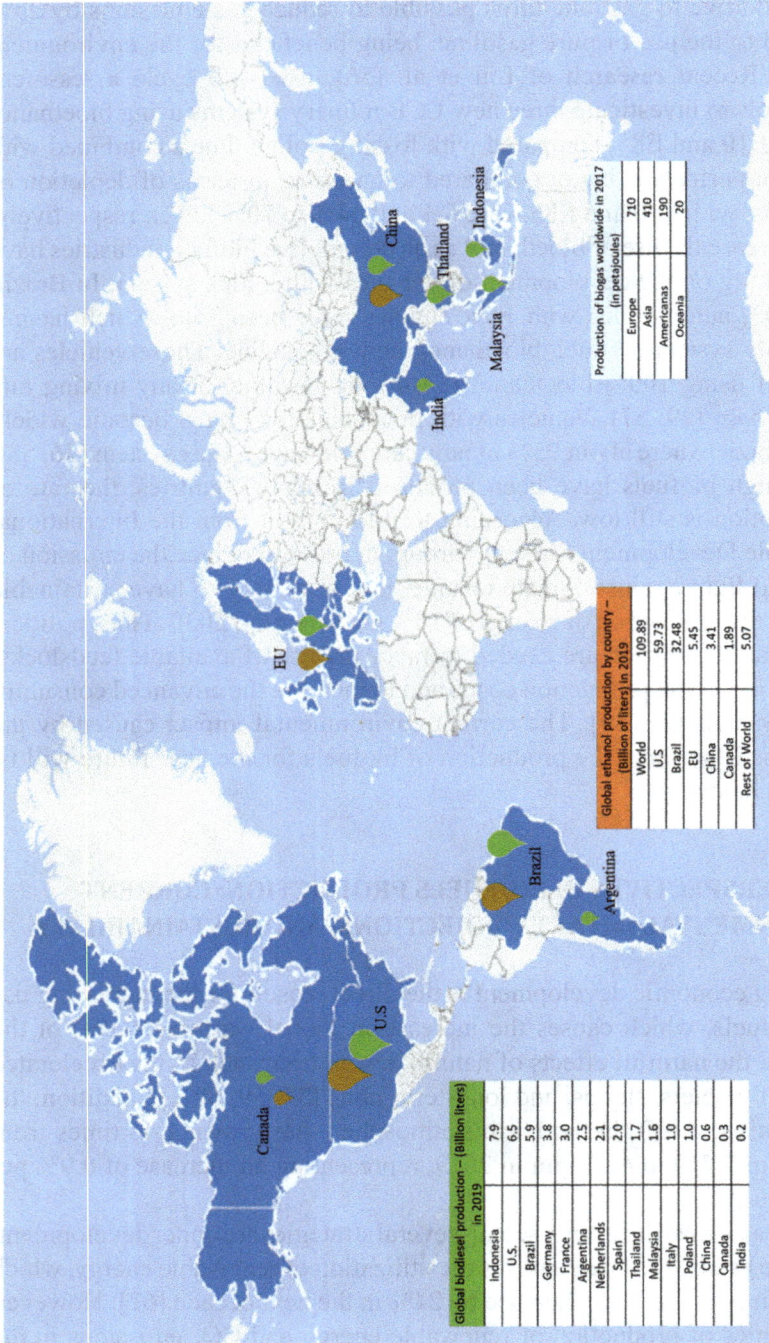

Production of biogas worldwide in 2017 (in petajoules)	
Europe	710
Asia	410
Americanas	190
Oceania	20

Global ethanol production by country – (Billion of liters) in 2019	
World	109.89
U.S	59.73
Brazil	32.48
EU	5.45
China	3.41
Canada	1.89
Rest of World	5.07

Global biodiesel production – (Billion liters) in 2019	
Indonesia	7.9
U.S.	6.5
Brazil	5.9
Germany	3.8
France	3.0
Argentina	2.5
Netherlands	2.1
Spain	2.0
Thailand	1.7
Malaysia	1.6
Italy	1.0
Poland	1.0
China	0.6
Canada	0.3
India	0.2

FIGURE 17.1   Global distribution of the main biofuels.

Source: Biodiesel and biogas production data taken from the Statista database (production of biogas worldwide in 2017, by region); (leading biodiesel producers worldwide in 2019, by country (in billion liters); [186].

world may vary according to economic, political, or social global events [63]. For example, the CO_2 emissions slowdown in 2019 was of 0.5%, but it would need to be analyzed in the context of the strong increase in carbon emissions (2.1%) due to the energy production boost in 2018 led especially by Russia, India, and USA [64]. Another issue that can be considered when analyzing future trends would be the pandemic COVID-19 crisis, which occurred in 2020. This health crisis brought momentary environmental benefits, such as the drastic decrease of global carbon emissions (expected to be 8%), but also challenges as a decrease in global investment and energy demand (6%). These factors together with economic issues, such as land distribution and higher food prices, can also influence the production of biofuels [65, 66]. It has been estimated that in the first period of 2020, global use of renewable energy and electricity increased about 1.5% and 2%, respectively, compared to the same period in 2019. However, for the first time in decades, it is expected to decrease during this year [67].

It is important to highlight that modern renewable energy share has increased only around 10% of the final global consumption of energy for the last 10 years [3]. Thus, to achieve our global goals for the use of cleaner energies and environmental improvement, the efforts must still focus, more than ever, to improve the production and use of renewable energies as biofuels.

17.3.1 CURRENT BIOFUELS PRODUCTION AND FUTURE TRENDS

In the world, the biggest contributors to global energy growth in 2019 were China, followed by Indonesia and India. China led the expansion of growth and energy consumption, accounted for about 75% of total growth, and achieved a greater increase in demand for each energy source [64]. Recent reports show that with the growth of renewable energies in the world, fossil fuel consumption fell for the first time to its lowest level. At the same time, an increase of more than 60% in the global production of biofuels has been observed since 2010, with a forecast to reach a consumption of 298 million tons of oil equivalent (Mtoe) by 2030 [2, 3, 64]. Countries with the highest production of biofuels are expected to double or triple their consumption of biofuels in the next 10 years compared to their current production [2].

Biofuels such as bioethanol, biodiesel, and others (biomethane and biohydrogen) are the most promising renewable energies to be produced on a large scale and for being applied in the transport sector. For this reason, countries have been increasing their investments to improve the production of these biofuels to address the economically, environmentally, and socially worldwide perspectives [30, 68]. Advantageously, conditions and feedstocks for their production can be adapted according to local characteristics like environmental parameters and available biomass [3, 30, 69].

In the first place, bioethanol with a production of 120,000 million liters is estimated to increase more than 1% during the next decade [68]. As seen before, USA and Brazil account for about 90% of bioethanol production and approximately 85% of the global production of liquid biofuels [33, 69]. Both countries have policies that have encouraged the production and use of bioethanol for decades. Many of these initiatives motivated other developing countries of America like Mexico, Colombia, and Argentina, and also Asia, to introduce smaller scale bioethanol programs that are so beneficial and necessary to achieve the international goals of sustainability [49]. As a consequence, Argentina (forecasted total production growth of more than 46% by 2030) and other Asian countries (forecasted total production growth between 19% and 50% by 2030) become large bioethanol producers. Nowadays, developed countries produce 58% of bioethanol in the world, and they could increase their production by 6% until 2028 [68]. Other important top producers are China (actual production of 10 billion liters), India (actual production of 2 billion liters), and the EU (actual production of almost 7 billion liters), as is shown in Table 17.1.

Another liquid biofuel with positive impacts on the environment is biodiesel. Nowadays, developed countries are responsible for 57% of the world's production of biodiesel, and it is expected by 2028 their production should increase by 9% [68, 69]. Table 17.1 shows actualized data of production and use of biodiesel around the world, having as greater producer USA, followed by the EU. In addition, it was calculated that global biodiesel production and consumption would increase by around 2% until 2030 [68, 69]. However, some of the developed countries do not produce and/or commercialize biofuels, thus they do not appear in future production statistics.

Additionally, reports show that biogas is another potential biofuel, for example, Dos Santos et al. [189], assessed energy potential and reduced emissions by applying biogas energy formed from the biodigestion of different organic wastes. It was found that potential power was between 4.5 and 6.9 GW, which would reduce CO_2 emissions at a rate of 4.93% per year.

It is worth to note there are some specifications for each biogas application or use. For example, for heat and electricity production usage biogas requires the removal of water vapor and H_2S. Its use in transportation, on the other hand, requires an upgrading process that consists of impurities removal (mainly CO_2), conserving high energy content to reach each country's quality standards [39, 70–72]. Fuels produced from biogas, such as

TABLE 17.1 Bioethanol and Biodiesel Projections: Production, Use, Price, and Emission Saving

	Bioethanol						Biodiesel					
	P^a (ML)	P^a (ML)	EP^a (ML)	EPG (%)	AG^a (%)	EGF (%)	P^a (ML)	P^a (ML)	EP^a (ML)	EPG (%)	AG^a (%)	RG (%)
	2018	2019	2028	2018–2028	2019–2020	2018–2028	2018	2019	2028	2018–2028	2019–2020	2018–2019
United States	60,911	59,718	62,062	1.8	1.85	4.7[b]	7,700	6,500	8,693	12.90	-1.85	-4.8[b]
Brazil	32,085	30,432	37,155	15.8	1.52	16.1[c]	5,400	5,358	5,916	9.56	1.10	8.5[c]
China	10,000	10,185	14,773	47.7	0.85	142.6[d]	1,140	1,153	1,279	12.19	1.13	4.8[d]
European Union	6,860	6,822	7,406	7.9	-1.88	-1.62[e]	13,522	13,605	12,871	-4.81	-0.73	3.0[e]
India	2,637	2,637	3,132	18.7	-3.41	-0.31[f]	184	199	238	29.35	2.15	2.4[f]
Canada	2,030	2,026	2,047	0.8	-1.05	-5.67[g]	550	558	611	11.09	1.10	5.2[g]
Thailand	1,923	2,000	2,810	46.1	-14.03	60.61[h]	1,490	1,517	2,320	55.70	4.84	12.8[h]
Argentina	1,235	1,206	1,802	45.9	-0.36	46.05[i]	2,450	2,505	2,768	12.98	1.40	12.82[i]
Global Total	127,584	126,302	143,111	12.1	-0.51	–	40,251	43,131	43,931	9.14	0.10	–
Global Consumption	125,864	127,250	143,190	13.7	–	–	39,927	43,556	44,227	10.77	–	–
World[l] Emission Saving (%)	–	–	2.3[j]	–	–	–	–	–	1.7[j]	–	–	–
Word Price (US Dollar per Hectoliter)^a	39.41	40.93	52.63	–	–	–	89.40	87.31	94.68	–	–	–

Overview of the annual and expected production, consumption, and price projections of bioethanol and biodiesel between the period 2018–2028, for larger producers and global (P: production, EP: expected production, EPG: expected production growth, AG: annual growth, EGF: expected growth in use as fuel, RG: on-road use growth).

Note: 1. Provides a Well-to-Wheel (WTW) emission savings using the baseline of the 2018 OECD-FAO Agricultural Outlook for the period 2015–2017 and by 2030 for major biofuel consuming countries.

Sources: Refs. [69, 75–83].

liquefied biogas (LBG) and compressed natural gas (Bio-CNG) have been commercially produced via biogas liquefaction and compression, respectively [73]. Currently, there is a global production of around 3.5 Mtoe of biomethane (obtained from biogas upgrading). Actually, biomethane represents about 0.1% of natural gas demand [74].

Regardless of biogas has been produced, consumed, and applied in several fields, its use in transportation is not well-established. However, IEA [74] forecasted that power and heat usage of biogas is going to increase from 70 to 85% between 2018–2040.

17.3.2   SUSTAINABILITY AND ENVIRONMENTAL BENEFITS OF BIOFUELS

Even with government efforts for the constant increase of biofuels production and promising future trends, there still are several challenges to be faced to achieve the Sustainable Development Scenario (a major transformation of the global energy systems, which is derived or aligned from other energy scenarios). According to IRENA and IEA, the increase of approximately 3% in the use of biofuels and renewables energies per year is a goal that would allow environmental and energy sustainability in the next decades [2, 3]. For the Sustainable Development Scenario, renewable energy share in the final energy supply would increase by 17% by 2030 and 25% by 2050. There is also a Transforming Energy Scenario that would bring sustainability in a shorter time-lapse and reduction of total carbon emissions. To achieve this goal, it would need an increase of six-fold and almost twice the actual trends to achieve 28% of the renewable energy share in the final energy supply by 2030 and 66% by 2050 [3].

In the context of the Sustainable Development Scenario and taking as an example the largest producers of biofuels and their forecast annual production growth during the next decade, all these countries need at least duplicate their predicted growth to achieve sustainable goals. For example, in the case of USA, its actual production is of 35 Mtoe of biofuels represents the 5% of its transportation energy sources [53], and have to increase to 95 Mtoe to get in the Sustainable Development Scenario, which means, this country has to get its forecasted annual production of 1.9% to 7% of annual production growth every year during the next decade. Other countries as China and India would need to increase their actual production 8-fold to fulfill the expected sustainable consumption of biofuels by 2030. According to statistics, the EU would be the top producer with the hardest challenge, it would have to increase 18-fold the annual growth of production to achieve a sustainable scenario (Table 17.2) [2].

TABLE 17.2 Expected Biofuel Production Growth to Achieve Sustainable Development Scenario by 2030

	Biofuel Production 2019 (Mtoe)	SDS Biofuel Consumption (Mtoe)	Forecast Annual Production Growth (%)	SDS Annual Production Growth (%)
United States	35	95	1.9	7
Brazil	22	37	1.7	5
China	3	26	15.3	19
European	16	51	0.5	9
India	1	8	11.8	22

Source: International Energy Agency (IEA) [2]; Transport biofuels, IEA, Paris.

Another important goal involved in sustainability is the mitigation of GHG emissions. In this context, the transportation sector is responsible for a large portion of the global use of fossil fuels, which contributes to a quarter of global CO_2 emissions producing 10 Gton/year of CO_2. Moreover, the number of cars on the road (4.6 ton CO_2/year/car) and the kilometers flown by airplanes (918 Mton in 2018 just for commercial operations) will be close to doubling by 2040 [84–87]. Hence, this sector will need to reduce 50% of its CO_2 emissions by the year 2050 to acquire the Planned Energy Scenario (perspective based on countries' current energy plans under the Paris Agreement) or 75% for the Transformation Scenario (perspective based on the renewable energy to keep GW below 1.5–2.0°C this century). In both scenarios, the production and use of renewable sources represent around 50% of the mandatory improvement [3, 190].

Forecasts from IEA [74] anticipate that clean energy technology investment would grow to 40% of total investment, due to the great decrease in fossil fuel investment. This scenario would mean an important increment since it has been stagnant at 33% in recent years. In this context, a transition to more electrified forms of transport and heat, together with greater production of renewable energy as biofuels, would reduce 60% of the CO_2 emissions related to the energy needed by 2050 [3].

Based on the actual panorama for getting the sustainable goals, there are some other limitations on the biofuels supply chain, that involves political, economic, and social activities, besides, to biomass logistic and technological operations.

17.4 CHALLENGES IN THE APPLICATION OF BIOFUELS IN TRANSPORT SECTOR

17.4.1 POLICIES FOR BIOFUELS AROUND THE WORLD

Due to environmental, social, or economic issues, many nations started to create political strategies to reduce dependence on petroleum and increase the share of biofuels in the transport sector [88]. Regulations and guidelines are created to encourage producers and manufacturers to develop this form of sustainable energy. These documents set goals on the quantities of biofuels that must be produced, the levels of GHG emissions, in addition to the restriction of biomass cultivation in areas intended for food planting [89].

Policies involving the production and use of biofuels have been under development for many years in different countries. For example, Brazil in 1975, developed the national fuel alcohol program (ProÁlcool), which aimed the production of bioethanol from sugarcane, in response to the oil crisis [90], while the National Biodiesel Production and Use Program (PNPB) was launched in 2004 [191]. The development of these programs led to the creation of the RenovaBio program, which integrated the production of the most used biofuels (bioethanol, biodiesel, biogas), aiming at the decarbonization of the Brazilian energy matrix and creating methodologies that promote the stability of biofuels prices [91, 92].

China was another country that developed important programs for biofuels production and application. In 1986, the 863 plan was created with the National High Technology Research, which aimed to develop bioethanol production from sweet potatoes and industrialize the biomass liquid fuel [8, 93, 94, 188]. Later in 2005, the Bioethanol Use Plan and the Renewable Energy Law were created, which focused on the importance of biofuel, promoting resources, and tax subsidies for its production [95, 96]. As part of the bioethanol implementation plan, China has adopted different national standards (GB 18350 and GB 18351) related to the production and use of biofuels. A GB 18351 standard allows a mixture of 10% (E10) of ethanol with gasoline [97]. In 2018, this mixture was already available in six provinces, such as Liaoning, Heilongjiang, Henan, Jilin, Guangxi, and Anhui [98].

In 2006, the Ministry of Finance of China created financial support policies for biofuels that encourage the development of biodiesel. These policies provided subsidies for biodiesel plants that used non-food raw materials [99–101]. The government strictly prohibited the use of used oil as cooking oil, but illegal use persists due to the high profit [98]. The purchase price of biodiesel based on residual oil is lower when compared to biodiesel based on cooking oil. In 2010, the development program started using B5 with help from China National Offshore Oil Corporation (CNOOC). However, there is no mandatory blending requirement for biodiesel in China [99, 102, 103].

USA government has already elaborated several political projects involving the production of different biofuels. The Clean Air Act was the first and the most influential environmental law to regulate air quality, established in 1963. From this policy, others such as the US Energy Independence and Security Law of 2007 and the Biomass Program in 2008 were created targeting to reduce gasoline consumption to 70% by 2030. As consequence, biofuels emerged as a potential alternative for transportation [92]. Currently, there is the standard renewable fuel program (RFS), which was established by the energy policy act (EPA) in 2005, to increase the volume of renewable fuel that is mixed with transport fuels. In 2007, this program was revised to meet the requirements of the energy independence and security act (EISA), that established specific targets for increase the production of cellulosic biofuels, biomass diesel, and advanced biofuel in increasing volumes until 2022 [55, 65, 192].

In European Union (EU) countries, policies to encourage biofuels application have evolved over the years. In 2008, was passed in the EU a legislation that allowed the use of biofuels in the transport sector. One year later, in 2009, as part of its "Climate Change Package," the EU adopted the Renewable Energy Directive (DRE 2009/28/EC). In this directive, each Member State must achieve a minimum usage quota of 20% of energy from renewable sources, reduce GHG emissions by 20% and increase energy efficiency by 20% [104, 193]. In 2018, an agreement on the new Renewable Energy Directive (REDII) was determined by the Council of the EU, which will be in force between 2021 and 2030. This new directive includes the use of advanced biofuels, and estimates consumption of total renewable energy in the European energy matrix in 2030 should be 32% and in the transport sector 14% [105, 106].

As seen above with the top biofuel producer countries, global governments have had initiatives to enhance the development of the biofuels market. Among the most important initiatives created are:

1. **Global Bioenergy Partnership (GBEP):** Intergovernmental initiative, with more than 70 members, including members of civil society, government, and the private sector. It aims to support "the deployment of biomass and biofuels, particularly in developing countries where the use of biomass is predominant" [107].

2. **Biofuture Platform (BFP):** An initiative that supports global actions to combat climate change, fostering solutions in low carbon transport and the bioeconomy [108].

3. **Innovation Mission-Challenging Innovation in Sustainable Biofuels (SBIC):** A global initiative with the objective of promoting the drastic acceleration of innovation in clean energy, which includes 23 countries. Through this initiative, participating countries pledged to double their investments in research and development of clean energy by the government over five years (2015–2020).

4. **Low-Carbon Technology Partnerships Initiative (LCTPi) – Below 50:** The main objective is to accelerate the development and diffusion of low carbon technologies. New fuel production technologies are starting to be deployed at new production facilities around the world.

All these projects aim to intensify the use of clean energy in the form of biofuels; however, other limitations involve the economic barriers between fossil and renewable fuels.

17.4.2   ECONOMIC CONSIDERATIONS

Even with fluctuations in the price of oil, biofuels still cost more than conventional fossil fuels. The productive process of biofuels is the main barrier to making these technologies economically viable. For this purpose, it is required public support to encourage feedstock supply, cultivation of biomass, and reduce the capital cost of biofuel processing [109, 110]. Moreover, the cost of biofuel production depends mainly of process involved, industrial logistics, obtaining biomass, storage, and distribution of the final product [111–114].

In the case of 1G and 2G biofuels the most important economic consideration is related to the dependence on biomass [10]. The share in cost of raw materials is 40% to 70% of the total production costs. Like most of biofuels available, 1G biofuels depend on obtaining biomass, involved in the food industries, which account for more than two-thirds of total production costs [10]. Meanwhile, costs of 2G biofuels are also dependent on additional steps (pretreatments) for the biomass involved in the production processes [115]. For example, in the production of 2G bioethanol, the highest costs are observed in the biomass conversion stage [116, 193]. This step involves the use of enzymes for the hydrolysis process, and these are responsible for about 20% to 40% of the total cost of one liter of second-generation bioethanol [194].

When comparing the total cost of liquid biofuels (bioethanol and biodiesel) with fossil fuels cost, as is shown in Figure 17.2, it is observed that biofuels demonstrated higher cost. However, USA, and Brazil have been the top producer countries with the lower costs during the last years, because their production is at an advanced stage (from the processing of the biomass involved until the industrial infrastructure) [7, 8].

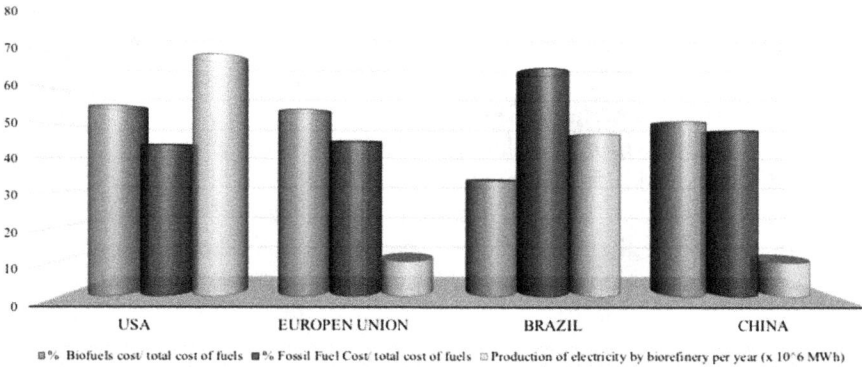

FIGURE 17.2 Comparative cost analysis of main biofuels vs fossil fuels.
Source: [114, 117–121, 195].

In this context, all processes related to biomass have an important impact on the cost of the biofuel supply chain that can vary according to regional availability, especially for those regions with the need of importation [2, 190].

17.4.3 BIOMASS FEEDSTOCK AVAILABILITY FOR BIOFUELS PRODUCTION

Many of the political projects developed to date encourage the production and increase availability of biofuels in society. Research carried out by special-ized agencies, such as the International Energy Agency, shows that in 2030, biofuels will contribute approximately 4% to 10% of the total energy involved in the planet's road transport. Thus, it will be necessary to use between 3.8% and 4.5% of the arable land available worldwide for the cultivation of crops directed to the production of biofuels, against only 1% of the total used [122].

According to the Global Renewable Energy Roadmap [30], the use of biomass in several sectors in the world grows approximately 3.7% per year between 2010 and 2030. It is estimated that in 2030 a total of 108 EJ

(Energy Joules) of biomass will be harvested, of which 31 EJ (28% of the total biomass) will be used to obtain liquid biofuels. From the total amount of available biomass, is expected a global production of approximately 650 billion liters of liquid biofuels per year until 2030 [30].

The use of biomass to produce first-generation biofuels has certain limitations due to its concomitant use as feed, causing great impacts on food prices [10]. In addition, the increase in demand of these feedstocks generates deforestation of natural reserves [123, 124]. These limitations have opened opportunity for the development of other generation biofuel technologies. It is estimated that by 2050, approximately 2150 Mtoe of agro-industrial waste, 955 Mtoe of forest materials, 2388 Mtoe of grains, 1194 Mtoe of waste, and 478 Mtoe of algae will be available worldwide, favoring the production of 2G and 3G biofuels [125, 196].

To get the expected demand for biofuel production in 2028, it will be necessary to increase the availability of feedstocks involved in the production process (Table 17.3). For example, in USA to obtain 59.72 ML of bioethanol in 2019, about 374.02 kton of maize were used and to achieved the forecasted amount for 2028, 388.70 kton will be required. Thus, countries that produce and use biofuels must take action to ensure biomass availability, such as the S2Biom Study, a project carried out in the EU that aims to provide sustainable supplies (non-food biomass) to support the development of the bioeconomy [6].

Lignocellulosic biomass (LCB) stands out among the several available feedstocks for the production of biofuels. This is an abundant organic resource in many countries around the world, with the produce of an estimated 181.5 Gton annually [126–129]. Currently, about 8.2 Gton are used, of which approximately 7 Gton of biomass are obtained from agricultural land, grass, and forest, and 1.2 Gton are derived from agricultural waste [128]. During 2018/2019, Brazilian mills harvested around 620 Mton of sugarcane, generating about 180 Mton of wet bagasse and 43 Mton of cane straw [6, 197].

LCB represents a suitable low cost and eco-friendly alternative for the production of biofuels and byproducts, however some aspects make difficult the conversion of this material at large-scale that corresponds to its recalcitrance, heterogeneity, and technical problems in its bioprocessing [130].

17.4.4 TECHNOLOGICAL CHALLENGES FOR LCB CONVERSION TO BIOFUELS: BIOGAS AND BIODIESEL

Bioprocessing of feedstocks is an early step of the biofuel production process and an important stage for the biorefinery value chain. The improvement of

TABLE 17.3 Feedstock Used for Bioethanol and Biodiesel Production: Production, Use, and Projections

	Bioethanol				Biodiesel			
	Biomass Used as Feedstock	Production (Kton) 2016–2019	Expected Use as Biofuel Feedstock 2030 (%)	Production Estimate for Biofuel Production (2028)	Biomass Used as Feedstock	Production (Vegetable Oil-Kton) 2016–2019	Expected Use as Biofuel Feedstock 2030 (%)	Production Required for Biofuel Production (2028)
United States	Maize	374,019	10	388.699	Soybean oil	13,432	25	11.915
Brazil	SugarCane	33,891	85	41,378	Soybean oil	9,312	40	10.282
China	Maize	260,005	8	377.128	Waste oils	28,254	–	–
European Union	Maize	63,635	8	69,082	Rapeseed oil	14,474	40	13.693
Thailand	Sugarcane	28,600[a]	68	40,183	Palm oil	3,410	50	5.215
Argentina	Maize	49,157	9	73,450	Soybean oil	9,289	70	10.264
Global Consumption as Biofuel Feedstock	–	121.9 mln.L	–	–	–	37.6 mln.L	–	–

Overview of the biomass used as feedstocks and production required of biomass for production of bioethanol and biodiesel between the period 2018–2028.

Source: Refs. [69, 131]; OECD-FAO Agricultural Outlook 2019–2028.

this key stage would mean a future large-scale application of biofuels in the transport sector since it impacts in the global cost of the process. Furthermore, other benefits would be market competition with fossil fuels, enhancement of technology for energy transition, and reduction of negative environmental impacts. Even thought, biorefineries still face technical/process challenges for biomass (LCB) conversion and utilization. The technical challenges involved in the use of these biomass as feedstocks to produce biofuels will be discussed in this section.

LCB can be used for biogas and biodiesel production through different processes. In the case of biogas, anaerobic digestion (AD) is the core process where occur all syntrophic mechanisms for biomass conversion. However, the high carbon and nitrogen ratio (C:N) of LCB negatively affects methane production, and the mono-digestion of this substrate is not an adequate strategy [126, 127, 129]. There are several methods to improve the use of lignocellulosic feedstock as substrate in the AD process, such as anaerobic co-digestion, bioaugmentation, and nutrient supplementation established to optimize the C:N to 20–30:1 [132]. For example, Kainthola et al. [133] evaluated the methane production of the anaerobic co-digestion of rice straw and food waste. It was shown that methane yield increased around 94% with respect to control in the optimized process (C:N 30, pH 7.32, food/microorganism ratio 1.87).

In biodiesel production, LCB can be hydrolyzed and used for lipids synthesis by oleaginous yeast, such as *Meyerozyma guilliermondii* and *Pichia kudriavzevii* or filamentous fungus-like *Mortierella isabellina* for subsequent obtaining of biodiesel [19, 134, 135]. Those microorganisms could accumulate up to 60–80% lipids of their cell dry weight and can grow exponentially while utilizing cheap substrates, becoming suitable candidates for biotechnological purposes [136, 137]. The use of cheap carbon sources like crop byproducts, significantly diminishes the cost of yeast oil production [109].

Due to the complexity of the lignocellulosic matrix, the crystallinity of the cellulose and the inhibition of metabolic operations by the lignin by-products, other strategies have been established to improve the use of LCB in the production of biofuels. Thus, pretreatments aim to modify the structural and compositional barriers of this biomass, in order to improve digestibility by enzymes to provide fermentable sugars for microorganisms [132, 138–140].

LCB pretreatment methods are widely known and can be classified in physical, chemical, physicochemical, and biological processes, and can be used alone or in combination.

The most common physical pretreatments are mechanical. One advantage of their use is the increase of the biochemical potential for biofuel production. For example, when pretreatments, such as roll milling, extrusion, pelletization, and hammer milling were applied to wheat straw, it was observed an increment in the maximum daily methane production and enzymatic hydrolysis yield related to the amount of released glucan content [141]. In another research, higher methane production was obtained from rice straw in AD as the result of particle size reduction, which improves the cellulose degradation rate [142, 143]. In this context, the most common mechanical method to increase the surface area and reduce the crystallinity of LCB is milling. The particle size is an important parameter that could lead to supplementary costs in the overall process of bioenergy production from lignocellulose. For example, Moiceanu et al. [144] used *Miscanthus* biomass in a 10 mm orifice sieve, which means energy consumption of 50–67 kJ/kg, that led to the increase of processing costs [144]. In addition, the energy consumption and susceptibility of milling machines in the presence of large materials such as stones and metals mixed with LCB, are some inconveniences [145]. Milling is also a currently used pretreatment on biodiesel production, but its purpose is centered on the release of lipids from marine algae species. A study with *Chlorella vulgaris* shows that from 208 mg/L of biomass could be obtained a lipid release of 10% when using bed milling [146].

On the other hand, chemical pretreatments like acid, alkaline, oxidative, and organo-solvent processes have also been extensively studied for pretreatment of LCB. Acid pretreatment causes the rupture of chemical bonds present in the biomass matrix, mainly removing the hemicellulose fraction [132]. One example of this pretreatment was shown by Taherdanak et al. [147]. In this study was observed the decrease of the crystallinity index of the wheat biomass accompanied by the solubilization of surface layers' structure. Moreover, Amnuaycheewa et al. [148] showed that biogas production and saccharification from rice straw can be enhanced by pretreatment with organic acids like oxalic and citric. Also, Huang et al. [149] used this pretreatment to obtain sulfuric acid-treated rice straw hydrolysate (SARSH) for microbial (*Trichosporon fermentans*) oil synthesis with potential for the production of biodiesel. During fermentation using SARSH without detoxifying, 1.7 g/L of lipid was obtained, a low yield compared to that obtained by fermentation with detoxified hydrolyzate, generating a 40.1% lipid content. Despite of the acid process is an efficient method for LCB, the main disadvantages are the high cost of acids, drastic operation conditions (pressure and temperature), and the formation of potential biological inhibitors [127, 150].

Looking to face the limitations of the most used pretreatments, researchers have found that chemical pretreatments assisted by alternative technologies are a promising way to unlock technological bioprocessing challenges.

Unlike acid pretreatment, alkaline pretreatment utilizes an alkali solution capable of promoting the solubilization of lignin present in biomass [127, 129]. As mentioned above, this pretreatment can be combined with the hydrogen peroxide pretreatment [151, 152], ultrasound [153], and hydrodynamic cavitation (HC) [154–157]. HC technology is a new and promising option for the pretreatment of LCB. Cavitation is the phenomenon of the formation of vaporous microbubbles or cavities growing and collapsing in a flowing liquid [156–158]. These collapsing cavities generate localized hot spots with very high temperatures and pressures in addition to intense shear [159]. Consequently, highly reactive oxygen species (ROS) named hydroxyl radicals (OH) are formed which produce oxidative degradation of LCB, turning more suitable for subsequent enzyme action and for microbial metabolization, which could result in higher yields of biofuels [158, 160–163].

17.4.5 CHALLENGES FOR BIOETHANOL TECHNOLOGIES: 1st, 2nd, AND 3rd GENERATION

With the implementation of first, second, and third-generation bioethanol production, also appeared some biotechnical and economic challenges to make these technologies viable, aiming to enhance their use in the transport sector. In this topic, some issues that companies have been facing on a large-scale bioethanol production will be discussed (Figure 17.3).

17.4.5.1 FIRST-GENERATION BIOETHANOL

Despite the first-generation bioethanol is a bioprocess well implemented in some countries such as USA, Brazil, and India, the industrial production of bioethanol by this technology has some challenges to be overcome. The main problems in 1G bioethanol production are the following:

1. **Climatic Problems for Crops:** With changes in climate dynamics in recent years, energy crops have been showing deficits due to drought. In addition to harming energy crops, this factor also interferes with the food and biofuel production chain related to them [164]. For example, in 2019, the severe drought that hit central-south Brazil for

more than four months lit the warning signal for the 2020 sugarcane harvest. Sugarcane has an annual cycle and, after its harvest, depends on a regular water regime for the full development of the plant until the next harvest [165]. As a consequence, this fact affected feedstock availability and led to a reduction of bioethanol production and an increase in price level.

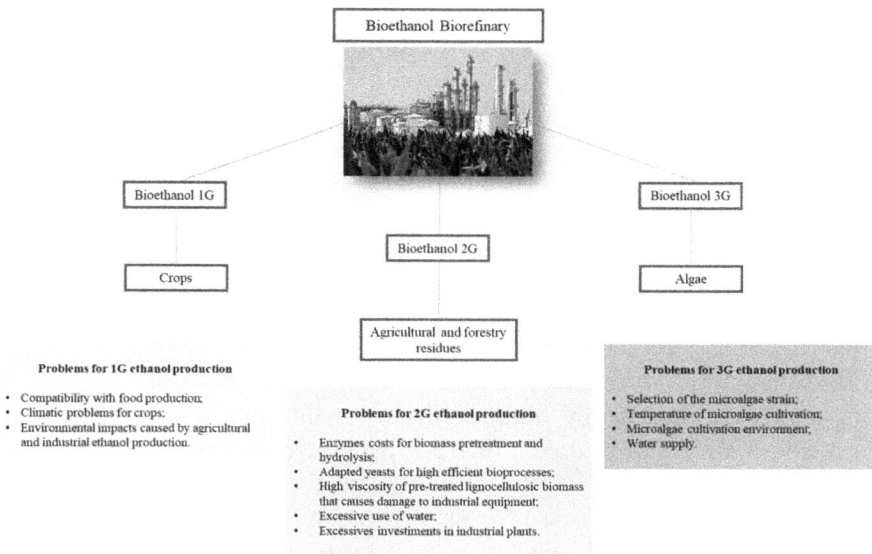

FIGURE 17.3 Bottlenecks for bioethanol production.

2. **Environmental Impacts Caused by Agricultural and Industrial Bioethanol Production:** Bioethanol production can also cause soil, water, and air pollution, a fact unknown to most of the population. The main environmental impacts encountered by agricultural activities are soil compaction by tractors and cultivation implements, contamination of water and soils by intensive use of fertilizers and pesticides, uncontrolled application of vinasse (as fertilizer), filter cake residues, and air pollution resulting from burning straw and other wastes. Furthermore, during industrial processes, electricity is generated from the burning of bagasse, excess straw, and other residues. Electricity generation also produce air and soil contamination by accumulation of GHGs and fly ash. Besides, the inadequate disposal of feedstocks residues and equipment washing water, condenser, and multi-jet and domestic wastewater also cause water pollution [166].

3. **Devaluation of the Price of Oil:** Besides, in some countries, low government investment in the sector and the increase in taxes could make the survival of some bioethanol plants even more difficult. If government interventions in the fuel price table are not effective, the competitiveness between gasoline and bioethanol may decrease, which would cause losses to biofuel plants [167]. Besides, in some countries, low government investment in the sector and the increase in taxes could make the survival of some bioethanol plants even more difficult.

17.4.5.2 SECOND-GENERATION BIOETHANOL

The problems reported related to 2G bioethanol production are based on global industrial experiences. Industries such as Raízen, Granbio, Poet, Beta Renewables, Abengoa (Synata Bio), and DuPont have found difficulties in the production phase regarding the pretreatment processes, microorganisms used, even in the mechanics and equipment costs of the unit operations involved in the process [168].

Substantially, the production of 2G bioethanol is based on the use of lignocellulosic and starchy biomass, which main challenges are pretreatment and hydrolysis processes that are magnified on an industrial scale. According to specialists working in 2G bioethanol plants, none of the pretreatments and hydrolytic processes available today are more advantageous than the other, and each varies in the production process, depending on the type of raw material and several other factors [168]. For these stages, enzymatic processes are preferred since they do not form microbial inhibitor compounds, such as furans and phenolics. However, the enzymatic cocktails used can make the process more expensive, due to their prices and the high enzymatic load required. Thus, it is essential to improve the efficiency of enzymes for reducing the price of this biofuel [168]. As stated by Schubert [169], cellulase cocktails cost around 15–20 cents per gallon bioethanol as compared to only 2–4 cents per gallon bioethanol for amylases used in the bioethanol production process by starch (1G bioethanol). The adoption of each type of enzyme in the pretreatment and enzymatic hydrolysis stages is totally dependent on the structure of feedstocks.

It is well documented that the high crystallinity of cellulose, high levels of lignin, hemicellulose, acetyl group, and degree of polymerization (DP) are undesirable characteristics of biomass that directly interfere in the performance of different enzyme cocktails [19, 170–175]. In an attempt to solve such problems, companies such as Novozymes and Genencor International have been studying and improving cellulose cocktails obtained by *Trichoderma reesei* and *Cellulomonas fimi* [173].

Recently, Santos et al. [176], modified the enzyme β-glycosidase, of the family 1 of the glycoside hydrolases (GH1) expressed by *Trichoderma harzianum*. With the site-directed mutation, the active site of the engineered enzyme became smaller and similar in size to that of β-glycosidases GH1. The results of the analyzes indicated that the modified enzyme showed a catalytic efficiency 300% higher than that of wild protein and became more tolerant of glucose, providing a significant increase in the release of sugar from all sources of plant biomass tested and also increase of thermal stability during fermentation. This engineered β-glycosidase can be used to supplement the enzymatic cocktails commercialized today for the degradation of biomass and the production of second-generation biofuels. Besides, this study shows that this enzyme can be produced by a homologous expression, which makes possible its production on a large scale and with low costs.

In the case of the fermentation stage, the main challenge is the selection of an appropriate microorganism to achieve the expected yield of bioethanol production on an industrial scale. There are microorganisms more efficient in the metabolism of hexoses from cellulose, than in pentoses from hemicellulose hydrolysis [177]. In this context, there are several studies to select wild yeast and produce engineered microorganisms also capable of ferment pentoses.

Furthermore, other important difficulties are operational and mechanical processes. GranBio reported the corrosion of the equipment in one of its plants that used the steam explosion (SE) pretreatment in which the LCB was subjected to very high temperature and pressure conditions. When the system was suddenly decompressed, the biomass hit the walls of the equipment causing routine damage and possibly interrupting the production. The solution to this problem was the use of the liquid hot water (LHW) process associated with the mechanical treatment of the fibers [177].

Raízen plant shows other problems related to the high level of dirt in the biomass, which required a pre-cleaning step. It was also necessary to redesign the process for the separation of sugars and lignin, introducing more filters and centrifuges. In this 2G bioethanol plant, the following problems were also presented: (I) the corrosion of pretreatment equipment, due to the silica present in the sugarcane straw, (II) the excessive use of water when working with wheat straw as feedstock, (III) the high water absorption by the sugarcane bagasse, making it a high viscosity material, clogging of pipes and valves [177].

Also, according to Lynd [178], the success of the industrial implementation of 2G bioethanol depends on investments in research and development.

One of the errors noticed in current biorefineries is the excessive investment in plant infrastructure, instead of concerning with supporting technological advances capable of reducing production costs.

17.4.5.3 THIRD-GENERATION BIOETHANOL

As analyzed previously in 1G and 2G technologies, the third-generation also presents bottlenecks. The main steps to 3G bioethanol production are microalgae cultivation, cell collection, pretreatment of biomass, fermentation, and recovery of bioethanol by distillation [179]. Algal biomass could be pretreated by mechanical or enzymatic methods to break cell walls, making carbohydrates available. After this step, yeasts are added to convert fermentable sugars into bioethanol [180–182].

Some issues in the production of 3G bioethanol are reported as follows.

The selection of the microalgae strain is one of the most important parameters. The carbohydrate composition of the cells must be considered since the glycosidic portion acts as a substrate for the fermentation process. As reported by Ueno [183], *Chlorella vulgaris* species is commonly used due to its capacity of carbohydrate accumulation higher than 65%. According to Kose and Oncel [184], the main parameters should be considerers for microalgae selection including (I) tolerance to oxygen gas levels, (II) resistance to photo-inhibition, (III) high rates of cell growth, (IV) low need for nutrients to develop, (V) high photosynthetic activity, (VI) easy adaptation to harsh environmental conditions (for open systems), (VII) high survival rates, and (VIII) genetic stability.

Temperature of microalgae cultivation is also an important factor that influences the production of 3G bioethanol. As reported by Costa and De Morais [180], it was observed that when *Chlorella sp.* was cultivated at 30°C, showed higher yields (448.0 $\mu mol.g^{-1}$) of dry mass, in comparison with cultures grown under 20°C (196.0 $\mu mol.g^{-1}$). It is emphasized that the drastic reduction in yield was also observed when the culture was performed at 35°C and 45°C.

The microalgae cultivation mode also influences 3G bioethanol production. Despite open lagoons are popular for being simple projects, contamination, temperature control, sun exposure, and other factors, cannot be controlled, highlighting the need for a careful selection of the bioreactor type [184].

Supply and physical and chemical characteristics of water are of fundamental importance for the seaweed cultivation during bioethanol 3G production. A large volume of water is required to produce algae. Thus, this resource can be optimized using wastewater, brackish, and with high levels of organic nutrients [184].

17.5 CONCLUSION AND FUTURE PERSPECTIVES

Biofuels are the main alternative to reduce environmental impact and cope with the energy crisis, and their global production has significantly increased in the last decade. In agreement with global sustainable goals, it is expected to achieve a biofuels production growth by about 10% by 2030, using 28% of available biomass as feedstock. Thus, it is extremely important to continue investing and encouraging the development of technologies to make biofuels an achievable option with a better cost-benefit for those countries without oil reserves and refineries. Policies and programs, like RenovaBio in Brazil, have been established to improve biofuel production and application in the transportation sector. Consequently, biofuels developing countries could support their future policies based on top producers' experiences. Despite the large number of biofuels manufactured in some countries, the actual industrial production does not reach sustainable goals and still faces environmental, economic, and technological challenges. At this moment, several strategies for biofuel technologies need to be applied by the integration of different biorefinery modes, such as 1G-1G integration (e.g., sugarcane-corn), 1.5G (e.g., 1G-2G), and other possibilities based on 2G and 3G. In this context, technological advances will be the core to boost biofuels to achieve a secure transition from fossil fuels to an energy sustainable panorama.

KEYWORDS

- biofuels
- biomass conversion
- European union
- global warming
- greenhouse gas
- life cycle assessment
- technological challenges
- transport sector

REFERENCES

1. Rajagopal, D., & Zilberman, D., (2007). *Review of Environmental, Economic and Policy Aspects of Biofuels*. The World Bank.
2. International Energy Agency (IEA), (2020a). *Transport Biofuels*. Retrieved from: https://www.iea.org/reports/transport-biofuels (accessed on 28 October 2021).
3. International Renewable Energy Agency (IRENA), (2020). *Global Renewable Outlook, Energy Transformation 2050*. Retrieved from: https://www.irena.org/-/media/Files/IRENA/Agency/Publication/2020/Apr/IRENA_Global_Renewables_Outlook_2020.pdf (accessed on 28 October 2021).
4. Raboni, M., Viotti, P., & Capodaglio, A.G. (2015). A comprehensive analysis of the current and future role of biofuels for transport in the European Union (EU). *Environment & Water Magazine, 10*, 9–21.
5. Beschkov, V., (2017). Biogas, biodiesel and bioethanol as multifunctional renewable fuels and raw materials. In: JacobLopes, E., & Zepka, L. Q., (eds.), *Frontiers in Bioenergy and Biofuels* (pp. 185–205).
6. International Energy Agency (IEA), (2020d). *Advanced Biofuels–Potential for Cost Reduction*. Retrieved from: https://www.ieabioenergy.com/publications/new-publication-advanced-biofuels-potential-for-cost-reduction/ (accessed on 28 October 2021).
7. Eisentraut, A., (2010). Sustainable Production of Second-generation Biofuels: Potential and Perspectives in Major Economies and Developing Countries, 2010/01, 9–221. Retrieved from: https://doi.org/10.1787/5kmh3njpt6r0-en.
8. Achinas, S., Horjus, J., Achinas, V., & Euverink, G. J. W., (2019). A pestle analysis of biofuels energy industry in Europe. *Sustainability, 11*(21), 5981.
9. Demirbas, A., (2011a). Competitive liquid biofuels from biomass. *Applied Energy, 88*(1), 17–28.
10. Azad, A. K., Rasul, M. G., Khan, M. M. K., Sharma, S. C., & Hazrat, M. A., (2015). Prospect of biofuels as an alternative transport fuel in Australia. *Renewable and Sustainable Energy Reviews, 43*, 331–351.
11. Ahlgren, E. O., Börjesson, H. M., & Grahn, M., (2017). Transport biofuels in global energy-economy modelling-a review of comprehensive energy systems assessment approaches. *GCB Bioenergy, 9*(7), 1168–1180.
12. Azad, A. K., & Ameer, U. S. M., (2013). Performance study of a diesel engine by first-generation bio-fuel blends with fossil fuel: An experimental study. *Journal of Renewable and Sustainable Energy, 5*(1), 013118.
13. Lin, T., Rodríguez, L. F., Shastri, Y. N., Hansen, A. C., & Ting, K. C., (2014). Integrated strategic and tactical biomass-biofuel supply chain optimization. *Bioresource Technology, 156*, 256–266.
14. Demirbas, M. F., (2009b). Biorefineries for biofuel upgrading: A critical review. *Applied Energy, 86*, S151–S161.
15. Naik, S. N., Goud, V. V., Rout, P. K., & Dalai, A. K., (2010). Production of first and second-generation biofuels: A comprehensive review. *Renewable and Sustainable Energy Reviews, 14*(2), 578–597.
16. Lee, R. A., & Lavoie, J. M., (2013). From first-to third-generation biofuels: Challenges of producing a commodity from a biomass of increasing complexity. *Animal Frontiers, 3*(2), 6–11.

17. Lü, J., Sheahan, C., & Fu, P., (2011). Metabolic engineering of algae for fourth-generation biofuels production. *Energy & Environmental Science, 4*(7), 2451–2466.
18. Abomohra, A. E. F., & Elshobary, M., (2019). Biodiesel, bioethanol, and biobutanol production from microalgae. In: *Microalgae Biotechnology for Development of Biofuel and Wastewater Treatment* (pp. 293–321). Springer, Singapore.
19. Kour, D., Rana, K. L., Yadav, N., Yadav, A. N., Rastegari, A. A., Singh, C., & Saxena, A. K., (2019). Technologies for biofuel production: Current development, challenges, and future prospects. In: *Prospects of Renewable Bioprocessing in Future Energy Systems* (pp. 1–50). Springer, Cham.
20. Azad, A. K., Uddin, S. A., & Alam, M. M., (2012). A comprehensive study of di diesel engine performance with vegetable oil: An alternative bio-fuel source of energy. *International Journal of Automotive and Mechanical Engineering (IJAME), 5*, 576–586.
21. Pimentel, D., & Patzek, T., (2008). Ethanol production using corn, switchgrass and wood; biodiesel production using soybean. In: *Biofuels, Solar and Wind as Renewable Energy Systems* (pp. 373–394). Springer, Dordrecht.
22. Demirbas, M. F., (2011b). Biofuels from algae for sustainable development. *Applied Energy, 88*(10), 3473–3480.
23. Azad, A. K., Rasul, M. G., Khan, M. M. K., & Sharma, S. C., (2014). Review of biodiesel production from microalgae: A novel source of green energy. In: *Int. Green Energy Conf.* (pp. 1–9).
24. OECD Library, (2019). *OECD Library-Enhancing Climate Change Migation through Agriculture*. OECD Organization for Economic Cooperation and Development (database). Retrieved from: https://www.oecd-ilibrary.org/sites/dce06785-en/index.html?itemId=/content/component/dce06785-en (accessed on 28 October 2021).
25. IEA, (2018). *Renewable-2018: Analysis and Forecast to 2023*, International Energy Agency, Paris. International Energy Agency. Biofuels. Retrieved from: http://www.iea.org/topics/biofuels/ (accessed on 28 October 2021).
26. International Energy Agency (IEA), (2019a). *Renewables 2019*. Retrieved from: https://www.iea.org/reports/renewables-2019/transport#abstract (accessed on 28 October 2021).
27. OECD, (2018). *Relatórios Econômicos OCDE Brasil, 2018*. Retrieved from https://epge.fgv.br/conferencias/apresentacao-do-relatorio-da-ocde-2018/files/relatorios-economicos-ocde-brasil-2018.pdf (accessed on 28 October 2021).
28. International Energy Agency (IEA), (2019b). *World Energy Model*. Retrieved from: Retrieved from: https://www.iea.org/reports/world-energy-model/sustainable-development-scenario (accessed on 28 October 2021).
29. Nogueira, L. A. H., Souza, G. M., Cortez, L. A. B., & De Brito, C. C. H., (2020). Biofuels for transport. In: *Future Energy* (pp. 173–197). Elsevier.
30. International Renewable energy Agency (IRENA), (2014). *Global Bioenergy Supply and Demand Projections: A Working Paper for REmap 2030, 1*, 1–88.
31. Cremiato, R., Mastellone, M. L., Tagliaferri, C., Zaccariello, L., & Lettieri, P., (2018). Environmental impact of municipal solid waste management using life cycle assessment: The effect of anaerobic digestion, materials recovery and secondary fuels production. *Renewable Energy, 124*, 180–188.
32. Callegari, A., Bolognesi, S., Cecconet, D., & Capodaglio, A. G., (2020). Production technologies, current role, and future prospects of biofuels feedstocks: A state-of-the-art review. *Critical Reviews in Environmental Science and Technology, 50*(4), 384–436.

33. Panchuk, M., Kryshtopa, S., Sładkowski, A., & Panchuk, A., (2020). Environmental aspects of the production and use of biofuels in transport. In: *Ecology in Transport: Problems and Solutions* (pp. 115–168). Springer, Cham. Retrieved from: https://link. springer.com/chapter/10.1007/978-3-030-42323-0_3 (accessed on 28 October 2021).

34. Sładkowski, A., (2020). *Ecology in Transport: Problems and Solutions.* Springer. Retrieved from: https://link.springer.com/book/10.1007%2F978-3-030-42323-0 (accessed on 28 October 2021).

35. Statista, (2020a). *Production of Biogas Worldwide in 2017, by Region.* Retrieved from: https://www.statista.com/statistics/481828/biogas-production-worldwide-by-region/ (accessed on 28 October 2021).

36. Scarlat, N., Dallemand, J. F., & Fahl, F., (2018). Biogas: Developments and perspectives in Europe. *Renewable Energy, 129*, 457–472.

37. Eurostat, S. A., (2017). Your Key to European Statistics. Retrieved from: https:// ec.europa.eu/eurostat (accessed on 8 December 2021).

38. Lampinen, A., (2013). Development of biogas technology systems for transport. *Tekniikan Waiheita, 31*(3), 5–37.

39. International Renewable Energy Agency (IRENA), (2018). *Biogas for Road Vehicles: Technology Brief.* International Renewable Energy Agency, Abu Dhabi. https://www. irena.org/-/media/Files/IRENA/Agency/Publication/2017/Mar/IRENA_Biogas_for_ Road_Vehicles_2017.pdf (accessed on 28 October 2021).

40. Van, F. F., (2012). *Perspectives for Biogas in Europe.* Oxford Institute for Energy Studies.

41. Gebremariam, S. N., & Marchetti, J. M., (2018). Biodiesel production economics. *Energy Conversion and Management,168*, 74–84.

42. Burton, R., & Biofuels, P., (2008). An overview of ASTM D6751: Biodiesel standards and testing methods. *Alternative Fuels Consortium.*

43. Statista, (2020b). *Leading Biodiesel Producers Worldwide in 2019, by Country (in billion liters).* Retrieved from: https://www.statista.com/statistics/271472/biodiesel-production-in-selected-countries/ (accessed on 28 October 2021).

44. European Commission Progress Reports, (2015). Retrieved from: https://ec.europa. eu/neighbourhood-enlargement/system/files/2018–12/20151110_report_turkey.pdf (accessed on 8 December 2021).

45. Mahapatra, M. K., & Kumar, A., (2017). A short review on biobutanol, a second-generation biofuel production from lignocellulosic biomass. *J. Clean Energy Technol., 5*(1), 27–30.

46. Yun, Y., (2020). Alcohol fuels: Current status and future direction. In: *Alcohol Fuels-Current Technologies and Future Prospect.* IntechOpen.

47. Cheng, J. J., & Timilsina, G. R., (2011). Status and barriers of advanced biofuel technologies: A review. *Renewable Energy, 36*(12), 3541–3549.

48. Cardona, C. A., & Sánchez, Ó. J., (2007). Fuel ethanol production: Process design trends and integration opportunities. *Bioresource Technology, 98*(12), 2415–2457.

49. Cortez, L. A., Griffin, M. W., Scaramucci, J. A., Scandiffio, M. I., & Braunbeck, O. A., (2003). Considerations on the worldwide use of bioethanol as a contribution for sustainability. *Management of Environmental Quality: An International Journal.*

50. Aydemir, E., Demirci, S., Dogan, A., Aytekin, A. Ö., & Sahin, F., (2014). Genetic modifications of *Saccharomyces cerevisiae* for ethanol production from starch fermentation: A review. *Journal of Bioprocessing & Biotechniques, 4*(7), 1.

51. Statista, (2020c). World production of fuel ethanol in 2019, by country (in millions of gallons). Retrieved from: https://www.statista.com/statistics/281606/ethanol-production-in-selected-countries/ (accessed on 28 October 2021).

52. Energy information administration (EIA), (2020a). https://www.eia.gov/energyexplained/use-of-energy/transportation.php (accessed on 28 October 2021).

53. Energy Information Administration (EIA), (2020b). *Use of Energy Explained - Energy Use for Transportation.* Retrieved from: https://www.eia.gov/energyexplained/use-of-energy/transportation.php (accessed on 28 October 2021).

54. Ramage, M., & Katzer, J., (2009). *Liquid Transportation Fuels from Coal and Biomass: Technological Status, Costs, and Environmental Impacts.* America's Energy Future Study, National Academy Panel on Alternative Liquid Transportation Fuels.

55. Robertson, G. P., Hamilton, S. K., Barham, B. L., Dale, B. E., Izaurralde, R. C., Jackson, R. D., & Tiedje, J. M., (2017). Cellulosic biofuel contributions to a sustainable energy future: Choices and outcomes. *Science, 356*(6345).

56. Liu, F., Short, M. D., Alvarez-Gaitan, J. P., Guo, X., Duan, J., Saint, C., & Hou, L. A., (2020). Environmental life cycle assessment of lignocellulosic ethanol-blended fuels: A case study. *Journal of Cleaner Production, 245*, 118933.

57. Cruz, C. H. B., Souza, G. M., & Cortez, L. A. B., (2014). Biofuels for transport. In: *Future Energy* (pp. 215–244). Elsevier.

58. Belincanta, J., Alchorne, J. A., & Teixeira Da, S. M., (2016). The Brazilian experience with ethanol fuel: Aspects of production, use, quality and distribution logistics. *Brazilian Journal of Chemical Engineering, 33*(4), 1091–1102.

59. Karmaker, A. K., Rahman, M. M., Hossain, M. A., & Ahmed, M. R., (2020). Exploration and corrective measures of greenhouse gas emission from fossil fuel power stations for Bangladesh. *Journal of Cleaner Production, 244*, 118645.

60. Wang, S., Li, F., Wu, D., Zhang, P., Wang, H., Tao, X., Ye, J., & Nabi, M., (2018). Enzyme pretreatment enhancing biogas yield from corn stover: Feasibility, optimization, and mechanism analysis. *Journal of Agricultural and Food Chemistry, 66*(38), 10026–10032.

61. NOAA-National Oceanic and Atmospheric Administration (2020). *Mauna Loa Atmospheric Baseline Observatory.* Retrieved from: https://www.noaa.gov/news/carbon-dioxide-levels-in-atmosphere-hit-record-high-in-may (accessed on 28 October 2021).

62. C2ES, (2020). Center for climate and energy solution. *Renewable Energy.* https://www.c2es.org/content/renewable-energy/#:~:text=Renewables%20made%20up%2017.1%20percent,come%20from%20wind%20and%20solar (accessed on 28 October 2021).

63. World Bank, (2020). *Global Economic Prospects.* Washington, DC: World Bank.

64. BP, (2020). *Statistical Review of World Energy.* Retrieved from: https://www.bp.com/en/global/corporate/energy-economics/statistical-review-of-world-energy.html (accessed on 28 October 2021).

65. Environmental Protection Agency (EPA), (n.d). *Economics of Biofuels.* Retrieved from: https://www.epa.gov/environmental-economics/economics-biofuels#main-content (accessed on 28 October 2021).

66. Anderson, W., (2019). *Biofuels Effects: Social, Economic, Political & Environmental.* Schoolworkhelper Editorial Team. Retrieved from SchoolWorkHelper, https://schoolworkhelper.net/biofuels-effect-on-social-economic-political-environmental/ (accessed on 8 December 2021).

67. International Energy Agency (IEA), (2020b). *The Impact of the Covid-19 Crisis on Clean Energy Progress.* Retrieved from https://www.iea.org/articles/the-impact-of-the-covid-19-crisis-on-clean-energy-progress (accessed on 28 October 2021).

68. OECD/FAO, (2019a). OECD-FAO Agricultural Outlook 2019–2028, OECD Publishing, Paris/Food and Agriculture Organization of the United Nations, Rome. Retrieved from

http://www.fao.org/3/CA4076EN/CA4076EN_Chapter9_Biofuels.pdf (accessed on 28 October 2021).

69. OECD/FAO, (2019b). *OECD-FAO Agricultural Outlook*. OECD Agriculture statistics (database). doi: Dx.doi.org/10.1787/agr-outl-data-en http://www.fao.org/3/ca4076en/ca4076en.pdf (accessed on 28 October 2021).

70. Johansson, N., (2008). Production of Liquid Biogas, LBG, with Cryogenic and Conventional Upgrading Technology. *Description of Systems and Evaluations of Energy Balances, 1,* 1–92.

71. Angelidaki, I., Xie, L., Luo, G., Zhang, Y., Oechsner, H., Lemmer, A., & Kougias, P. G., (2019). Biogas upgrading: Current and emerging technologies. In: *Biofuels: Alternative Feedstocks and Conversion Processes for the Production of Liquid and Gaseous Biofuels* (pp. 817–843). Academic Press.

72. Panahi, H. K. S., Tabatabaei, M., Aghbashlo, M., Dehhaghi, M., Rehan, M., & Nizami, A. S., (2019). Recent updates on the production and upgrading of bio-crude oil from microalgae. *Bioresource Technology Reports, 7,* 100216.

73. Yang, L., Ge, X., Wan, C., Yu, F., & Li, Y., (2014). Progress and perspectives in converting biogas to transportation fuels. *Renewable and Sustainable Energy Reviews, 40,* 1133–1152.

74. International Energy Agency (IEA), (2020c). *Outlook for Biogas and Biomethane: Prospects for Organic Growth in World Energy Outlook Special Report.* https://www.iea.org/reports/outlook-for-biogas-and-biomethane-prospects-for-organic-growth/an-introduction-to-biogas-and-biomethane (accessed on 28 October 2021).

75. Independent Statistics Analysis (EIA), (2020). *July 2020 Monthly Energy Review US.* Retrieved from: https://www.eia.gov/totalenergy/data/monthly/pdf/mer.pdf (accessed on 28 October 2021).

76. United States Department of Agriculture-(USDA), (2019c). Brasil Biofuels Annual 2019. Retrieved from: https://apps.fas.usda.gov/newgainapi/api/report/downloadreportbyfilename?filename=Biofuels%20Annual_Sao%20Paulo%20ATO_Brazil_8-9-2019.pdf (accessed on 28 October 2021).

77. United States Department of Agriculture-(USDA). (2019d). *Biofuels Annual, China Will Miss E10 by 2020 Goal by Wide Margin.* Retrieved from: https://apps.fas.usda.gov/newgainapi/api/report/downloadreportbyfilename?filename=Biofuels%20Annual_Beijing_China%20-%20Peoples%20Republic%20of_8-9-2019.pdf (accessed on 28 October 2021).

78. United States Department of Agriculture-(USDA), (2019e). *EU Biofuels Annual 2019.* Retrieved from: https://apps.fas.usda.gov/newgainapi/api/report/downloadreportbyfilename?filename=Biofuels%20Annual_The%20Hague_EU-28_7-15-2019.pdf (accessed on 28 October 2021).

79. United States Department of Agriculture-(USDA), (2019f). *India Biofuels Annual 2019.* Retrieved from: https://apps.fas.usda.gov/newgainapi/api/report/downloadreportbyfilename?filename=Biofuels%20Annual_New%20Delhi_India_8-9-2019.pdf (accessed on 28 October 2021).

80. United States Department of Agriculture-(USDA), (2019g). *Canada Biofuels Annual 2019, 1–613.*

81. United States Department of Agriculture-(USDA), (2019h). *Thailand Biofuels Annual 2019.* Retrieved from: https://apps.fas.usda.gov/newgainapi/api/Report/DownloadReportByFileName?fileName=Biofuels%20Annual_Bangkok_Thailand_11-04-2019 (accessed on 28 October 2021).

82. United States Department of Agriculture-(USDA), (2019i). *Argentina Biofuels Annual 2019*. Retrieved from: https://apps.fas.usda.gov/newgainapi/api/report/downloadrepo rtbyfilename?filename=Biofuels%20Annual_Buenos%20Aires_Argentina_8-9-2019. pdf (accessed on 28 October 2021).

83. OECDilibrary (2019j), "OECDilibrary-Enhancing Climate Change Mitigation through Agriculture," OECD Organization for Economic Cooperation and Development (database).

84. Environmental Protection Agency (EPA), (2020). *Green Vehicle Guide: Greenhouse Gas Emissions from a Typical Passenger Vehicle*. https://www.epa.gov/greenvehicles/ greenhouse-gas-emissions-typical-passenger-vehicle (accessed on 28 October 2021).

85. Mofijur, M. G. R. M., Rasul, M. G., Hyde, J., Azad, A. K., Mamat, R., & Bhuiya, M. M. K., (2016). Role of biofuel and their binary (diesel-biodiesel) and ternary (ethanol-biodiesel-diesel) blends on internal combustion engines emission reduction. *Renewable and Sustainable Energy Reviews, 53*, 265–278.

86. Smith, M. N., (2016). *The Number of Cars Worldwide is Set to Double by 2040*. World Economic Forum. Published online at weforum.org. Retrieved from: https://www. weforum.org/agenda/2016/04/the-number-of-cars-worldwide-is-set-to-double-by-2040 (accessed on 28 October 2021).

87. Graver, B., Zhang, K., & Rutherford, D., (2019). *Emissions from Commercial Aviation,* 2018.

88. Pathak, L., & Shah, K., (2019). Renewable energy resources, policies and gaps in BRICS countries and the global impact. *Frontiers in Energy, 13*(3), 506–521.

89. De Fátima, V. M., (2019). *Northeastern Sugar and Ethanol Production, 129*, 1–5.

90. Hoogwijk, M., Faaij, A., Van, D. B. R., Berndes, G., Gielen, D., & Turkenburg, W., (2003). Exploration of the ranges of the global potential of biomass for energy. *Biomass and Bioenergy, 25*(2), 119–133.

91. Grassi, M. C. B., & Pereira, G. A. G., (2019). Energy-cane and RenovaBio: Brazilian vectors to boost the development of Biofuels. *Industrial Crops and Products, 129*, 201–205.

92. Acharya, R. N., & Perez-Pena, R., (2020). Role of comparative advantage in biofuel policy adoption in Latin America. *Sustainability, 12*(4), 1411.

93. Griliches, Z., (1957). Hybrid corn: An exploration in the economics of technological change. *Econometrica, Journal of the Econometric Society*, 501–522.

94. Su, Y., Zhang, P., & Su, Y., (2015). An overview of biofuels policies and industrialization in the major biofuel producing countries. *Renewable and Sustainable Energy Reviews, 50*, 991–1003.

95. Utkulu, U., & Seymen, D., (2004). Revealed comparative advantage and competitiveness: Evidence for Turkey vis-à-vis the EU/15. In: *European Trade Study Group 6th Annual Conference, ETSG* (pp. 1–26).

96. Lapan, H., & Moschini, G., (2012). Second-best biofuel policies and the welfare effects of quantity mandates and subsidies. *Journal of Environmental Economics and Management, 63*(2), 224–241.

97. Saravanan, A. P., Pugazhendhi, A., & Mathimani, T., (2020). A comprehensive assessment of biofuel policies in the BRICS nations: Implementation, blending target and gaps. *Fuel, 272*, 117635.

98. Hao, H., Liu, Z., Zhao, F., Ren, J., Chang, S., Rong, K., & Du, J., (2018). Biofuel for vehicle use in China: Current status, future potential and policy implications. *Renewable and Sustainable Energy Reviews, 82*, 645–653.

99. Qiu, H., Sun, L., Huang, J., & Rozelle, S., (2012). Liquid biofuels in China: Current status, government policies, and future opportunities and challenges. *Renewable and Sustainable Energy Reviews, 16*(5), 3095–3104.

100. Chang, S., Zhao, L., Timilsina, G. R., & Zhang, X., (2012). Biofuels development in China: Technology options and policies needed to meet the 2020 target. *Energy Policy, 51*, 64–79.

101. Lu, Y. H., & Yang, Y. Q., (2019). Sugarcane Biofuels Production in China. In: *Sugarcane Biofuels* (pp. 139–155). Springer, Cham.

102. Schuman, S., & Lin, A., (2012). China's Renewable Energy Law and its impact on renewable power in China: Progress, challenges and recommendations for improving implementation. *Energy Policy, 51*, 89–109.

103. Nahm, J., (2017). Exploiting the implementation gap: Policy divergence and industrial upgrading in China's wind and solar sectors. *The China Quarterly, 231*, 705–727.

104. European Commission, (2020). *Biofuels.* Retrieved from: https://ec.europa.eu/energy/topics/renewable-energy/biofuels/overview_en (accessed on 28 October 2021).

105. EU, (2018). *Council-Council of the European Union.* Proposal for a directive of the European Parliament and of the council on the promotion of the use of energy from renewable sources (recast). Brussels. Available at: http://data.consilium.europa.eu/ (accessed on 28 October 2021).

106. European Commission, (2019). *Renewable Energy – Recast to 2030, (RED II).* Retrieved from: https://ec.europa.eu/jrc/en/jec/renewable-energy-recast-2030-red-ii (accessed on 28 October 2021).

107. Global bioenergy. (GBEP), (2016). Retrieved from: http://www.globalbioenergy.org/fileadmin/user_upload/gbep/docs/ENGLISH_Background_note_GBEP_Setember_2016_FINAL.pdf (accessed on 28 October 2021).

108. Biofuture Platform, (2020). Retrieved from: http://www.biofutureplatform.org/ (accessed on 28 October 2021).

109. Demirbas, A., (2009c). Political, economic and environmental impacts of biofuels: A review. *Applied Energy, 86*, S108–S117.

110. Demirbas, A., (2017). The social, economic, and environmental importance of biofuels in the future. *Energy Sources, Part B: Economics, Planning, and Policy, 12*(1), 47–55.

111. Van, R. R., Sanders, J., Bakker, R., Blaauw, R., Zwart, R., & Van, D. D. B., (2011). Biofuel-driven biorefineries for the co-production of transportation fuels and added-value products. In: *Handbook of Biofuels Production* (pp. 559–580). Woodhead Publishing.

112. Cremonez, P. A., Feroldi, M., Feiden, A., Teleken, J. G., Gris, D. J., Dieter, J., & Antonelli, J., (2015). Current scenario and prospects of use of liquid biofuels in South America. *Renewable and Sustainable Energy Reviews, 43*, 352–362.

113. Lucena, A. F., Clarke, L., Schaeffer, R., Szklo, A., Rochedo, P. R., Nogueira, L. P., & Kober, T., (2016). Climate policy scenarios in Brazil: A multi-model comparison for energy. *Energy Economics, 56*, 564–574.

114. Guimarães, L. N., (2020). Is there a Latin American electricity transition? A snapshot of intraregional differences. In: *The Regulation and Policy of Latin American Energy Transitions* (pp. 3–20). Elsevier.

115. Srivastava, N., Kharwar, R. K., & Mishra, P. K., (2019). Cost economy analysis of biomass-based biofuel production. In: *New and Future Developments in Microbial Biotechnology and Bioengineering* (pp. 1–10). Elsevier.

116. Soratana, K., Harden, C. L., Zaimes, G. G., Rasutis, D., Antaya, C. L., Khanna, V., & Landis, A. E., (2014). The role of sustainability and life cycle thinking in US biofuels policies. *Energy Policy, 75*, 316–326.

117. Tozer, L., (2019). The urban material politics of decarbonization in Stockholm, London and San Francisco. *Geoforum, 102*, 106–115.
118. Silva, N. D. L. D., (2010). Biodiesel production = process and characterizations, 1–192. Retrieved from: http://www.repositorio.unicamp.br/handle/REPOSIP/266978.
119. Lu, C., (2018). When will biofuels be economically feasible for commercial flights? Considering the difference between environmental benefits and fuel purchase costs. *Journal of Cleaner Production, 181*, 365–373.
120. Markel, E., Sims, C., & English, B. C., (2018). Policy uncertainty and the optimal investment decisions of second-generation biofuel producers. *Energy Economics, 76*, 89–100.
121. Li, B., Li, Y., Liu, H., Liu, F., Wang, Z., & Wang, J., (2017). Combustion and emission characteristics of diesel engine fueled with biodiesel/PODE blends. *Applied Energy, 206*, 425–431.
122. Alisson, E., (2013). *Biofuels Face Challenges for Expansion*, 1–5. Retrieved from: https://agencia.fapesp.br/biocombustiveis-enfrentam-desafios-para-expansao/18259/ (accessed on 28 October 2021).
123. Gerbens-Leenes, P. W., (2017). Bioenergy water footprints, comparing first, second and third-generation feedstocks for bioenergy supply in 2040. *European Water, 59*, 373–380.
124. Khanna, M., Wang, W., Hudiburg, T. W., & DeLucia, E. H., (2017). The social inefficiency of regulating indirect land-use change due to biofuels. *Nature Communications, 8*(1), 1–9.
125. Espi, E., Ribas, I., Diaz, C., & Sastron, O., (2020). Feedstocks for advanced biofuels. In: *Sustainable Mobility*. IntechOpen.
126. Paul, S., & Dutta, A., (2018). Challenges and opportunities of lignocellulosic biomass for anaerobic digestion. *Resources, Conservation and Recycling, 130*, 164–174.
127. Baruah, J., Nath, B. K., Sharma, R., Kumar, S., Deka, R. C., Baruah, D. C., & Kalita, E., (2018). Recent trends in the pretreatment of lignocellulosic biomass for value-added products. In: *Frontiers in Energy Research* (Vol. 6, p. 141).
128. Dahmen, N., Lewandowski, I., Zibek, S., & Weidtmann, A., (2019). Integrated lignocellulosic value chains in a growing bioeconomy: Status quo and perspectives. *GCB Bioenergy, 11*(1), 107–117.
129. Cheah, W. Y., Sankaran, R., Show, P. L., Ibrahim, T. N. B. T., Chew, K. W., Culaba, A., & Jo-Shu, C., (2020). Pretreatment methods for lignocellulosic biofuels production: Current advances, challenges and future prospects. *Biofuel Research Journal, 7*(1), 1115.
130. Sharma, R., Joshi, R., & Kumar, D., (2020). Present status and future prospect of genetic and metabolic engineering for biofuels production from lignocellulosic biomass. In: *Genetic and Metabolic Engineering for Improved Biofuel Production from Lignocellulosic Biomass* (pp. 171–192). Elsevier.
131. Russel, T. H., & Frymier, P., (2012). Bioethanol production in Thailand: A teaching case study comparing cassava and sugar cane molasses. *The Journal of Sustainability Education*.
132. Abraham, A., Mathew, A. K., Park, H., Choi, O., Sindhu, R., Parameswaran, B., & Sang, B. I., (2020). Pretreatment strategies for enhanced biogas production from lignocellulosic biomass. *Bioresource Technology, 301*, 122725.
133. Kainthola, J., Kalamdhad, A. S., & Goud, V. V., (2020). Optimization of process parameters for accelerated methane yield from anaerobic co-digestion of rice straw and food waste. *Renewable Energy, 149*, 1352–1359.
134. Ruan, Z., Zanotti, M., Wang, X., Ducey, C., & Liu, Y., (2012). Evaluation of lipid accumulation from lignocellulosic sugars by *Mortierella isabellina* for biodiesel production. *Bioresource Technology, 110*, 198–205.

135. Ananthi, V., Siva, P. G., Chang, S. W., Ravindran, B., Nguyen, D. D., Vo, D. V. N., La, D. D., et al., (2019). Enhanced microbial biodiesel production from lignocellulosic hydrolysates using yeast isolates. *Fuel, 256*, 115932.
136. Patel, A., Arora, N., Sartaj, K., Pruthi, V., & Pruthi, P. A., (2016). Sustainable biodiesel production from oleaginous yeasts utilizing hydrolysates of various non-edible lignocellulosic biomasses. In: *Renewable and Sustainable Energy Reviews* (Vol. 62, pp. 836–855). Elsevier Ltd.
137. Martani, F., Maestroni, L., Torchio, M., Ami, D., Natalello, A., Lotti, M., & Branduardi, P., (2020). Conversion of sugar beet residues into lipids by *Lipomyces starkeyi* for biodiesel production. *Microbial Cell Factories, 19*(1), 1–13.
138. Mosier, N., Wyman, C., Dale, B., Elander, R., Lee, Y. Y., Holtzapple, M., & Ladisch, M., (2005). Features of promising technologies for pretreatment of lignocellulosic biomass. *Bioresource Technology, 96*(6), 673–686.
139. Hendriks, A. T. W. M., & Zeeman, G., (2009). Pretreatments to enhance the digestibility of lignocellulosic biomass. *Bioresource Technology, 100*(1), 10–18.
140. Dziekońska-Kubczak, U. A., Berłowska, J., Dziugan, P. T., Patelski, P., Balcerek, M., Pielech-Przybylska, K. J., & Domański, J. T., (2018). Comparison of steam explosion, dilute acid, and alkali pretreatments on enzymatic saccharification and fermentation of hardwood sawdust. *BioResources, 13*(3), 6970–6984.
141. Victorin, M., Davidsson, Å., & Wallberg, O., (2020). Characterization of mechanically pretreated wheat straw for biogas production. *Bioenergy Research*, 1–12.
142. Dai, X., Hua, Y., Dai, L., & Cai, C., (2019). Particle size reduction of rice straw enhances methane production under anaerobic digestion. *Bioresource Technology, 293*, 122043.
143. Dell,ʼO. P. P., & Spena, V. A., (2020). Mechanical pretreatment of lignocellulosic biomass to improve biogas production: Comparison of results for giant reed and wheat straw. *Energy, 203*, 117798.
144. Moiceanu, G., Paraschiv, G., Voicu, G., Dinca, M., Negoita, O., Chitoiu, M., & Tudor, P., (2019). Energy consumption at size reduction of lignocellulose biomass for bioenergy. *Sustainability, 11*(9), 2477.
145. Dahunsi, S. O., (2019). Mechanical pretreatment of lignocelluloses for enhanced biogas production: Methane yield prediction from biomass structural components. *Bioresource Technology, 280*, 18–26.
146. Zheng, H., Yin, J., Gao, Z., Huang, H., Ji, X., & Dou, C., (2011). Disruption of *Chlorella vulgaris* cells for the release of biodiesel-producing lipids: A comparison of grinding, ultrasonication, bead milling, enzymatic lysis, and microwaves. *Applied Biochemistry and Biotechnology, 164*(7), 1215–1224.
147. Taherdanak, M., Zilouei, H., & Karimi, K., (2016). The influence of dilute sulfuric acid pretreatment on biogas production from wheat plant. *International Journal of Green Energy, 13*(11), 1129–1134.
148. Amnuaycheewa, P., Hengaroonprasan, R., Rattanaporn, K., Kirdponpattara, S., Cheenkachorn, K., & Sriariyanun, M., (2016). Enhancing enzymatic hydrolysis and biogas production from rice straw by pretreatment with organic acids. *Industrial Crops and Products, 87*, 247–254.
149. Huang, C., Chen, X. F., Xiong, L., & Ma, L. L., (2012). Oil production by the yeast *Trichosporon dermatitis* cultured in enzymatic hydrolysates of corncobs. *Bioresource Technology, 110*, 711–714.

150. Jönsson, L. J., & Martín, C., (2016). Pretreatment of lignocellulose: Formation of inhibitory by-products and strategies for minimizing their effects. In: *Bioresource Technology* (Vol. 199, pp. 103–112). Elsevier Ltd.

151. Yilmaz, F., Kökdemir, Ü. E., & Perendeci, N. A., (2019). Enhancement of methane production from banana harvesting residues: Optimization of thermal-alkaline hydrogen peroxide pretreatment process by experimental design. *Waste and Biomass Valorization, 10*(10), 3071–3087.

152. Zhang, H., Huang, S., Wei, W., Zhang, J., & Xie, J., (2019). Investigation of alkaline hydrogen peroxide pretreatment and tween 80 to enhance enzymatic hydrolysis of sugarcane bagasse. *Biotechnology for Biofuels, 12*(1), 107.

153. Wahid, R., Romero-Guiza, M., Moset, V., Møller, H. B., & Fernández, B., (2020). Improved anaerobic biodegradability of wheat straw, solid cattle manure and solid slaughterhouse by alkali, ultrasonic and alkali-ultrasonic pretreatment. *Environmental Technology, 41*(8), 997–1006.

154. Bimestre, T. A., Júnior, J. A. M., Botura, C. A., Canettieri, E. V., & Tuna, C. E., (2020). Theoretical modeling and experimental validation of hydrodynamic cavitation reactor with a Venturi tube for sugarcane bagasse pretreatment. *Bioresource Technology, 311*, 123540.

155. Terán H. R., Kamoei, D. V., Ahmed, M. A., Da Silva, S. S., Han, J. I., & Santos, J. C. D., (2018). A new approach for bioethanol production from sugarcane bagasse using hydrodynamic cavitation assisted-pretreatment and column reactors. *Ultrasonics Sonochemistry, 43*, 219–226.

156. Terán H. R., Dionízio, R. M., Sánchez, M. S., Prado, C. A., De Sousa, J. R., Da Silva, S. S., & Santos, J. C., (2020). Hydrodynamic cavitation-assisted continuous pretreatment of sugarcane bagasse for ethanol production: Effects of geometric parameters of the cavitation device. *Ultrasonics Sonochemistry, 63*.

157. Nalawade, K., Saikia, P., Behera, S., Konde, K., & Patil, S., (2020). Assessment of multiple pretreatment strategies for 2G L-lactic acid production from sugarcane bagasse. *Biomass Conversion and Biorefinery, 1–14*.

158. Nagarajan, S., & Ranade, V. V., (2020). Pretreatment of distillery spent wash (vinasse) with vortex based cavitation and its influence on biogas generation. *Bioresource Technology Reports, 100480*.

159. Gogate, P. R., & Pandit, A. B., (2005). A review and assessment of hydrodynamic cavitation as a technology for the future. *Ultrasonics Sonochemistry, 12*(1, 2 SPEC. ISS.), 21–27.

160. Zieliński, M., Dębowski, M., Kisielewska, M., Nowicka, A., Rokicka, M., & Szwarc, K., (2019). Cavitation-based pretreatment strategies to enhance biogas production in a small-scale agricultural biogas plant. *Energy for Sustainable Development, 49*, 21–26.

161. Terán H. R., Dos, S. J. C., Ahmed, M. A., Jeon, S. H., Da Silva, S. S., & Han, J. I., (2016). Hydrodynamic cavitation-assisted alkaline pretreatment as a new approach for sugarcane bagasse biorefineries. *Bioresource Technology, 214*, 609–614.

162. Patil, P. N., Gogate, P. R., Csoka, L., Dregelyi-Kiss, A., & Horvath, M., (2016). Intensification of biogas production using pretreatment based on hydrodynamic cavitation. *Ultrasonics Sonochemistry, 30*, 79–86.

163. Huang, C., Zong, M. H., Wu, H., & Liu, Q. P., (2009). Microbial oil production from rice straw hydrolysate by *Trichosporon fermentans*. *Bioresource Technology, 100*(19), 4535–4538.

164. Morrow, III. W. R., Gopal, A., Fitts, G., Lewis, S., Dale, L., & Masanet, E., (2014). Feedstock loss from drought is a major economic risk for biofuel producers. *Biomass and Bioenergy, 69,* 135–143.
165. Gomes, (2018). Drought lights up alert for Brazil's sugarcane crop in-2019. Retrieved from: https://www.novacana.com/n/cana/safra/seca-alerta-safra-cana-brasil-2019-200718 (accessed on 28 October 2021).
166. Pugliese, L., Lourencetti, C., & Ribeiro, M. L., (2017). Environmental impacts on Brazilian ethanol production: A brief discussion from field to industry. *Multidisciplinary Brazilian Journal, 20*(1), 142–165.
167. Costa, C. C. D., & Burnquist, H. L., (2016). Impacts of gasoline price controls on biofuel ethanol in Brazil. *Economic Studies* (São Paulo), *46*(4), 1003–1028.
168. Nova Cana, (2017). Dilemma of Ethanol 2G: The challenges of cellulosic feedstock. Retrieved from https://www.novacana.com/n/etanol/2-geracao-celulose/dilema-etanol-2g-desafios-materia-prima-celulosica-260117 (accessed on 28 October 2021).
169. Schubert, C., (2006). Can biofuels finally take center stage?. *Nature Biotechnology, 24*(7), 777–784.
170. Nazhad, M. M., Ramos, L. P., Paszner, L., & Saddler, J. N., (1995). Structural constraints affecting the initial enzymatic hydrolysis of recycled paper. *Enzyme and Microbial Technology, 17*(1), 68–74.
171. Chang, V. S., & Holtzapple, M. T., (2000). Fundamental factors affecting biomass enzymatic reactivity. In: *Twenty-First Symposium on Biotechnology for Fuels and Chemicals* (pp. 5–37). Humana Press, Totowa, NJ.
172. Zhu, J. Y., Wang, G. S., Pan, X. J., & Gleisner, R., (2008). The status of and key barriers in lignocellulosic ethanol production: A technological perspective. In: *International Conference on Biomass Energy Technologies: Guangzhou, China, December 3–5, 2008* (Vol. 1, pp. 1–12). [Guangzhou, China]: Guangzhou Institute of Energy Conversion, The Chinese Academy of Science.
173. Banerjee, S., Mudliar, S., Sen, R., Giri, B., Satpute, D., Chakrabarti, T., & Pandey, R. A., (2010). Commercializing lignocellulosic bioethanol: Technology bottlenecks and possible remedies. *Biofuels, Bioproducts and Biorefining: Innovation for a Sustainable Economy, 4*(1), 77–93.
174. Chen, H., Han, Q., Venditti, R. A., & Jameel, H., (2015). Enzymatic hydrolysis of pretreated newspaper having high lignin content for bioethanol production. *BioResources, 10*(3), 4077–4098.
175. Abo, B. O., Gao, M., Wang, Y., Wu, C., Ma, H., & Wang, Q., (2019). Lignocellulosic biomass for bioethanol: An overview on pretreatment, hydrolysis and fermentation processes. *Reviews on Environmental Health, 34*(1), 57–68.
176. Santos, C. A., Morais, M. A., Terrett, O. M., Lyczakowski, J. J., Zanphorlin, L. M., Ferreira-Filho, J. A., & Souza, A. P., (2019). An engineered GH1 β-glucosidase displays enhanced glucose tolerance and increased sugar release from lignocellulosic materials. *Scientific Reports, 9*(1), 1–10.
177. Marques, F., (2018). FAPESP Magazine: Obstacles on the way. Retrieved from: https://revistapesquisa.fapesp.br/obstaculos-no-caminho/ (accessed on 28 October 2021).
178. Lynd, L. R., (2017). The grand challenge of cellulosic biofuels. *Nature Biotechnology, 35*(10), 912.
179. Özçimen, D., Koçer, A. T., İnan, B., & Özer, T., (2020). Bioethanol production from microalgae. In: *Handbook of Microalgae-Based Processes and Products* (pp. 373–389). Academic Press.

180. Costa, J. A. V., & De Morais, M. G., (2011). The role of biochemical engineering in the production of biofuels from microalgae. *Bioresource Technology, 102*(1), 2–9.

181. Amin, S., (2009). Review on biofuel oil and gas production processes from microalgae. *Energy Conversion and Management, 50*(7), 1834–1840.

182. Kumar, N., Sonthalia, A., & Pali, H. S., (2020). Next-generation biofuels—opportunities and challenges. In: *Innovations in Sustainable Energy and Cleaner Environment* (pp. 171–191). Springer, Singapore.

183. Ueno, Y., Kurano, N., & Miyachi, S., (1998). Ethanol production by dark fermentation in the marine green alga, *Chlorococcum littorale. Journal of Fermentation and Bioengineering, 86*(1), 38–43.

184. Kose, A., & Oncel, S. S., (2017). Algae as a promising resource for biofuel industry: Facts and challenges. *International Journal of Energy Research, 41*(7), 924–951.

185. Dos, S. I. F. S., Vieira, N. D. B., De Nóbrega, L. G. B., Barros, R. M., & Tiago, F. G. L., (2018). Assessment of potential biogas production from multiple organic wastes in Brazil: Impact on energy generation, use, and emissions abatement. *Resources, Conservation and Recycling, 131*, 54–63.

186. NREL (Transforming Energy), (2018). *Renewable Energy Data Book*. Retrieved from https://www.nrel.gov/docs/fy20osti/75284.pdf (accessed on 28 October 2021).

187. U.S. Energy Information Administration (USEIA). Total Petroleum and Other Liquids Production. (2014). Retrieved from: https://www.eia.gov/forecasts/steo/reDort/global oil.cfin.

188. Kumar, D., & Singh, V. (2019). Bioethanol production from corn. In *Corn* (pp. 615–631). AACC International Press.

189. International Energy Agency (IEA). (2019c). IEA convenes 2019 meeting of the Renewable Industry Advisory Board. Retrieved from: https://www.iea.org/news/iea-convenes-2019-meeting-of-the-renewable-industry-advisory-board.

190. OECD/FAO (2020), OECD-FAO Agricultural Outlook 2020–2029, FAO, Rome/OECD Publishing, Paris, https://doi.org/10.1787/1112c23b-en. https://www.fao.org/3/ca8861en/CA8861EN.pdf.

191. Laursen, K. (2015). Revealed comparative advantage and the alternatives as measures of international specialization. *Eurasian Business Review, 5*(1), 99–115.

192. Congressional Research Service (CRS). (2019). The Renewable Fuel Standard (RFS): An Overview. Retrieved from: https://crsreports.congress.gov/product/pdf/R/R43325

193. Subramaniam, Y., Masron, T. A., & Azman, N. H. N. (2020). Biofuels, environmental sustainability, and food security: A review of 51 countries. *Energy Research & Social Science, 68*, 101549.

194. Centro Nacional de Pesquisa em Energia e Materiais. (CNPEM) (2017). Retrieved from: http://pages.cnpem.br/2gbioethanol/wp-content/uploads/sites/85/2017/06/Jose_ Bressiani_ Granbio.pdf (accessed on 8 December 2021).

195. Carriquiry, M., Elobeid, A., Dumortier, J., & Goodrich, R. (2020). Incorporating sub-national Brazilian agricultural production and land-use into US biofuel policy evaluation. *Applied Economic Perspectives and Policy, 42*(3), 497–523.

196. International Energy Agency (IEA). (2019). *Bioenergy*. Annual Report 2018. Paris: IEA; 2019. Retrieved from: ieabioenergy.com/wp-content/uploads/2020/05/IEA-Bioenergy-Annual-Report-2019.pdf (accessed on 8 December 2021).

197. CONAB (2019). *Monitoring of the Brazilian crop: coffee. Brasilia: Conab.* Retrieved from: https://www.conab.gov.br (accessed on 8 December 2021).

Index

For Product Safety Concerns and Information please contact our EU
representative GPSR@taylorandfrancis.com
Taylor & Francis Verlag GmbH, Kaufingerstraße 24, 80331 München, Germany